Rudenko International Conference "Methodological Problems in Reliability Study of Large Energy Systems" (RSES 2022)

E3S Web of Conferences Volume 384 (2022)

Alushta, Russia
19-23 September 2022

Editors:

Sergey Senderov
Alexander Michalevich
Husein Guliev

ISBN: 978-1-7138-7247-4

Printed from e-media with permission by:

Curran Associates, Inc.
57 Morehouse Lane
Red Hook, NY 12571

Some format issues inherent in the e-media version may also appear in this print version.

For additional information, please contact EDP Sciences – Web of Conferences
at the address below.

EDP Sciences – Web of Conferences
17, Avenue du Hoggar
Parc d'Activité de Courtabœuf
BP 112
F-91944 Les Ulis Cedex A
France

Phone: +33 (0) 1 69 18 75 75
Fax: +33 (0) 1 69 28 84 91

contact-edps@webofconferences.org

Additional copies of this publication are available from:

Curran Associates, Inc.
57 Morehouse Lane
Red Hook, NY 12571 USA
Phone: 845-758-0400
Fax: 845-758-2633
Email: curran@proceedings.com
Web: www.proceedings.com

TABLE OF CONTENTS

Statement of Peer review

In submitting conference proceedings to *Web of Conferences*, the editors of the proceedings certify to the Publisher that

1. They adhere to its **Policy on Publishing Integrity** in order to safeguard good scientific practice in publishing.
2. All articles have been subjected to peer review administered by the proceedings editors.
3. Reviews have been conducted by expert referees, who have been requested to provide unbiased and constructive comments aimed, whenever possible, at improving the work.
4. Proceedings editors have taken all reasonable steps to ensure the quality of the materials they publish and their decision to accept or reject a paper for publication has been based only on the merits of the work and the relevance to the journal.

Title, date and place of the conference

Methodological problems in reliability study of large energy systems (RSES) 2022
19-23 September 2022, Alushta, Russia

Proceedings editor(s):

1. Academician of Russian Academy of Science Valeriy Stennikov, Melentiev Energy Systems Institute, Russia.
2. Prof. Sergey Senderov, Melentiev Energy Systems Institute, Russia.
3. Prof. Alexander Michalevich, Energy Institute of Belarus NAS, Belarus.
4. Prof. Husein Guliev, Azerbaijan Research and Design and Exploration Institute of Power Engineering, Azerbaijan.

Date and editor's signature 15. 01. 2023.

Electrodynamic withstand and effects of short-circuit currents on high-voltage electrical equipment

Alexander Khrennikov [1]*, *Robert* Shulga[2], and *Nikolay* Alexandrov[3]

[1]JSC "STC FGC UES" Rosseti, 123007 Kashirskoe Highway, 22, Moscow, Russia
[2]VEI-branch of FSUE " RFNC-VNIITF im. Akademik E. I. Zababakhina", 105118 Krasnokazarmennaya st., 12/38, Moscow, Russia
[3]SPE Dynamics, 428000 Anisimova str, 6, Cheboksary, Russia

Abstract. The analysis of simplified expressions allows us to estimate the electrodynamic effects of short-circuit currents on the main elements of a power electrical station and substation, which include busbars, high voltage circuit-breakers, disconnectors, electrical machines and transformers. Examples of calculations of magnetic field induction, mechanical forces and pressures that determine the electrodynamic withstand of electrical equipment are given. Simplified qualitative calculations allow us to assess the impact of key parameters on the selection of appropriate elements and the possible range of their changes during the design, operation and modernization at the preliminary design stage.

1 Introduction

The mechanical effects of short-circuit currents (SC) determine the resistance of electrical equipment to electrodynamic influences. The power electrical circuit of electric power station and substation contains electrical machines (EM) in the form of a generator or motor, a busbar, high voltage circuit-breakers (CB) and a disconnector, a power transformer and an overhead transmission line or cable line. Transient processes in these elements are quite widely described in the literature, for example, in [1, 2] for EM, in [3,4] for power transformer, and the calculated modes of switching on and short-circuit are given in [5,6].

Currently, there is a modernization and improvement of electrical equipment to improve energy efficiency through the use of new designs, technologies and materials [6 - 8].

Despite the availability of digital programs for calculating overcurrents and mechanical effects, for example, EMTP, ETAP, REST, ELAX - 2D, ELINDST 2.0, etc., there is a need to conduct high-quality calculations of the rigidity and stability of transformer windings during short-circuit: calculation of the magnetic field, strength of winding conductors during bending by axial and radial forces, axial vibrations of transformer windings, allowable electric field strengths of oil barrier insulation, value of insulation safety factors.

Block diagram of the connections of power elements at electric power station or substation shows on fig. 1, where they are indicated: EM - electric machine, busbar, CB - high voltage circuit-breaker and D - disconnector, T - power transformer, L - overhead transmission (cable) line.

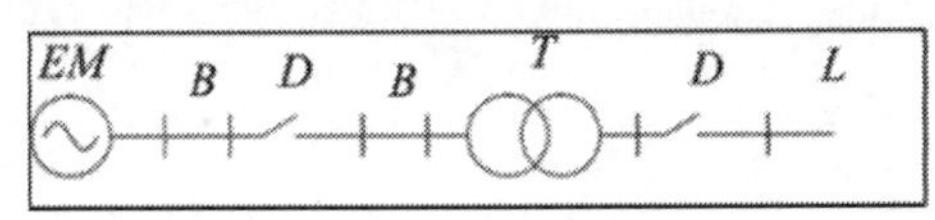

Fig. 1. Block diagram of the connections of power elements at electric power station or substation.

2 BUSBAR

When calculating the strength of execution and termination on busbar insulators (W), in accordance with Ampère's law, the magnetic induction B is determined by the formula

$$B = \mu_0 I / 2\pi d, \quad (1)$$

where $\mu 0 = 4\pi\times 10\text{-}7$ *H/m*, *I*, *d* are respectively the amplitude of the current and the distance between the busbars, *l* is the length of the busbar.

In the case of the opposite direction of current *I* in two parallel busbar trunkings of length *l*, repulsive forces *F* arise, which are equal to

$$F=2\times 10^{-7}\times I2\times l/d \quad (2)$$

Figure 2a shows the arrangement of 2 parallel busbars, and fig. 2b shows the vectors of currents *I*, magnetic inductions *B*, and repulsive forces *F*.

* Corresponding author: ak2390@inbox.ru

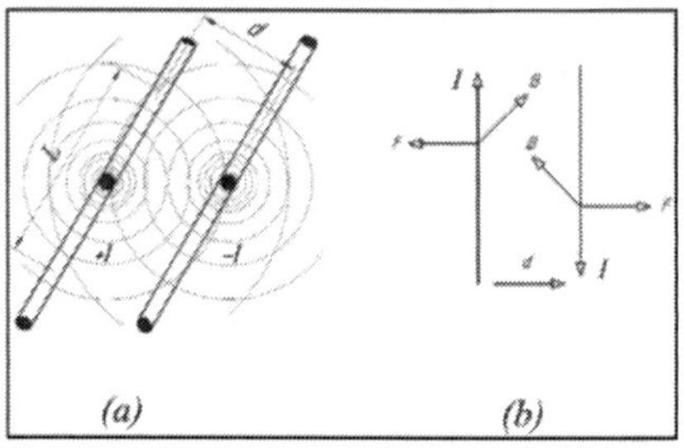

Fig. 2. Arrangement of parallel busbars with current *I* of length *l* (a) and vectors of counter currents *I*, magnetic inductions *B* and repulsive forces *F* (b).

Example 1. Given the values of the short-circuit current equal to $I=10^5$ *A*, *d*=0.5 *m*, = 1 *m*, in accordance with (2) the value *F*=4000 *N* or 400 *kg* of force that the switchgear busbars must withstand at a length of 1 m.

Example 2. For a test circuit with a tubular busbar for a 1200 MW turbo generator of Nuclear Power Station with a short-circuit current of 300 kA at a distance of *d* = 5 m, the repulsive force is 3600 N over a length of 1 m.

When exposed to a sinusoid of short-circuit current with amplitude I_k, dissipation factor σ, mains frequency ω, attenuation coefficient α in formula (2), the value of I_2 should be taken equal to

$$I_2 = I_K^2/\sigma(\cos(\omega t)-e^{-\alpha t})2= =(I_s/2)^2(1/2+e^{-2\alpha t}-2e^{-\alpha t}\cos(\omega t)+1/2\cos(2\omega t)) \quad (3)$$

As a result, the short-circuit current leads to the appearance of three components of mechanical effects: a constant force and two variable forces with frequencies that are multiples of ω and 2ω. When switching on the EM and power transformer, due to the asymmetric turn-on current, the current contains frequencies that are multiples of 1, 3, 5 harmonics.

3 Circuit-breaker (disconnector).

The equivalent circuit of the circuit breaker and disconnector (CB and D, respectively) is shown in Fig. 3,a. Due to the flow of current I in opposite directions of the horizontal branches, repulsive forces F arise between them. The same force causes the knives CB and D to separate from the traverse. Figure 3 shows the directions of the forces of interaction F in the circuit breaker and disconnector along the gap *x* of length *d*, the radius of the conductors and connections is *r*.

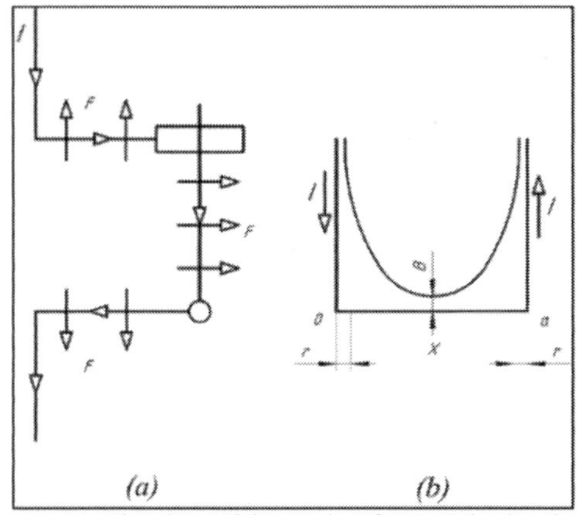

Fig. 3. Directions of interaction forces F in the circuit breaker and disconnector: a) interaction scheme of *F*, b) distribution of induction *B* (Tl) over the gap x of length *l*.

The force of interaction F for circuit breaker (disconnector) is determined

$$F = 2\cdot 10^{-7}\ \ln\left(\frac{d}{r}-1\right)\cdot l \quad (4)$$

Example 3. Given the values of the short-circuit current equal to $I=10^5$ A, *d* = 0.3 m, *r*= 0.01 m, *l* =1m, in accordance with (4) $F=2\times10^{-7}\ln(0.3/0.01^{-1})\times10^{10}$ = 6870 N, which is more than 1.5 times the impact force of Example 1 for parallel busbars.

4 Single-layer winding of an electrical machine.

The layout of conductors in a single-layer winding of electric machine is shown in fig. 4, where *l* is the length of the conductor, *x* is the current length of the EM slot, A is the linear current density (A/m).

On fig. 4b shows the case when the direction of induction B is perpendicular to the plane l_x.

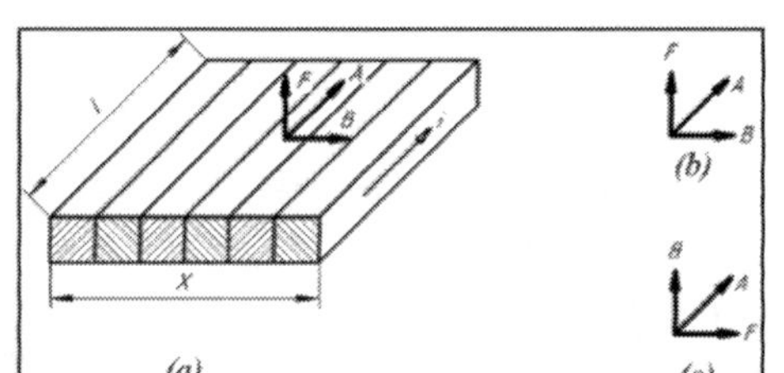

Fig. 4. Layout of conductors in a single-layer winding of electric machine (a), vectors of induction *B*, force *F* and linear current density *A* (b, c).

Linear current density A is

$$A = \sum {}^{I}/_{x}\ ,\ \text{A/m} \quad (5)$$

The force of interaction of the magnetic field on conductors with current is

$$F = 10^{-6}\ \text{BA}\ lx\ ,\text{N} \quad (6)$$

The specific pressure p per unit area *l* x is equal to

$$p = 10^{-6}\text{BA}\ \ \text{N/m}^2 \quad (7)$$

The effect of force on the windings of EM and power transformer from the magnetic field should be considered using the example of a power transformer.

5 Windings of the power transformer

The distribution of the magnetic field of induction B in the channel between the cylindrical windings of the power transformer is shown in fig. 5a. Ampère's law to the elementary circuit Adx in fig. 5a has the form of a circular integral

$$\int Hdx = Bdx = 4\pi Adx, \quad (8)$$

where linear current density $A = \sum I / x$ (A/m).

Neglecting the terms to the other three sides of the closed contour Adx, one can obtain

$$B_0 = 4\pi A \quad (9)$$

The distribution of induction B over the thickness of the winding is shown as a trapezoid in the lower part of fig. 5a.

The magnetic flux B, crossing the conductors in the axial direction, acts on the tension (expansion) of the winding outward (due to the opposite signs of the current and the linear current density A).

The average induction due to the trapezoidal distribution of B is B0/2. The radial pressure (force) pr from the axial component Bo is directed outward of the winding and is equal to

$$p_r = 10^{-7}\, 4\pi A_a^2, \quad (10),$$

where $A_a = \sqrt{2}$ A.

When a short-circuit current flows through the transformer, the inner winding is subjected to a compressive force due to the direction of forces inward towards the magnetic core, and the outer winding experiences a tensile stress due to outward forces [5].

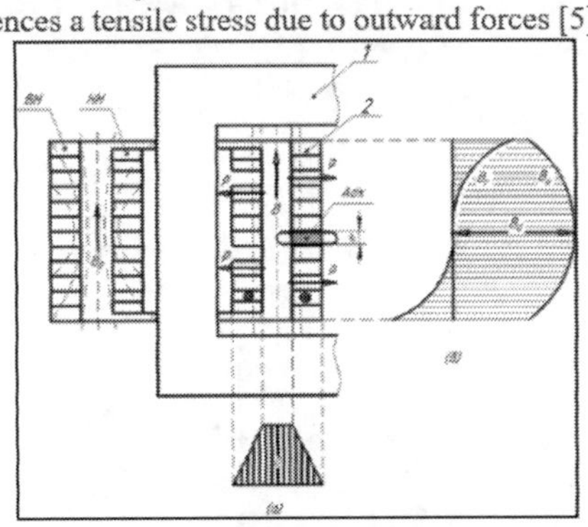

Fig. 5. The magnetic field of induction B in power transformer channel (a) and the components of the magnetic induction: B_r-radial, creates pressure pa, B_a-axial, creates pressure p_r (b); 1-core, 2- HV and LV windings.

Typically, tensile stress does not pose a risk of damage to the outer winding [2, 5]. The compressive stress of the radial forces causes damage to the internal winding - a wave of radial deformations (hoop buckling) (Fig. 6) [6 - 8].

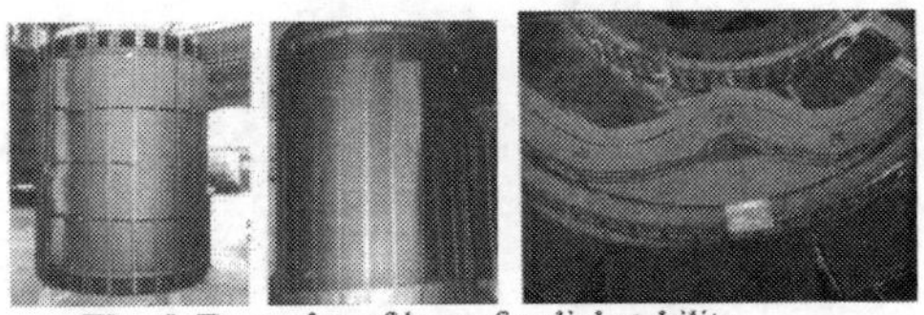

Fig. 6. Examples of loss of radial stability

6 Mathematical model of winding radial buckling

Next, a mathematical model is given for the radial deformation (example in the fig. 6) of one turn of a transformer winding in the form of an equation, which is more accurate than the previously presented models. Due to the use of trigonometric functions when describing the shape and volume of deformation and smaller error, this model more realistically reflects the processes occurring in the windings of power transformers in the case of short circuit.

Assumptions adopted in the preparation of the model:

The maximum size of the bulge of transformer winding is taken equal to the size of the concavity: a;

2. The convexity range is assumed to be equal to the concavity range: $\angle \Phi / 2$;

As an equation describing the concave-convex deformation of the current-conducting coil turn, the function $\varepsilon(\phi)$, described in polar coordinates, is chosen (Fig. 7):

$$\varepsilon(\phi) = a \cdot sin(2 \cdot \pi \cdot \phi / \Phi), with\ 0 \le \phi < \Phi, \quad (11)$$

where a - maximum deformation, the maximum deviation of the distorted section from the ideal state;

$\angle\Phi$ - deformation range, the central angle within which the deformation is observed;

ϕ - is the function argument, the angle that is set clockwise from the vertical semiaxis.

The equation of radial deformation:

$$e(\phi) = a * sin(2 * \pi * \phi/\Phi), with\ 0 \le \phi < \Phi. \quad (12)$$

The equation describing the shape of the turn of coil:

$$r(\phi) = R + e(\phi) = R + a * sin(2 * \pi * \phi/\Phi), with\ 0 \le \phi < \Phi; \quad (13)$$

R - winding radius, with $\Phi \le \phi < 2 * \pi$

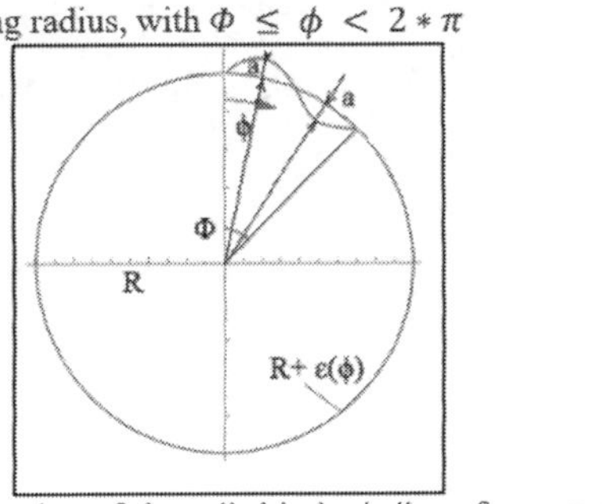

Fig. 7. Top view of the cylindrical winding of a power transformer with a concave-convex deformation after short-circuit with radial buckling.

https://doi.org/10.1051/e3sconf/202338401001

The expression for the radial concave-convex deformation (11), for reasons of changing the position of the middle of the gap between the windings, we have for radial force:

$$F_r = \frac{\mu_0 (I_{max}\omega)^2 \rho_1 \pi (D_{12} + a \cdot \sin(2\pi\phi/\Phi))}{2h} \quad (14)$$

where μ_0 - is magnetic conductivity;

I_{max} - short circuit current amplitude;

ω - the number of turns of the winding through which short-circuit current flows;

ρ1 - the coefficient of reduction of the ideal leakage field to the real (Rogowski coefficient);

*D*12 - the distance of the tank wall to the middle of the gap between the windings;

h - winding height.

Φ - deformation range, the central angle within which the deformation is observed;

ϕ - is the function argument, the angle that is set clockwise from the vertical semiaxis.

The expression (14) for the radial force acting on the transformer windings in the event of a radial hoop buckling in case of a short circuit has greater accuracy than other expressions.

This expression (14) for the radial force more accurately reflects the processes occurring in the windings of power transformers in case of short circuit and the occurrence of radial hoop buckling, due to the use of trigonometric functions in describing the shape and volume of deformation, *Φ* - deformation range, the central angle within which the deformation is observed, and h - winding height [24-30].

7 Axial deformations and axial forces

An example of the axial deformation of the LV winding of 250MVA/220 kV transformer, i.e. the movement of the winding in the axial (vertical) direction under the action of electrodynamic forces during the flow of short-circuit currents (Fig. 8) [31-33].

Axial forces tend to bend the turns of the coils towards each other, which leads to the effect of tensile forces on the insulation layers between the turns of the inner and outer windings [5, 8]. Axial forces also exert pressure on the insulation of the winding concentrator. The change in short-circuit impedance in the HV-LV mode was *Δ* Zk = + 20% [7, 8].

Fig. 8. Photo of phase "B" of the low voltage winding of 250MVA / 220 kV transformer, illustrating the loss of the axial stability of the winding, i.e. the movement of the winding in the axial (vertical) direction under the action of electrodynamic forces during the flow of short-circuit currents.

Own axial forces F_b (in newton) acting on the windings tend to reduce the height of the windings, therefore, the derivative of the magnetic field energy must be taken along the height of the winding h, i.e.

$$F_{b(2)} = \frac{\partial W_M}{\partial h} = \frac{1}{2} I_{max}^2 \frac{\partial L_k}{\partial h} = -\frac{\mu_0 (I_{max}\omega)^2 \rho_1 \pi D_{12} \delta'}{2h^2} \quad (15)$$

where $F_{b(2)}$ - own axial force acting on both windings.

If we consider the axial deformation as a change in the height of the winding, then the axial deformation formula will look like:

$$\Delta = h_{HV} - h_{LV}. \quad (16)$$

It follows that:

$$h_{HV} = h_{LV} + \Delta \quad (17)$$

Calculation of internal axial forces acting on the windings of the transformer is presented in fig. 9.

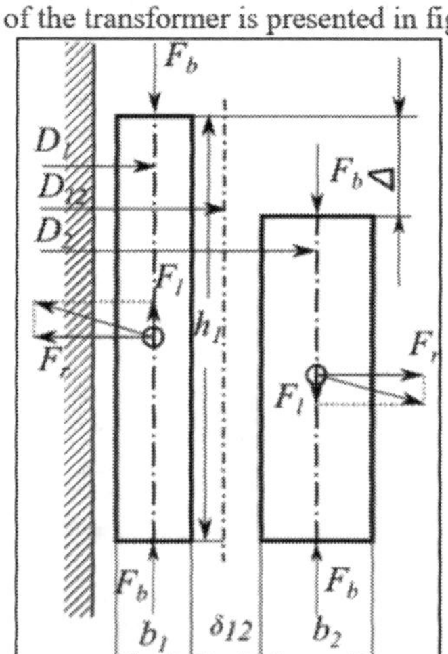

Fig. 9. Calculation of internal axial forces acting on the windings of the transformer.

You can substitute (17) in the formula (16) and then we get the expression for the axial force acting on the winding of the power transformer:

$$F_{axial} = -\frac{\mu_0 (I_{max}\omega)^2 \rho_1 \pi D_{12} \delta'}{2(h_{LV} + \Delta)} \quad (18)$$

 https://doi.org/10.1051/e3sconf/202338401001

Where $\delta' = \delta_{12} + \frac{b_1+b_2}{3}$, this is the maximum width of the main leakage channel.

8 Elliptical deformation of winding.

Here we study the case where one or more winding segments have been slightly deformed from the ideal circular form to an elliptic form [by 11] (Fig. 10).

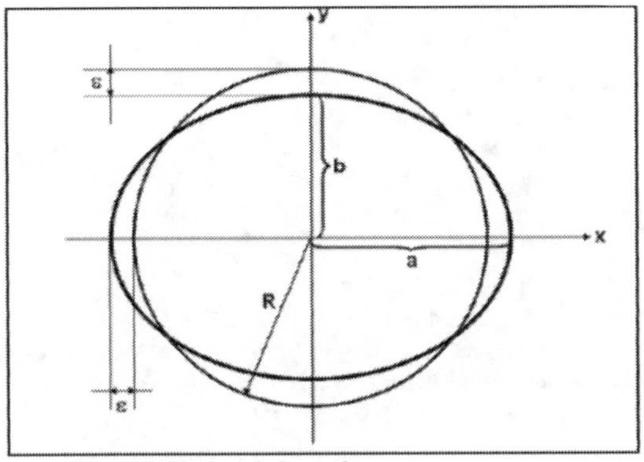

Fig. 10. Elliptical deformation.

To the first order of approximation in the small parameter ε/R, we obtain the following radial deformation $\delta\ (r,\varphi)$ of the elliptic winding compared to the unperturbed circular winding:

$$\delta(r,\varphi) = \delta(\varphi) = r - R = \varepsilon\cos(2\varphi) \quad (19)$$

where R - the radius of the undeformed winding, r - the curve describing the elliptical deformation, ε=R-r(90°).

For small values of the ratio ε / R, the condition of maintaining the length of the coil before and after deformation is satisfied:

$$l = 2\pi R(1 + \varepsilon^2/(4R^2)) \approx 2\pi R \quad (20)$$

Substituting the equation of elliptic deformation into the force formula for radial deformation, we get:

$$F_r = (\mu_0\,(I_{max}\,\omega)^2\,\rho_1\,\pi(D_{12} + \varepsilon\cdot\cos(2\varphi)))/2h \quad (21)$$

The whole refinement of the formula for radial forces is reduced to the fact that the distance D12 changes, which is the middle between the primary and secondary windings. Initially, we assume that only one of the windings is deformed, and then the formula correctly reflects the processes occurring in the transformer winding. In the literature about the second winding in the elliptical deformation is not mentioned.

The second assumption with elliptical deformation is that we mean that the ellipse is as the y axis perpendicular to the tank wall, but this may not be the case and will have to enter another angle into the equation. But this is a topic for further research.

9 Residual deformations with twisting of the winding and lodging of the turns on the edge.

Example 4. Setting for a powerful transformer by (4) with a normal linear current density *A*=75000 A/m and a short-circuit current equal to 20 *Inom.* (nominal current), the radial force is equal to p_r= 10-7×4π×(75000×20)2 =28260 N/m2

Example 5. For a power transformer with a power of 30,000 kVA, we take the following values (4):

a=1.25 m, b=0.04 m, c=1.3 m, d=0.1 m, A=75000 A/m, Ik=20×Inom., value of ln(a/b ×d/c) is 0.875. Then the axial force will be

p_a = 4×10-7×(75000 ×20)2 ×0.875 = 8000 N/m2 or 8 kg/cm2

As a result, the axial force is 4 times lower than the radial force. Axial forces from winding layers add up to the middle of the winding and can reach hundreds of tons. As long as the ratios a/b and c/d are almost equal, the axial forces remain acceptable. If the ends of one winding protrude relative to each other, then logarithm ln of these ratios increases sharply, and with it the axial force increases sharply.

It is also possible the occurrence of tangential winding twisting forces under the action of axial forces, which are much greater in the LV winding than in the HV. At a 500 kV substation, the autotransformer 250 MVA/500 kV/110 kV was disconnected from the action of gas protection. There were deformations of the HV winding with twisting and lodging of the turns on the edge under the action of tangential forces (Fig. 11) [6-8].

Fig. 11. Residual deformations of the HV winding of 250 MVA/500 kV/110 kV autotransformer with twisting of the winding and lodging of the turns on the edge.

10 220 kV SF_6 power transformer.

220 kV SF6 power transformer, produced by Russian Federation manufacturer, with a capacity of 63 MVA is a group of three single-phase 220 kV transformers with SF_6 insulation and cooling for normal operation, outdoor installation and at an underground substation is showed at fig. 12.

The design of the main units is determined: the core, the LV, HV and regulating windings, the LV, HV and regulating taps, the active part, the transformer tank and the on-load tap-changer.

The following calculations were carried out: electromagnetic calculation, insulation calculation, calculation of the electrodynamic withstand of windings in case of short-circuit, gas-dynamic calculation of the cooling system, thermal calculation of the windings and magnetic system, calculation of the transformer tank and the tap changer tank for mechanical strength [9].

Fig. 12. 63 MVA/220 kV SF6 power transformer, produced by Russian Federation manufacturer.

Critical conductor subsidence forces, maximum axial forces and compressive stresses in spacers, average areas of insulation elements over which axial forces are distributed, and coefficients of rigidity of secant concenters were calculated using the ELDINST-2.0 program [9-12].

11 Conclusions.

1. Calculations of the mechanical impact of short-circuit current on high voltage circuit-breakers, disconnectors, electrical machines and power transformers, calculations of magnetic field induction, mechanical forces are given.
2. Received and investigated mathematical models: radial, elliptical, axial deformation of power transformer windings.
3. Calculations of the electrodynamic stability of the windings of SF_6 gas-insulated transformer 63 MVA/220 kV, produced by Russian Federation manufacturer, are given.

References

1. R. Ryudenberg, *Transient processes in electric power systems*, Moscow,712 p. (1955)
2. *Power transformers*. Reference book / Edited by S. D. Lizunov, A. K. Lokhanin. Moscow: Energoizdat, 616 p. (2004)
3. S. I. Lur'e *Stability of transformers during short circuit*. Moscow: Znak, 520 p. (2005)
4. R. N. Shulga, *Turn-on currents, short-circuit currents of transformers and synchronous generators*, Energoekspert No. 2, pp. 44-48. (2021)
5. S.B. Vasyutinskiy, *Calculation and design of transformers*. Leningrad: LPI (1976)
6. A.Yu. Khrennikov, *High-voltage equipment in electrical systems: diagnostics, defects, damageability, monitoring*: textbook, Moscow: INFRA-M, 186 p. (2019)
7. A.Yu. Khrennikov, *Technical diagnostics and accident rate of electrical equipment*, Moscow: LITRES, 306 p. (2021)
8. A. Yu. Khrennikov, V. V. Vakhnina, A. A. Kuvshinov, N. M. Alexandrov, *Transformers at power facilities: tests, diagnostics, damageability, monitoring*. Moscow: Direct-Media,336 p. (2021)
9. V.I. Lazarev, I.V. Lazarev, *On the reason for the decrease in the forces of axial pressing of the windings of transformers in case of short-circuit*, Electrical engineering and electric power industry, Zaporozhye, No. 1, p.18 - 22 (2005)
10. A. Yu. Khrennikov, *Diagnostics of Electrical Equipment Faults and Power Overhead Transmission Line Condition by Monitoring Systems (Smart Grid): Short-Circuit Testing of Power Transformers*, Nova publishers, New York, USA, p. 165 (2016)
11. A. Yu. Khrennikov, A. A. Kuvshinov, I. A. Shkuropat, *Providing of reliable operation of electrical networks*, New York: NOVA PUBLISHERS, 296 p. (2019)
12. A. Yu. Khrennikov, English Original Reader for Technical Students. *Power transformers: short-circuit testing, monitoring systems (Smart Grid)*, Moscow, LITRES, 230 p. (2023)

Improving Belarusian energy system reliability in the context of Belarusian NPP commissioning

Mikhail Kashin[1,*], *Nikolay* Novikov[2] and *Alexander* Novikov[3]

[1] Belenergosetproject RUE, Republic of Belarus
[2] STC FGC UES JSC, JIHT RAS, NRU MPEI, Russia
[3] STC Energobezopasnost, Russia

Abstract. The article presents the results of computational experiments confirming the efficacy of the phase-shifting transformer (PST) as a type of a phase-shifter, which is to be installed at the backbone 750/330 kV substation Belorusskaya, in the repair and post-emergency states of the Belarusian power system, as well as an analysis of technical and economic performance indicators.

Introduction

The commissioning of a nuclear power plant in the Republic of Belarus in 2021 caused a radical transformation in the power system, as the installed capacity of the Belarusian nuclear power plant is about 40% of the annual maximum load.

The backbone grid of the Belarusian power system, as well as interconnections with adjacent power systems are formed at 330-750 kV. At present, along with changes in the internal 330 kV connections, the composition of the external 330 kV interconnections is changing. In this regard, maintaining the standard backup capacity in the Belarusian power system poses a question of how to regulate the flow of active power between the grids of 330 kV and 750 kV voltage classes.

In connection with the changes in the topology of the 330-750 kV backbone grid as was outlined above, given the limited interconnection redundancy capabilities, the Belarusian power grid comes close to the need to build an active-adaptive 330-750 kV grid, allowing for real-time grid control using phase shifters (PS).

The article considers the possibility of regulating active power flows along the 750 kV transmission line Smolensk NPP - Belorusskaya, which connects the Belarusian power system and the UES of Russia.

As is known, the active power flow in an AC power line is determined by expression (1).

$$P = \frac{U_1 \cdot U_2}{X_l} \cdot \sin \delta_{21} \qquad (1)$$

where U_1 and U_2 are the voltage moduli of the power source at the beginning of the line and the voltage of the electric load at the end of the line;

X_l - line reactance;

δ_{21} - the phase angle between the voltage vector of the power source and the load.

The voltage moduli on the power source and load buses (including at different points of power systems) are formed based on specified parameters of power flows, equipment reliability, reactive power balances, and other basic conditions, and lack a significant regulation range to address the issue of active power flow control. It follows that the amount of active power transmitted through the transmission line can be controlled in two ways:

- by changing the line reactance,
- by changing the phase angle between the source and load voltage vectors.

The AC line resistance is predominantly inductive in nature. Therefore, one of the engineering means of regulating the flow of active power along a transmission line is series compensation devices (SCD), which are batteries of capacitors included in series in the transmission line to compensate for part of its inductance. However, in the repair and emergency states of the Belarusian power system, as will be shown below, it is necessary to limit the flow of active power along the 750 kV transmission line Smolensk NPP - Belorusskaya.

Therefore, for the Belarusian power system regulating the power flow in the 750/330 kV grid by changing the phase angle between the vectors of voltages of the beginning and end of the transmission line proves relevant.

To regulate the flows of active power in three-phase AC grids, one uses the phase shifter, a specialized modification of the power transformer. The phase shifter consists of two transformers: a regulated transformer, which is connected in parallel with the line, and a series transformer, the secondary winding of which is connected in series with the line. Due to the winding connection scheme, the voltage vector on the series winding is directed at an angle of 90 electrical degrees to the phase voltage of the grid.

By changing the voltage on the series winding with a regulated transformer, it is possible to rotate the vector of total voltage at the beginning of the grid and control the shift angle between the voltage vectors at the

* Corresponding author: m.kashin@besp.by

 https://doi.org/10.1051/e3sconf/202338401002

beginning and at the end of the transmission line by changing the flow of active power transmitted through it relative to the natural load flow [1].

There are two types of phase shifters:

- phase-shifting transformers (PST): they the ability to continuously regulate quadrature voltage;
- quadrature booster: they regulate step voltages.

The quadrature booster (QB) is a three-phase transformer built into a single-tank; it has no OLTCs. QB is a simple, cheap, and reliable phase shifter used in power flow control in a 220-500 kV grids of elaborate topologies.

At 750 kV substations (SS), the functions of the PST can be performed by a quadrature booster installed at the neutral point of a 750/330 kV autotransformer (AT).

It should be noted that at present, the output voltage phase shift is changed mainly by means of traditional mechanical on-load tap changers (OLTC). OLTCs have a relatively low response time (a few seconds) and a relatively low reliability. However, if it is necessary to increase the speed of the phase shifter, thyristor switches can be used, which perform a similar function of switching transformer windings using thyristors or triacs [1].

Phase shifters have been widely used globally since the second half of the XX century. Phase shifters have been used in the British power grid since 1969. [2]. In the IPS of the North-West at the Leningrad NPP, 750/330 kV AT1 is equipped with a phase shifter. In 2019, a 500/220 kV phase shifter with a capacity of 195 MVA was commissioned at the Volzhskaya HPP [3]. Since 2009, the 500 kV Ulken substation in Kazakhstan has been using a 500/220 kV phase shifter with a capacity of 400 MWA [4].

1 Special considerations relating to the application of phase shifters in the Belarusian power system

The Belarusian power system is connected to adjacent power systems through the 750 kV power line Smolensk NPP - Belorusskaya and a number of 330 kV transmission lines, the load of which is formed on the basis of power flows of all power systems involved. The lack of control of the distribution of active power between transmission lines of 330 kV and 750 kV voltage classes limits the utilization of the design capacity of grid elements. These limits do not allow going through all maintenance and emergency states without changing the composition of generating and consuming facilities in the Belarusian power system, which compromises the security of the power system.

The substation that serves as a junction connecting 750 kV and 330 kV grids is the 750 kV Belorusskaya substation. A unique feature of the 750 kV Belorusskaya substation circuit is that the 750 kV and 330 kV grids are connected through a single element being the 750/330 kV AT.

The analysis of the experience of operating PSTs in CIS countries demonstrated that the most appropriate way to regulate the power flows through the 750/330 kV AT of the 750 kV Belorusskaya substation is to use a standard single-phase regulating transformer of the ODTsNP 92000/150 type, designed specifically to work with the 750/330 kV AT and already proven in use.

Our study attests to the fact that the use of this engineering solution, both in normal and in emergency states, provides the ability to control the loading of the 750/330 kV autotransformer, and, consequently, that of the only 750 kV transmission line Smolensk NPP - Belorusskaya.

Currently, the Belarusian power system is connected to the adjacent power systems of Russia, Lithuania, and Ukraine through eight 330 kV transmission lines and one 750 kV transmission line. The 330 kV power transmission lines previously connected to the Ukrainian power system were disconnected due to Ukraine's joining ENTSO. In the case of the withdrawal of the Lithuanian power system from parallel operation with the Belarusian power system, only four 330 kV and 750 kV transmission lines of the interconnection with the UES of Russia will remain in operation (Figure 1):

1. 750 kV overhead line Smolenskaya NPP - Belorusskaya;
2. 330 kV overhead line Krychaw - Roslavl;
3. 330 kV overhead line Viciebsk - Talashkino;
4. 330 kV overhead line Polatsk - Novosokolniki.

Fig. 1 330 kV and 750 kV transmission lines connecting the Belarusian power system with the UES of Russia.

2 Calculations of power flows and voltages

The expansion of power grids is accompanied by an increase in additional power and electricity losses caused by the growth of equalizing power due to their heterogeneity. In this case, grids of different voltage classes are loaded according to their impedances and topology, and oftentimes the capacity of the entire closed network is limited because of its single element. Managing the power flow of such a network makes it possible to optimize active power and electricity losses, as well as to ensure the required level of capacity of such a closed network by redistributing power flows between grids of different nominal voltages.

As noted above, the standard solution for regulating the flows between the 330 kV and 750 kV grids is to install a phase-shifting transformer (PST) with a regulation range of ± 68 kV (± 20x5%) in the primary winding of the existing 750/330 kV AT , which performs quadrature regulation of the AT voltage.

To perform longitudinal voltage regulation at the neutral point of the AT of the 750 kV substation Belorusskaya, an on-load tape-changer (OLTC) with a ±64 kV (±20x5%) additional EMF regulation range is currently installed.

The additional quadrature EMF of the phase-shifting transformer is directed at a 90° angle to the main phase voltage. The longitudinal and quadrature voltage control is carried out at the neutral point of the 750/330 kV autotransformer. Therefore, the total transformation ratio of the autotransformer is a complex number and is defined by the expression:

$$K_T = \frac{\left((U_{nom.MV} \pm \Delta U_{OLTC})/\sqrt{3} + j\Delta U_{PST}\right)}{\left((U_{nom.HV} \pm \Delta U_{OLTC})/\sqrt{3} + j\Delta U_{PST}\right)} = K_{TR} + jK_{TI}, \quad (2)$$

where

$U_{nom.MV}$ - nominal voltage of the MV winding of the autotransformer, equal to 330 kV;

$U_{nom.HV}$ - nominal voltage of the HV winding of the autotransformer, equal to 750 kV;

ΔU_{OLTC} - the value of additional longitudinal EMF generated by the OLTC (± 64 kV);

ΔU_{PST} - the value of the additional quadrature EMF generated by the PST (± 68 kV).

K_{TR} - the real part of the complex transformation ratio;

K_{TI} - the imaginary part of the complex transformation ratio.

To evaluate the technical performance of the use the PST at the 750 kV substation Belorusskaya, we carried out a series of electrical analysis calculations of the backbone grid of the IPS of Belarus in its normal and maintenance states under different PST taps (no quadrature regulation in case when PST tap No. 22 is selected) for winter maximum and summer minimum power flows. In the calculations of power flows, we assumed that the condition of the self-balancing of the Belarusian power system holds in terms of active power.

Analysis of the calculation results showed that the regulation ranges of active power flows between the 330 kV and 750 kV grids are determined by the grid configuration and do not depend on the level of loads. The configuration of the grid of the Belarusian power system does not depend on the season of the year, so below are the results of calculations of winter maximum power flows.

The profiles of active power flows over the 750/330 kV AT and 330 kV transmission lines (Krychaw - Roslavl, 330 kV Viciebsk - Talashkino overhead line, Polatsk - Novosokolniki) with different taps of the PST selected in the normal, maintenance (n-1) and maintenance-and-emergency (n-2) states are shown in Figures 2-4.

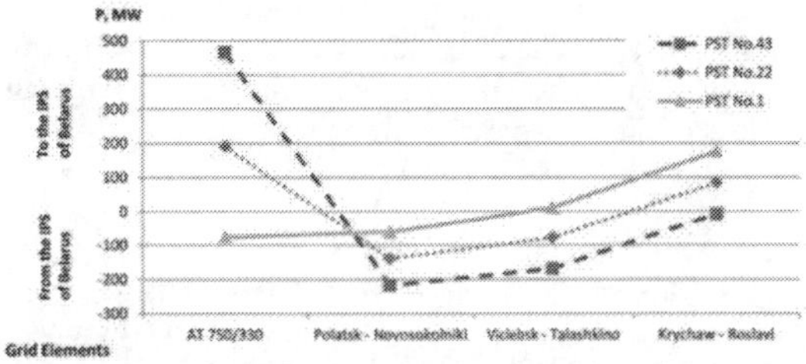

Fig. 2. Profile of active power flows on 330 kV transmission lines and the 750/330 kV AT of the IPS of Belarus with different taps of the PST selected under normal operating conditions.

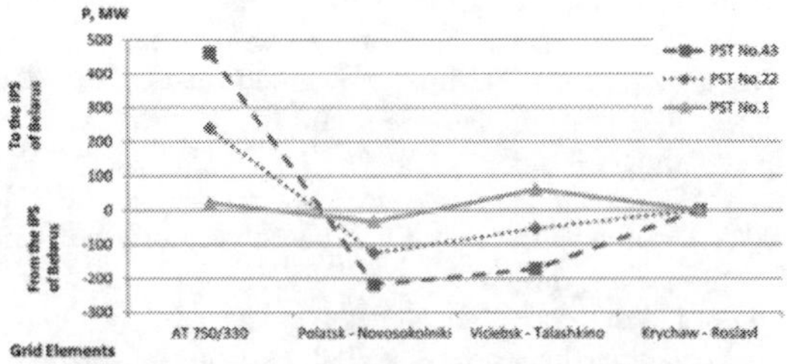

Fig. 3. Active power flow profiles along the 330 kV transmission line and the 750/330 kV AT of the IPS of Belarus with different taps of the PST selected in the maintenance state (disconnection of the 330 kV Krychaw-Roslavl overhead line).

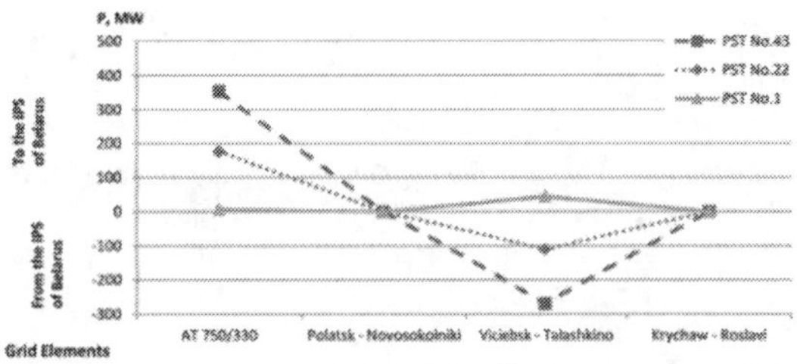

Fig. 4. Active power flow profiles along the 330 kV transmission line and the 750/330 kV AT of the IPS of Belarus with different taps of the PST selected in the maintenance and emergency state (disconnection of 330 kV overhead lines Krychaw-Roslavl and Polatsk - Novosokolniki).

The analysis of Figures 2-4 shows that the possible range of change in the active power flow through the 750/330 kV AT of the 750 kV substation Belorusskaya is as follows:

- in the normal operating state, it is 540 MW,
- in the repair state - 440 MW;
- in the maintenance and emergency state - 350 MW.

Consequently, when using a longitudinal and quadrature voltage regulation at the 750 kV substation Belorusskaya it is possible to transfer the flow of active power from the 330 kV grid of the cross-border section of the IPS of Belarus - IPS of the Center to the 750 kV overhead line Smolenskaya NPP - Belorusskaya and back, in the range of 540-350 MW.

The security of the power system is understood as the ability to withstand sudden power flow changes, preventing cascading accidents with mass disruption of

https://doi.org/10.1051/e3sconf/202338401002

power supply to consumers. The deterministic criterion n-1 is widely used to assess the security of a power grid. The criterion means that in case of emergency involving the loss of any independent element of the grid, the grid continues to perform its functions in full [5].

As part of this study, one of the tasks was to ensure the security of the power system by providing the required transmission capacity of 750 kV and 330 kV interconnections in the event of an emergency shutdown of the Belarusian NPP.

Performed power flow analysis with a shortage of active power in the Belarusian power system (in case of an emergency shutdown of the Belarusian NPP) show that the bottleneck element for receiving backup power is the 750/330 kV AT at the 750 kV substation Belorusskaya, whereas the transmission capacity of the 330 kV transmission line proved adequate. In these states, the use of the PST allows eliminating the overloading of the 750/330 kV AT by increasing the load of 330 kV transmission lines by about 200-250 MW. Thus, when receiving backup power from the UES of Russia, there is no overloading of grid elements and no need to disconnect consumers.

Managing the loading of 750 kV and 330 kV grid elements using the PST also makes it possible to go through maintenance states in the backbone grid of the Belarusian power system without compromising the required parameters.

3 Additional considerations related to the use of the PST

As mentioned above, during an emergency shutdown of the Belarusian NPP, the PST allows increasing the interconnection transmission capacity of the Belarusian power system by up to 250 MW, and, consequently, the power system does not incur losses for electricity not served. The economic effect from the installation of the PST, according to aggregate estimates, will be 110% of the cost of its installation.

In addition, a similar effect of eliminating the imbalance of active power in the power system can be achieved by constructing a new cross-system 330 kV overhead line or a shunting generating unit of similar capacity. The cost of PST installation is 10-20 times less than the cost of either of the above options.

It should be noted that the possibility of regulation of active power flows between the 750 kV and 330 kV grid is determined by the permissible voltages for these grids (787 kV and 363 kV, respectively). Voltages at the nodes of the 330-750 kV power system are determined by reactive power balances, which largely depend on the active power load of consumers and, consequently, the load with respect to the active power of grid elements.

Figure 5 shows an example of an alignment chart of the voltage on the buses of 750 kV substation Belorusskaya at different positions taps of the tap changer and PST.

Since the change in the power system load is a dynamic process with oftentimes sharp fluctuations in active power (in case of emergencies), to determine promptly the tap of the PST and OLTC appropriate for the current state (with the aid of an alignment chart similar to Figure 5) in the future will require the creation of a digital twin of the longitudinal and quadrature regulation device [2] and its integration into the model of the digital twin of the power system.

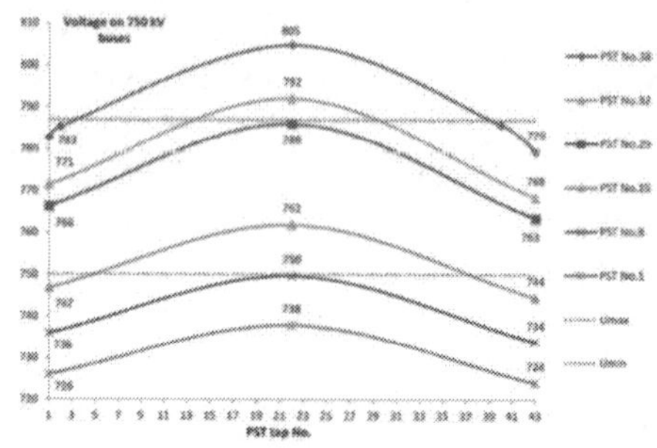

Fig. 5. Alignment chart of voltage on 750 kV buses of the 750 kV substation Belorusskaya at different positions of taps of the OLTC and PST.

Conclusion

1. The control ranges of active power flows between the 330 kV and 750 kV grid using the PST are determined by the grid configuration and do not depend on the magnitude of loads.
2. Load control of 750 kV and 330 kV grid elements using the PST allows to go through all maintenance, maintenance-and-emergency, and emergency states without constructing new transmission lines.
3. When adopting the PST, it is advisable to create its digital twin and integrate it into the digital twin model of the power system.

References

1. Astashev M.G., Novikov M.A., Panfilov D.I. *Application of phase shifters with thyristor switches in active-adaptive power grids,* Energiya edinoi seti, E **5**, 70 (2013)
2. Bushuev V.V., Novikov N.L., Novikov A.N. *Digitalization of economy and energy: prospects and challenges,* Ekonomicheskie strategii, E **6**, 96. (2019)
3. Brilinsky A.S., Evdokunin G.A., Kritsky V.A., Matvienkov Yu.V., Sidelnikov A.P., Smirnova L.S. *Phase-shifting transformer in the power supply circuit of a large hydropower plant,* Proceedings of the STC of the Unified Energy System, E **1**, 6 (2019)
4. Evdokunin G. et al. *Phase-shifting transformer put to use in Kazakhstan for the first time in the CIS,* Novosti elektrotekhniki, E **6**, 12 (2008)
5. Astashev M.G., Novikov M.A., Panfilov D.I., Rashitov P.A., Remizevich T.V., Fedorova M.I. *On power flow analysis of power transmission lines with regulated phase shifters,* Izvestiya RAN "Energetika", E **1**, 15 (2016)

Once again on loading and capacity selection of autotransformers in transmission grids

Nikita Vasilenko[1,*], *Misrikhan* Misrikhanov[1], and *Andrey* Shuntov[1]

[1] Moscow Power Engineering Institute, 17 Krasnokazarmennaya str., Moscow, Russia

Abstract. The aspects which have a bearing upon selection of capacity of autotransformers, such as load graphs, permissible temperature characteristics, n-1, n-2 modes, are considered. On the basis of statistical analysis suggestions for improving the current method of capacity selection of autotransformers in transmission grids are given.

1 Introduction

In previous work [1], authors considered aspects of choosing the power of autotransformers (AT) installed on 220-750 kV substations. It was noted that the capacity of these autotransformers is selected under uncertainty, since we need to predict the operation conditions for the period of 10 years and more (referring to scheme of power system development).

The following problems have been identified:

The scope of regulation of current standard [2] does not include loading capacity of most of the transformer equipment installed in transmission grids, namely 220-750 kV autotransformers with 125-800 MVA capacity and OF cooling system. Thus, its lack of the methodical base for taking a correct decision.

Lastly, it has become customary to consider every possible condition (mode) in the power system for capacity selection of autotransformers during design of substations. Not only do we include n-1modes (with failure of one grid element), but also n-2 modes (with failure of two concurrent grid elements). And mode with autotransformer loading higher than in n-1 mode can always be found.

On top of that, the tendency for industrial load decreasing from the 1990s, the limited number of unit sizes, and cautious attitude of the operating organizations to transformer overloading are added [5].

Under these circumstances a low load of costly transformer equipment appears.

In this paper, the authors set a goal to update the previous study, taking into account a wider sample of data, to confirm or refute the previously obtained results, and propose a scientifically substantiated recommendation for the capacity selection of 220 - 750 kV autotransformers in transmission grids.

2 Accounting for load graphs

The real daily measurement data of check measurement days in 2015, 2020, and 2021 years for more than 300 substations according with their voltage type were analyzed and the values of characteristic equivalent two-stage load graphs of 220-500 kV substations from different regions of the country were obtained. Results are given in the Table 1.

Defined:

– load factor f_{load} (the average load divided by the peak load);

– load curve irregularity factor f_{irreg} (the minimum load divided by the peak load).

– f_{max} factor, calculated as the winter peak load ($S_{max(w)}$) divided by the summer peak load ($S_{max(s)}$);

The mean value of f_{max} is still equal to 1.2. Likewise, for about 30% substations the summer peak load is higher than the winter peak load. Geographically these substations are located in all regions of the country (not only in southern regions). It is harder design condition, since rated autotransformer capacity conform to ambient temperature of 20°C, i.e. permissible loading is decreasing while growth of temperature and conversely.

Mean values of load factor and load curve irregularity factor are f_{load}=0.829 and f_{irreg}=0.644. Therefore, mean pre-load - K_1 of equivalent two-stage load graph is in relatively thin range from 0.644 to 0.829 (further arithmetical averages were taken).

Daily load graph characteristics in the Table 1 organized by four groups: 1 group – industrial load, presented in oil production regions of the country enhanced by high share of hydro power plants in the generation mix in some cases; 2 group – regions with large industrial centers and cities; 3– regions with mixed type of load with measurable share of domestic consumer

*Corresponding author: vasilenkone.ird@gmail.com

and service sector demand; And the 4th group – the same as the third group, but with high share of agricultural industry.

As shown by the Table 1 the K_1 values of equivalent load graphs could be different. These values are varying from 0.69 to 0.81 with almost constant duration of overload h≈8–9 hours. The deviation of the f_{load}, f_{irreg} for 2015, 2020 and 2021 years is about 1.5%, so the prevalent type of load in the regions under consideration has not changed over the years.

Thermal model of ageing of transformer insulation, top-oil and hot-spot temperature limitations and number of other boundary conditions [2,3] are the foundation of Russian and international methods of definition of loading capacity of autotransformers. In assessment of insulation ageing the known 6°C principle is used. However, permitted temperatures in Russian and international standards [2,3] are different. The comparison is shown in Table 2. The international standards are more sparing. The new standard IEC 60076 – 7 – 2018. "Power transformers – Part 7: Loading guide for mineral-oil-immersed power transformers" was published in 2018, which, however, in terms of permissible loading characteristics complies with the previous standard IEC 60076 – 7:2005. Thus, maximum permissible winding hot-spot temperature from [2] - 160°C, while in [3] – 140°C (Table 2).

Examples of loading of autotransformers are given in Fig. 1: solid lines – normal cyclic loading, dashed lines - emergency loading. Curves in Fig. 1 are given for equivalent ambient temperature of +20°C, curves in Fig. 1(a) are given for –10°C and for +20°C in Fig. 1(b). It is common for Moscow region for winter and summer periods thoroughly. Calculations were settled by the procedure [3]. Normal cyclic loadings were calculated with daily loss of life is less than 1. Numbers with daggers in Fig. 1 - duration (h) of overloading K_2. The current limitation from Table 2, which is $K_2 \leq 1.3$, is disregarded in Fig. 1

Thus, the conclusions obtained earlier in [1] continue to be relevant.

Table 1 Characteristics of Equivalent two-stage load graph

Prevalent type of load	f_{load}, pu.	f_{irreg}, pu.	K_1, pu.	K_2, pu.	*h*, h
1 group – industrial load (regions of Siberia)	0.874	0.747	0.811	1	8.1
2 group – regions with large industrial centers and cities (regions of Ural)	0.823	0.643	0.733	1	8.1
3 group – mixed load (Central regions)	0.810	0.605	0.708	1	8.4
4 group – mixed load with high share of agricultural industry (South regions)	0.807	0.583	0.695	1	8.8
Mean value	0.829	0.644	0.736	1	8.3

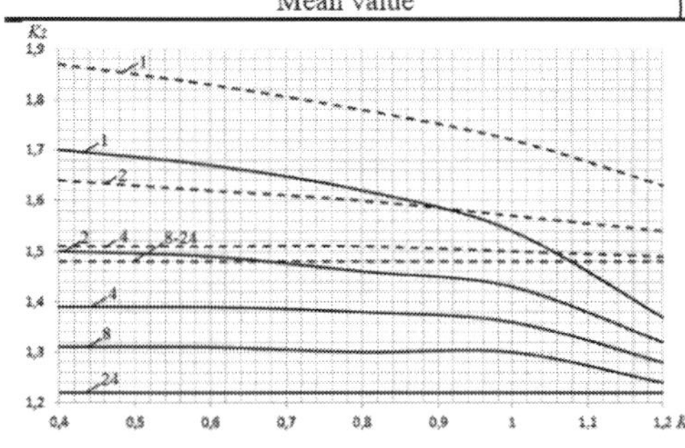

Fig. 1.a Loading of AT which rating over 100 MV·A with OF cooling system for –10 °C.

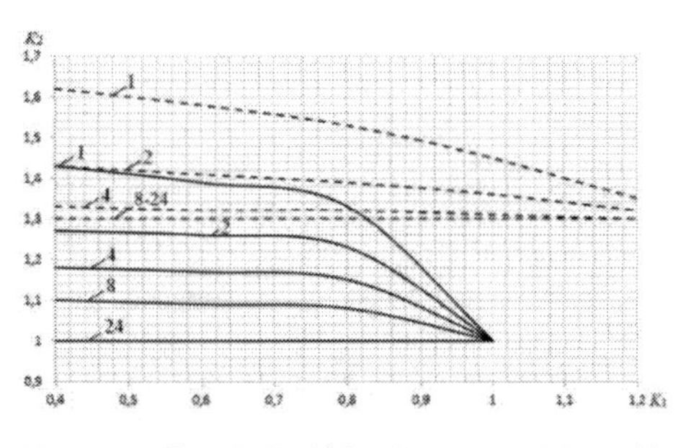

Fig. 1.b Loading of AT which rating over 100 MV·A with OF cooling system for +20°C.

Namely:

- loading of autotransformers for the real values K_1 from 0.6 to 0.8 and h≈8–9 h in transmission grid in Russia (presented in Table 1) are almost constant

- Normal cyclic loads can be limited not by daily loss of life, but the maximum permissible winding hot-spot temperature (120 °C – Table 2).

Permissible cyclic loading of autotransformers with 125-800 MVA capacity and OF cooling system is given in Table 3 for equivalent ambient temperature +20 °C (numerator) and corresponding daily loss of life (denominator).

The comparison of data from Table 1 and Fig. 1,b shows: with removing constraints from daily loss of life under other factors being equal the permissible loading K_2 is growing from 1.0 to 1.16. In this case daily loss of life for duration of 8 hours doesn't exceeded 3.98. Thus, in calendar day insulation is really "aged" on 3.98 days. If annual mean outage doesn't exceed, assume, a month then insulation will be aged on 120 days for this period.

In a steady state mode when two AT are in service and the loading of each is $(1.16/2)S_{rated}$, the daily loss of life is less than 0,01 (the winding hot-spot temperature is 57 °C), so it can be neglected. This state is common to temperatures up to 80 °C [4]. Thus, determination of permissible cyclic loading based on winding hot-spot temperature is proven design condition for increasing of AT loading. Moreover, the newly introduced transformer equipment is equipped with digital control devices that allow calculating the actual insulation depreciation along the process.

Less stringent requirements are imposed by the industry instruction [6] for the loading of AT with a capacity of more than 100 MV·A with an OF cooling system (Table 4).

Table 2 Maximum permissible characteristics of autotransformers

Characteristic	autotransformers capacity, MV·A		
	below 2.5	from 2.5 to 100	above 100
Normal cyclic load:			
Current, pu.	1.5/1.5/1.5	1.5/1.5/1.5	**/1.3/1.3
Winding hot-spot temperature, °C	140/140/120	140/140/120	**/120/120
Top-oil temperature, °C	95/105/105	95/105/105	**/105/105
Long-time emergency loading:			
Current, pu.	2.0/1.8/1.8	2.0/1.5/1.5	**/1.3/1.3
Winding hot-spot temperature, °C	160/150/140	160*(140*)/140/140	**/130/140
Top-oil temperature, °C	115/115/115	115/115/115	**/115/115

Note: the first number – from GOST 14209 – 85; the second – from IEC 60354 – 91, the third – from IEC 60076 – 7:2018; * the first number for transformers up to 110 kV, the second – above 110 kV ** – not rated

Table 3 Cyclic loading and daily insulation loss of life for winding hot-spot temperature of 120°C

Overload duration, h	Permissible overload K_2 for K_1			
	K_1=0.4	K_1=0.6	K_1=0.8	K_1=1.0
1.0	1,43 / 0,60	1,39 / 0,64	1,33 / 0,77	1,25 / 1,89
2.0	1,27 / 0,61	1,26 / 0,67	1,23 / 0,78	1,20 / 1,90
4.0	1,19 / 1,05	1,18 / 1,06	1,18 / 1,35	1,17 / 2,47
8.0	1,16 / 2,44	1,16 / 2,57	1,16 / 2,85	1,16 / 3,98
24.0	1,16 / 11,10	1,16 / 11,20	1,16 / 11,46	1,16 / 11,77

Permissible emergency overload K_2 = 1.1-1.5 in Table 4, while from the positions of the permissible current load according to [3] from Table 2 $K_2 \leq 1.3$ (more stringent design condition). These differences, apparently, should be taken into account by operating organizations and manufacturing plants of transformer equipment.

Table 4 Permissible emergency overload of autotransformers

Overload duration, h	Permissible overload K_2 at ambient temperature, °C							
	−25	−20	−10	0	+10	+20	+30	+40
1.0	1.6	1.5	1.5	1.4	1.4	1.3	1.2	1.2
2.0	1.5	1.5	1.5	1.4	1.3	1.3	1.2	1.1
4.0	1.5	1.5	1.4	1.4	1.3	1.2	1.2	1.1
8.0	1.5	1.5	1.4	1.4	1.3	1.2	1.2	1.1
24.0	1.5	1.5	1.4	1.4	1.3	1.2	1.2	1.1

3 Accounting for n-1 modes

As it is known, the nominal parameters of electrical devices are selected according to the normal mode and checked for maintenance and postfault mode. In this regard, 153 two-transformer substations were selected: 220/110 (103 substations) and 500/110-220 kV (50 substations). For them, the power flow calculations in the grid were carried out in the n–1 mode (when one of the autotransformers was turned off). The electric grid model was also formed based on actual measurements on regime days 2014–2016 yrs. Accumulation curves of the load distribution of autotransformers in normal mode in relation to their rated capacity in 2020 is determined as the arithmetic mean value of the power of autotransformers per day (winter or summer regime day) in relation to the rated capacity according to 124 substations (107 - 220 kV substations, 24 - 330 kV substations, 49 - 500 kV substations)

 https://doi.org/10.1051/e3sconf/202338401003

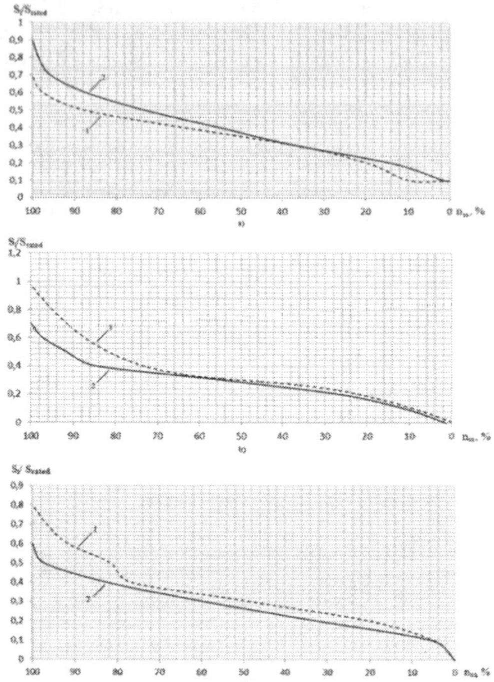

Fig. 2 Integral load distribution of AT S_{normal}/S_{rated} according to data from 2015 (a),2020 (b), and 2021(c) years: 1 – for 220 kV substations, 2 – for 500 kV (a) and 330-500 kV (b,c)

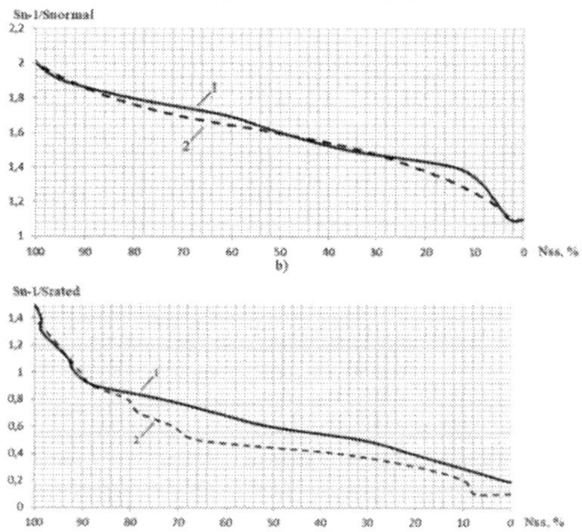

Fig. 3 Integral load distribution of AT S_{n-1}/S_{rated} (a) and S_{n-1}/S_{normal} (b): 1 – for 220 kV substations, 2 – for 500 kV

The distribution of loads of autotransformers in normal mode (S_{normal}) and in n – 1 mode (S_{n-1}), referred to their rated power (S_{rated}) and normal load (S_{normal}) are shown in Fig. 2-3. The predominant loads of autotransformers in normal mode are concentrated in the area S_{normal}/S_{rated} << 0.6. In the n–1 mode, the load rises to $S_{n-1}/\ S_{rated}$<< 1.5. In this regard, the $S_{n-1}/\ S_{rated}$ distribution curve is more indicative. Here the load of autotransformers increases by 1.1-2.0 times compared to that for normal operation; the average value of S_{n-1}/S_{rated} = 1.5, and the maximum value is 2.0. The last value is typical for single-ended substation, when the entire load falls on the autotransformer remaining in operation. The average value of 1.5 in the n–1 mode characterizes the generalized effect of grids of adjacent voltage levels on the resulting power flow in the electrical grid.

In other words, when one of the autotransformers is out-of-operation, its load is approximately equally distributed over the grid between the shunt connections of other voltage levels and the autotransformer that remains in operation. Thus, the connection scheme (configuration) of electrical grids is a significant influencing factor in the justification and selection of the power of autotransformers in the transmission grids of power systems.

In 90% of autotransformers, the relative load in the n–1 mode does not exceed 1.0, i.e. rated capacity of the corresponding autotransformers. In this regard, it can be assumed that their rated capacity is selected in the transmission grids of the country's power systems is not entirely rational, since it practically does not take into account the available overload capacity of transformer equipment.

To confirm this conclusion, a correlation analysis was carried out between the years of commissioning of autotransformers (Fig. 4) and their loading in the n–1 mode (during the analysis, 108 substations 220 and 500 kV were selected).

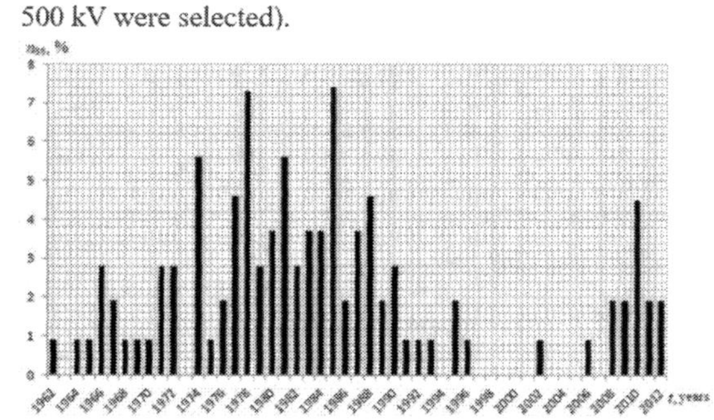

Fig. 4 Distribution of commissioning of the considered autotransformers by years.

Calculations have shown that the correlation coefficient between the load of autotransformers and their year of commissioning is 0.09, which indicates the absence of cause-and-effect relationships here.

4 Accounting for n-2 modes

In [1] there were calculated AT loadings S_{n-1} and S_{n-2} on 73 two-transformer substations with voltage level from 220 kV to 500 kV in n–1 and n–2 modes in all regions of the country for peak loading scheme of 2015 year. N-1 modes were modeled with failure of one autotransformer. In n-2 modes all OHL 110 kV and above in a corresponding region were orderly and sequential failured in addition to failure of one autotransformer (more specifically – with one autotransformer out-of-operation due to scheduled maintenance) n-1 modes. For one particular substation, the number of modes was in the range of 100 to 400.

Obtained results:

– from 16073 *n*–2 modes - 4919 modes (about 30%) lead to excess $S_{n-2}>S_{n-1}$;

– from these 4919 modes - 4718 design conditions (95.9%) lead to excess of autotransformer loading less than 0.1 in comparison with the same in *n*–1 mode and 4718+115=4833 (98.3%) – less than 0.2;

– the number of modes, where $S_{n-2}/S_{n-1}>1.1$, decreases rapidly with increasing of power surge;

– between 16073 design modes there was found one extreme event with $S_{n-2}/S_{n-1}=1.81$.

It should be noted that the ratio $S_{n-2}/S_{n-1} >> 1.0$ only in 20% of cases was caused by the failure of the overhead line from the substation under consideration. Another 30% were added by failure of overhead lines, which evacuated the power of power plants. The remaining 50% are other failures.

As was shown, ceteris paribus, permissible emergency overloads are 0.2–0.3 higher than systematic loads. Thus, if capacity of autotransformers is selected according to n-1 mode, then while verify it in n-2 mode we can exclude 98.3% design conditions from consideration, $S_{n-2}/S_{n-1}<1.2$. Next, in [1] reliability evaluation of n-1 modem, but authors could not found even one for reliability evaluation. In a point of fact, in the majority of cases rated capacity of autotransformers installed are exceed power flows through autotransformers not only in n-1 mode, but also in n-2 mode. In other words, rated capacities of given autotransformers due to various reasons turn out to be such uprated, that it's impossible to estimate real limitations for doing reliability evaluation.

Further, in [1], the reliability evaluation of n–2 modes was conducted using the example of a real 220/110 kV substation with S_{rated}=250 MV·A.. According to the results of the calculation of modes, it turned out that in the most severe n-2 mode, there is a limitation on power transmission in the amount of 37 MW. Further, the calculation of the resulting value of the undersupply of electricity was made, taking into account the reliability parameters of the grid elements for the 3 most severe n-2 modes.

It is shown that in order to eliminate restrictions on power flows in n–2 modes in the scheme of the substation under consideration, one can go for the option of increasing the autotransformers' capacity from 125 to 200 MV·A. With the maximum recommended value of specific damage of 240 rub./(kW·h) economic consequences due to the unreliability of the scheme for $\Delta W \approx 4.0$ MWh will amount to 10.8 million rubles. while the costs associated with an increase in the rated power of the AT, 134 million rubles, i.e. about an order of magnitude higher

Conclusion

1. It was revealed that since 2015 the average load of autotransformers installed in transmission grids of the country has retained its value, as well as the characteristic load curves by region, foreign approaches to determining the load capacity of AT have not fundamentally changed.

2. The parameters of the actual equivalent load graphs of substations in transmission grids may differ significantly across the regions of the country. The revealed preload is within a wide range of 0.6 - 0.9 with an overload duration of 8-9 h. However, the permissible loading of autotransformers in this case turns out to be practically unchanged, which is fundamentally important when assessing it for a given perspective.

3. The existing cumbersome practice of justifying and choosing the capacity of autotransformers in transmission grids, taking into account not only the n–1 modes, but also hundreds and thousands of n-2 modes, is redundant. It is not possible to substantiate the advisability of increasing the capacity of autotransformers in n–2 modes even under the most severe design conditions and initial data.

4. The power of autotransformers should be selected based on the maximum power flow in the n–1 mode, i.e. when one of them is disconnected at the substation, taking into account the permissible cyclic loading. -2 modes should be considered only to determine the requirements for emergency automatics.

5. There is an urgent need to revise GOST 14209-85 and expand its provisions to power autotransformers of all standard sizes and increased voltages. One of the options is harmonization with the International Electrotechnical Commission standard, which offers very gentle design conditions for justifying the loading of autotransformers.

References

1 Vasilenko NE, Plotnikov AV, Shuntov AV (2017) About loading of autotransformers in main electrical grids and conclusions for designing, 3S Web Conf, Vol. 25, № 01012 (DOI: 10.1051/e3sconf/20172501012)

2 State Standard (GOST) 14209 – 85. General-purpose oil-immersed power transformers. Permissible loads, Standartinform Publ., Moscow.

3 Power transformers – Part 7: Loading guide for mineral-oil-immersed power transformers: IEC 60076-7-2018. -. IEC, 2018, 94 p.

4 Neklepaev BN (1986) Electric Part of Power Stations and Substations, Energoatomizdat, 640 p.

5 Hublarov NN (2004) On the development of a new manual for the loads of operated autotransformers 220 – 750 kV, Vestnik VNIIE, pp. 90-102.

6 Standart of PJSC FGC UES Operating instructions for transformers: STO 56947007 – 29.180.01.116-2012", 2010, 52 p.

Improvement of Algorithms of the Topological Model of the Steady Mode of Electric Power Systems

Dauren Akhmetbaev [1], *Abdygali* Dzhandigulov [2,*], *Arman* Akhmetbaev[3], *Svetlana* Bystrova[4], *Shakizat* Begezhanova[2]

[1]S. Seifullin Kazakh Agrotechnical University, Astana, 010011, Kazakhstan
[2]L.N. Gumilyov Eurasian National University, Astana, 010000, Kazakhstan
[3]JSC Kazakhtelecom, Almaty, 050057/A15H9D6, Kazakhstan
[4]Ekibastuz Engineering and Technical Institute named after academician K. Satpaev, Ekibastuz, 141208, Kazakhstan.

Abstract. The work improves the algorithms for the formation of a steady state mode of electric power systems, developed on the basis of a topological model. The topological model is developed on the basis of the theory of directed graphs using a matrix of current distribution coefficients. Algorithms for the formation of a steady state provide for conducting calculations with respect to all independent nodes of a complex network of a power system, which leads to an increase in the number of operations of iterative processes. The paper proposes a transformation of the topological model of the system, in which, in the process of solving the transformed model, iteration is performed only over nodes that have non-zero load or generation powers. The rest of the nodal stresses are simply calculated through the stresses of the above nodes. The resulting transformation makes it possible to reduce not only the number of operations in each iteration, but also the number of iterations itself. In particular, when studying a 5-node test circuit [1], instead of four complex equations of the original system, it is sufficient to iterate only one equation (reduction by 25%), and for the New England test 39-node circuit [2], instead of 38 equations It is enough to iterate over 26 equations (32% reduction).

Introduction

Methodological problems of the analysis of electric power systems (EPS) cover a wide range of issues and tasks related to the control of its modes, first of all, this is a study on the formation of equations of steady-state modes of EPS. The complexity of solving the problem of controlling the EPS, taking into account the restrictions that take into account economic and reliability indicators, is explained by the large dimension of the nonlinear equations that describe its mode and composition of telemetry and the variety of mathematical models of modes associated with this [3]. The application of the matrix calculation method in the theory of electric networks, developed in [4] and further developed in [5, 6], unambiguously determined the direction of research on the formation of the equations of the steady state mode of the EPS based on the generalized parameters of its equivalent circuit. In some papers, it is proposed to reduce complexity by replacing the nonlinear model with linear models of integer programming [7]. The topological content of the $\underline{Z}$-nodal generalized parameters made it possible to take it as the basis of mathematical models and methods for calculating complex EPS circuits.

The foundations of the topology of the equivalent circuit of electrical networks were laid down in the classical works of Kirchhoff and Maxwell. They first introduced the concepts of trees and obtained topological expressions for the determinants of the node conductance matrix) $\underline{Y_y}$ and the circuit resistance matrix $\underline{Z_k}$.

In modern literature, these topological expressions for the determinants $\underline{\Delta_y}$ and $\underline{\Delta_k}$, written in terms of the values of undirected trees and complements, are known as derivatives of Kirchhoff's laws [8].

The practical application and development of topological methods in the analysis of electrical circuits became possible after the publication of the works of Percival, Sechu, Mason, Coates and others. In the works [9,10] of Percival V.S. the concepts of a common branch representing a group of trees and a pair of generalized branches representing a group of 2 trees of a graph are derived, and theorems are given for determining groups of trees of a graph of a circuit composed of subcircuits connected by a tree in parallel and in the form of a contour. This theorem served as the basis for the development, Ionkin P.A., Sokolov A.A. [11,12], ways to find trees by decomposing the original diagram graph into node pairs, by branches and by node. The further development of the topological method, in relation to the problems of analyzing steady-state modes of EPS, was received in the works of O.T. Geraskin [13, 14].

To find the current distribution coefficients of electrical networks, the ratios of the sum of the weights of specific trees to the sum of the weights of all possible trees of the graph are calculated [15]. The complexity of the formation of matrices of current distribution coefficients lies in determining the numerators of

*Corresponding author: abeked@mail.ru

https://doi.org/10.1051/e3sconf/202338401004

topological expressions by dividing the network into two parts in order to find two graph trees. Akhmetbaev D.S. proposed an analytical approach to determining the topological content of the matrix of current distribution coefficients based on the properties of possible graph trees, without dividing the network into two parts [16,17]. In [18], an efficient algorithm for directed search and determination of the weights of possible graph trees is implemented without involving previously defined trees.

Studies conducted to analyze the methods and algorithms for searching for possible trees of a complex graph led to the development of optimal algorithms based on the principles of diakoptics [18]. The basis of the idea of optimization is the construction of special classes of trees. At the same time, in the process of grouping, parts of future graphs that are "parent" for groups of graphs are allocated and the corresponding graphs of a significantly lower dimension are constructed.

Based on the topological algorithm of the current distribution coefficients, a matrix equation of the steady state is formed.

1 Topological algorithm for the formation of a steady state

Nodal voltages are determined by the expression obtained by analytical transformation of the known equations of the electrical state of networks [19,20]:

$$U = U_0 + C^T Z_B C \overline{U_D^{-1}} \overline{S(U)}, \qquad (1)$$

where C is a rectangular complex matrix of current distribution coefficients; Z_B is the diagonal matrix of branch resistances; U_D is the diagonal matrix of nodal conjugate stresses; $S(U)$ is a column vector depending on the power of nodal loads and generators and conductivities of transmission lines; T - matrix transposition sign.

The resulting system of nonlinear equations for the steady state of electric power systems is developed on the basis of the topological model of the matrix of current distribution coefficients. This system was proposed and applied by the authors and represents $(n-1)$ complex nonlinear equations with respect to $(n-1)$ complex values of nodal stresses. Where n is the total number of nodes in the network. The resulting system is solved by iterative methods. As the design scheme of the EPS becomes more complex, the number of operations of the iterative process increases significantly, which reduces the efficiency of obtaining calculation results. An improved algorithm for organizing the iterative process is proposed below.

2 Improving the algorithms for calculating the steady state.

System (1) is solved by iterative methods. In this paper, a transformation of the system under consideration is proposed, in which, in the process of solving the transformed system, iteration is performed only over nodes that have non-zero load or generation powers. The rest of the nodal voltages are simply calculated through the voltages of the above nodes.

In system (1), the column vector $S(U)$ is expressed by the formula:

$$S(U) = S_0 - j \cdot A \cdot |U|^2, \qquad (2)$$

where S_0 is a column vector of given (constant) load and generation powers, A is a diagonal matrix with real elements, calculated through given conductivities b.

Denote by $M = C^T Z_B C$. Substitute expression (2) into (1)

$$U = U_0 + M\overline{U_D^{-1}}(S_0 - j \cdot A \cdot |U|^2).$$

Let's open the brackets and carry out transformations:

$$U = U_0 + M\overline{U_D^{-1}}S_0 - j \cdot M \cdot A \cdot U,$$

$$(E + j \cdot M \cdot A)U = U_0 + M\overline{U_D^{-1}}S_0,$$

$$U = (E + j \cdot M \cdot A)^{-1}(U_0 + M\overline{U_D^{-1}}S_0). \qquad (3)$$

System (3) can be rewritten as

$$U = V_0 + G\overline{U_D^{-1}}\bar{S}_0, \qquad (4)$$

where $V_0 = (E + j \cdot M \cdot A)^{-1} U_0, G = (E + j \cdot M \cdot A)^{-1} M$.

Component wise, it can be written as:

$$U_k = (V_0)_k + \sum_l G_{kl} \cdot {(\bar{S}_0)_l}/{\overline{U_l}}. \qquad (5)$$

From the system of equations (4-5), we see that when solving by iterative methods, in the right part of the system, only those values of nodal voltages are used, in which there are non-zero generation or load powers. That is, system (4-5) can be divided into two subsystems: equations with non-zero generation and/or load powers and equations without them. Only the first part is solved by iterative methods. And the second part is only formulas for calculating the remaining nodal stresses.

The resulting transformation makes it possible to reduce not only the number of operations in each iteration, but also the number of iterations itself. In particular, when studying a 5-node test circuit, instead of four complex equations of the original system, it is sufficient to iterate only one equation (75% reduction).

3 Implementation of the algorithm

Consider a 5-node test circuit.

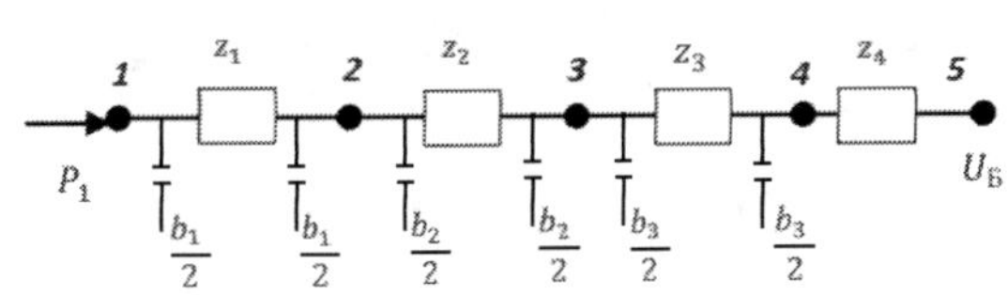

Fig.1. Design scheme

This scheme is a test circuit rucase_5_4 and the source data for is taken from open sourcesx [1].

Table 1. Initial data by nodes.

Nodes №	U nom	Voltage		Weighting mode, S, MVA		Generation power	
	kV	d, degree	Modulus, kV	P, MW	Q, MVA	P, MW	Q, MVA
1	500	68.73	484.53	0.00	0.00	5400	0.00
2	500	45.05	482.05	0.00	0.00	0.00	0.00
3	500	21.97	489.35	0.00	0.00	0.00	0.00
4	500	0.12	499.90	0.00	0.00	0.00	0.00
5	500	0.00	500.00	0.00	0.00	0.00	457

Table 2. Line design parameters

Lines №	Complex resistances, Z, Ohm	Conductivity b, cm
1	0.7000 + 17.5000i	8.3200
2	0.7000 + 17.5000i	8.3200
3	0.7000 + 17.5000i	8.3200
4	0.0000 + 0.1000i	0.0000

The matrices M and the column vector S_0 in this case have the form:

$$M = \begin{pmatrix} 2.1 + 52.6j & 1.4 + 35.1j & 0.7 + 17.6j & 0.1j \\ 1.4 + 35.1j & 1.4 + 35.1j & 0.7 + 17.6j & 0.1j \\ 0.7 + 17.6j & 0.7 + 17.6j & 0.7 + 17.6j & 0.1j \\ 0.1j & 0.1j & 0.1j & 0.1j \end{pmatrix},$$

$$A = \begin{pmatrix} 4.16 & 0 & 0 & 0 \\ 0 & 8.32 & 0 & 0 \\ 0 & 0 & 8.32 & 0 \\ 0 & 0 & 0 & 4.16 \end{pmatrix} \cdot 10^{-3},$$

$$S_0 = \begin{pmatrix} 5400 \\ 0 \\ 0 \\ 0 \end{pmatrix}, (E + j \cdot M \cdot A)^{-1} U_0 =$$

$$= \begin{pmatrix} 1232 - 64.767j \\ 01142 - 56.465j \\ 886.8 - 33.289j \\ 502.4 - 0.102j \end{pmatrix},$$

$$(E + j \cdot M \cdot A)^{-1} M \cdot diag(\bar{S_0}) =$$

$$= \begin{pmatrix} (0.4754 + 5.676j) \times 10^5 & 0 & 0 & 0 \\ (0.3865 + 4.319j) \times 10^5 & 0 & 0 & 0 \\ (0.2161 + 2.335j) \times 10^5 & 0 & 0 & 0 \\ 69.948 + 1.33j \times 10^3 & 0 & 0 & 0 \end{pmatrix}.$$

Thus, the first part of system (5) in this case is one complex equation for the voltage of the first node, which we write in the form:

$$U_1 = (V_0)_1 + G_{11} \cdot {(\bar{S}_0)_1}/{\overline{U_1}}. \quad (6)$$

If we write equation (6) component by component, using the following notation,

$$U_1 = x + j \cdot y, \ (V_0)_1 = x_0 + j \cdot y_0 = 1232 - 64.767j,$$

$$G_{11} \cdot (\bar{S}_0)_1 = \alpha + j \cdot \beta = (0.4754 + 5.676j) \times 10^5,$$

then we get the system of equations

$$\begin{cases} x = x_0 + \dfrac{\alpha x - \beta y}{x^2 + y^2}, \\ y = y_0 + \dfrac{\alpha y + \beta x}{x^2 + y^2}, \end{cases}$$

which reduces to a quadratic equation with respect to, for example, the variable x, solving which we find two solutions:

$$x_1 = 175.64, x_2 = 1.105 \times 10^3,$$

the corresponding imaginary parts are equal:

$$y_1 = 451.48, y_2 = 402.64$$

The second solution is eliminated due to the constraints $|U_1| \leq 500$. The corresponding phase and modulus will be: $\arg(U_1) = 68.74, \ |U_1| = 484.44$.

Table 3. The total results for all nodes will have the following values:

№	Generation power		Voltage				Voltage Relative difference, %	
			Test's results		Results for the simplified algorithm			
	P, MW	Q, MVA	d, degree	Modulus, kV	d, degree	Modulus, kV	d, %	Modulus,%
1	5400	0.00	68.74	484.53	68.73	484.44	0,0%	-0,02%
2	0.00	0.00	45.05	482.05	45.73	481.86	1,5%	-0,04%
3	0.00	0.00	21.98	489.35	21.98	489.23	0,01%	-0,02%
4	0.00	0.00	0.12	499.90	0.12	499.90	0,0%	0,00%
5	0.00	457.0	0.00	500.00	0.00	500.00	0	0

Thus, in the problem under consideration, there is no need to apply iterative methods of solution, and the solution is found exactly.

Conclusion

The proposed transformation significantly reduces the number of calculations performed in the calculations of the steady state.

Studying the question of conversion efficiency in a more general case is of interest and is the subject of further research.

References

1. 5-node test circuit rucase_5_4 *http://energy.komisc.ru/downloads/schemes/rucase_5_4.txt*
2. New England 39-node test circuit https://electricgrids.engr.tamu.edu/electric-grid-test-cases/new-england-ieee-39-bus-system/
3. L.A. Melentiev System Research in Energy. Moscow, Nauka, 1974.
4. N.A. Melnikov Matrix method of analysis of electrical circuits. 2nd revised edition, Moscow, Energia, 1972.
5. V.S. Khachatryan. Method and algorithm for calculating the steady state regimes of large energy systems. Izvestiya AN SSSR. Energy and transport. 1973.- No. 4.

6. A.D. Krayushkin. Software implementation of the Newton method with a variable step on the ES computer for calculating the normal electrical modes of a large-scale network // Automated Control Systems in the Energy Industry. - Moscow, 1982.
7. S.Karimi, Hazhar&Natarajan, Balasubramaniam. (2021). Joint Topology Identification and State Estimation in Unobservable Distribution Grids. IEEE Transactions on Smart Grid. PP. 1–1. 10.1109/TSG.2021.3102179.
8. O. T. Geraskin. Topological content of knot and contour electric network determinants and calculation of their values by means of a computer // Izvestiya AN USSR. Energetics and transport, 1966 (2)
9. W.S. Percival. Solution of passive electrical networks by means of mathematical trees // Proceedings IEE (London. - 1953, v.100(III, 65)
10. W.S. Percival. Implored matrix and determinant methods of solving networks // Proceedings IEE (London). - 1954.- v.101, (IV, 7).
11. P.A. Ionkin, A.A. Sokolov. Fundamentals of construction and transformation of graphs for the calculation of electrical circuits // Electricity. 1964 (5).
12. P.A. Ionkin. General equations for calculating electrical circuits using graphs // Electricity. 1964(8)
13. O.T. Geraskin. Graphs of an electrical network and topological formulas // Izvestiya AN USSR. Energy and transport. 1971(4).
14. O.T. Geraskin Topological analysis of current distribution coefficients in electrical networks. Izvestiya vuzov. Energetics, 1967 (5).
15. D.S. Akhmetbaev. Topological method of analysis of stationary regimes of complex electrical networks // Electrical equipment: operation and repair. Moscow. 2010.
16. D. S. Akhmetbaev. Method for calculating steady-state modes of electric networks based on current distribution coefficients. Electricity, 2010 (11)
17. D.S. Akhmetbayev, A.R. Dzhandigulov. Finding all spanning graphs of a given graph. Computer program. Certificate of entering information into the state register of rights to objects protected by copyright. No. 1551 dated January 31, 2019
18. D.S. Akhmetbayev, A.R. Dzhandigulov, S.V. Bystrova. Topological system method of formation of transformer transformation coefficients. E3S Web of Conferences. 216.01087.10.1051/e3sconf/202021601087. (2020)
19. D.S. Akhmetbayev, A.D. Akhmetbayev, A. Berdygozhin. The Topological Method for Producing the Inverse Form Nodal Equations Electric Power System. Electricestvo. 5. 18-27. 10.24160/0013-5380-2018-5-18-27.
20. D.S. Akhmetbaev, A.R. Dzhandigulov Development of algorithms for a new topological method for calculating current distribution coefficients in complex electrical networks//Eurasian journal of mathematical and computer applications. ISSN 2306–6172, Volume 7, Issue 3 (2019) 4 – 12

Real-time Power System Topology Recognition through Convolutional Neural Networks

Natalja Gotman[1]*, *Galina* Shumilova[1]

[1]Federal Research Center "Komi Scientific Center of the Ural Branch Russian Academy of Sciences", ISE and EPN, 167000, Russia

Abstract. This paper investigates the status of a transmission line (on/off) using a 140-bus Northeast Power Coordinating Council (NPCC) test system model [1]. The application of the software package ANDES [2] to obtain a database when solving this problem in the transient process of the power system is considered. The Deep Learning Neural Networks (DLNN) [3] were proposed to solve the problem, in particular, a convolutional neural network (CNN), the input variables of which are voltage and current phasors obtained from phasor measurement units (PMU). Calculations to determine the state of lines were performed using a program developed in the Julia language using the Flux package (a machine learning library that includes functions for creating CNN models). The results of the studies are presented.

1 Introduction

The task of detecting changes in a topology of an electric power system in transient states is one of most popular and complex tasks to be solved in the operational management of the power system. The electrical network topology is determined using information about the position of circuit breakers and disconnecting switches (open/closed), which is transmitted via telemechanics channels to the dispatch control centers. As a result of disturbances, this information can be distorted, leading to errors in determining connections of network components. Also sensors that transmit and receive information may be malfunctioning due to equipment failures or cyber attacks.

Modern PMUs also provide a line state information (on/off) due to the telesignalization function implemented in them, but not placed in all buses of the power system PMUs do not allow to get information about the states of all lines. In this paper, this problem is solved using convolutional neural networks based on data measured by the PMUs. Deep neural networks are becoming one of the most popular machine learning techniques for creating artificial intelligence systems in various fields due to their enhanced performance and scalability. With respect to solving energy problems, DLNN methods are widely used for the detection of false data injection attacks [4-5], load forecasting [6], load modeling [7], fault location [8], detecting defects in damaged equipment due to short circuits [9], etc.

There are various DLNN architectures (multilayer perceptron, neocognitron, autoencoders, convolutional neural networks, limited Boltzmann machine, deep trust networks, etc.). In this paper, a convolutional neural network based classifier, similar to the one developed earlier by the authors for the 14-bus electrical network [10], but focused on the 140-bus network, is used to detect changes in the topology of the 140-bus electrical network in real time.

The remaining parts are organized as follows: Sections 2, 3, and 4 briefly describe the architecture of the convolutional neural network, the ANDES software package used to obtain the database, and the 140-bus power system under study, respectively. Numerical results with the study of the influence of various factors on the accuracy of topology detection are given in Sections 5, 6. Then Section 7 concludes results and discusses future research.

2 Convolutional Neural Networks

To solve the problem of determining the change of line status in transient states, especially in the case of large volumes of input parameters, of all the above DLNN architectures the most suitable are CNNs. CNN receives input data, transforms them using a number of interrelated layers, and outputs a set of probabilities (estimates). The CNN architecture (Fig. 1) consists of three main groups of layers: 1) an input layer; 2) feature extraction layers; 3) classification layers. The input layer accepts 3D signals. Feature extraction layers have a repeating structure: convolution (filter) $\rightarrow$ activation (ReLU) $\rightarrow$ pooling. The convolution layer is a set of feature maps, which have a scanning kernel (filter). Fully connected layers are the layers of the usual multilayer perseptron, which determine the final classification decision. Gradient-based optimization method (error back-propagation algorithm) is used to trained CNN. A more detailed description of CNN is presented by the authors in [10].

The authors used the ANDES software package to obtain a database for the topology detection problem decision of a 140-bus test system.

* Corresponding author: gotman@energy.komisc.ru

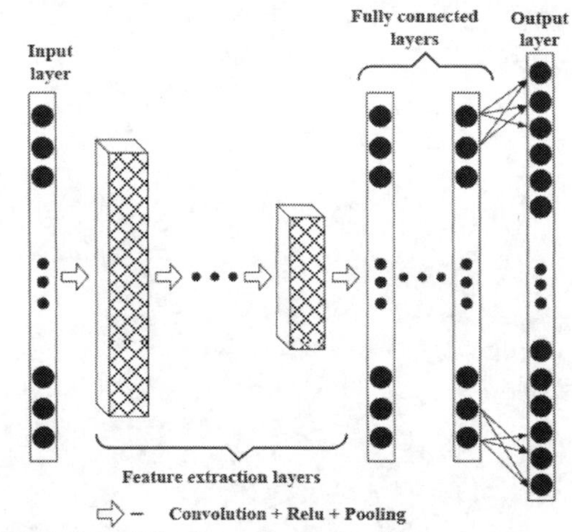

Fig. 1. Convolutional neural network architecture.

3 ANDES Software Package

ANDES is a free open-source software package developed based on the Python programming language for modeling and numerical analysis of power system operation [2]. The package has a hybrid symbolic-numeric structure for solving differential algebraic equations (DAEs) describing the energy system model [11].

The ANDES package allows you to develop abstract models in the symbolic layer independently of the test scheme. The computation time for processing equations depends on the number and complexity of model types, not on the number of devices in a particular test scheme.

Thus, ANDES is much easier to use than other DAEs modeling tools for dynamic power system simulation while maintaining high computational efficiency: ANDES can perform a 20-second transient simulation of a 2000-bus system in a few seconds on an ordinary desktop computer.

The ANDES calculations were performed by the authors in Google Colab, a free interactive cloud environment for working in Python.

4 Investigated NPCC 140-bus System

The 140-bus Northeast Power Coordinating Council (NPCC) system [1] is the equivalent of the Northeast Region of the Eastern Interconnection of the United States and Canada power system.

The scheme under study is shown in Fig. 2. The system includes 48 generators, 233 lines and 92 load buses. In Fig. 2 the generator buses are shown as squares and the load buses as circles. The base bus (78) is shown as a triangle.

As mentioned above, the implementation of the algorithm for detecting the topology of the electrical network uses the data transmitted by the PMU. In the NPCC scheme, the PMU is installed in 57 buses [12].

The database for solving the problem of determining the network topology are phasor measurements of voltage and current in the transient process, but the package ANDES in the calculation file of the necessary values provides only the values of the modulus and phase of the voltage in the buses. The following calculation formulas were used to determine the missing phasor current values in the branches:

$$\dot{y} = 1/(r + jx), \tag{1}$$

$$\dot{I}_{н} = (\dot{U}_{н} - \dot{U}_{к}) \times \dot{y} + j0,5 \times \dot{U}_{н} \times b, \tag{2}$$

$$\dot{I}_{к} = (\dot{U}_{к} - \dot{U}_{н}) \times \dot{y} + j0,5 \times \dot{U}_{к} \times b, \tag{3}$$

where $\dot{U}_{н}$ - the bus voltage phasor at the beginning of the line, $\dot{U}_{к}$ - the bus voltage phasor at the end of the line, r and x - respectively, the active and reactive resistances of the line, $\dot{y}$ - complex conductivity of the line, b - capacitive conductivity of the line, $\dot{I}_{н}$ - the current phasor at the beginning of the line, $\dot{I}_{к}$ - the current phasor at the end of the line.

5 Numerical Experiments

The data base, calculated with the ANDES package, includes 600 modes obtained by means of the load changes in all load buses in the range 75–125 percent of the base level and adding a random value equal to 0–20 percent of the base load of the bus to the obtained values. For each bus, transient calculations associated with turning off one of the lines and turning it back on after three seconds with an auto-reclosing device have been performed. In the experiment, 100 lines turned off/in is considered (these lines are highlighted by a dotted line in Fig. 2). Half of the 600 calculated modes was used to train the CNN, and the other half was used for testing. Voltage and current measurement values were used with the addition of a random noise: the phase change of voltage by ±0.5° and the modulus change of current by ±0.5%. The superposition of noise is due to the PMU errors and voltage and current instrument transformers.

Calculations to determine the topology were performed using the program written in the Julia language using the Flux package.

6 Influence of Various Factors on the Accuracy of Topology Change Detection

Variants of the CNN input parameters. 10 variants of input parameters are considered.

The first option consists of changes in measurement values of one (first) time slice after the beginning of the transient mode. In this case, the number of CNN input parameters is 256 (voltage phases in the PMU placement buses and current modules in the branches incident to the PMU placement buses).

Changes in measurement values for one time slice are calculated as the difference between the

 https://doi.org/10.1051/e3sconf/202338401005

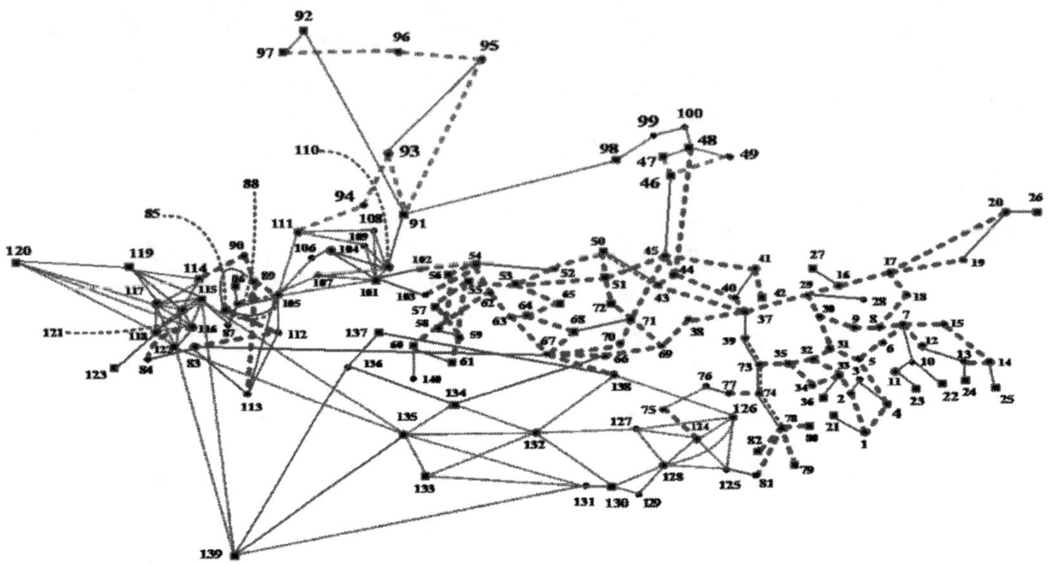

Fig. 2. NPCC 140-bus power system.

measurements of the current and the preceding time slice. The time interval between time slices is 0.02 s, which corresponds to the PMU sampling rate.

Since to solve the problem, an CNN is used, the input layer of which, as mentioned above, accepts three-dimensional signals, the first version of the input parameters is a three-dimensional 1x256x1 matrix. For each variant of input parameters a different CNN architecture is used.

The results of the calculations according to the variants proposed above are shown in Table 1.

Table 1. Average accuracy of topology change detection for different data options when disconnecting/connecting one line.

Data option	Dimension of input parameters	Average calculation accuracy (%)
1	1x256x1	91,48
2	2x256x1	92,19
3	3x256x1	92,81
4	4x256x1	92,71
5	5x256x1	92,79
6	6x256x1	95,24
7	7x256x1	94,92
8	8x256x1	95,08
9	9x256x1	95,03
10	10x256x1	95,11

The accuracy of the calculations shows how correctly the CNN has determined the state of the lines in all the tested samples. The lowest accuracy corresponds to the first variant of the input data (91.48%). Starting from the sixth variant accuracy in detecting changes in the network topology in question in the event of switching-off / in one of the lines is over 95%. Thus, the accumulation of the PMU data from the time of the emergency improves the topology detection accuracy of the electric network.

Rough error in PMU measurements. The effect of coarse measurement error associated with a change in the sign of the current modulus was investigated for 6 - 10 variants of the input data. The essence of the study was that in one of the time slices (after 0.02 s and 0.14 s after the start of the transient) the sign of the current modulus for one of the two lines (25th or 28th) changed, and the indicated change remained constant until the end of the time interval in question.

The interval of current modulus values in all 600 modes for Line 25 is [0.0295;4.8855] (mean value 1.072 p.u.); for Line 28 is [0.7323;5.8693] (mean value 2.7079 p.u.). Training was performed with data without coarse error. The test results are shown in Table 2.

Based on the results obtained, a conclusion was made about a sharp decrease in the accuracy of detecting topology changes in the case of switching-off one of the lines (in Table 2, these values are highlighted in bold). For line 25 with a smaller value of the current modulus, the accuracy in this case is higher than for line 28. When the line was turned on, the classifiers correctly identified the topology state in most cases.

Due to the low accuracy of topology detection, a classifier using the 10th variant of input parameters was tested. The training sample included 20% of the data with a coarse error. The results of the calculations are shown in Table 3, which shows that the accuracy for the data without coarse error decreased slightly, but the accuracy for the data with coarse error increased.

https://doi.org/10.1051/e3sconf/202338401005

Table 2. Calculation results for data with gross errors.

Data size		Error	Time when the error occurred, c	Accuracy (%) when the line outage occurs	Accuracy (%) when the line is turned on
6x256x1	Changing the sign of the current modulus	No gross errors	-	92,07	98,4
		when the line 25 outage occurs	0,02	**28,935**	95,92
			0,14	**49,905**	98,4
		when the line 25 is turned on	0,02	92,07	98,39
			0,14	92,07	79,33
		when the line 28 outage occurs	0,02	**6,375**	98,4
			0,14	**19,3**	98,4
		when line 28 is turned on	0,02	92,07	99,42
			0,14	92,07	67,43
7x256x1		No gross errors	-	91,68	98,16
		when the line 25 outage occurs	0,02	**43,09**	98,16
			0,14	**40,04**	98,16
		when line 25 is turned on	0,02	91,68	99,31
			0,14	91,68	98,24
		when the line 28 outage occurs	0,02	**7,92**	98,16
			0,14	**11,26**	98,16
		when line 28 is turned on	0,02	91,68	99,78
			0,14	91,68	79,31
8x256x1		No gross errors	-	91,69	98,46
		when the line 25 outage occurs	0,02	**27,26**	98,46
			0,14	**14,6**	98,46
		when line 25 is turned on	0,02	92,58	99,39
			0,14	91,69	98,91
		when the line 28 outage occurs	0,02	**11,48**	98,46
			0,14	**2,95**	98,46
		when line 28 is turned on	0,02	91,69	99,66
			0,14	91,69	99,59
9x256x1		No gross errors	-	92,29	97,76
		when the line 25 outage occurs	0,02	**31,28**	97,76
			0,14	**24,38**	97,76
		when line 25 is turned on	0,02	92,29	99,48
			0,14	92,29	99,45
		when the line 28 outage occurs	0,02	**7,34**	97,76
			0,14	**4,26**	97,76
		when line 28 is turned on	0,02	92,29	99,82
			0,14	92,29	99,8
10x256x1		No gross errors	-	91,95	98,26
		when the line 25 outage occurs	0,02	32,84	98,26
			0,14	**21,0**	98,26
		when line 25 is turned on	0,02	91,95	95,3
			0,14	91,95	93,79
		when the line 28 outage occurs	0,02	**13,5**	98,26
			0,14	**9,51**	98,26
		when line 28 is turned on	0,02	91,95	98,3
			0,14	91,95	99,18

7 Conclusion

In this work, studies were performed to detect the power transmission line state (switching-off/in) using the test model of the 140-bus NPCC system (USA, Canada). The application of the ANDES software package to obtain a database when solving this problem in the transient mode of the power system is considered.

The developed classifier based on convolutional neural network processes a large volume of input data in real time. Phasor measurements of phases and effective

Table 3. Accuracy of calculations for 10 variants of input parameters when including data with gross error in the training sample.

Data		Accuracy when the line outage occurs, (%)	Accuracy when the line is turned on, (%)
No gross errors		90,13	98,17
Changing the sign of the current modulus	when the line 25 outage occurs in 0,02 s	91,45	98,16
	when the line 25 outage occurs in 0,14 s	89,63	98,16
	when the line 28 outage occurs in 0,02 s	86,48	98,16
	when the line 28 outage occurs in 0,14 s	81,07	98,16

values of voltages and currents (modules) come from 57 PMU installed in the power system. The accuracy of the line status detection is more than 95% when the phases of voltages and the effective values of currents of 6-10 time slices are used as input parameters. Studies on the effect of coarse error (reversal of the sign of the current modulus) on the accuracy of topology detection have shown that to increase the accuracy it is necessary to include data with a coarse error in the training sample. In the future, we will test this work on real-data (as opposed to synthetically generated data).

References

1. Y. Lei, Y. Liu et al., "A Study on Wind Frequency under High Wind Penetration on an NPCC System Model", in Proceedings of the IEEE PES General Meeting, National Harbor, MD, USA, July 2014
2. ANDES. Python software for symbolic power system modeling and numerical analysis, [online] Available: https://docs.andes.app/en/stable/index.html
3. Josh Patterson and Adam Gibson, Deep Learning. A Practitioner's Approach / Transl. from Eng., Moscow, DMK Press (2018), 418 p. (in Russian)
4. Y.B. He, G.J. Mendis, J. Wei. "Real-time detection of false data injection attacks in smart grid: a deep learning-based intelligent mechanism", *IEEE Trans. on Power Systems*, vol. 8, no.5, pp. 2505–2516, (2017)
5. H.Z. Wang, J.Q. Ruan, G. Wang, et al., "Deep learning-based interval state estimation of AC smart grids against sparse cyber attacks", *IEEE Trans. Ind. Inf.*, vol. 14, no. 11, pp. 4766–4778, (2018)
6. H. Shi, M.H. Xu, R. Li, "Deep learning for household load forecasting-A novel pooling deep RNN", *IEEE Trans. Smart Grid*, vol. 9, no. 5, pp. 5271–5280, (2017)
7. C. Zheng, S.H. Wang, Y.L. Liu, *et al.*, "A novel RNN based load modeling method with measurement data in active distribution system",

Electr. Power Syst. Res., vol. 166, pp. 112–124, (2019)

8. W.T. Li, D. Deka, M. Chertkov, M. Wang, "Real-time faulted line localization and PMU placement in power systems through convolutional neural networks", IEEE Transactions on Power Systems, vol. 34, no. 6, pp. 4640–4651, (2019)
9. A. Muhammad, J.M. Lee, S.W. Hong, S.J. Lee and E.H. Lee, "Deep learning application in power system with a case study on solar irradiation forecasting", in Proc. Int. Conf. Artif. Intell. Inf. Commun. (ICAIIC), Feb. 2019, pp. 275–279
10. N.E. Gotman, G.P. Shumilova, "Power electrical network topology detection using changes of phasor measurements in transient states", in Proceedings of Methodological Problems in Reliability Study of Large Energy Systems, Vol. 139 (E3S Web of Conferences, Tashkent, Uzbekistan, 2019)
11. H. Cui, F. Li, K. Tomsovic, "Hybrid Symbolic–Numeric Framework for Power System Modeling and Analysis", ar Xiv:2002.09455v2 [eess.SY], Aug 2020, pp. 1–13
12. M. Khokhlov, I. Golub, "The Unified Approach for Optimal PMU Placement to Ensuring the Reliability of Power System Observability", in Proceedings of Methodological Problems in Reliability Study of Large Energy Systems, vol. 65 (At: Saint Petersburg, Russia, 2014) (in Russian)

Method for determining the critical parameters of the sustainability of an energy system with integrated renewable energy sources

Huseyngulu Guliyev*

Azerbaijan Technical University, Baku, Azerbaijan

Abstract. This paper proposes a method for modeling the regime of a power system with integrated variable renewable sources - wind power units and solar PV installations, the generation of which is continuously changing in accordance with the random variability of wind speed and solar radiation. Wind speed and solar radiation data were obtained for real regions in which a 240 MW wind farm and a 230 MW solar PV power plant are currently under construction. A method is proposed for calculating the value of the critical stress in the system for indefinitely changing wind speed and the level of its fluctuations, which determine the stochastic variability of the system regime. For each interval of change in wind speed, a set of probable modes of the system and the corresponding critical voltages in the network of the system are determined. The computational experiments were performed for the 30-node IEEE scheme.

Introduction

In the electric power systems (EPS) over the past 30 years, there have been significant changes, both in the structure of generation and in electrical transmission and distribution networks.

Modern EPS contain mode control devices that make it possible to make power facilities (generators, buses) flexible in controlling the generation of active and reactive power and their distribution, as well as control the demand for consumed power, thereby creating the best conditions for covering the power balance under various random disturbances.

In these cases, voltage instability is considered one of the main threats to the safety of the operation of modern EPS. In recent years, renewable energy sources (RES) have been widely used in energy systems to improve the efficiency of functioning and reduce the impact of harmful emissions on the environment. The increase in the generated power of RES in the power system began to have a significant impact on voltage stability [1, 2]. The article investigates the influence of the growth in the power of integrable wind generators on the change in the limits of voltage stability, the assessment of the possibility of violation of stability during shutdowns of system elements.

The risk of violating the stability of the power system is determined by the probability that one of the main elements of the system (generator, line or section of the power supply network) will continue to function without failures for a certain period of time. The growth of renewable energy (primarily solar and wind) is expected to increase in order to reduce the use of fossil fuels. However, solar and wind energy have risks and uncertainties of deteriorating voltage stability limits, which must be taken into account when controlling the flow distribution of the electrical network of the power system [3].

Compared to power systems using synchronous generators (SGs), power systems with significant installed capacity renewable energy sources (RES) (the so-called inverter power systems) [4, 5] do not have sufficient inertia, which requires more in-depth studies of sustainability problems in such systems with RES. For these purposes, appropriate analytical methods are needed, on the basis of which it is possible to obtain voltage stability estimates in a power system with a large integrated power of variable renewable sources, which are controlled by a scheme with inverters.

Voltage stability in a power system with RES has become important from the point of view of the continuity of trouble-free power supply to consumers. Compared to traditional energy systems, the ability to control voltage in an electrical distribution network with connected renewable sources is limited [6].

In most cases, the integration of high-capacity RES into the power system is carried out by connecting them through electrical networks of distributed generation. To directly connect renewable energy sources to a distributed generation network, power inverters (SI) are used. Generating wind and solar installations equipped with SI are usually called inverter generators (IG).

* Corresponding author: huseyngulu@mail.ru

 https://doi.org/10.1051/e3sconf/202338401006

At present, in power systems, a significant proportion of the generated active power of IG installations operated in power systems can be large (up to several hundred megawatts) and have large mechanical inertia. The large inertia of the rotors of these IGs is prevented by a sharp change in the rotor speed, which contributes to the stabilization of the system mode during disturbances.

Unlike SG, IG installations are located closer to the load connected to the DG system, and in cases where a large amount of RES connection with variable generation is used in DG, they create problems associated with controlling power flows and voltage stabilization in the system [7-9].

The concepts and methods of sustainability studies used for conventional power systems are not sufficient to analyze the sustainability of power systems containing a large capacity of renewable energy sources with variable output (power plants with Wind and PV power plants).

Voltage stability in a power system with RES is an important parameter, the control and management of which will allow assessing the stochastic variability of the current state of the power system, from the limit state (maximum flows, critical voltages at the nodes). Voltage instability occurs in the form of a gradual decrease in its value in some nodes of the system. This is due to the inability of DG systems to provide enough generated electricity to cover the growing demand in the system [10].

At all possible levels of RES generation stochasticity, the integration of renewable sources will improve the voltage profile by reducing power losses [11, 12].

The paper investigates the influence of connecting a high-power wind farm on the limits of voltage stability in the power system.

With a deterministic setting of the power generation value of a wind farm for fixed wind speeds, the voltage stability margin can be estimated in a simplified way from the curves [13].

1 Methods for estimating the probability of distribution of critical stress values during stochastic mode changes in a power system with a wind farm operating in the range of wind speed variability up to the nominal

1.1 Methods of sequential weighting of the power system mode

The essence of these methods is to identify the share of the contribution of each source to cover a given amount of load growth and the corresponding part of the losses from the transmission of this power through the network. Let K_{Gi} be an indicator of the growth of $P_{Gi,0}$ active power generation by the i generator and be the base (nominal) value of the power generation of this source. In this case, the generation power P_{Gi} at the point after the increase in power can be written:

$$P_{Gi} = P_{Gi,0}\left(1 + K_{Gi}\right) \tag{1}$$

where $i = 1,2,...,n$ for all participating generators.

The value K_{Gi} shows the degree of participation of the $i-$th generator in violation of the voltage stability limit.

1.2 Load limit maximization method

This method is based on a surface approximation that describes the space of states of the power system in the coordinates of the generating capacities of the stations. The essence of the approach based on state surface modeling is outlined below. The possibility of separability of the limit load condition is shown. A method is proposed for the maximum load limit for power transmission over the network (dedicated transmission line). As a result, the application of this method makes it possible to build a model of the value of the maximum load of the system, depending on the variability of the increase in sources. By approximating the surface of the maximum power transfer in the form of Functions depending on the increase in the power of generators (GD_i), it is possible, using well-known optimization methods, to find the maximum point of the curve of the function, $P_{np}^{\max}$. The identification of the model $P_{np}(GD_i)$ and the determination of its maximum can be made by carrying out the following computational procedures:

1. Determination of the dependence $P_{np}(GD_i)$ on the power variability of each $i-$th generator. The set of values $P_{np}(GD_i)$ is determined from a series of curves $P-U$ for different generators GD from a series of selected ones. In this case, the model $P_{np}(GD_i)$ is described in the form of a first-order dependence for each individual generator, except for the balancing one. If we construct continuous curves $P_{np}(GD_i)$ for all i generators, then we can represent the expression for the general case as the following polynomial:

$$\left[P_{np}(GD)\right]_j = B_{j,0} + B_{j,1}GD_j + \cdots + B_{j,n}GD_j^n = B_{j,0} + \sum_{p=1}^{n} B_{j,p}GD_j^p \tag{2}$$

where $\left[P_{np}(GD)\right]_j$ – is a polynomial approximation of the dependence $P_{np}(GD)$ for the case of studying the interaction between two generators (a basic generator and another generator connected to buses j); $B_{j,p}$ - polynomial coefficients; n – number of coefficients; B_0 – permanent member is equal for all $B_0 = B_{j,0}$.

2. The surface $P_{np}(GD_i)$ is approximated for the multidimensional case, subject to the conditions of separability of the state. For all n-generators, the surface $P_{np}(GD)_j$ is defined as:

$$P_{np}(GD) = B_0 + \sum\left[P_{np}(GD)\right]_j,\ j = 1,...,n \qquad (3)$$

Equation (3) describes the surface for all possible options from among the selected generators. A combination of all polynomial equations (2) with the same constant B_0 is determined, the value of which is taken as in paragraph 1.

3. The best vector $P_{np}(GD)$ is determined by the maximum value P_{np}. This best vector $P_{np}(GD)$ is found as the maximum of function (3) under the following constraints: $\Sigma GD_j = 1$ and $0 \le GD_j \le 1$.

Stage 1 studies can be performed using the ETAP software package to determine the voltage stability limit in terms of fulfilling commercial conditions, while stages 2 and 3 require the use of software in solving problems of analysis and operational control. The proposed method has been tested for various versions of the IEEE test scheme.

2 Simulation results

Fig. 1 shows a fragment of a 30-node IEEE single-line test circuit, on the basis of which the studies were performed. There are 4 generator buses in the system, one of which represents a common connection point for the wind farm (buses B2 and B3). There are 10 buses and 34 branches, with the sum of loads connected to the component $P_n = 283{,}4\,MBm$; $Q = 126{,}2\,MBAp$. The total installed capacity of the wind farm (stage 1) with a total of 60 wind turbines (2 MW each).

On Fig. 2 shows the curves for changing the power limit in the system for the options for the participation of pair generators 1-2; 1-6; 1-8 when regulating the mode of transmission of the total load in the event of an emergency failure of one of the lines (case N-1).

At the same time, it is clear from the picture that in the case of G1-4, regulation with the 4th source is possible up to the price of generation, but it is not possible for the prices of generation after that. In other words, it is not possible to adjust the entire range. However, in the case of G1-2, it is known that the adjustment margin is large, and it is possible to ensure the stable mode of the system in the entire range.

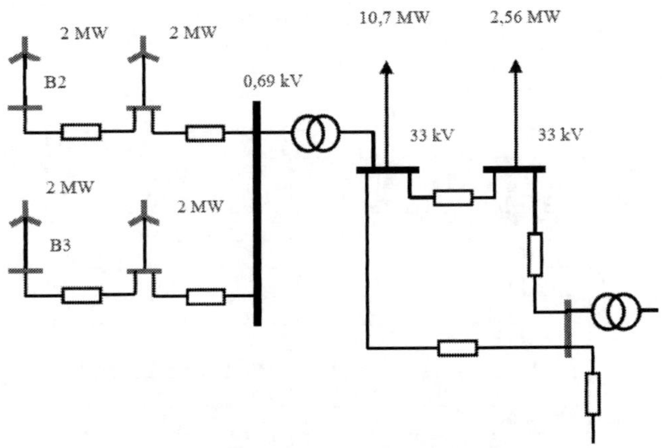

Fig. 1. Fragment of 30 nodal IEEE single-line test circuit

The case of G1-2 is not suitable for adjustment. It should be noted that the obtained results were adapted to the real conditions as much as possible, as the most probable accidents in the electrical system for the Azerenergy system during one year were investigated and damage to the 110 kV power transmission line was found. In this case, the case of accidental opening of the electric transmission line was also considered. According to the N-1 criterion, the power transmission line is considered to be sequentially opened from protection.

Options for integration in the RES system for the Azerenergy system were also considered. The results were obtained using the ETAP 19.5 software package. The results obtained at wind speeds of 14 m/s, 10 m/s and 5 m/s are shown in Tables 1 and 2.

Note that the wind speed values given in the tables reflect real values obtained from local metrological stations for distant regions of Azerbaijan.

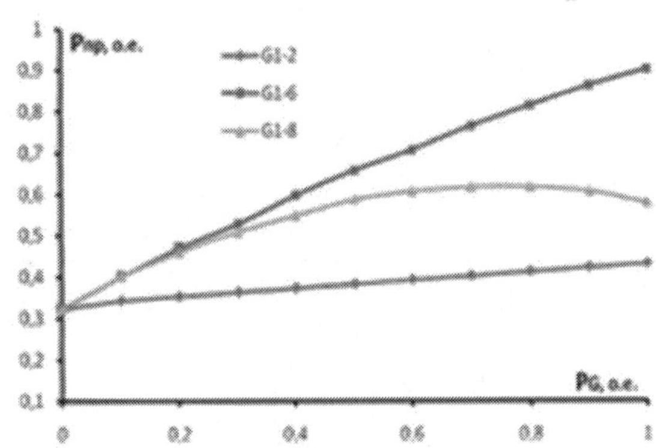

Fig. 2. Changing the limit of transmitted power in the case of one regulating source and emergency shutdown of one element

 https://doi.org/10.1051/e3sconf/202338401006

Table 1. Limiting values of transmitted power and critical voltages in the node of connection of the wind farm to the power system for the values of the load curve in different periods and different values of the network load in terms of active power

Wind speed 14 m/sec		
Load	Critical voltage, %	Power from buses 3_30
0	0	0
106	52,84	170,6
20	90,64	30,6
10	94,98	20,6
Operating point	98,97	10,6
Additional points		
100	55,85	110,6
70	68,18	80,6
50	76,95	60,6
30	86,11	40,6
Wind speed 10 m/sec		
0	0	0
175	30,82	185,6
30	84,08	40,6
20	88,9	30,6
10	93,56	20,6
Operating point	97,89	10,6
Additional points		
170	32,51	180,6
160	35,33	170,6
100	53,66	110,6
70	65,48	80,6
50	74,49	60,6
Wind speed 5 m/sec		
0	0	0
178	26,74	188,6
30	82,6	40,6
20	87,6	30,6
10	92,45	20,6
Operating point	96,99	10,6
Additional points		
170	27,48	180,6
160	30,99	170,6
100	50,98	110,6
70	63,32	80,6
50	72,66	60,6

Table 1 shows the limit values of the transmitted power and critical voltages at the node of the wind farm connection to the power system, depending on the network load in terms of active power, and in Table 2 - the limit values of the transmitted reactive power and critical voltages for reactive power values. Based on the values given in the table, it is possible to estimate the stability limit of the system with the increase of the share of renewable energy sources in the daily consumption.

Table 2. Limit values of transmitted reactive power and critical voltages in the wind farm connection point to the power system for reactive power values

Wind speed 14 m/sec		
Q, load	Critical voltage, %	reactive power from buses 3_30
72	53,58	73,9
70	55,04	71,9
60	60,37	61,9
50	65,42	51,9
30	76,62	31,9
20	83,14	21,9
10	90,52	11,9
Operating point	98,97	1,9
Wind speed 10 m/sec		
178	31,55	179,9
100	48,76	101,9
50	66,37	51,9
30	76,56	31,9
20	82,71	21,9
10	89,76	11,9
Operating point	97,89	1,9
Wind speed 5 m/sec		
0	0	0
278	18,55	279,9
200	29,43	201,9
178	32,63	179,9
100	48,36	101,9
50	65,7	51,9
30	75,81	31,9
20	81,91	21,9
10	88,96	11,9
Operating point	96,99	1,9

3 Analysis of the influence of an integrated wind farm on the limits of voltage stability

The integration of a wind station into the power system depends on many factors, such as data on wind speeds for a long period of observation, the presence of a connection with a system capable of providing the maximum power generation by the wind station, the presence of a source of active power generation in the absence of the possibility of generating a wind station, and many others. Thus, it may not be possible to integrate a wind farm by connecting to a "weak" system bus or through a weak area of a system distribution network closer to the wind farm to be connected. In addition, the load of the system also changes over time. Therefore, the work considered the influence of load variability, as well as the connection scheme of a wind farm on the stability of the system voltage. It should be noted that the operational and short-term forecasting of the daily load schedule for the day and a longer period ahead, in this work, is taken as given.

On Fig. 3 shows the voltage values depending on the wind speed of 5, 10 and 14 m/s at a load of 20 and 30%. As can be seen, at the point of connection of the wind turbine, the voltage value improves with increasing speed, i.e. with an increase in power output from a wind generator.

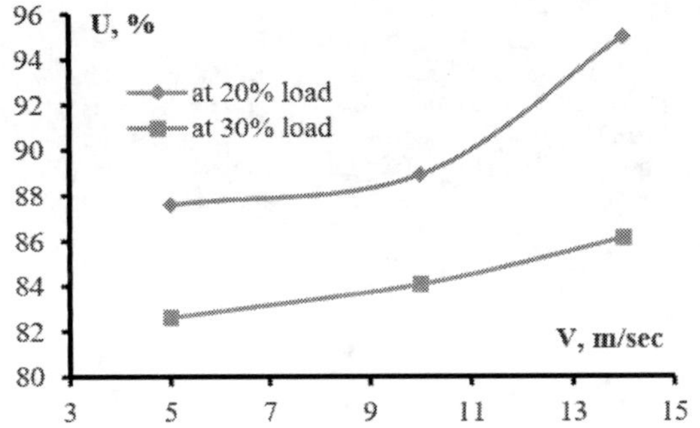

Fig. 3. Dependence of the voltage value on the wind speed at 20% and 30% load.

Conclusion

1. In the electrical system, with the growth of wind power generation and solar PV power plants, the risks of voltage instability may increase, which have deeper consequences for post-failure conditions caused by the shutdown of lines and generators. The influence of integrating a large share of wind energy generation on the stability of the voltage of the power system is considered. On the basis of computer simulation on IEEE test circuits, it was found that the magnitude of the generation power of a wind farm, its connection scheme to the power system, the position of the load curve in the studied interval have a significant impact on the stability of the system voltage.

2. A preliminary analysis of the results of modeling the modes of power systems with wind turbines shows that wind farms connected to some nodes of the system can be used as a set of additional tools to increase voltage stability.

References

1. Rakhmanov N.R., Kurbatsky V.G., Guliyev H.B., Tomin N.V. Mammadov Z.A. *Probabilistic assessment of the regime of an energy system with a different degree of integration of wind source,* Methodological issues of researching the reliability of large energy systems. Collection of scientific articles. Issue. 68. Research and ensuring the reliability of energy systems. Bishkek, 2017, p. 232-241.
2. Rakhmanov N.R., Kurbatsky V.G., Guliyev H.B., Tomin N.V., Mustafaev A.A., Ibragimov F.Sh. *Method for estimating the parameters of the stability limit of power systems in case of emergency failures of its elements*, Problems of energy and resource saving, Tashkent, No. 3-4, 2019, pp. 69-78.
3. Shakirov V.A., Kurbaysky V.G., Tomin N.V., Guliyev H.B. *Siting and sizing of wind farms taking into account stochastic nature of generation*, E3S Web Conf., Rudenko International Conference "Methodological problems in reliability study of large energy systems" (RSES 2020), Vol. 216, 2020, pp. 1-6.
4. Cutsem, T.V. *Voltage stability, phenomena, countermeasure and analysis method*, Proc, IEEE, 2000, Vol.18, No2, pp. 208-127.
5. Eremia, M, Shahidepour, M. *Handbook of Electrical Power System Dynamics: Modeling stability, and Control John Wiley and Sons,* Hoboken, NJ 2013, 946 p.
6. Kundur, P, Paserba, J., Ajjarapu, V., et.al: *Definition and classification of power system stability*, IEEE Trans. Power System, 2014, Vol.19, No.3, pp.1387-1401.
7. Abe, A., Fukunaga, Y., Isono., A., et.al. *Power System Voltage stability*, IEEE Trans. Pow. Apparat. Syst., 1982, Vol.101, No.10, pp.3830-3840.
8. Ajjarapu, V., Lee, B. *Bifurcation theory and its application to nonlinear dynamical phenomena in an electrical Power syst.*, IEEE Trans. Power Syst., 1992, Vol.7, No.1, pp.124-131.
9. Zahidi, R.A., Abidin, I.Z., Hashim, A., et.al. *Study of static voltage stability index G S an indicator for undev voltage load shedding schemes*, Proceedings of the international Conference on Energy and Environment (ICEE), INSPEC Accession Number: 11089129, Malacca, Malaysia, 7-8 December 2009, pp.1-6.
10. Kamel, M., Karrar, A.A., Eltoni, A.H. *Development and application of new voltage stability index for on line monitoring and shedding,* IEEE Trans. Power. Syst., 2018, pp.1231-1241.
11. Zimerman, R.D., Murillo-Sanchez, C.E., Thomas, R.J. *Manpower: Steady state operations, planning and analysis tools for power systems research and education*, IEEE Trans. Power Syst., 2011, Vol.16, No.1, pp.12-19.
12. Blackont of 2003. *Descriptions and Responses (Online) Available:* http: //www.pserc.wise edu.
13. B.H.Lee and K.Y. Lee. *A study on voltage collapse mechanism in electric power systems,* IEEE Transactions Power Syst., 1991, Vol.6, No. 3, pp. 966-974.

Ensuring the reliability of the technological equipment of the cable company through the use of renewable energy sources

V.V. Tsypkina[1] *, V.P. Ivanova[1]

[1]Tashkent State Technical University named after I. Karimov, Tashkent, Uzbekista

Abstract. To date, a "green" transformation of the industry is being carried out in the global energy sector, which will inevitably affect the decline in the level of consumption of natural energy resources, as well as the accompanying high financial and technological risks. The article discusses the issues of accelerating the possible energy transfer, in relation to enterprises producing cable and wire products. The analysis of possible methods and technologies that are aimed at making the transition and increasing the energy efficiency of cable machines is carried out. Improving the efficiency of technological processes will increase the profitability of the cable industry, as well as obtain significant savings in electrical energy. The solution of the task is focused on the use of an additional unit included in the electromechanical system of technological equipment having increased alternating loads (runout). The theoretical possibility of carrying out the improvement of cable machines has been studied by developing a mathematical model describing the physics of the process that ensures the energy efficiency of the proposed method. It is noted that the application of the considered method is possible not only for cable units and machines, but also for technological equipment with a certain shock load.

Introduction

The "green" economy is essentially a new trend, which until recently was a fashionable trend in the economic policy of any developed state. It should be noted that recently this direction has formed the basis of measures for the global green transformation of the economies of states whose policies are focused on the growth of welfare and social equality, while significantly reducing natural (environmental) risks and environmental deficits [1]. In the near future, the "green" economy will become an integral part of many industries and services.

The use of "green" technologies in the industrial sector is usually focused on increasing the energy and resource efficiency of the production process, which is inextricably connected with changes in energy and transport infrastructures operating on a low-carbon basis. The use of renewable energy sources in industry will eventually lead to a "green" transformation of the entire industrial sector, which will inevitably affect the decline in the level of consumption of natural energy resources, as well as the reduction of concomitant high financial and technological risks.

The orientation of "green" technology to individual elements of the economy with structural adjustment will eventually lead to a way out of the current environmental crisis in the world. Therefore, industrial enterprises shall play a leading role in solving the current environmental situation by introducing "green" technologies into the production process, given that in developed countries businesses own copyrights to most "green" technologies [2].

One of the ways to introduce "green" technologies is the "ecological modernization" of the production infrastructure. However, it is impossible to solve this problem at the global level, which means that it is necessary to consider a point orientation to separate technological equipment. The work on the modernization of the working mechanisms of production machines in the long term implies a complete restructuring of all production technological chains and the creation of a "green" transformation of the enterprise. Thus, this direction is relevant for every country, for every region and for every company.

Given the existing interconnections to the decarbonization process, the economy of any state is not ready for the introduction of clean energy, because the percentage of its production is not high enough, while the share of electricity worldwide in the structure of production costs will grow to 20-30% [3]. For instance, the forecast for the production of steel by electrometallurgy in the period 2050-2070 is 50-70%, the rest relates to the hydrogen method and electrolysis production. At the same time, the cost of electricity for the production of one ton of steel will be 3.5-4 times higher than at present. This suggests that the economies of the states will not be ready for this growth.

Currently, there is an active integration of environmental factors with the management strategy of companies in the world. Some large manufacturers, for example BMW (Germany), when concluding contracts with steel suppliers for volumes of more than €20 million tons per year, focus only on the purchase of "green" steel [3]. The introduction of "green" technologies into other industries is becoming a stable trend, which is accompanied by the use of both

*Corresponding author: veronika.tsipkina@yandex.com

 https://doi.org/10.1051/e3sconf/202338401007

fundamentally new approaches and modernization of technological equipment installed at enterprises in order to increase the unit capacity of the main equipment.

The "green" transformation of the cable industry is also of great importance, because the technological process of cable and wire products manufacture includes a large number of energy-consuming technological equipment: melting, drawing, extrusion and stranding equipment, the work of which shall be focused on improving the reliability of its operation and reducing the cost of finished products through the introduction of innovative technologies.

Methods and technologies to improve the reliability of cable equipment.

The conducted research of the Association "Electrocable" (Moscow) shows that the cable industry is currently in a stage of intensive growth [4], which is ensured by the introduction of innovative scientific developments that have found their application both in the process and in the development of new designs of cable and wire products using modern materials. All these aspects have allowed the industry to maintain a high technological level. At the same time, most cable companies have technological equipment, the average service life of which lies in the range from 5 to 20 years. The program of technical re-equipment of the enterprise is carried out as part of the development of new types of products and is implemented through the purchase of modern technological lines. In relation to the established technological park of cable machines, due to the impossibility of its abrupt renewal, measures are being taken to modernize them by improving the design of equipment and technology operating at the enterprise [5, 6, 7, 8, 9].

One of the possible ways to solve the problem of modernization of technological equipment is the introduction of alternative energy sources, i.e. the "energy transition" from the existing system to the innovative one due to renewable energy sources (RES), which is based on the creation of a flexible energy system of both a cable enterprise and a cable machine by organizing "smart networks" that allow optimizing electricity consumption and to increase the efficiency of RES application in the manufacturing technology of cable and wire products [9].

Wires and cables are always in great demand in the economy of any state. At the same time, the traditional manufacturing technology is based on the use of copper, less often aluminum conductors, which make up the cable nomenclature of various purposes. Engineering and technical services are constantly working to improve both the operational characteristics of cable products, as well as strength, flexibility, mass, corrosion and chemical resistance, fire resistance, electromagnetic compatibility, etc. [10].

The technology of manufacturing cable products is based on standard technological operations, which are usually performed on cable machines with high energy intensity. Therefore, the issue of improving the efficiency of equipment is the main task, the solution of which will create great opportunities to improve the energy efficiency of technology and reduce the cost of finished products. The introduction of alternative energy sources will improve the reliability of the equipment and will contribute to reducing energy consumption and cost of finished cable products.

The results obtained

Scientific research in the field of using alternative energy sources at cable enterprises was carried out on technological machines with an increased level of vibration and shock load, namely:

- stranding equipment – general stranding machines, braiding machines, armoring machines, "Drum twister" stranding machine, rigid stranding machine, cable armoring line, cable screening line;
- rewinding equipment;
- cable production waste recycling lines;
- hydraulic presses for cable aluminum sheathing.

As an object of research, the type of general stranding machines was determined, which strand both the current-conducting conductors and the workpiece of cables and wires from insulated cores (Fig.1).

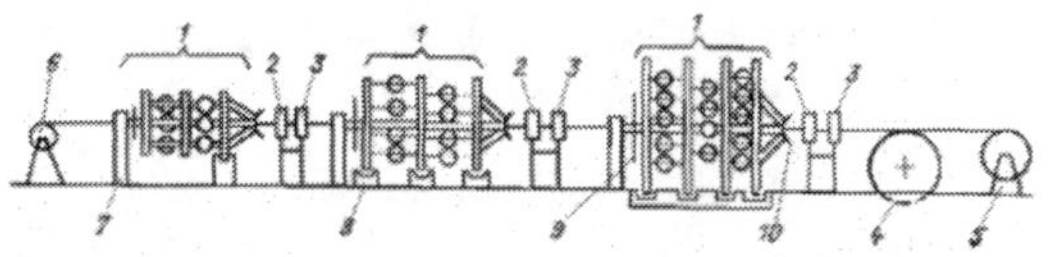

Fig. 1. General stranding machine:
1 — carriage – stranding pay-off; 2 — calibers; 3 - sealing rollers; 4— pull-off device; 5 — take-up device; 6 —fixed pay-off device; 7 — support stand; 8 — support roller; 9 — unscrewing device; 10 — distribution socket

The composition of the working part of the stranding machine (Fig.1) contains units that have a certain technological load and provide quality parameters of the finished product, namely:

Carriage .... 4 pcs.
Wheel-type pull-off device..............................1 pc.
Take-up device ....1 pc.

The choice of the stranding machine loading option is made depending on the production task, i.e. the type of core strand (insulated, bare) of the cable product order, taking into account technological factors:

- number of core strand;
- core cross-section;
- dimensions of the pay-off drums with a core;
- requirements for the strand workpiece (back-twist stranding / stranding without back-twist);
- required performance of this operation.

At the same time, for machines of this type, the energy consumption level is taken in the power range

(Table 1) according to the equipment passport, and the actual power consumption depends on the multitasking of the equipment, i.e. the use of additional options in the technology: a taping head, a scroll, etc.

Table 1. Power consumption of the stranding machine

Working unit	Quantity, pcs.	Full maximum power, kW	Actual power consumption, kW
Carriage	4	300	250
Take-up device	1	15	10
Wheel-type pull-off device	1	110	100
Total:		**425**	**360**

It should be noted that the discrepancy in the values of the full maximum power and the actual power consumption of the upgraded cable machine directly depends on the volume of work and the level of equipment load. According to the equipment passport, the range of the initial loaded wire in the carriage is for the wire diameter in the range of Ø 35-300 mm^2. Depending on the design of the cable product, the carriages of the stranding machine are loaded with coils with wire, and depending on the number of wires in the layer of the workpiece (conductive core), the amount of load and the number of unused working carriages are determined. Thus, the power consumption of the stranding machine has a maximum value at full load of all working carriages with a wire diameter of Ø 300 mm^2 and a minimum value at Ø 35 mm^2, as well as not fully loaded carriages (to 80%, 60%, 40%, 20%, 0%). Therefore, when calculating power consumption, the average load of technological equipment is taken, taking into account the carriages that are not involved.

The improvement of the operation of the stranding machine (Fig.2) by installing additional units with alternative energy sources will create conditions that allow accumulating additional energy with its return to the power system of technological equipment. Thus, to ensure an increase in the level of reliability of the equipment both in the event of a power outage in the workshop, and the return of stored energy from the batteries to the electromechanical system, which will increase the energy efficiency of the upgraded cable equipment.

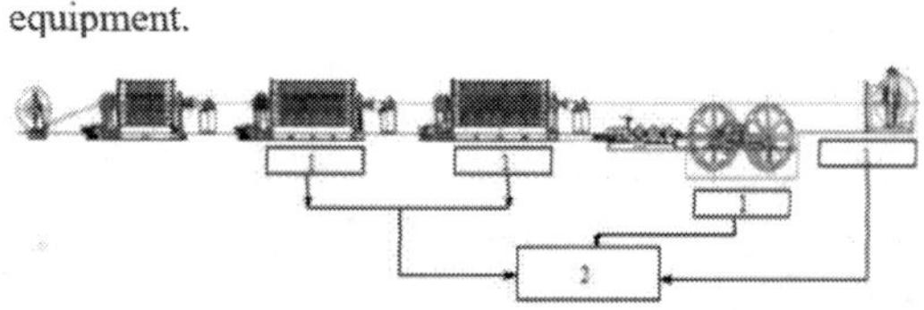

Fig. 2. Block diagram of an improved stranding machine: 1 - unit with alternative energy sources, 2 – accumulating battery

The presented functional diagram of the improved stranding machine (Fig. 2) explains the experimental part of the scientific research conducted at the Departments of "Electromechanics, Electrical Technology" and "Electric Machines" of TASHSTU (Tashkent, Uzbekistan) and JV LLC "HPS" (Navoi, Uzbekistan). According to the results obtained, an application for a patent of the Republic of Uzbekistan for a utility model has been filed. The solution of the research task is focused on the use of an additional unit included in the electromechanical system of technological equipment – a stranding machine with increased alternating loads (runout).

As a result of the research work carried out, the theoretical possibility of improving cable machines was studied by developing a mathematical model that describes the physics of the stranding process of the workpiece of a cable product (conductive core, core).

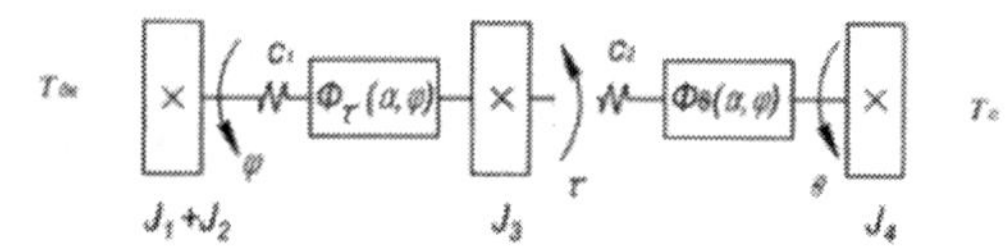

Fig. 3. Dynamic model of a stranding machine

The equation of motion of the stranding mechanism was compiled by developing a dynamic three–mass model (Fig. 3), where J_1+J_2 are the moments of inertia of the first carriage; φ is the generalized coordinate of the first carriage, J_3 and J_4 are the moments of inertia of subsequent carriages; τ and θ are redundant generalized coordinates of subsequent carriages. The results of solving the mathematical model are shown in Fig. 4.

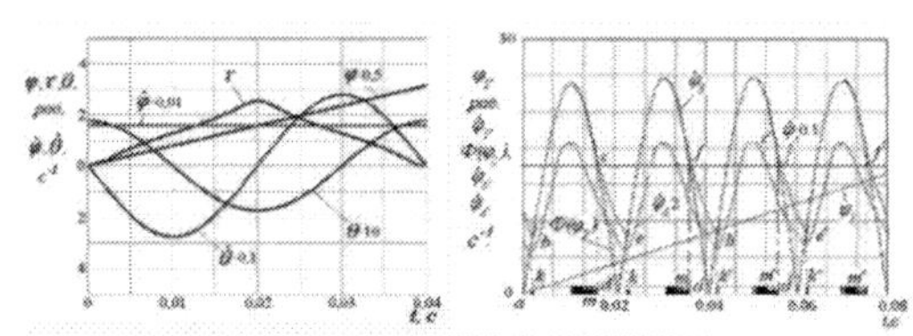

Fig. 4. Results of solving mathematical models

The verification of the adequacy of the mathematical model of the improved stranding machine was based on the comparison of the results of experimental studies with a working cycle that meets the requirements of technology and regulatory documentation for cable equipment operating at the JV LLC "HPS" (Navoi, Uzbekistan).

Conclusion

The research work carried out to improve the energy efficiency of the stranding machine by upgrading it by installing working mechanisms in the electromechanical system: carriages, pull-off and take-up devices, additional units assembled on the basis of alternative energy sources allowed to draw the following conclusions:

The results of the experimental part are shown in Table 2.

Table 2. Stranding machine modernization results

Working unit	Quantity, pcs.	Energy efficiency	
		kW	%
Carriage	4	0,83-1	2,5-3
Take-up device	1	5,33	0,8
Wheel-type pull-off device	1	1,36	1,5
Total		**7,52-7,69**	

1. The introduction of an additional unit into the electromechanical system of the stranding equipment made it possible to ensure the reliability (in 1,5 times) of the technological equipment of the cable enterprise through the use of renewable energy sources.

2. The application of the considered method is possible not only for cable units and machines, but also for technological equipment with a certain shock load.

References

1. Towards a "green" economy: ways to sustainable development and poverty eradication. UNEP report, 2011. p.17.
2. World Economic and Social Survey 2011. The Great Green Technological Transformation. UN Department of Economic and Social Affairs. New York, 2011. P. IX.
3. S. Zinchenko. New challenges of green transformation. https://gmk.center/opinion/novye-vyzovy-zelenoj-transformacii/
4. M.V. Tretyakov, 79th GENERAL MEETING OF THE PARTICIPANTS OF THE ASSOCIATION "ELECTROCABLE", Magazine "Cables and wires", No. 2 (394) 2022.
5. V. Tsypkina, D. Madrakhimov, Method of resource saving of cable production recycables E3S Web Conf., Rudenko International Conference "Methodological problems in reliability study of large energy systems" (RSES 2019), E3S Web of Conferences 139:01058, Follow journal, DOI: 10.1051/e3sconf/201913901058, License CC BY, https://doi.org/10.1051/e3sconf/201913901058
6. V.V. Tsypkina, A. A. Ivanov, Development of generalized requirements for the modernization of the electric drive of cable drawing machines // Universum: Technical sciences. 2017. No.4 (37). URL: https://cyberleninka.ru/article/n/razrabotka-obobschennyh-trebovaniy-k-modernizatsii-elektroprivoda-volochilnyh-mashin-kabelnogo-proizvodstva (Reference date: 31.07.2022).
7. N. Pirmatov, V. Ivanova Improving Energy Efficiency of Drawing Equipment through Energy and Resources Saving, ISSN: 2350-0328International Journal of AdvancedResearch in Science,Engineering and TechnologyVol. 6, Issue 3, March 2019, https://www.researchgate.net/publication/
8. V.P. Ivanova, B.E. Yakubov, Development of an information model of a resource-saving drawing method // Universum: Technical sciences. 2017. No.3 (36). URL: https://cyberleninka.ru/article/n/razrabotka-informatsionnoy-modeli-resursosberegayuschego-sposoba-volocheniya (Reference date: 31.07.2022).
9. I.B. Peshkov, Actual problems of the domestic cable industry, Magazine "Cables and Wires", No. 5 (391) 2021.
10. Yu. Svetikov, Modern trends in the development of cable production, Magazine "Components and Technologies", No. 8 '2007. file:///G:/SOFT/sovremennye-tendentsii-razvitiya-kabelnogo-proizvodstva.pdf

https://doi.org/10.1051/e3sconf/202338401008

The influence of the market type in the electric power industry on its reliability on the example EPS of Russia

Mikhail Chukreyev[1,*]

[1] Institute for Socio-Economic & Energy Problems of the North Federal Research Center of the Komi Science Centre of the Ural Branch of the Russian Academy of Sciences, Syktyvkar, Russia

Abstract. In modern conditions of competition in international markets, the cost of electricity has a huge impact on the ability of state to produce a competitive product. The percentage of electricity costs to goods manufactured in Russia is 2-2.5 higher than European countries, which undoubtedly affects competitiveness, and therefore leads to a decrease in Russia's growth rate. Therefore, work on the effectiveness of the energy industry is very important. The article discusses the mechanisms of energy markets, their features are given. The influence of selected mechanism on necessary full reserve of power is shown, the author's vision on further actions in these matters is given.

1 Introduction

In December 2000, RAO UES of Russia submitted for approval by the Russian Federation Government concept of restructuring the industry. The transition to market relations was supposed, but their form was subjected to serious discussion and criticism from scientific and production organizations. Despite this, with minimal changes, the concept was adopted and in February 2003 the State Duma of the Russian Federation adopted the Law "On Electricity".

If you see the adopted law shows that not one of the goals set in it is aimed at reducing electricity prices, ensuring energy security, sustainable functioning, and attracting investments. As a result of the electric power industry reform a Russian power market was created. Wholesale trade in electric energy in the federal wholesale electricity and power market (FWEPM) was organized. FWEPM was a system of contractual relations of many participants.

In the desire of mankind for competition, the development of markets in the electric power industry is inevitable, but it in view of its special nature and significance for the state cannot be uncontrolled. Over the years of the development of market relations, it is customary to distinguish between four types of markets [1,2,3 and so on]: an adjustable natural monopoly, the only buyer, competition in the wholesale market and retail markets.

The article [4] considers the influence of these market models on the final price. In this, we consider the effect of the influence of various market models on the level of reliability, all other things being equal.

2 Market models

Consider each of the models from the position of ensuring their reliability and predictability the energy system.

Adjustable natural monopoly. This type of vertically integrated companies usually subject to state regulation. In its absence, rising prices would occur. This type of markets is very natural for electric power, due to the characteristics of the manufactured products and the need to maintain proper reliability. This option increases its positive qualities as the system enlarges.

Monopoly companies plan the development of the EPS (both power plants and electric networks) on its territory, based on ensuring the reliability and efficiency of power supply to consumers, environmental requirements, etc. Plans for the construction of new power plants and power lines are subject to coordination with the regulatory authority, after which Construction costs are included in the form of an investment component in electricity tariffs. Thus, the costs of the EPS development is guaranteed to be paid by consumers.

The only buyer. This market model differs from the previous separation of the generation sphere into several economic independent electrogenerating companies (EGCs), which compete with each other for the supply of electricity to a single procurement agency. New electricity manufacturers (NEM) may also appear. The remaining areas remain integrated within the framework of one company, which is still a monopolist in relation to consumers. Naturally, this company (Purchase agency) should, as before, be regulated by the state.

The company "Purchase Agency" retains the obligation to ensure uninterrupted supply of consumers. What implies early planning and implementation the development of EPS on its territory in order to avoid

* Corresponding author: mchukreyev@gmail.com

electricity deficiency. Power supply contracts are individual in nature aimed at creating conditions for the development of uninterrupted deliveries. The procurement agency, in advance plans the development of EPS, predicts the power consumption, makes up promising balance and energy balance sheets, etc.

Competition in the wholesale market. This model is significantly different from the previous ones. Instead of one Purchase agency, several distribution and sales companies (DSC) appear. These companies, as a rule, do not have their own generating capacities and monopoly supply consumers with electricity on their territory. They are responsible for reliable power supply and remain subject to regulation by regional or municipal authorities. They own the distribution networks of low voltage, which they must develop if necessary.

In this market model, control over the development of energy system and maintaining the proper level of balance reliability are entrusted to an independent system operator (SO). At the same time, the presence of a wholesale market deprives consumers of a privileged position, which gives market power to suppliers and negatively affects the capabilities of the system to predictly develop and maintain a proper level of reliability. Now for this it is necessary to create economic capabilities, which significantly affects the increase in the tariff for power.

Competition in wholesale and retail markets. In addition to Model 3, the possibility of competition in retail markets is created here. Electricity consumers can now be supplied from different DSCs or sales companies (SC). Since the distribution of electricity remains a monopoly sphere, the state regulation of DSC is preserved.

The model provides for direct supply of electricity from manufacturers to consumers, bypassing DSC or SC. This means either the direct access of consumers to the wholesale market, or some combination of wholesale and retail markets. At the same time, of course, the procedures for accounting, mutual settlements, etc. are complicated.

At the same time, the market is preserved and even aggravated by problems with the construction of new power plants. There is still no body responsible for the development of generating capacities and the prevention of deficiency in the wholesale electricity market.

3 The influence of the market model on the formed reserve value

Consider the influence of the presented market models on formed full reserve value. In a few words about the task of substantiating a power reserve. The full reserve consists of three components, the most difficult to determine is an operational power reserve. The other two components are repair and strategic reserves, which, in conditions of redundancy the energy system, loses its significance. The repair reserve has a minimal impact on the indicators of balance reliability.

The determination of the value of the operational reserve of power is two-stage. At the first, for a given composition of generating sources, and, consequently, the values of the operational power reserve, the task of assessing the indicators of balance reliability (IBR) for all territorial EPS zones is solved. On the second, depending on the results of the IBR and the accepted regulatory values, the specified initial parameters the operational reserve of power are corrected.

In relation to our country, the most acceptable in the practice of planning the development of the EPS remains IBR in the form of the integral probability of the appearance ($J_л$) or the absence ($\rho = 1 - J_л$) of the power deficiency in the territorial zones of the EPS design scheme. As normative parameters of this indicator in the methodological recommendations [5], the values of 0.004 and 0.996 are used, respectively. In the countries of Western Europe and the United States, several different, but also probabilistic IBR and normative values for them. In our country, many works are devoted to the development of IBR assessment models, for example [5-8].

The value of the power reserve is primarily affected by the predicted demand and the accepted indicators of balance reliability. The second for one country is constantly, and the first parameter changes depending on the selected market model in the energy system. Consider in order each of the models from this position.

With an adjustable natural monopoly as with the "only buyer" model, the only managing body is responsible for the development and safety of the energy system. He plans to input power on its territory based on the actual necessity and economic feasibility. Since the cost of input power is included in the form of an investment component in tariffs for electricity, the work on introducing capacities will always be paid, and therefore there are a lot of people who want to invest in new generating objects, which removes a number of uncertainties, primarily the need to interest a potential investor. The possibility of point input of capacities also contributes to a smaller volume of their input, which in the market will significantly affect the tariff.

The model based on competition in the wholesale market implies control over the development of the energy system through an independent system operator. His power includes decision making in the volume of the necessary input of capacities for uninterrupted operation of the energy system. At the same time, these solutions are often dictated by the need to create an economically reasonable investment base. This is a very important point, since it must be understood that in this market model market power on the side of suppliers, which means they dictate the conditions under which investments are interesting to them. Initially, in order to remove part of such risks and attracting those who wish to build a new generation, a program of contracts for providing power (CPP) was created, which guaranteed the refund of invested funds, since without it the existing tariff for electricity and strange risks did not attract those who wanted to participate in the investment program.

In the model based on competition inside the wholesale and retail markets, there are the same problems as in the previous one, only the scale of the problems is even more. There are still no control of the

 https://doi.org/10.1051/e3sconf/202338401008

unified managing body on the issues of promising development and compliance with the proper level of reliability.

4 The tariff effect on the formed volume of installed power for different market models in the electric power industry

Consider the influence of the presented market models on the required volume of installed capacity. Suppose that for all market models, in the formation of a model of promising development, criteria were used to maintain the reliability of the energy system according to the STD. Suppose (from [4]) that in the EPS there are sixteen power plants of power suppliers (table 1). Conditional costs of suppliers are random numbers in monetary units (units).

Table 1. Characteristics of power plants

№	Passed power, MW	Conditional constant costs, units.	Specific conditional constant costs, units/MW
1	1100	11000	10
2	1400	42000	30
3	500	19000	38
4	1700	68000	40
5	465	19065	41
6	550	24750	45
7	1200	55200	46
8	300	16500	55
9	405	23490	58
10	500	30000	60
11	160	9760	61
12	1200	73200	61
13	240	15600	65
14	35	2345	67
15	750	56250	75
16	500	45500	91
Sum	10725	1136120	

The results of solving the problem of determining value for power suppliers the considered models are shown in Fig. 1.

According to [4], the first two models will be workers at a tariff value lower than for model 3 and 4. So for 1 and 2 models, the level of installed power is 8280 MW, while for two other models 9755 MW. This is primarily due to the departure from the programs for creating CPP's in favor of the prices obtained through a competitive power selection (CPS), which should have the necessary investment attractiveness for options with a purely market mechanism of relations between subjects in the electric power industry. All this, coupled with an increased duration the CPS, and therefore the increase in the error of planning the maximum load when forecasting demand leads to an increase in the predicted demand.

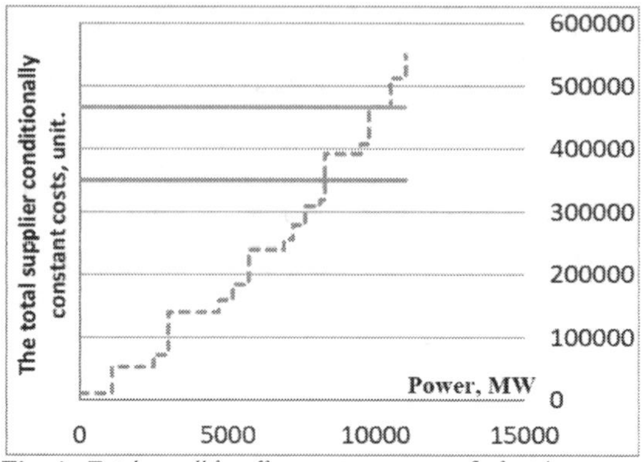

Fig. 1. Total conditionally constant costs of electric power suppliers and recommendations received according to the level of power value. 1 and 2 market type (red line), 4th market type (gray line).

The procedure for determining demand is carried out in accordance with Order No. 431 according to which it is formed from the value of the combined maximum load and the planned reservation coefficient. The analysis given in [9] shows that the value of the combined maximum load grows with an increase in the advance period. Moreover, this growth is in no way associated with the analysis of retrospective information about a change in the actual parameters [6]. Given the data of Scheme and development program EPS on 2021-2027, it can be seen that an increase in the forecasting period of up to six to seven years will lead to at least a 4 percent increase in the maximum load in both price zones of the EPS of Russia. All this directly affects both the high demand and the determined power reserve, and therefore the necessary volume of installed power.

5 The influence investment component on the level required power reserve

In [1], the process of obtaining all the necessary indicators is presented in detail, here we will present the influence of these indicators on the reserve size.

In Russia, countries of Western Europe, the United States and a number of other countries, the co-renewal of the investment component of tariffs and prices for the competitive market is 0.171, while the "only buyer" for markets is 0.0625, and for the regulated monopoly-0.052. Which by 2.7-3.3 times increases the influence of the investment component on the price of electricity. And the higher the country's industrial development, the greater the effect has it. In fact, this value indicates the necessary for new power plants excess of wholesale prices over the costs of similar existing power plants. This excess creates a price barrier to enter the competitive electricity market for new manufacturers.

In regulated monopolies and in the "only purchase" market, the investment component of the tariff will

 https://doi.org/10.1051/e3sconf/202338401008

depend on the composition and proportions of new (under construction) power plants. The final investment component will depend on the optimal composition (and capacities) of new power plants.

Since the investment attractiveness is primarily economic, the guaranteed value for electricity most strongly affects it. And it also depends on production costs that are different in different types of power plants. With adjustable markets (mechanisms 1 and 2), the weighted average generation costs will be determined by the structure of existing power plants. If you use the data presented and transfer it to the structure of the introduced generating capacities, then with the same payback periods, regulated markets will reduce capital costs 2.8 times. Which for our example corresponds to a decrease in the required installed capacity by 3-5%.

6 Full power reserve for different market models

We analyze the information about the predicted and actual demand and the necessary coating. To do this, we turn to Scheme and development program EPS of Russia [10]. Table 2 shows excerpts from Scheme and development program 2011 to 2022.

Table 2. Power Balance EPS of Russia

Year	Maximum consumption, GW	Change the year by the year, %	Coating, GW	Change the year by the year, %
Predictable values				
2014	168809		236850,4	
2015	173096	2,54	240725,2	1,64
2016	177475	2,53	244637	1,62
2017	181478	2,56	242291,3	-0,96
2018	165151	-9	232516,4	-4,03
2019	166262	0,67	232335,4	-0,08
2020	160484	-3,47	232620,4	0,12
2021	161311	0,51	225686,2	-2,98
2022	162011	0,43	228545	1,27
Actual values				
2011	153973	1,61	200346,6	-0,6,4
2012	158988	3,25	207362	3,5
2013	158659	-0,21	207700,5	0,16
2014	157219	-0,91	213875,1	2,97
2015	158871	1,05	223733	4,61
2016	154116	-2,99	215379,9	-3,73
2017	157355	2,1	221166,2	2,69
2018	156570	-0,5	215904,4	-2,38
2019	159360	1,78	218713,4	1,3
2020	155003	-2,73	217433,8	-0,58
2021	157543	1,64	217082,6	-0,16
2022	156946	-0,38	221925,1	2,23

Table 3. Full reserve for various market models

Year	Full reserve predicted (model 4), %	Full reserve actual (model 4), %	Full reserve predicted (model 2), %
2011	-	30,12	16,4
2012	-	30,43	16,23
2013	-	30,91	16,07
2014	40,31	36,03	14,9
2015	39,07	40,83	15,82
2016	37,84	39,75	15,04
2017	33,51	40,55	15,55
2018	40,79	37,9	14,69
2019	39,74	37,24	14,42
2020	44,95	40,28	15,07
2021	39,91	37,79	14,83
2022	41,07	41,4	15,33

Table 2 shows that the forecast is always highly overstated, despite the presence of huge excess power (table 3). This is primarily due to the need to ensure attractiveness for potential investors, which requires a comfortable tariff level for them, i.e. At the forefront, it is not the optimization of the system and a decrease in the tariff in favor of the consumer, but the work on the supplier. This is the main difference between models based on competition without direct control of the state and models that implies this control. For comparison, the use of models 1 and 2 allows you to focus on forecast demand, taking into account 2-3 years in the future, which reduces the impact of the forecast error and does not depend on economic attractiveness, which is achieved by direct control over the introduced capacity and guaranteed payment for its construction. With this in mind, the "Orion-M" program was simulated necessary the level of a full reserve taking into account the same level of normative reliability as in other market models. The results are presented in table 3. It is clearly visible more than two multiple advantage of models with state control over purely market mechanisms.

7 Conclusions

Electric energy is a resource that does not have an alternative, so the creation of any kind of markets raises questions. The idea of market works when there are various alternatives to the same product, in the case of electricity, this is not so. Therefore, the adopted model of the energy market directly affects the method of monitoring the reliability of the energy system its ability to predictably develop.

The studies show that models 1 and 2 make it possible to achieve predictions of the energy system, to simplify the task of maintaining due reliability. In addition, the adjustable market will reduce the problem of the price barrier for new generating capacities. With the current structure of the energy industry, the construction of a new generation requires a significant increase in price. Partly because of this, the price of selected capacities as a result of a CPS to strive for the value that allows to finance new objects without using the CPP mechanism. This drawback of the electric power market in the long term can be overcome only through state regulation.

References

1. L. S. Belyaev, S. V. Horseskalnikov The market in the electric power industry: problems of development of generating power (Science, Novosibirsk, 2004), 220 p.
2. Hunt S., Shuttleworth G. Competetion and Choice in Electricity. - Chichester: John Wiley, 1996.
3. E. P. Volkov, V. A. Barinov, A. S. Manevich Problems and prospects for the development of electric power industry in Russia (Energoatomizdat, Moscow, 2001), 432 p.
4. M. Yu. Chukreyev "The influence of various factors on power price" in Methods / Methodical issues of studying the reliability of large energy systems Vol. 72. (Publishing House ISEM SB RAS, Irkutsk, 2021), pp. 244-254.
5. Global Demand Response Capacity is Expected to Grow to 144 GW in 2025, URL: https://www.navigantresearch.com/news-and-views/global-demand-response-capacity-is-expected-to-grow-to-144-gw-in-2025
6. Methodical recommendations for the design of the development of power systems. (Approved by the Order of Russia Ministry of Energy June 30, 2003, No. 281). Moscow, Russian Ministry of Energy from 153-34.20.118 2003.
7. I. M. Markovich Modes of energy systems (Energy, Moscow, 1969), 351 p.
8. Yu. Ya. Chukreyev Models for ensuring the reliability of energy power systems (Komi SC UB RAS, Syktyvkar, 1995), 176 p.
9. Yu. Ya. Chukreyev, M. Yu. Chukreyev "Components of the power balance consumption and their influence on the procedure of competitive power selection UES of Russia" in Methods / Methodical issues of studying the reliability of large energy systems Vol. 71. (Publishing House ISEM SB RAS, Irkutsk, 2020), pp. 169-179.
10. URL: https://www.so-ups.ru/future-planning/dev-sch

Ensuring balance reliability in managing the development of electric power systems with renewable energy sources

Iyrii Chukreyev[1,*]

[1] Institute for Socio-Economic & Energy Problems of the North Federal Research Center of the Komi Science Centre of the Ural Branch of the Russian Academy of Sciences, Syktyvkar, Russia

Abstract. The issues of substantiating balance reliability - reserves of the generating power of the EPS in the presence of renewal power in the standpoint of achieving the required normative indicators of balance reliability and the execution planned volumes of electricity generation at the wind power plants (WPP) and solar power plants (SPP) in the balance sheet. The results of substantiation reservation tools are given for various scenarios for the development of SPP and WPP both in power and the production of electricity.

1 Introduction

Planning for the development of electric power systems in modern conditions requires the accounting features of the work renewable power sources (RPS), mainly wind (WPP) and solar (SPP) power plants. The process of commissioning these types of stations is at a higher rate in countries with limited traditional energy resources. In Russia, the development of environmentally friendly generation touched a lesser extent, which is quite understandable. Nevertheless, the development of the industry under the program CPP RPS in the power balance of the South union power system (UPS) by 2025 provides for increasing such capacities up to 3.5 GW.

It is clear that the installations RPS do not radically solve the problems of reliability, due to the features of their work, depending on random natural factors associated with their energy supply, especially for the period of passing maximum loads. An analysis of the current RPS shows that the share of the SPP in the coating of maximum loads is insignificant small (less than 10%), the WPP share can fluctuate from several percent of the percent to tens. Nevertheless, when justifying the operational capacity power reserve of power systems for a promising period, it is necessary to take into account the features operation of these types power plants.

The article discusses the methodological problems of evaluating the indicators of balance reliability and justifying the value of the operational power reserve in the management development of electric power systems if there is a receipt WPP and SPP in the structure of the power balance. At the same time, the approach is considered to be an unquestioned that only those whose solution is associated with the need to take into account the specified restrictions on the supply of energy resources and the expected unplanned breaks in the operation of its elements - failures of the system due to emergency damage is considered to be an indisputable reliability. This aspect provides algorithms for embedding the SPP and WPP, taking into account the emergency failures of their attitudes and power restrictions caused by the insufficiency of energy resources (sun, wind).

2 Methodological approaches to assessing the balance reliability of the EPS

When developing mathematical models for evaluating the balance reliability of multi-zone EPS, approaches that are based on the use of analytical methods, statistical modeling methods or their combined are used. The last two approaches were most widely used due to the large opportunities for taking into account various kinds of restrictions [1, 2]. The methodology for solving the problem of evaluating balance reliability indicators when using these approaches, regardless of the control principles in electroenergy industry, is based on the solution of two interconnected stages:

- the formation load levels and random states of generating power caused by unscheduled conclusions in emergency repair of power plants;

- assessments the formed conditions for ensuring the load of territorial zones.

In the developed methodology [1, 2], as in the methods proposed in [3, 4] determination the indicators of balance reliability is carried out sequentially for all time intervals T_t, during which the structure of generating capacities and load remain unchanged. The formation of random states is carried out by methods of statistical modeling on the analytically obtained probabilities of reducing the generating capacities in

* Corresponding author: ychukreyev@gmail.com

 https://doi.org/10.1051/e3sconf/202338401009

individual zones of the EPS due to an unscheduled output of equipment into emergency repairs (regardless of the type of equipment). To evaluate the formed accidental condition of the system, approaches based on the use of methods for permanent current idealization are used [1, 4]. In this case, either a linear or nonlinear setting of the problem can be used. Summing up of balance reliability in the form of power deficits or probability of their existence in all the allocated time intervals allows you to determine their integral values. The most widespread in our country was a probabilistic indicator of balance reliability in the form of an integral probability of power deficiency ($J_{д}$) [2, 5].

From the point of view of evaluating the indicators of balance reliability from the entire multitude of states of generating power and load formed by randomly, only those conditions in which power deficiency are observed are of interest. There are extremely few such conditions, less than one percent at the largest steps in the steps of discrete daily load (usually not more than 8 out of 24) and their practical absence on other steps. As a matter of fact, this is not caused by the accounting for evaluating the indicators of balance reliability EPS the specification by types of generating equipment involved in consumer coating. Crimeless states do not affect the indicators of balance reliability and in existing models, the generation power can be redistributed between the different types of generating equipment (HPP, TPPs, NPPs, RPS). This moment was somewhat accompanied by the development of a methodology for evaluating balance reliability indicators taking into account the limited resource for the production of electricity at HPP [6]. This approach can be extended to the WPP and SPP.

The methodological principles of substantiating the operational power reserve value remain unchanged since the distant 60s of the last century. In modern conditions, a regulatory indicator of balance reliability for territorial zones ($J_{д}^{н}$= 0.004) is used to make decisions on the justification of reservation [5, 7], approved by order of the Russian Federation Ministry of Energy (№321 30.04.2021). The accepted value of the normative indicator of balance reliability has been used for more than 40 years (since the 80s of the last century). In the countries of Western Europe, the normative indicator of balance reliability $LOLH^{н}$ = 3-8 hours/day, in North America - $LOLE^{н}$ = 0.1 times a year.

All of the listed regulatory indicators of balance reliability is characterized by approximately the same methodological basis for their receipt with different information component, especially according to the regimes of electrical consumption [1]. The European standard for the indicators of balance reliability is focused on hourly electricity consumption schedules for all 8760 hours of the year, the North American provides for accounting for the load only the maximum hour of the day (365 values). The domestic regulatory indicator of balance reliability is focused on only one medium - hour daily schedule of December the month in the assumption of its action within 250 working days of the year.

3 The possibilities of accounting for the WPP and SPP when evaluating the indicators of balance reliability EPS

The use of statistical information on the climatic features of the region. The easiest way to take into account the WPP and SPP in model support the task of evaluating the indicators of balance reliability is to obtain probabilistic functions of reducing the generating power of these stations types caused on the one hand by their unreliable, on the other - the unsecured by their energy carriers (sun, wind). To implement this approach, it is necessary to have statistical information about the power of wind or solar activity by the hour of the year for a sufficiently large temporary segment (10 or more years). Having these two probabilistic rows, it is possible to build a resulting series of reducing the generating power for these types of stations [8]. A significant disadvantage of this approach is the loss of temporary and territorial dependencies of wind loads and activity of the sun when constructing the resulting rows of power reduction.

The problem can be solved by constructing the probabilistic series of reducing the generating capacity for all time periods (at the limit of 8760 hours) and special correlation matrices taking into account the change in probabilities in the territorial section of the power system. It should be noted that this is a very difficult procedure in the aspect of the information filling of the task. They encountered such a problem when assessing the effect of changes in temperature on the load power in different territorial zones EPS of Russia. The decision was the creation of a correlation matrix at the level of Russia UPS [9].

Using the planned electricity production at SPP and WPP. Recall that information about the values of electricity planned during the year at all types of power plants is given in the annual work scheme and development program of Russia UPS for the seven year period. Focusing on these indicators, which in such documents are considered a high degree of probability of reliable, it is possible to determine the distribution of electricity production by hours of the estimated period. At the same time, the corresponding shares of the clock period can be obtained in reporting documents on the work of current WPP and SPP.

As noted, a feature WPP and SPP is the unequal of their production of electricity caused by natural factors. This feature is proposed to take this feature by randomly setting the values of electricity production at WPP and SPP for the same periods of time T_t, according to which the indicators of balance reliability are evaluated. In a model for evaluating the balance reliability indicators, taking into account the participation of these types of stations in the coverage of consumers in defective states, it is possible to carry out by separate for them and all other types of power plants for the formation of capacities reducing the equipment caused by the conclusions of the equipment into emergency repairs [6]. At the same time, of course, mathematical models of assessment of each random state formed by power should be transformed.

 https://doi.org/10.1051/e3sconf/202338401009

4 Transformation the problem of assessing a random condition when taking into account the production of electricity at SPP and WPP

The task of streaming in the models of evaluating the indicators of balance reliability is called the task of distributing power deficiency. There are many productions of this task [2, 3]. Consider the simplest linear production, as the target function of which uses the functionality of maximizing the load or minimize power deficiency in the EPS, consisting of n territorial zones [1, 2]:

$$\sum_{j=1}^{n} c_j^{н}(\overline{P}_j^{н} - P_j^{н}) \rightarrow \min, \qquad (1)$$

The restrictions in these expressions are compliance with capacity balance, as well as restrictions on the power of loads and generating capacities in relation to all J-M territorial zones [1]. Accounting for the energy supply of WPP and SPP somewhat changes this target function, due to the introduction of two additional terms into it, characterizing the loading of these $c_j^{в}(\overline{P}_j^{в} - P_j^{в})$ and all other types of stations $c_j^{др}(\overline{P}_j^{др} - P_j^{др})$. In the introduced additional terms in expression (1) $c_j^{н}$- the loading coefficients are taken equal to one; $c_j^{др}, c_j^{в}$- coefficients reflecting the need for loading in defective states of other stations types ($P_j^{др}$) and renewable ($P_j^{в}$); $P_j^{н}, \overline{P}_j^{н}, P_j^{др}, \overline{P}_j^{др}, P_j^{в}, \overline{P}_j^{в}$-accordingly covered with the need and power of the load demand, used and existing powerful power of other types of stations and RPS j - th territorial zone.

In order to exclude the influence of the introduced components of power in the deficit states of the EPS, the values of the coefficients of the target function with component generating power ($c_j^{в}$ and $c_j^{др}$) are accepted much less (100 or more times) of the coefficient with the mainly plain of optimized functionality (1).

5 Principles of determining the production of electricity

For specified in the model of evaluating indicators balance reliability of time periods T_t, random values of electricity production at WPP and SPP are formed. The amount of electricity production for all the allocated periods of time should be equal to the specified planned value, for example, the SDP UPS. When solving the problem of assessing defective states, this requirement is carried out by the use of a special approach based on dynamic variation of the coefficients with additional terms of the target function (1).

The time period T_t for which the production of electricity is divided into many equal parts (up to 100). It is known [1-3] that obtaining reliable results of assessing the indicators of balance reliability requires the formation of at least 25,000 random conditions at each time interval T_t. Thus, for each of the 100 broken areas, 250 states are played. For each selected section (up to 100), the methods of statistical modeling are formed by the states of the generating power. At the same time, the coefficients $c_j^{в}$ and $c_j^{др}$ additional terms for expression (1) in the allocated areas are unstable. Their values depend on how it is necessary to redistribute the generating power between all types of power plants and RPS. So that at the time under consideration, to approach the given volume of electricity production as close as possible.

For the ratio of the target function specified in the first section, the production of electric power is determined by the generation electric power on them and is compared with the one that should be on the site. If the production of electricity is lower than the planned, the coefficients are additional for the target function (1) of the terms for generating renewables $c_j^{в}$ are increased, and for other types of stations $c_j^{др}$ they decrease. Thus, when playing the next group of states (for example, 250 states) in the 2nd section there is an increase in the production of electricity at the WPP and SPP. When considering the last section, the estimated value of electricity production is approached the value specified for the period T_t.

The most characteristic case was discussed above when the modest value production of electricity on WPP and SPP reaches the given value using all the allocated sections of the time period T_t. When an accidentally specified electricity production on WPP and SPP is characterized by a low value, the effect is slightly different. The achievement given volume of electricity production at WPP and SPP is achieved in a smaller number the selected areas. This leads to the exclusion power of WPP and SPP when considering all subsequent sections. This condition is possible when in UPS power balance is insufficiency the generating power other types of plants and prevails WPP and SPP. This generally leads to an increase in balance reliability in the form of integral probability of power deficiency $J_д$. And, of course, to increase the operational reserve of power.

6 Experimental calculations

The influence of WPP and SPP on balance reliability indicators and value of power operational reserve was considered on the example South UPS. Information on promising capacity and electricity balance sheets for 2025 was taken from the SDP UPS for 2021-2027. Important characteristics South UPS, from the standpoint the issues of evaluating balance reliability, are the established capacities of all stations types (28687 MW) and WPP with SPP (3645 MW), as well as the production of electricity for them (112317 and 7700 million kWh). WPP and SPP used 2112 hours. The maximum load combined with Russia UPS is 16995 MW. The calculated scheme Russia UPS (Fig. 1) and information content the task of assessing the balance reliability indicators were obtained when performing joint research with the «SO UPS» JSC. From the same work in Fig. 1 shows the values the maximum

Table 1. Comparison the values of operational power reserve for various parameters of electricity production at WPP and SPP

№	Power WPP and SPP, MW	Power used in balance, MW		Electricity production of WPP and SPP, million kW · h		Operational power reserve	
		WPP and SPP	other	Plan, million kW · h., %	Calculation	MW	%
1	2	3	4	5	6	7	8
Without restrictions on electricity production at WPP and SPP						1725	10,15
1.	910	506	18229	1925 / 1,72	1786,8	1740	10,24
2.	1820	956	17893	3850 / 3,43	3583,0	1854	10,91
3.	3645	1685	17453	7700 / 6,86	7099,7	2143	12,61
4.	5465	2672	17106	11550 / 10,28	11374,3	2783	16,37
5.	7290	3812	17181	15400 / 13,71	14250,9	3998	23,52
Without restrictions on electricity production at WPP and SPP						790	4,65
6.	910	604	16476	1925 / 1,72	1792,8	795	4,68
7.	1820	866	16239	3850 / 3,43	3588,6	830	4,88
8.	3645	1682	15498	7700 / 6,86	7254,8	950	5,59
9.	5465	2510	14966	11550 / 10,28	11224,3	1210	7,12
10.	7290	3290	14667	15400 / 13,71	14852,1	1648	9,70

permissible power with neighboring UPS: 1572 MW (issuance), 2610 MW (receipt).

When using the balance reliability indicators annual load in model with 8760 clock changes, you can play the volume of electricity production for each hour estimated period. The formation of only one random parameter of electricity production for an hour is not entirely correct (not representative). Therefore, it is proposed for each stage of the annual load graphics, at least 4-multiple repetition of calculations during the recruitment of other randomly obtained electricity production values.

When applying the indicators of the balance reliability of the average daily graph load on December [5], implemented in the model [1, 2], the effects probabilistic nature of electricity production on WPP with SPP for 250 working days of the year is applied. Here, representativeness is fulfilled, and this approach is implemented in software "Orion-M" [2].

Research was carried out for isolated work South UPS and when it was part of the Russia UPS (Fig. 1). Due to the lack of information on the distribution electricity production at WPP and SPP on the calculation hours in the calculations below, the assumption of its uniform distribution was made. This made it possible to determine the volume of electricity produced at each selected time interval T_t. At this stage, an uniform law of distribution of electricity production at a time interval T_t was applied. It is clear that both the first and second assumptions are not entirely correct and used only to show the performance of the proposed methodological approach. The receipt of more reliable information on

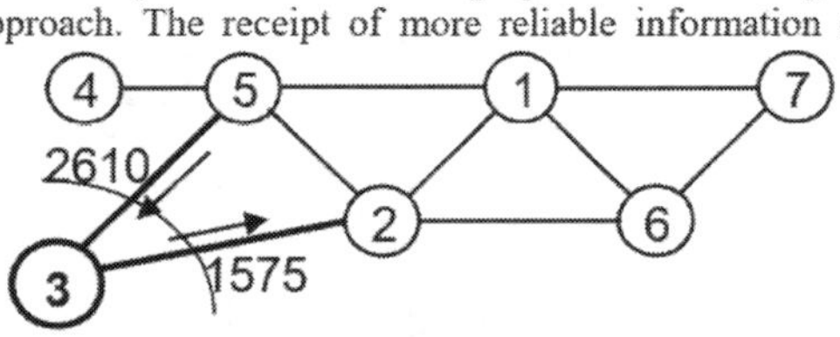

Fig. 1. South UPS and Russia UPS:
1 – Ural UPS; 2 – Middle VolgaUPS; 3 – South UPS; 4 – North-West UPS; 5 – Centre UPS; 6 – Kazakhstan; 7 – Siberia UPS.

the distribution of electricity production at WPP and SPP, as well as the laws of their changes problems in modern conditions, cannot cause them.

The scenarios of electricity production at WPP and SPP South UPS at 2025 were subject to consideration the planned value of 7700 million kWh, its decrease of 2 and 4 and an increase of 1.5 and 2 times with the corresponding change in the installed capacity. In table 1 for these scenarios, results optimization the value of operational power reserve in South UPS are provided, subject to compliance the probabilistic normative indicator of balance reliability is $J_{л}$ =0.004. For a more visible presentation of the proposed algorithm for accounting the work WPP and SPP, the work of South UPS was considered in isolated mode and as part of Russia UPS. For the 5 considered scenarios changes the power WPP and SPP (column 2 in the table), according to the ratio of hours use, the planned values production of electricity on them (column 5) are determined. The producer of the column 5 shows the percentage electricity production at WPP and SPP to the planned values. The approximate equality values the production of electricity at WPP and SPP and their values planned for a year (columns 5 and 6) for all considered scenariors.

The task in the model of an additional condition for

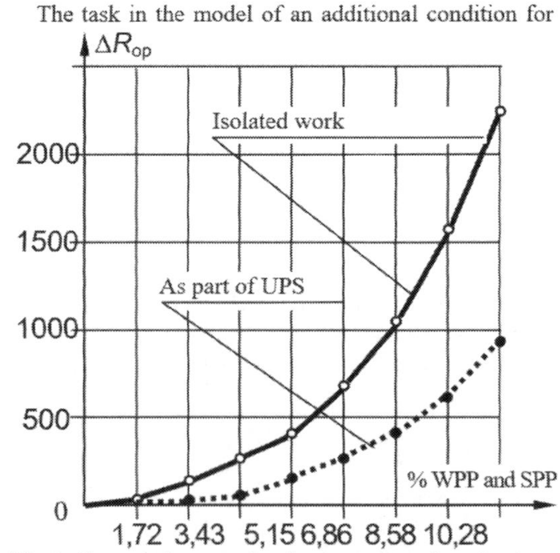

Fig. 2. Change value operational power reserve from the share (%) of electricity production at WPP and SPP.

redistribution of capacities between renewables and other types of stations leads to the fact that at some time intervals allocated randomly, the WPP and SPP are ready to provide a systemic power deficit, but cannot do this due to restrictions on the electricity produced on them. This leads to an increase in the calculation indicator of balance reliability in the form of an integral probability of the appearance of a power $J_д$ deficiency at this time interval. This can be explained that for the conditions of isolated operation South UPS, the power used in the balance sheet for all types of stations exceeds the maximum load (16995 MW) according to all scenarios of electricity production at WPP and SPP (column 4, Table 1). During the operation of South UPS, this principle is violated as part of Russia UPS. The reason for this is quite simple - the lack of its own generating power to cover the deficiency is covered with excessive Russia UPS. The capacity of 2610 MW (Fig. 1) allows this to be done.

The principles of redistribution generating capacity between the immediately personal types of power plants in defective states (section 4) and the probabilistic random task of the electricity production at WPP and SPP in different periods of time (section 5) made it possible to evaluate the influence values of electricity on the operational power reserve (column 7 and 8). In Fig. 2 shows the characteristics of a change in the operational power reserve in relation to its value, obtained without taking into account any restrictions, for various shares of electricity production at WPP and SPP. It can be seen that with a small proportion of electricity production at WPP and SPP (up to 3.5 %), which is characteristic of current state of the Russian electric power industry, an increase in the operational reserve of power is slightly (about 125 MW / 0.7 %). It is clear that taking into account the peculiarities of the work WPP and SPP in the task of substantiating the operational power reserve when working as part of Russian UPS in this case is not required.

Calculations showed that with the value of 7700 million kW h (6.86 %) and assumed assumptions planned by 2025, an increase in the operational capacity reserve by 160 MW (0.94 %) when working in Russia UPS and 420 MW (420 MWs (2.47 %) with isolated work. This already requires the preparation of more detailed information about the work WPP and SPP and the development of software that allows you to determine the necessary additives to the value of operational capacity reserve at the development stage of Russia UPS. These ratios on the influence of WPP and SPP on the reliability of energy systems are quite well consistent with the results presented in the work [10] by reviewing foreign sources, in relation to operating conditions.

References

1. Yu. Ya. Chukreyev, M. Yu. Chukreyev Models for evaluating indicators of balance reliability when managing the development of energy power systems(Komi SC UB RAS, Syktyvkar, 2014), 207 p.
2. Yu. Ya. Chukreyev Models for ensuring the reliability of energy power systems (Komi SC UB RAS, Syktyvkar, 1995), 176 p.
3. Kovalev G.F., Lebedeva L.M. Nadyezhnost' system elektroenergetiki. Novosibirsk: Nauka, 2015, 224 p.
4. Billinton, R. Reliability Evaluation of Power Systems. Second Edition / R. Billinton, R.N. Allan. New York and London: Plenum Press, 1996, 509 p.
5. Methodical recommendations for the design of the development of power systems. (Approved by the Order of Russia Ministry of Energy June 30, 2003, No. 281). Moscow, Russian Ministry of Energy from 153-34.20.118 2003.
6. Chukreev Yu.Ya., Chukreev M.Yu. Methodological principles of accounting for hydroelectric power plants in the task of ensuring the balance reliability UES of Russia // Methodological Problems in Reliability Study of Large Energy Systems: Volume 70. Methodical and practical problems of power systems reliability / otv. ed. N.I. Voropai / Book 1. Irkutsk (2019), 54-63 p.
7. Volkenau I.M., Zeiliger A.N., Khabachev L.D. Ekonomika formirovaniya elektroenergeticheskikh sistem. M.: Energoatomizdat, 1981, 320 p.
8. Krupenev D.S., Perzhabinskij S.M. Ocenka nadezhnosti jelektrojenergeticheskih sistem s vetrovymi jelektrostancijami // Izvestija RAN. Jenergetika, 2017, № 1. s. 39-47.
9. Chukreev Ju.Ja. Chukreev M.Ju., Chuprov V.S. Harakteristiki rezhima jelektropotreblenija primenitel'no k zadacham balansovoj nadezhnosti pri upravlenii razvitiem jelektrojenergeticheskih sistem rezul'taty / Metodicheskie voprosy issledovanija nadezhnosti bol'shih sistem jenergetiki. Vyp. 68. Irkutsk: Izd-vo ISJeM SO RAN, 2017, s. 328-337.
10. Il'enko A.V. «Ogranichenie vyrabotki SJeS i VJeS javljaetsja normal'noj praktikoj» // Jenergetika v Rossii i mire. Peretok.ru, 2021. https://peretok.ru/opinion/24103/

Development of Methods for Calculation the Load Angle of a Synchronous Generator Based on PMU Data

Y.E. Tepikin[1]*, *V.R.* Rafikov[1], and *F.N.* Gaydamakin[1]

[1]«AP Soft LLC», Ekaterinburg, Russia

Abstract. The development of the Wide Area Measurement Systems (WAMS) in the Unified Electric Power System (UES) of Russia and using phasor measurement technology (or PMU technology) open up new prospects in the electric power system (EPS) operational dispatch and automatic control. PMU devices installation on synchronous generators (SG) allows to obtain information not only about currents and stator voltages phasors, but also information about the excitation current and excitation voltage. This set of high-precision measurements of the network state parameters, in the presence of generators mathematical models, provides the possibility of calculating such parameters, the direct measurements of which are very difficult. One of these parameters, which can be determined using PMU, is the load angle of the SG (the angle between the generalized voltage phasor and the generator EMF phasor in the *dq* axis system) in the steady states and transients. The value of the load angle can be used to analyze the static and dynamic stability, as well as to quantify the damping properties of the SG at low-frequency oscillations (LFO) of the power systems electrical parameters. This study analyzes the possibilities for calculating the load angle based on a typical set of generator PMU signals: stator current and voltage vectors, excitation current and voltage. Two methods for calculating the load angle in steady states and transients have been developed and implemented. The first method is based on solving a system of nonlinear algebraic equations, and the basis of the second method is the mathematical model of the SG. The presented methods were tested using the SimInTech software package. Further, using the developed methods, the load angles of two turbogenerators installed at the plants of the Russian UES were calculated according to the PMU data of the parameters of the electromechanical transient process.

1 Introduction

The use of the PMU technology within the Wide Area Measurement Systems (WAMS) allows solving many urgent problems, contributing to an increase in the reliability of the operation of electric power systems (EPS) [1-3]. PMU devices are able to provide information about the voltages and currents phasors in steady states and transients with high discretization and accuracy [4,5]. Also, PMUs installed on synchronous generators (SG) make it possible to obtain the values of excitation current and excitation voltage [4,5]. Such a set of high-precision data opens up broad prospects in the field of verifying models of EPS objects, identifying the actual parameters of equivalent circuits, monitoring changes in the power systems electrical parameters, as well as analyzing transient processes in EPS [1]. The availability of PMU data in conjunction with the mathematical models of the generator provides the possibility of calculating such parameters, the measurement of which is impossible, or is associated with great difficulties. One of these parameters is the SG load angle (the angle between the generalized voltage phasor and the generator EMF phasor in the *dq* axis system). Information about the actual nature of the change in the load angle in various states can help in solving problems of analyzing the dynamic behaviour of generators and equivalents of power systems during transients.

An urgent problem inherent in the EPS is low-frequency oscillations (LFO) of the parameters of the electric power state [6, 7]. Arising due to the presence of various kinds of disturbing influences in the EPS, as well as due to the lack of sufficient damping, LFO reduce the reliability of the power system and can lead to a disruption of stability [6-9]. Therefore, it is important to ensure the stability of the EPS to solve the problems of identification, monitoring of LFO, as well as minimizing the negative impact of fluctuations on the power system.

It is possible to determine whether a specific SG is a source of oscillations, as well as to characterize its quantitative participation in the damping of the LFO, by calculating the values of the synchronizing power or synchronizing energy of the generator [10–12].

To assess the synchronizing power and energy, it is important to have as accurate information as possible about the nature of the change in the load angle during the electromechanical process [10]. Direct measurements of the load angle, as well as synchronized measurements of the SG rotor speed, are almost always unavailable or difficult to implement in practice. Thus, the task of developing methods that could provide information on the nature of the change in the load angle of the SG both in real time, increasing

*Corresponding author: tye@ap-soft.ru

the steady states and in transients with acceptable accuracy is an urgent task. The presence of synchrophasors measurements will allow determining the load angles almost in efficiency of monitoring and identification of LFO, as well as the analysis of transient processes and the stability of the EPS.

2 Methods for calculating the load angle

2.1 Method based on the solution of nonlinear algebraic equations

This method allows to determine the load angle of the SG in the steady states and transients according to the following equations, which are the dependences of the generator active and reactive powers on the load angle:

$$P_g = \frac{E'_q \cdot U_g}{X'_d} \cdot \sin\theta - U_g^2 \cdot \frac{X_q - X'_d}{X_q \cdot X'_d} \cdot \sin\theta \cdot \cos\theta; \quad (1)$$

$$Q_g = \frac{E'_q \cdot U_g}{X'_d} \cdot \cos\theta - U_g^2 \cdot \left(\frac{\sin^2\theta}{X_q} + \frac{\cos^2\theta}{X'_d} \right), \quad (2)$$

where U_g is the positive sequence phasor modulus of the SG linear voltage; P_g and Q_g are the respectively active and reactive three-phase SG powers; X_q is the synchronous reactance along the q axis; X'_d is the transient reactance along the d axis; E'_q and θ are the respectively transient EMF along the q axis and SG load angle.

This method requires values of phase voltages and powers, which can be obtained with PMU. The modulus of the linear voltage positive sequence phasor is calculated based on information about the voltage phasors of the stator phases. In the system of nonlinear algebraic equations (1-2), the transient EMF and the load angle SG are unknown. The solution of the presented nonlinear equations makes it possible to obtain the dependence of the change in the value of the load angle during the transient.

The presented method for calculating the load angle requires only measurements from the side of the SG stator, a minimum of passport SG parameters (only synchronous and transient resistances). Also, this method is not demanding on computing power. It allows to start the calculation from any measurement point, including from any moment during the transient, which makes it possible to use the described method to solve tasks in real time. However, the disadvantage of the method is that it does not take into account the influence of damper circuits, which can affect the accuracy of determining the load angle, and, as a result, the accuracy of calculating the synchronizing power and energy. Therefore, it is required to develop an approach that can most accurately take into account transients in all SG circuits.

2.2 Method based on the mathematical model of a synchronous generator

For a more complete account of the process in all SG circuits and, consequently, a more accurate the generator load angle determination, a method has been developed based on the "full" mathematical model of the SG based on the Gorev-Park equations [13,14]. This model takes into account transients in the stator circuit, in the excitation winding, as well as in damper circuits along the d and q axes.

$$U_d + R_a \cdot I_d + \frac{d}{dt}\psi_d + \omega \cdot \psi_q = 0; \quad (3)$$

$$U_q + R_a \cdot I_q + \frac{d}{dt}\psi_q - \omega \cdot \psi_d = 0; \quad (4)$$

$$U_f = R_f \cdot U_f + \frac{d}{dt}\psi_f; \quad (5)$$

$$0 = R_{1d} \cdot I_{1d} + \frac{d}{dt}\psi_{1d}; \quad (6)$$

$$0 = R_{1q} \cdot I_{1q} + \frac{d}{dt}\psi_{1q}; \quad (7)$$

$$\psi_d = X_d \cdot I_d + X_{ad} \cdot I_f + X_{1ad} \cdot I_{1d}; \quad (8)$$

$$\psi_q = X_q \cdot I_q + X_{1aq} \cdot I_{1q}; \quad (9)$$

$$\psi_f = X_f \cdot I_f + X_{ad} \cdot I_d + X_{1f} \cdot I_{1d}; \quad (10)$$

$$\psi_{1d} = X_{1d} \cdot I_{1d} + X_{1f} \cdot I_f + X_{1ad} \cdot I_d; \quad (11)$$

$$\psi_{1q} = X_{1q} \cdot I_{1q} + X_{1aq} \cdot I_q; \quad (12)$$

$$I_q = I_g \cdot \cos(\varphi + \theta); \quad (13)$$

$$I_d = I_g \cdot \sin(\varphi + \theta); \quad (14)$$

$$U_q = U_g \cdot \cos(\theta); \quad (15)$$

$$U_d = U_g \cdot \sin(\theta); \quad (16)$$

where U_d, U_q, I_d and I_q are the respectively voltage and current phasors d- and q- components; ψ_d and ψ_q are the respectively stator flux linkage d- and q- components; R_a is the resistance of the stator circuit; I_{1d} and I_{1q} are the respectively damper circuit currents d- and q- components; U_f is the field voltage; R_f is the resistance of the field winding; R_{1d} and R_{1q} are the respectively damper circuits resistances d- and q- components; ψ_{1d} and ψ_{1q} are the respectively damper circuits flux linkages d- and q- components; X_{1d} and X_{1q} are the respectively damper circuits reactances d- and q- components; X_{1f} is the mutual reactance between the field winding and the damper circuit along the d-axis; X_{1ad} is the mutual reactance between the stator circuit and the damper circuit along the d-axis; X_{1aq} is the mutual reactance between the stator circuit and the damper circuit along the q-axis; φ is the angle between positive sequence voltage and current phasors; I_g is the positive sequence current phasor modulus.

Calculation using this method requires values of phase voltages and currents, excitation voltage and excitation current, which can be obtained with PMU. In equations (3-16), the derivatives are written in terms of finite differences, and at each step, an optimization problem is solved, where the unknowns determined in the calculation process are the flux linkages of the generator circuits, the currents of the damper circuits, and the load angle of the SG. When using the described method, the calculation starts from the previous steady state.

The algorithm for calculating the load angle using the system of equations (3-16) is as follows:

1. Downloading the required data from the PMU. If necessary, pre-processing and filtering of data is carried out.
2. Calculation of the positive sequence of stator current and voltage. Recalculation of electrical parameters to relative unit values.
3. Recalculation of generator parameters. In technical data sheets, passports and other documentation, synchronous, transient, subtransient reactances, as well as transient and subtransient time constants are given. In turn, the mathematical model based on the Gorev-Park equations uses the resistances and reactances of the generator circuits. Therefore, at this stage, it is required to recalculate the parameters in accordance with one of the existing approaches [15].
4. Calculation of the initial conditions according to the steady state data preceding the transient.
5. Solution of the optimization problem, which results in the dependence of the change in the load angle during the process.

Thus, this method requires a large number of measurements, including the parameters of the rotor circuit (excitation currents and excitation voltages), as well as a full set of generator parameters (calculated reactances, time constants, no-load excitation current and voltage). The calculation in accordance with this method must be started strictly from the steady state. However, the described approach most fully reflects the transients in the rotor circuits (excitation winding and damper circuits), as a result, providing a more accurate result of calculating the SG load angle.

3 Case studies

3.1 Experiments with model data

Testing the performance of the developed methods, as well as assessing the accuracy of calculating the load angle during electromechanical transients, was carried out using dynamic simulation environment of the SimInTech software package.

Using the model of a two-machine power system, data were obtained for currents, stator voltages, excitation currents and voltages during an electromechanical transient, which is a swing with increasing amplitude and a frequency of 0,86 Hz. Also, the actual (reference) values of the load angle were obtained from the model. Then the calculated values by the proposed methods were compared with the reference values. Figure 1 illustrates the block-schema of test dynamic model.

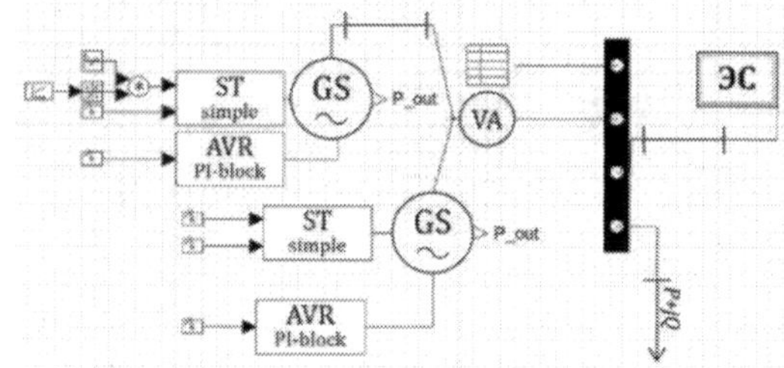

Fig.1. Block-schema of test dynamic model

Figure 2 shows a graph of the change in the active power of the modal generator and the results of calculating the load angle in comparison with the reference value during LFO simulation.

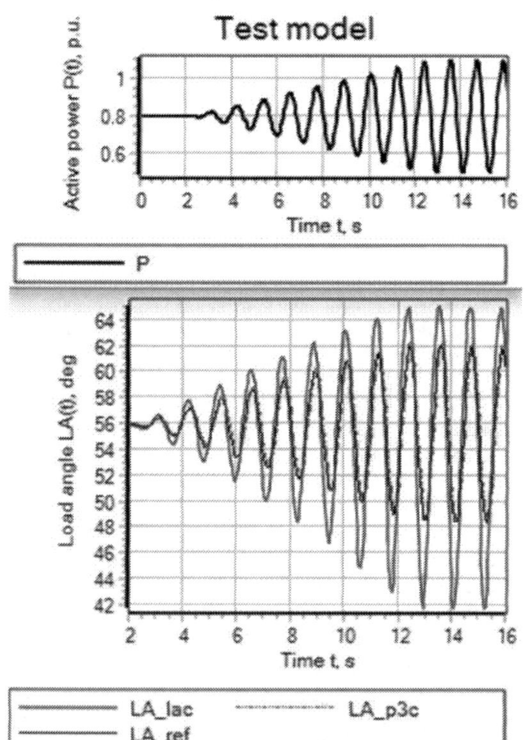

Fig. 2. Active power and load angle during test LFO simulation

In this figure and further, P is active power, LA_lac is load angle calculating by means of (1) and (2) equations; LA_p3c is load angle calculating by means of (3) - (16) equations; LA_ref is reference load angle from model.

Experiments on model data have shown that, in steady state, the methods accurately determine the load angle. In the transient, although both methods give an acceptable result, the accuracy of calculating the load angle using the SG «full» mathematical model is higher. Thus, the average difference between the calculation result and the standard at each measurement point was less than 0,5 degrees for the method based on the «full» mathematical model of the SG, and about four degrees for the method based on the solution of a system of algebraic equations.

3.2 Experiments with real-field data

Experiments with model data proved the efficiency of the developed methods and the correctness of the results of calculating the load angle in transients. At the next stage, these methods were used to calculate the load angle of real generators in the presence of oscillations in the EPS. Two turbogenerators were considered: one from the generators of the Power plant 1 and one from the generators of the Power plant 2. In both cases, LFO took place.

Figure 3 illustrates the graph of the change in the active power of the generator of the Power plant 1 and the part of the load angle calculating results during LFO.

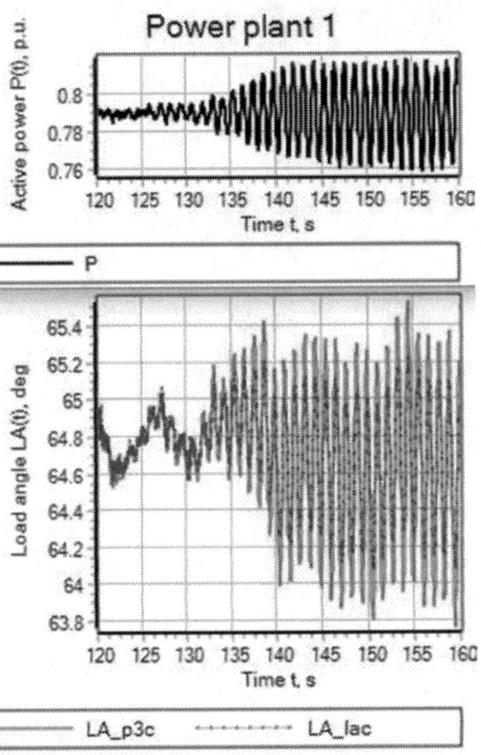

Fig. 3. Active power and load angle during LFO recorded at the Power plant 1

Figure 4 shows a graph of the change in the active power of the Power plant 2 generator and the part of the load angle calculating results during LFO.

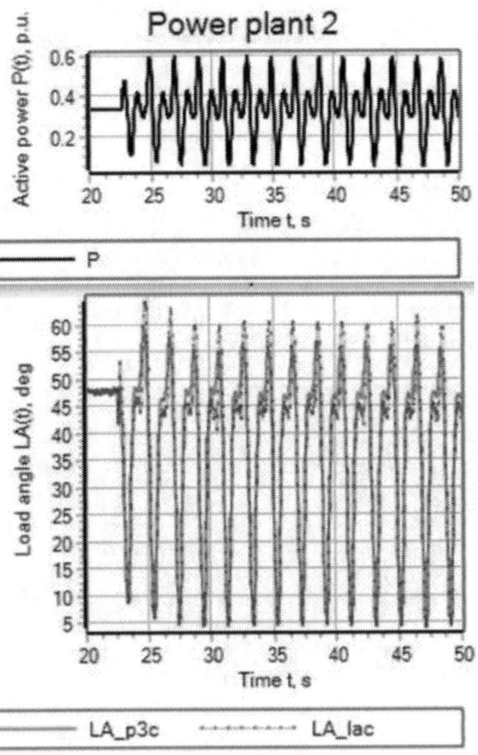

Fig. 4. Active power and load angle during LFO recorded at the Power plant 2

Table 1 shows the time for calculating the load angle of generators by different methods.

Table 1. Comparison of calculation duration.

Plant	Method	Calculation duration, s	Time of transient, s	Number of measuring points
Power plant 1	First method	87,7	220	11000
	Second method	177,7		
Power plant 2	First method	136,7	240	12000

An analysis of the results of calculating the load angle showed that the developed methods almost equally reflect the very nature of the angle change during the transient, however, there is a quantitative difference (primarily in amplitude) between the results, which is at least 10% in the transient. In steady state, the results are absolutely identical. Also, the duration of the calculation by the method based on the solution of a system of algebraic equations is almost two times less than by the method based on the use of the SG mathematical model.

Based on the results, we can conclude that if it is necessary to speed up the calculations, as well as in the presence of only stator measurements, a method based on solving a system of nonlinear algebraic equations should be used. However, in this case, not taking into account the influence of damper circuits can lead to a decrease in the accuracy of calculating the load angle. If a more complete consideration of the influence of all SG circuits is required, the load angle should be calculated using the «full» mathematical model of the generator.

4 Conclusion

The article proposes two methods for calculating the load angle. The first method is based on solving a system of nonlinear algebraic equations, requires only measurements of the stator circuit, has a short calculation time and can be used in real time. However, it does not take into account the influence of damper circuits. The second method is based on solving a system of equations representing the mathematical model of the SG, which takes into account transients in the stator circuit, excitation winding and generator damper circuits. This method requires not only measurements from the stator side, but also measurements of the excitation current and voltage. However, it gives a more accurate result.

The developed methods were successfully tested using the SimInTech software package. Further, using the presented methods, the generators load angles of the Power plant 1 and the Power plant 2 were calculated according to the transients data in the power system.

Further, the authors plan to continue research in the field of application of the calculated load angles in the tasks of developing methods for identifying the oscillations source, as well as methods for quantifying the participation of the SG in the LCO damping.

References

1. Zhukov A.V., Dubinin D.M. Development of the technology of phasor measurement unit data in the UES of Russia // Digital substation. – 2017. – No.8. – pp. 24-33.
2. Zhukov A.V., Satsuk E.I., Dubinin D.M., Opalev O.L., Utkin D.N. Experience in the development, implementation and operation of the monitoring system of transitional modes in the UES of Russia // Reports collection of the 5th International Scientific and Technical Conference "Modern trends of the relay protection systems and automation development in power systems". – Sochi, 2015.
3. Neber A.A. Applied issues of application of phasor measurement unit data of electrical mode parameters // Reports collection of the 3rd International Scientific and Technical Conference "Modern trends of relay protection systems and automation development in energy systems". – Saint Petersburg, 2011.
4. aUSS R 59364-2021. Relay protection and automation. Transient mode monitoring system. Norms and requirements; introduction. 2021-05-01. Moscow: Standartinform, 2021.
5. aUSS R 59365-2021. Relay protection and automation. Transient mode monitoring system. Phasor measurement unit. Norms and requirements; introduction. 2021-05-01. M.: Standartinform, 2021.
6. Zhukov, A.V. Monitoring of low-frequency oscillations in electro-energy systems / A.V. Zhukov, T. G. Klimova, A. I. Rascheplyaev // Electricity. - 2013. – No. 2. – pp. 20-27.
7. Methods for detecting fluctuations in the parameters of the electrical mode of the power system and their application for energy system management tasks / A.V. Zhukov, E. I. Satsuk, D. M. Dubinin [et al.] // Energetik. – 2018. – No. 12. – pp. 3-9.
8. Klimova, T. G. Analysis of a periodic disturbances influence on a low-frequency oscillations occurrence in the power system / T. G. Klimova, M. V. Savvatin // Gazette of the Moscow Power Institute. Bulletin of the MEI. – 2015. – No. 6. – pp. 56-62.
9. Opalev, O. L. Investigation of systemic interzonal low-frequency oscillations of the electrical mode parameters in the UES of Russia / O. L. Opalev // News of STC of UES. – 2018. – № 2(79). – Pp. 54-72.
10. S. Guo, F. Ouyang and F. Lv, A New Generator Control Device Level Disturbance Source Location Method of Power Oscillation Based on PMU. 2019 IEEE 3rd Conference on Energy Internet and Energy System Integration (EI2), 2019, pp. 1558-1562, doi: 10.1109/EI247390.2019.9061977.
11. Guo Siyuan, Zhang Shoushou, Li Li , Song Junying, Zhao Yongsheng. An Oscillation Energy Calculation Method Suitable for the Dis-turbance Source Location of Generator Control Systems. Journal of Phys-ics: Conference Series. 1518. 012081. 10.1088/1742-6596/1518/1/012081.
12. Kovalenko, P. Y. Estimation of synchronous generator participation in low-frequency oscillations damping based on synchronized phasor measurements / P. Y. Kovalenko, Y. P. Zakharov, A. S. Berdin // WIT Transactions on Ecology and the Environment. – 2014. – Vol. 190 VOL-UME 1. – P. 319-325. – DOI 10.2495/EQ140311.
13. Kundur P. Power System Stability and Control. – New York: McGraw-Hill 1994.
14. Peter W. Sauer, M. A. Pai, Joe H. Chow. Power System Dynamics and Stability: With Synchrophasor Measurement and Power System Toolbox. IEEE, 2017, doi: 10.1002/9781119355755.
15. Canay, I.M. Modelling of alternating-current machines having mul-tiple rotor circuits. IEEE Transactions on Energy Conversion, vol. 8, no. 2, pp. 280-296, June 1993, doi: 10.1109/60.222719.

On the functioning of capacitor banks in smart grids

Lidiia Kovernikova[1]*

[1]Melentiev Energy Systems Institute, Power Systems Department, 664033, Lermontov Str., Irkutsk, Russian Federation

Abstract. The amount of electronic equipment is increasing in created smart grids. This equipment introduces distortions into the electrical network, due to which the voltages at the network nodes, and hence the currents, become nonsinusoidal. The paper is devoted to the analysis of the effect of capacitor banks on the electrical network with nonsinusoidal mode parameters due to the resonant circuits they create with the electrical network at harmonic frequencies, as well as the influence of nonsinusoidal network modes on capacitor banks. The paper presents an overview of scientific publications on the problem of harmonic resonances in nonsinusoidal network modes, as well as the values of power quality indices that were measured in the study of the effect of the capacitor bank on the network mode. With the help of the "Harmonics" software package, the reduction in the service life of the capacitor bank in the nonsinusoidal mode, the frequency characteristics of admittance, conductance, susceptance of the network node with the capacitor bank, and the resonant powers of the capacitor bank at harmonic frequencies were calculated.

1 Introduction

Smart grids are created in Russia. They have distributed generation as well. Electronic equipment ensures the functioning of smart grids. It is the reason for the deterioration of the power quality in smart grids. The power quality in them does not meet the requirements of GOST 32144-2013 [1]. The results of measurements of the power quality indices confirm this fact [2]. Electronic equipment has non-linear current-voltage characteristics, and it consumes nonsinusoidal current from the supply network. The nonsinusoidal current flows in the electrical network and creates nonsinusoidal voltage drops on the network elements. The voltages at the network nodes become nonsinusoidal. A distorted voltage supplies the linear loads. Nonsinusoidal currents and voltages contain sinusoidal components of the fundamental frequency, i.e. with the frequency corresponding to the period of the original waveform or the first harmonic number (n=1), and frequencies that are multiples and non-multiples of the fundamental frequency with harmonic numbers higher than the first (n>1). The active power transmitted through the electrical network contains the active power of the first harmonic and the active power of the harmonic numbers n>1 [3, 4]. The active power of the first harmonic is useful because it does useful work. Active powers with harmonic numbers n>1 are harmful, since they do not perform useful work. The time of rapid development of industrial technologies using electronic equipment was already in the last three decades of the twentieth century. The power quality has deteriorated sharply due to the harmonics of currents, voltages, and powers. They caused additional power losses in electrical networks, false alarms of control systems, reduced service life of electrical equipment, resonant overvoltages and other consequences [5-19].

The present paper explores the effect of the capacitor bank on the electrical network and the electrical network on the capacitor bank in the nonsinusoidal mode. The paper presents the overview of publications, the results of measurements of the power quality indices, obtained during experimental study of the effect of the capacitor bank on the network mode, methodical approach for predicting harmonic resonance, the analysis of the parameters of the mode and of the network node with the capacitor bank using the "Harmonics" software package, assessments of resonant powers and the service life of the capacitor bank.

2 Relevance of harmonic resonances in electrical networks

One of the problems of the nonsinusoidal modes relates to capacitor banks in electrical networks. Capacitor banks regulate the voltage value in the network nodes and allow obtaining the normative value of the power factor. The electrical network and the capacitor bank at harmonic frequencies create resonant circuits [6, 7, 9-11].

If there are current harmonics in the electrical network, on which resonant circuits have formed, then currents and voltages increase on these harmonics. That is why the authors of [7] write that current resonances are especially dangerous for static capacitors. The authors of [12] note that current harmonics overload capacitor banks and damage them.

*Corresponding author: kovernikova@isem.irk.ru

Regulatory documents on the electric power industry for many years indicate the need to take into account resonant modes at harmonic frequencies. The 2003 regulation states [13]: "The nonsinusoidality of the voltage at the point of common connection can be created both by distorting consumer electrical receivers and by equipment of power supply organizations operating in modes that contribute to the appearance of resonant modes". The current document "On a unified technical policy in the power grid complex" indicates: "The use of the capacitor bank is allowed provided that resonance phenomena are excluded under all operating modes of the electrical network" [14].

The authors [10, 15-19] in their papers pay great attention to the prediction and detection of resonant modes in electrical networks at harmonic frequencies. The authors of [10] proposed to use the frequency scanning method to detect the resonant circuit. They used impedance charts to establish the resonant mode. Large impedance values characterize parallel resonances. They form peaks on the chart. Small impedance values characterize series resonances. They form troughs on the chart. The authors of [15] developed a model of the electrical network to predict the values of voltage and current harmonics. They made special measurements on this network to verify the model. The electrical network where took measurements consisted of the electrical system, a transformer, a 500 kV power line, which formed the resonant circuit at the 5-th harmonic. The authors of [16] studied the harmonics of voltages and currents in high-voltage networks under resonant phenomena. They used the theory of long lines in their research. The authors studied the influence of the network configuration on the input impedance near resonant frequencies. They found that when the network configuration changes, the increase in the harmonics of currents and voltages is possible. The authors of [17] presented the results of studies of parallel resonance with the controlled 12-stage capacitor bank. They found that each stage of the capacitor bank creates the resonance at the certain frequency. The authors of [18] note that interest in harmonic resonances has increased due to the use of power electronics in all areas of human activity. For example, damage to communication systems equipment occurs in resonant modes due to overvoltages and interference. Interference distorts information transmitted over power lines and recorded by smart meters. The authors of [19] studied harmonic resonances in urban networks. They note that they arise due to the electronic equipment of households. The authors point out those harmonic resonances in low voltage networks must be taken into account when planning networks.

3 Results of experimental studies of the effect of the capacitor bank on the power quality

Power quality indices were measured to evaluate the effect of the capacitor bank on voltage waveform distortion. The indices of the *n*-th voltage harmonic factor ($K_{U(n)}$) and the total harmonic distortion (K_U) were measured. The state standard [1] establishes for the values of power quality indices $K_{U(n)}$ and K_U for 95% and 100% of the measurement time the normative values – $K_{U(n)N95}$, $K_{U(n)N100}$, K_{UN95}, K_{UN100}.

The capacitor bank is installed in the 110 kV network at the substation of the network organization. The measurements were carried out on 110 kV buses of a railway traction substation. The 110 kV transmission line 200 meters long connects the buses of the traction substation with the 110 kV buses of the network organization, where the capacitor bank is connected.

Figure 1 shows a fragment of the network scheme, where the capacitor bank is connected at node 1980, and measurements were taken at node 1981. The notations used in the scheme are: PS – power system, EN – electrical network.

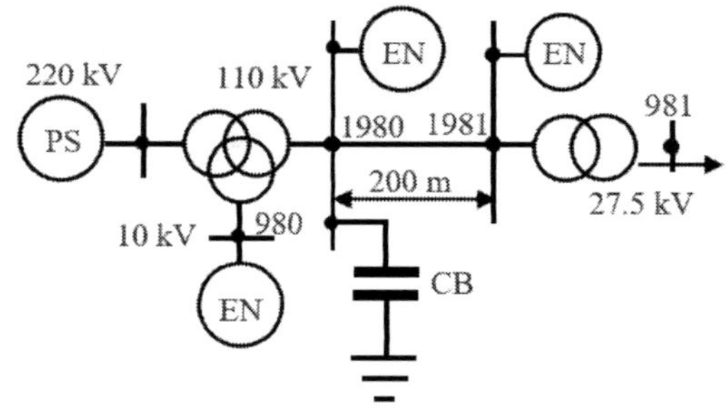

Fig.1. Fragment of the network scheme.

The measurement results are shown in Fig. 2-7. The capacitor bank switches on at the nineteenth minute. Charts of $K_{U(3)}$ and $K_{U(5)}$ for three phases (A, B, C) and the normative values established for them in [1] are shown in Fig. 2 and Fig. 3. Power quality indices $K_{U(3)}$ did not exceed $K_{U(3)N95}$ before switching on the capacitor bank. They have exceeded the normative value $K_{U(3)N100}$ after switching on the capacitor bank. Power quality indices $K_{U(5)}$ before switching on the capacitor bank exceeded the normative value $K_{U(5)N95}$, after switching on the capacitor bank they began to exceed $K_{U(5)N100}$.

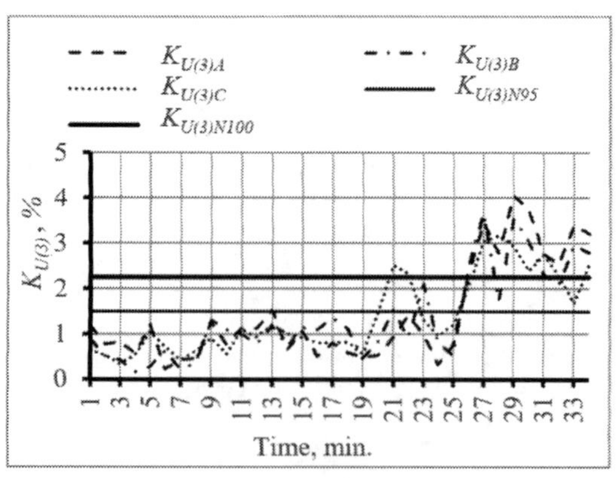

Fig. 2. Measured power quality indices of $K_{U(3)}$.

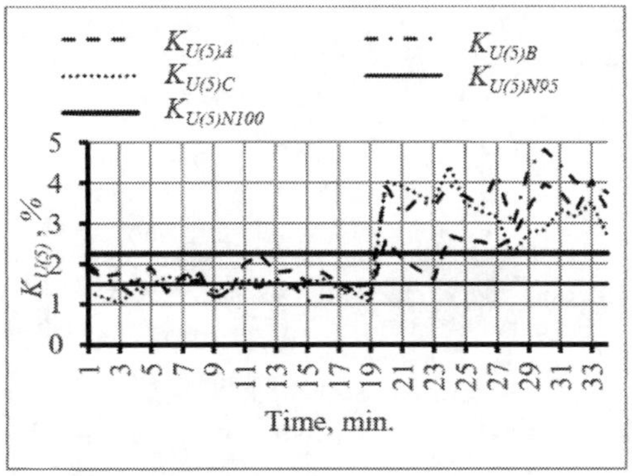

Fig. 3. Measured power quality indices of $K_{U(5)}$.

Figures 4 and 5 show charts of power quality indices for one of the phases. Different normative values in the range from 0.5% to 1% are established for the power quality indices in Fig. 4 [1]. The measured power quality indices did not exceed the normative values $K_{U(n)N95}$ before switching on the capacitor bank, their values decreased after the capacitor bank was switched on. The normative value $K_{U(n)N95}$ for the power quality indices in Fig. 5 are 0.4% [1]. They exceeded the normative

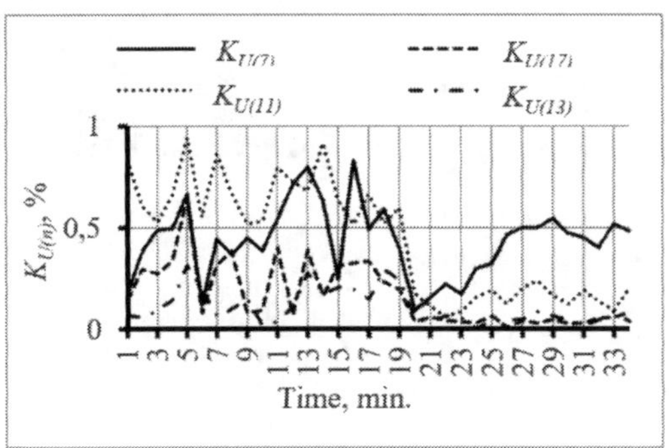

Fig. 4. Measured power quality indices of $K_{U(7)}$, $K_{U(11)}$, $K_{U(13)}$, $K_{U(17)}$.

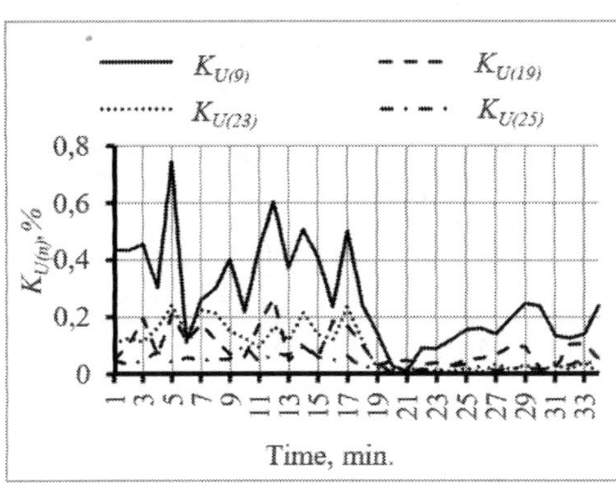

Fig. 5. Measured power quality indices of $K_{U(9)}$, $K_{U(19)}$, $K_{U(23)}$, $K_{U(25)}$.

value before switching on the capacitor bank only for the 9-th harmonic. They became less than the normative value at all harmonics after switching on the capacitor bank.

All power quality indices in Fig. 6 are less than the normative values $K_{U(n)N95}$ equal 0.2% and 0.4% both before switching on and after switching on the capacitor bank.

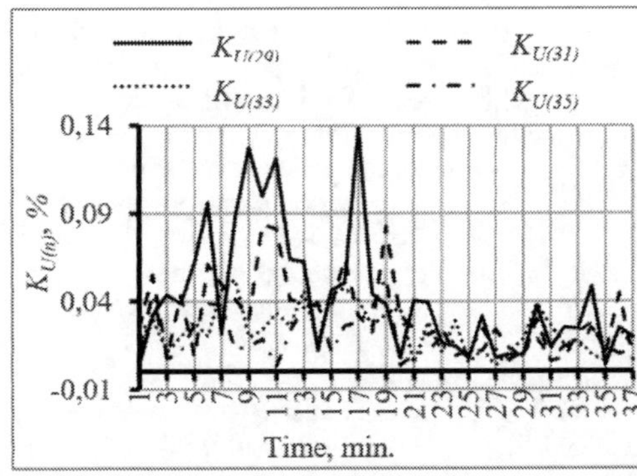

Fig. 6. Measured power quality indices of $K_{U(29)}$, $K_{U(31)}$, $K_{U(33)}$, $K_{U(35)}$.

The values of K_U in the three phases are shown in Fig. 7. They exceeded the normative value K_{UN95} before the capacitor bank was switched on and exceeded K_{UN100} after it was switched on.

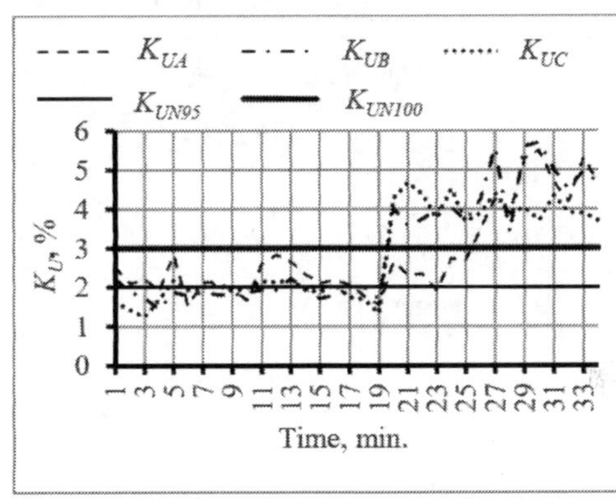

Fig. 7. Measured power quality indices of K_U.

4 Methodical approach to the analysis of harmonic resonances

A resonant circuit occurs when the values of the inductive and capacitive elements of the electrical circuit are equal [20]. The phase difference between voltage and current at the input of the resonant circuit is equal to zero. If the inductive and capacitive elements are connected in series, then a series resonance occurs, i.e. voltage resonance. The impedance of the resonant circuit is active, the current in the circuit is the largest, and large voltage drops occur on the inductive and capacitive elements. When the inductive and capacitive elements are connected in parallel, a parallel resonance occurs, i.e. current resonance. The susceptance of the circuit becomes equal to zero. The admittance is active and minimal. Parallel connection of inductive and capacitive elements creates the large resistance. The current in the common branch is the smallest, but it is possible to increase it in reactive elements. If there is the capacitor bank in the network, resonance occurs between the capacitance of the capacitor bank and the inductance of the electrical network at the connection point of the capacitor bank.

To analyse harmonic resonances, it is necessary to calculate the network mode at the harmonic frequency, as well as conductance, susceptance, and admittance of network nodes, voltage harmonics, power quality indices and power of capacitor banks at which resonance occurs. All the above parameters can be calculated using the system of matrix equations [21]

$$\mathbf{U_n}=\mathbf{Z_n}\mathbf{I_n}, \tag{1}$$

where $\mathbf{n}$ – the harmonic number; $\mathbf{U_n}$ – the column-matrix of voltage phasors at network nodes to be determined; $\mathbf{Z_n}$ – the square matrix of self- and mutual impedances of network nodes, wich is obtained as result of inversion of nodal admittance matrix $\mathbf{Y_n}$; $\mathbf{I_n}$ – the column-matrix of currents phasors at network nodes representing non-linear loads, which should be determined from the results of measurements of current harmonics [22].

Matrix $\mathbf{Y_n}$ is formed according to the known parameters of the electrical network elements [21].

To establish the presence of the resonant circuit, the frequency response of the following parameters of the network node is calculated: the admittance (y_n), the conductance (g_n), the susceptance (b_n).

$$y_n=1/z_n \tag{2}$$

$$g_n=r_n/z_n \tag{3}$$

$$b_n=x_n/z_n \tag{4}$$

The reactance x_n can be inductive, i.e. $x_n=\omega nL$, and capacitive i.e. $x_n=-1/(\omega nC)$, $\omega=2\pi f$, $f=50$ Hz. The impedance z_n is calculated from expression (5).

$$z_n=[(r_n)^2+(x_n)^2]^{1/2} \tag{5}$$

To assess the effect of harmonic resonance on voltage harmonics at the network nodes, the power quality indices $K_{U(n)}$ are calculated in accordance with [1].

The experience of studying harmonic modes shows that most often $K_{U(n)}$ exceed the normative values of 3, 5, 7, 9, 11, 13, 17, 19, 23, 25-th harmonics. Resonance circuits arise not only on harmonics, but also on interharmonics. The frequency scanning method is implemented in the "Harmonics" software package to determine the numbers of resonant harmonics and interharmonics using frequency response y_n, r_n, b_n [23]. The "Harmonics" software package determines the nodes of the network with resonant circuits, and calculates the parameters of the network and network modes.

The algorithm for analysing harmonic resonance in the connection node of the capacitor bank using the "Harmonics" software package is given below:

1. Input information on the electrical network.
2. Setting the interval of studied harmonics $[n1, n2]$ and harmonic scanning step Δn.
3. Formation of the matrix $\mathbf{Y_n}$.
4. Calculation of the elements of the matrix $\mathbf{Z_n}$.
5. Calculation of y_n, g_n, b_n for the network node in which the capacitor bank is connected.
6. Calculation of U_n, $K_{U(n)}$ for the network node in which the capacitor bank is connected.
7. Checking condition $n \leq n2$. If the condition is not met, then return to step 2. If the condition is met, then go to step 8.
8. Analysis of the calculated susceptance values b_n to determine the number of the resonant harmonic.
9. Calculation of the power of the capacitor bank at the resonant harmonic.
10. Calculation of the multiplicity of reduction in the service life of the capacitor bank at the resonant harmonic.
11. Output of calculation results.

Additional losses of active power appear in the insulation of the capacitor bank with the nonsinusoidal voltage at its terminals. They increase the operating temperature of the insulation, which shortens its service life and, as the result, the service life of the capacitor bank.

The authors of [7] propose to use the multiplicity of the service life reduction (y) to assess the effect of the nonsinusoidal voltage on the capacitor bank. It is calculated by expression (6).

$$y=1/z, \tag{6}$$

where $z=Z/Z_s$ is the relative reduction in the service life of the insulation, $Z=C\exp(-bt)$ and $Z_s=C\exp(-bt_s)$ - service life of the insulation at temperature differences between the hottest point of the capacitor and the environment for nonsinusoidal voltage t and for sinusoidal voltage t_s. After transformations, z takes the form

$$z=\exp(-b\Delta t), \tag{7}$$

where $\Delta t=t-t_s$, $b=0.086$.

The value of b is determined by the Montzinger rule, which establishes that "an increase in temperature by 8°C compared to the permissible temperature reduces the service life of the insulation by 2 times". The temperature drop Δt is proportional to the power losses from voltage harmonics and in [7] is calculated as

$$\Delta t = t_s\left(\sum_{n=2}^{40} nU_n^2\right). \tag{8}$$

As a result, the multiplicity of the reduction in the service life of the capacitor bank will be determined as

$$y = \exp\left[bt_s\left(\sum_{n=2}^{40} nK_U^2\right)\right]. \tag{9}$$

5 Example of analysis of harmonic resonance in the network node with the capacitor bank

The analysis of harmonic resonance was made for the node 1980 of the electrical network, the fragment of which is shown in Fig. 1. Calculation of parameters for the analysis of harmonic resonance with the help of the "Harmonics" software package was carried out in accordance with the algorithm. The network scheme for calculation has 174 nodes, 217 links, including reactors and capacitor banks, loads represented by asynchronous motors, active-inductive power loads, sources of harmonics – an aluminium plant and traction loads of the railway.

Figures 8 and 9 show the frequency responses of conductance, susceptance and admittance for the node 1980 with the capacitor bank switched off and on. The change in the nature of susceptance occurs in both cases. At the points of intersection of the susceptance curves with the horizontal axis, which determine the resonant harmonic, the susceptance changes from inductive to capacitive, which corresponds to parallel resonance [20]. Resonance exists at the 17-th harmonic without the capacitor bank (Fig. 8). The susceptance chart of the 1980 node shows that it is inductive at harmonics 3 to 17. The resonant circuit at the 17-th harmonic was formed due to the capacitive elements of the electrical network. Resonance occurs on the harmonic in the range of 3.75-4.00 after switching on the capacitor bank (Fig. 9).

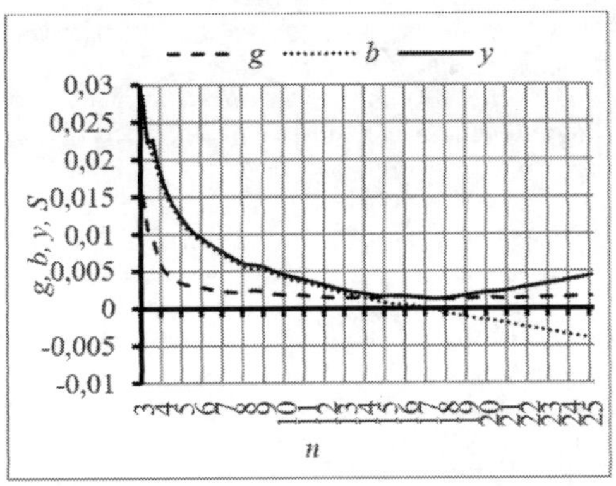

Fig. 8. Frequency responses of admittance, conductance, susceptance on the 3–25-th harmonics without the capacitor bank.

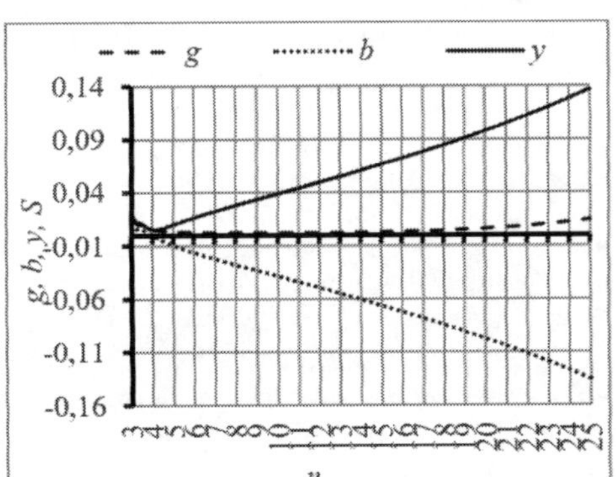

Fig. 9. Frequency responses of admittance, conductance, susceptance on the 3–25-th harmonics with the capacitor bank.

The power quality indices K_U and $K_{U(n)}$ in the 1981 node and the normative values set for them with the capacitor bank on (CBon) and off (CBof) in the 1980 node are shown in Fig. 10. The diagram shows that the values K_U, $K_{U(3)}$ and $K_{U(5)}$ exceed their normative values with the capacitor bank on, as was also obtained as the result of measurements in the 1981 node. The power quality indices do not exceed the normative values for other harmonics.

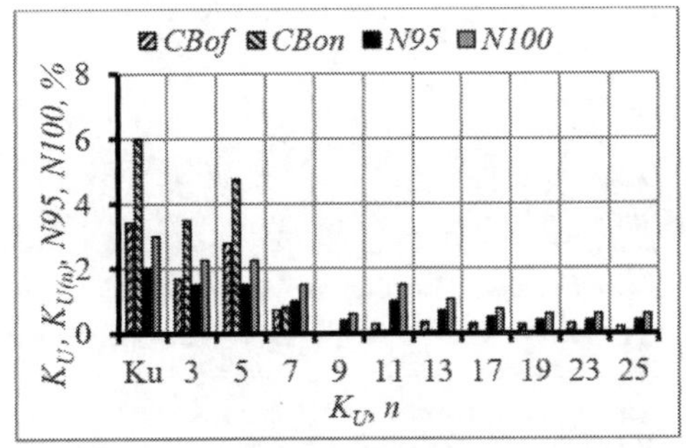

Fig. 10. Calculated power quality indices of K_U, $K_{U(n)}$.

Capacitor bank power (Q) for harmonics, on which the capacitor bank in node 1981 would create the resonant circuit with the electrical network, were calculated using the "Harmonics" software package. They are shown in Table 1.

Table 1. Resonance powers of the capacitor bank, MVAr

n	3	5	7	9	11	13	17
Q	75	49	27	13	8	6	2.5

To evaluate the effect of voltage harmonics on the service life of the capacitor bank in the 1980 node, the reduction factor of its service life was calculated using expression (9). In the calculation, the measured maximum values of the power quality indices $K_{U(n)}$ for 40 harmonics in three phases were used. For the capacitor bank with the capacity of 50 MVAr, the insulation heat resistance class A was adopted with the allowable temperature of 105°C and the ambient temperature of 45°C [24]. The multiplicity of the reduction in the service life of the capacitor bank is 1.25.

Conclusion

Harmonic resonances in electrical networks occur randomly with changes in the scheme and parameters of the electrical network, when the capacitor bank is switch on.

When forming the resonant circuit in the electrical network, power quality indices $K_{U(n)}$ and K_U exceed the normative values. Significant harmonics of currents and voltages create power losses in the capacitor bank, leading to the reduction in its service life.

The possibility of harmonic resonances should be taken into account when designing electrical networks.

The "Harmonics" software package and the proposed methodological approach to the analysis of resonant modes can be used in managing the power quality in smart grid to predict resonant modes and evaluate their effect on capacitor banks.

The research was carried out under State Assignment Project (No. FWEU-2021-0001) of the Fundamental Research Program of Russian Federation 2021-2030 using the resources of the High-Temperature Circuit Multi-Access Research Centre (Ministry of Science and Higher Education of the Russian Federation, project no 13.CKP.21.0038).

References

1. State Standard 32144-2013. *Electric Energy. Electromagnetic compatibility of technical equipment. Power quality limits in public power supply systems* (M.: Standartinform, 2014)
2. L.I. Kovernikova, V.V. Sudnova, R.G. Shamonov, et al., *Power quality: current state, outstanding issues, and proposals to their solutions* (Nauka, 2017)
3. L.I. Kovernikova, *Active powers of harmonics in the nodes of connecting non-linear loads to the high- voltage network*, Elektrichestvo, No. 3, 12-20 (2017)
4. V.I. Kochkin, *Reactive power in electrical networks. Technologies of controlled compensation*, Novosti elektrotekhniki, No. 3(51), 1-9 (2008)
5. J. Arrillaga, D. Bradley, P. Bodger, *Power systems harmonics* (Energoatomizdat, 1990)
6. I.V. Zhezhelenko, *Higher harmonics in power supply systems of industrial enterprises* (Energoatomizdat, 2010)
7. A.K. Shidlovsky, V.G. Kuznetsov, *Improving the power quality in electrical networks* (Nauk. Dumka, 1985)
8. E.F. Fuchs, D.J. Roesler, K.P. Kovacs, *Aging of electrical appliances due to harmonics of the power system's voltage*, IEEE Trans. Power Delivery, **1**(3), 301-307 (1986)
9. V.E. Tonkal, A.V. Novoseltsev, S.P. Denisyuk, V.Ya. Zhuikov, M.T. Strelkov, Yu.A. Yatsenko, *Energy balance in electrical circuits* (Nauk. Dumka, 1992)
10. IEEE power engineering society, *Tutorial on harmonics modeling and simulation*, IEEE Catalog number: 98TR125-0 (1998)
11. I.I. Artyukhov, I.I. Bochkarev, S.V. Molot, *Problems of the power quality in the power supply systems of gas transmission enterprises with the intensive introduction of frequency-controlled electric drives*, in Proceedings of the International Scientific and Practical Conference "Management of the power quality", Moscow, 23-25 November 2016, Moscow, Russian Federation (2016)
12. V.M. Zyryanov, N.A. Mitrofanov, *Operation of capacitor units under conditions of harmonic distortions of current and voltage curves*, in Proceedings of the International Scientific and Practical Conference "Management of the power quality", Moscow, 05-07 December 2018, Moscow, Russian Federation (2018)
13. RD 153-34.0-15.502-2002. *Guidelines for monitoring and analysing the power quality in public power supply systems. Part 2. Analysis of the power quality* (Moscow: Energoservice, 2003)
14. Regulations of PJSC FGC UES "*On the Unified Technical Policy in the Electric Grid Complex*", St. Petersburg, Russian Federation (2021)
15. D.K. Geddey, D.C. Smith, P.O.Wright, P.M. Hart, G.E. Littler, *Control of harmonic distortion on Australian Electricity supply networks*, In the book "The impact of high-voltage electrical installations on the environment", 51-59, (Energoatomizdat, 1988)
16. K. Hlava, I. Kabrhel, J. Smid, M. Blasko, *Experimental investigation of harmonic impedances and other electrical parameters of high-voltage network*, In the book "The impact of high-voltage electrical installations on the environment", 59-64, (Energoatomizdat, 1988)
17. T. Vinnal, T. Sakkos, K. Janson, M. Jarkovoi, *Resonances in LV industrial networks when using shunt capacitors for power factor correction*, in Proceedings of 23th International Conference on Electricity Distribution (CIRED), 15-18 June 2015, Lyon, France (2015)
18. L. Eggenschwiler, O. Galland, D. Chollet, F. Decorvet, D. Roggo, P. Favre-Perrod, *Frequency scans and resonance mode analysis for resonance problems identification in power networks in presence of harmonic pollution*, in Proceedings of 24th International Conference on Electricity Distribution (CIRED), 12-15 June 2017, Glasgow, United Kingdom (2017)
19. J. Meyer, R. Stiegler, P. Schegner, I. Röder, A. Belger, *Harmonic resonances in residential low voltage networks caused by consumer electronics*, in Proceedings of 24th International Conference on Electricity Distribution (CIRED), 12-15 June 2017, Glasgow, United Kingdom (2017)
20. L.A. Neiman, K.S. Demirchan, *Theoretical foundations of electrical engineering. Textbook for universities*, **1** (Energoatomizdat, 1981)
21. L.A. Zhukov, I.P. Stratan, *Steady-state operation modes of complex electrical networks and systems: Calculation methods* (Energiya, 1979)
22. L.I. Kovernikova, Van Chung Luong, *Algorithm for modeling non-linear loads of the high-voltage network by measured parameters*, Elektrichestvo, No. 10, 16-25 (2017)
23. L.I. Kovernikova, *"The "Harmonics" software package for analysis and normalization of harmonic modes in high-voltage networks"*, in Proceedings of the International Scientific and Practical Conference "Management of the power quality", 23-25 November 2016, Moscow, Russian Federation (2016)
24. Energy-T. Product Catalog. https://www.energy-t.ru/catalog/krm/batarei-staticheskix-kondensatorov.html

Coherence indicators of generators for express assessment of electric power system transient stability

Dmitry Efimov[1,2]*, and *Sergey* Stashkevich[1]

[1]Melentiev Energy Systems Institute, Electric Power System Department, 664033 Irkutsk, Russia
[2]Irkutsk National Research Technical University, Energy Institute, 664074 Irkutsk, Russia

Abstract. A brief overview of the generator coherence indicators is presented, the areas and limitations of their applicability are indicated. The application of the area method for the rapid assessment of the transient stability of complex multi-machine electric power system based on the features of the dynamics of the system behavior, determined by the heterogeneous structure of the electrical network, is considered. A distinctive feature of the proposed approach is the use of the characteristics of the heterogeneity of the power system structure, which determine the presence of weak links (or cut-sets) of the electrical network, which are critical from the point of view of a possible violation of the system stability.

1 Introduction

Due to the instability of the electricity market and the growing demand for electricity, modern EPSs (electric power systems) are forced to work closer and closer to the limits of their stability. This makes the system more vulnerable to internal failures and external disturbances and increases the risks of stability loss.

Conventionally, the assessment of transient stability comes down to detailed simulation of EPS responses to disturbances – that is, to calculations of transients (this is the numerical integration of a system of nonlinear differential equations of large dimensions). However, a disturbance is unpredictable, and the calculations should be carried out at least with a minimum, but in advance - the calculation itself takes a lot of time, and it still takes time to develop control actions.

The key points of the proposed approach to the analysis of the transient stability of EPS are the following:

1. Study of heterogeneities in electric power systems.
2. Revealing the coherence of the movement of generators under disturbances.
3. Reducing the models of dynamics of electric power systems.
4. Study of stability using reduced models.

In view of the foregoing, an express assessment of the transient stability of a complex EPS using the area method for a given studied conditions, including a given network topology, electric mode and disturbance, is performed as follows:

1) Using structural analysis algorithms, weak (dangerous from the point of view of a possible violation of the stability of the system under a specific disturbance) cut-sets of the studied network are determined and ranked according to the degree of weakness;
2) For each weak section, as the degree of weakness decreases, a two-machine equivalent is formed;
3) An assessment of the transient stability by the area method is carried out for each of the obtained two-machine equivalents as the weakness of the sections decreases. The evaluation ends at the section for which the applied area method no longer violates the stability of the system;
4) If for any of the network cut-set, according to the above algorithm, a violation of the stability of two-machine equivalents is not recorded, we can assume that the stability of the EPS in a given studied conditions is not violated. Otherwise, the estimates show a violation of the stability of the system for a given network topology, electric mode and disturbance.

2 Heterogeneity of EPS and coherence of movement of generators

Structural heterogeneity is a fundamental property of any systems with a complex structure. The structure of the electrical network of a complex EPS is always heterogeneous and includes strongly connected subsystems and weak connections between them. It is important to identify this heterogeneity, quantify it and use it in modeling the EPS, its study and control of electric modes [1].

During the operation of the EPS, it is subject to disturbances and reacts to them by changing the parameters of the system electric mode. This reaction is determined both by the magnitude and location of the disturbance, and by the internal properties of the system itself. The nodes and connections (sections) of the system, the electric mode parameters of which reach unacceptable values first of all, are called weak points.

*Corresponding author: d.n.efimov@mail.ru

When simulating electromechanical transients and evaluating the stability of EPS during disturbances, the presence of strong connections between the generators of a strongly coupled subsystem determines the coherence (identity, consistency in time) of the movement of generators in transients. Besides the presence of large reserves of throughput for connections between generators, which guarantees their mutual stability within the subsystem. On the contrary, weak connections between strongly connected subsystems create threats of stability violation. Due to the limited transmission capacity of weak links, it is along them that violations of the stability of the system usually occur during disturbances. Therefore, the identification and quantitative assessment of weak links in the structure of the electrical network are important tasks in the process of studying stability and determining control actions to ensure it [2].

The purpose of the study of heterogeneities is to identify weak links in the structure of the electrical network and thereby determine the strongly connected subsystems in this structure. Violations of the EPS stability and cascade development of emergency processes during regime changes will occur primarily through weak links (between subsystems) and less likely through stronger links (within subsystems). Therefore, the stability of the EPS must be analyzed primarily in relation to weak links.

Structural inhomogeneity of the EPS determines the specifics of the movement of the system generators in the transient electromechanical process, namely, their coherent movement. The coherence of the movement of generators is an objective basis for reducing the mathematical model of the EPS dynamics by aggregating (combining) generators included in the same subsystem.

The motion coherence of generators i and j is defined as

$$\delta_i(t) - \delta_j(t) = const \qquad (1)$$

where $\delta(t)$ are the angles of the rotors of generators i and j as functions of time in a single coordinate system.

Identification of the coherence of the movement of generators during disturbances is the identification of groups of generators, the mutual (relative to each other) movement of which in the transient is close to coherent.

3 Metrics and similarity/difference matrices for identifying coherent groups of generators

Coherence can be determined by directly comparing the motion curves of the generators (which requires numerical integration of the transient). So, in Fig. 1, in the studied conditions *a)* and *b)*, generators 1 and 2 are more coherent to each other than each of them is to generator 3. In case *c)*, generators 2 and 3 have the greatest mutual coherence.

In addition, coherence can be estimated by formal indicators of mutual similarity or difference in the movement of generators in the transient, based on the admittances of the equivalent network, inertia and acceleration of machines.

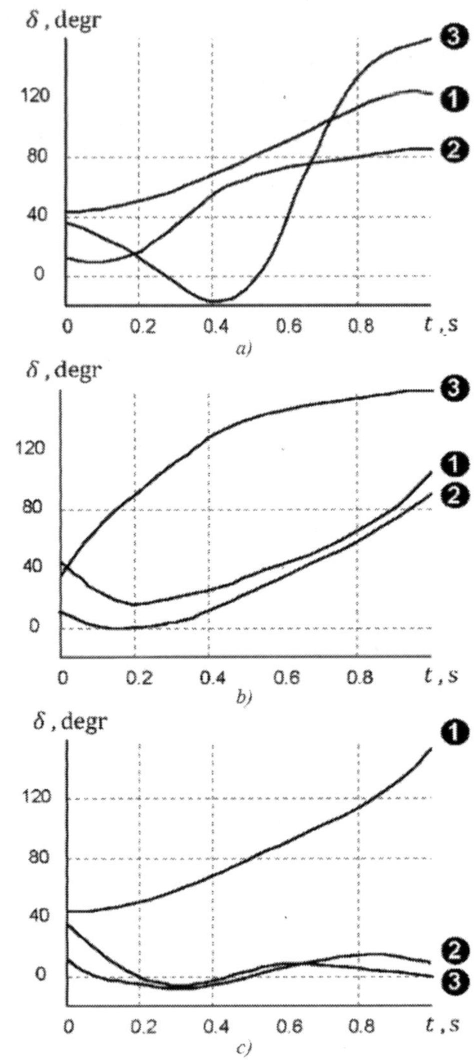

Fig. 1. An example of the change in time of the angles of the generators in three studied conditions.

The equivalent network of the investigated EPS for calculating the indicators of similarity (coherence) or difference (non-coherence) of generators, in structural analysis is obtained from the conventional network used to calculate steady-state electric modes, by:

- representation of loads by constant admittances, generators by a two-node equivalent (bus – transient impedance – transient electromotive force (EMF)), power of the primary engine (in the equations of motion of the generator rotors) – by a constant,
- exclusion of all nodes that do not contain EMF.

Any pair of equivalent network nodes turns out to be interconnected by an equivalent connection. From

https://doi.org/10.1051/e3sconf/202338401012

the point of view of interaction and mutual influence of generators, EPS can be represented as a complete graph, at the vertices of which generators are connected, and the edges represent the relationship between generators. Edges can be assigned some numerical characteristics that determine the degree of interaction and mutual influence of generators. Then the response of the system as a whole to the disturbance, instead of the curves of the movements of the generators, can be described by a matrix of similarity or difference indicators of the movements of the generators in pairs. On the basis of such matrices, a classification (identification of coherent groups) of generators can be made. The result of it is a set of nested subsystems (groups of generators of greater or lesser coherence) for a given studied conditions.

The simplest indicators of coherence are determined through the admittance of the electrical network, in other cases they take into account the electric mode parameters and dynamic parameters of the EPS. In this case, the force of mutual influence of two generators can be interpreted as the strength of the connection between them and as the similarity (coherence) of their movement in the transient.

Of the formal indicators of the similarity of generators, two have been conventionally and most widely used (see, for example, [3, 4]):

- *"Distance-admittance"*, defined as the admittance of connections of the equivalent network obtained from the initial network (for generator models "EMF behind transient impedance") with the exception of all nodes that do not contain EMF. These measures describe how closely two generators are electrically connected.
- *"Distance-reflection"* as the magnitude of the acceleration components (or active unbalance on the shaft) of one machine due to a change in the angle of the other. Unlike distance-admittance, this measure reflects the dynamic influence (synchronizing power) of one generator on others as a result of disturbance.

Note that here the historically accepted term "distance" is inaccurate, since both of the indicators are essentially measures of similarity (proximity).

In addition, the problem of determining coherence was solved using such features as:

- symmetry of generators relative to the system, initial mutual accelerations of the rotors [5, 6],
- differences in the angular deviations of the EMF of generators in a pair from their initial value at a given time interval, determined by a linearized model, [7, 8],
- equality of synchronizing powers, observance of stability conditions within the group, and a number of others [3, 4, 9–11].

In various sources, the following were considered as indicators of the similarity of generators:

- electrical connectivity (structural maximum as the capacity of an equivalent connection) [12];
- dynamic (taking into account inertia) connectivity [11] or dynamic connectivity, determined using pairs of eigenvalues of the linearized model of EPS dynamics [13];
- the severity of the disturbance for a pair of generators based on the values of the Lyapunov function, written as an integral of energy for the mathematical model of the EPS dynamics in positional idealization [14].

For a more accurate assessment of coherence, it is often recommended to numerically calculate the initial stage of the transient using a linearized or nonlinear model [4, 15]. An analytical algorithm for estimating motion coherence based on the area method for pairs of generators was developed in [10, 13] using the assumption of invariance in the transient of the mutual acceleration component of a pair of machines, determined by the change in the angles of the machines of the remaining part of the system. For the same purposes, indicators of the influence of disturbances on the behavior of EPS elements can be used, additionally taking into account the parameters of the disturbance and the dynamic characteristics of generators [9, 10].

Let's assume that the reference (accepted obviously correct) process of classification of generators is known, obtained on the basis of numerical integration of the electromechanical transient process and visual analysis of integral curves. Then, evaluating the similarity with it of classification processes based on matrices of formal indicators of similarity or difference of generators, it is possible to determine to what extent this or that indicator is adequate to the task of identifying groups of coherent generators - that is, whether and in what cases this indicator is a measure of coherence generator movements.

4 Energies of acceleration and deceleration

The indicators based on the values of the energies of the mutual acceleration and actually implemented in the transient braking of generators are determined by integrating the mutual acceleration over the mutual angle (Fig. 2).

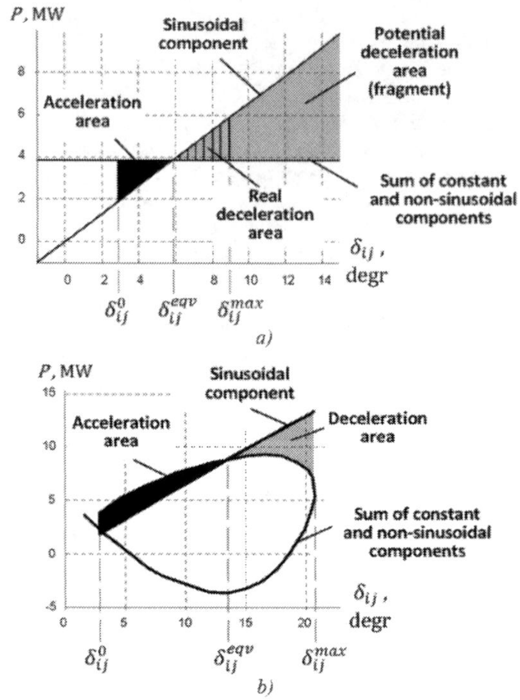

Fig. 2. Determination of the energies (areas) of mutual acceleration and deceleration of generators i and j for a simple disturbance by integrating over the mutual angle.

Fig. 2a illustrates the result of analytical integration with the assumption of the constancy of the non-sinusoidal component, and fig. 2b – the result of numerical integration without that assumption.

The integration result contains three components:

1) *Constant* component reflects the difference in the design and operating parameters of a pair of generators (turbine power, moments of inertia, EMF modules and modules of self-admittances of driving points),
2) *Sinusoidal* component reflects the influence of direct connection between generators,
3) *Non-sinusoidal* component reflects the influence of the remaining part of the system (asymmetry in the location of two generators relative to all other generators, manifested through the difference in power flows related to the moments of inertia).

5 Reducing the Power System Dynamics Models – Two-Machine Equivalent

Identification of dangerous sections from the point of view of a possible violation of the stability of the system under a specific disturbance is carried out by means of a cluster analysis of indicators that estimate the degree of coherence of the movement of generators in the transient [1, 2].

Classification of generators can also be made on the base of analysis and comparison of transient curves for various studied conditions. However, this way eliminates the very need for classification (since the detailed calculations necessary to obtain the curves in themselves provide an answer to most of the questions that arise in the study of a given conditions).

Hence, it is obvious that it is necessary to use the most simplified models of the movement of generators in the transient, or to search for the possibilities of classifying without resorting to numerical integration at all. To organize and visualize the hierarchical structure of the electrical network, dendrograms (classification trees) are used, i.e. a graphical method for presenting the results of hierarchical clustering, which shows the degree of proximity of individual energy objects and clusters, and also graphically demonstrates the sequence of their combination or separation. Dendrograms store information about further divisions (or associations) of the electrical network into smaller (larger) islands.

The transient stability of an EPS is conventionally estimated by numerically integrating its model offline for different studied conditions (network topologies, electric modes, and accident scenarios). The simplest classical positional model of EPS dynamics is written for the i-th generator in the form

$$\frac{d^2\delta_i}{dt^2} = \frac{\omega_0}{T_{Ji}} \cdot \frac{P_{Ti} - P_i}{S_{\text{НОМ}i}}, \quad i = \overline{1, N}. \tag{2}$$

Here t is time, δ_i is the angle of the generator rotor relative to the synchronous axis, P_{Ti} and P_i are electrical active powers (power, created by the accelerating torque of the turbine, and the generator power output to the network), $S_{\text{НОМ}i}$ is the rated power of the generator, T_{Ji} is the rotor inertia constant, ω_0 is the rated angular speed of the rotor, N is the number of generators in the EPS. The power supplied to the network is defined as

$$P_i = \sum_{j=1}^{N} E_i' E_j' y_{ij} \sin(\delta_{ij} - \alpha_{ij}), \quad i = \overline{1, N}, \tag{3}$$

where $\delta_{ij} = \delta_i - \delta_j$ is the mutual angle of the rotors of generators i and j, E_i' and E_j' are transient EMFs of generators i and j, y_{ij} and α_{ij} are the modules and complementing to 90° the angles of the complexes of intrinsic and mutual admittances of the electrical network.

Different elements of the EPS react to disturbances in different ways. However, the behaviors (reactions) of some elements are more similar to each other than to others. This creates the possibility of aggregating information in order to reduce computational costs.

Reducing the dynamics models of electric power systems is an aggregated representing of each of the coherent groups of generators by one equivalent generator. The conventional task of aggregation is to reduce the dimension of analyzed network (electromechanical equivalencing).

Reducing the EPS dynamics model during the express assessment of transient stability consists in constructing a two-machine equivalent – that is, a two-node network, each of the nodes of which is an equivalent generator (that is, it aggregates a group of coherently moving generators).

The most common models of equivalent generators are:

- The center of inertia, characterized by aggregated (summed over the group) parameters, when the power, constant of inertia, admittance of connections with the nodes of the external network are taken equal to the sums of the values of the corresponding parameters of the generators of the aggregated group, and the equivalent EMF – equal to the sum of EMF of equivalent branches weighted by their admittances [10].
- A representative generator that reflects the dynamic characteristics of a coherent group to the greatest extent (compared to other generators). The degree of representativeness is determined by the value of the "participation factor" (the contribution of the movement of the generator to the movement of the entire group). Such approaches are common when using selective modal stability analysis (see, for example, [16]).

The dynamic equivalent of REI (Radial Equivalent Independent), similar in implementation to the center of inertia model, but providing equality of power losses in the steady state for the initial and equivalent networks. This is achieved by introducing a temporary fictitious "network of zero power balance", electrically connecting the nodes of these networks with each other [17, 18].

By analogy with (2) and (3), the dynamics model of the two-machine (machine I and machine K) EPS equivalent can be written as

$$\frac{d^2\delta_{IK}}{dt^2}=\frac{d^2\delta_I}{dt^2}-\frac{d^2\delta_K}{dt^2}=\frac{\omega_0}{T_{JI}}\cdot\frac{P_{\text{т}I}-P_I}{S_{\text{ном}I}}-\frac{\omega_0}{T_{JK}}\times \times\frac{P_{\text{т}K}-P_K}{S_{\text{ном}K}}=\omega_0\left(\frac{P_{\text{т}I}-P_I}{T_{JI}S_{\text{ном}I}}-\frac{P_{\text{т}K}-P_K}{T_{JK}S_{\text{ном}K}}\right), \quad (3)$$

where

$$P_I=E_I'^2 y_{II}\sin(-\alpha_{IK})+ +E_I'E_K'y_{IK}\sin(\delta_{IK}-\alpha_{IK}), \quad (4)$$

$$P_K=E_K'^2 y_{KK}\sin(-\alpha_{KI})+ +E_I'E_K'y_{IK}\sin(\delta_{KI}-\alpha_{KI}). \quad (5)$$

6 Summary of the Results

The highest accuracy of coherence recognition is achieved when using the acceleration and deceleration energies, which require numerical integration of the transient. However, the suitability of these indicators for rapid recognition is limited by the complexity of the calculation. At the same time, determined from a simplified model of generator motion, these indicators are suitable for identifying groups of coherent generators in cases of complex disturbances.

For simple disturbances, the most accurate identification of groups of coherent generators is made by the matrix of absolute values of the initial mutual accelerations of their rotors. If it is necessary to use indicators of mutual similarity or difference not only for generators, but for any nodes of the studied network (which is required in a number of tasks), indicators such as mutual admittance modules, synchronizing powers and structural maxima are acceptable.

To identify groups of coherent generators under complex disturbances, the use of indicators based on numerically determined acceleration and deceleration energies is preferable to indicators of the quality of the transient (since energy-based indicators, firstly, provide a quantitative measure of stability and, secondly, do not require expert assignment of the integration interval).

The identification of groups of coherent generators based on energy indicators determined through the analytical integration of mutual acceleration over the mutual angle of generators is less reliable than the identification based on numerically determined energy indicators. The reason is the assumption about the constancy of the influence of the system on the mutual motion of the generators, introduced to create the possibility of analytical integration. In addition, these indicators (as well as the rest, determined without numerical integration) do not allow taking into account complex disturbances (sequences of several commutations that do not coincide in time).

Energy-based indicators can be used to identify groups of coherent generators, provided that among the mutual movements of the generators there are no movements that are unstable on oscillations greater than the first. With this limitation, the most universal indicators are the absolute and relative values of the acceleration pads. The main advantage of energy indicators in comparison with visual analysis of curves and indicators of the quality of the transient is in their completely formal (not requiring expert decisions) calculation.

The time interval of integration and the complexity of obtaining indicators of the quality of the transient and indicators based on numerically determined energies are comparable.

The study was carried out within the framework of the state assignment project "Theoretical foundations, models and methods for managing the development and operation of intelligent electric power systems," topic FWEU-2021-0001, reg. no. AAAA-A21-121012190027-4, using the resources of the High-Temperature Circuit Multi-Access Research Center (Ministry of Science and Higher Education of the Russian Federation of Russia, project no. 13.ЦКП.21.0038).

References

1. Voytov O.N., Voropai N.I., Gamm A.Z., Golub I.I., Efimov D.N. Analysis of non-homogeneities of electric power systems. Novosibirsk: Nauka (1999) (in Russian)
2. Abramenkova N.A., Voropai N.I., Zaslavskaya T.B. Structural Analysis of Electric Power Systems: in Problems of Modeling and Synthesis. Novosibirsk: Nauka (1990) (in Russian)
3. Lee S.T.Y., Schweppe F.C. *Distance measures and coherency recognition transient stability equivalents,* IEEE Trans. PAS-92, **92**, no. 5, pp.1550-1558 (1973)

4. Voropai N.I. Methods for equivalencing the electric power systems under large disturbances (literature review). Irkutsk: SEI, 124 p. (VINITI, no. 6521-73 dep.) (1973) (in Russian)
5. Xue Y., Wehenkel L., Belhomme R., Rousseaux P., Pavella M. et al. *Extended equal area criterion revisited,* IEEE Trans. PWRS-7, no. 3, pp.1012-1022 (1992)
6. Xue Y., Pavella M. *Critical cluster identification in transient stability studies,* IEE Proc. Pt.C, **140**, no.6, November, pp.481-489 (1993)
7. Germond A.J., Podmore R. *Dynamic aggregation of generating unit models,* IEEE Trans. PAS-97, **97**, no.4, pp.1060-1069 (1978)
8. Podmore R. *Identification of coherent generators for dynamic equivalents,* IEEE Trans. PAS-92, **92**, no. 4, pp.1344-1353 (1978)
9. Kartvelishvili N.A., Galaktionov Yu.I. Idealization of complex dynamic systems. Moscow: Nauka (1976) (in Russian)
10. Voropai N.I. Reduction of mathematical models of the dynamics of electric power systems. Novosibirsk: Nauka (1981) (in Russian)
11. Machowski J. *Dynamic equivalents for transient stability studies of electrical power systems*, Elec. Power&Energy Syst., **7**, no. 4, pp. 215-224 (1985)
12. Abramenkova N.A. *Determination of the structure of electric power systems in the analysis of static stability,* Izv. AS USSR. Energy and transport, no. 3, pp.33-40 (1985) (in Russian)
13. Barinov V.A., Sovalov S.A. *Application of modal theory for the analysis and synthesis of electric power systems,* Electron. Modeling, vol. 9, no. 5, pp. 72-77 (1987) (in Russian)
14. Abramenkova N.A., Voropai N.I., Zaslavskaya T.B. *Construction of a hierarchical system of models for studying the dynamic properties of electric power systems,* Simulation approach to managing the functioning of EPS, Irkutsk: SEI, pp. 11-20 (1989) (in Russian)
15. Pires de Souza E.J.S., Cardaso E.N. *Dynamic equivalent utilizing the REI approach to aggregate coherent generators,* IEEE Int. Symp. Circuits and Syst. Proc., Montreal, **2**, pp. 622-625 (1984)
16. Voropai N.I., Golub I.I., Efimov D.N., Iskakov A.B., Yadykin I.B. *Spectral and modal methods in studies of the stability of electric power systems and their control,* Automat. & Remote Control, no. 10, pp. 3–34 (2020) (in Russian)
17. Dimo P. Nodal Analysis of Power Systems, Abacus Press, Kent, England (1975)
18. Stadler J., Renner H. *Application of dynamic REI reduction,* 4th IEEE PES Innovative Smart Grid Technologies Europe, Copenhagen, Denmark, (2013)

On the feasibility of using distributed generation to tackle the issue of insufficient transmission capacity of power lines in urban area

Gabdushev Dmitry[1]*, *Buslov* Victor[1], *Gusev* Sergey[1], *Okhlopkov* Andrey[2], *Bitney* Vladislav[2]

[1]PJSC Rosseti Moscow region, Department of Electrical Modes, Moscow
[2]PJSC Mosenergo, Expertise and Technical Development Service, Moscow

Abstract. Issues related to the use of distributed generation for solving the problem of insufficient grid capacity in Moscow are analyzed. The problem area, where the current load of the power transmission line (TL) does not meet the requirements for the mode parameters is considered. The paper proposes an alternative way of tackling this problem by installing a low-capacity gas power plant (GPP) instead of reconstructing the power grid. Results obtained from this study can be used by electricity distribution companies as guidelines for solving the above problem.

1 Introduction

Having so many transmission lines that have reached the end of their standard service life poses a problem – insufficient transmission capacity because they were designed for smaller loads 40-50 years ago. Energy intensity and human population have greatly increased since then, thereby imposing a limit on the transmitted power due to technical limitations.

Higher energy efficiency is a characteristic feature of today's energy transition. Distributed generation is a key element of energy efficiency. In this paper, distributed generation refers to power plants located close to the place of energy consumption and connected either directly to the consumer or to the distribution grid [1].

Integration of distributed generation in urban areas is a promising direction for energy development and helps to cope with an intensive increase in load, which is especially characteristic of urban areas. This approach allows to reduce power flows through 35-220 kV distribution grids, which gives significant economic advantages when there is increased power consumption [2-3]. It should be noted that introduction of distributed generation facilities, in addition to postponing the upgrade of transmission lines (investments in such measures are limited) allows for more reliable power supply [4-5].

When laying power transmission lines in major cities, priority is given to cable lines (CL) due to high aesthetic requirements, reduced space and threat to human life in case of emergency. The problem of overcapacity is especially noticeable for cable lines, whose replacement is more costly economically and technically.

A developed power supply scheme for Moscow for 2030, prepared under a state contract by Inter RAO - Engineering Ltd., shows that insufficient transmission capacity is one of the main problematic aspects of the reliability of power supply to consumers in the Moscow region [6]. Megacities consume more than 75% of the world's primary energy. According to a study [7], this number will only grow. It is in megacities that one gets a high demand for electricity; building new large power plants or upgrading old ones is very expensive and limited by dense buildings.

2 Analysis of the Moscow power grid problem area

We consider a network with a lack of capacity. It consists of cable lines (Fig. 1). In one of the cases analyzed, emergency outage (EO) can lead to disconnection of consumers and limit output power.

Power flow distribution, voltages in the network nodes, as well as branch currents under in different EOs in the network are calculated:

Mode 1. Normal network operation mode.

Post-fault modes:

Mode 2. EO CL 220 kV CHP-Main–Raduga No.1 (branch 1-2).

Mode 3. EO, CL 110 kV Raduga–Oranzhevaya, with a tap-off to Zelenaya substation No. 1 (branch 5-12 and 12-8).

Mode 4. EO CL 110 kV Raduga–Oranzhevaya, with a tap-off to Zelenaya substation No. 2 (branch 4-13 and 13-9).

Mode 5. EO busbar section I, 110 kV Oranzhevaya substation.

Mode 6. EO busbar section II, 110 kV Oranzhevaya substation.

Mode 7. EO CL 110 kV Oranzhevaya – generator 2 (branch 9-11).

*Corresponding author: gdima512@yandex.ru

The computations are summarized in Table 1.

Table 1. Results.

TL	Emergency long current overload, %						
	Mode 1	Mode 2	Mode 3	Mode 4	Mode 5	Mode 6	Mode 7
1-2	64.3	0	65.3	65.3	64.6	89.3	65.1
1-3	64.3	128.1	65.3	65.3	64.6	89.4	65.1
4-6	25.9	22.9	25.4	26.4	26.1	26.8	25.9
5-7	23.7	26.8	24.2	23.2	23.5	22.8	23.7
5-12	21.7	21.8	0	45.2	30.0	102.4	24.7
4-13	21.5	19.9	45.2	0.0	32.2	30.0	24.7
13-9	22.7	25.9	43.8	0.0	43.6	0.0	24.4
12-8	20.8	20.0	0	43.8	0.0	73.4	22.5
9-10	54.6	56.3	54.8	54.8	54.7	0.0	88.7
9-11	56.7	58.5	56.9	56.9	56.7	0.0	0.0

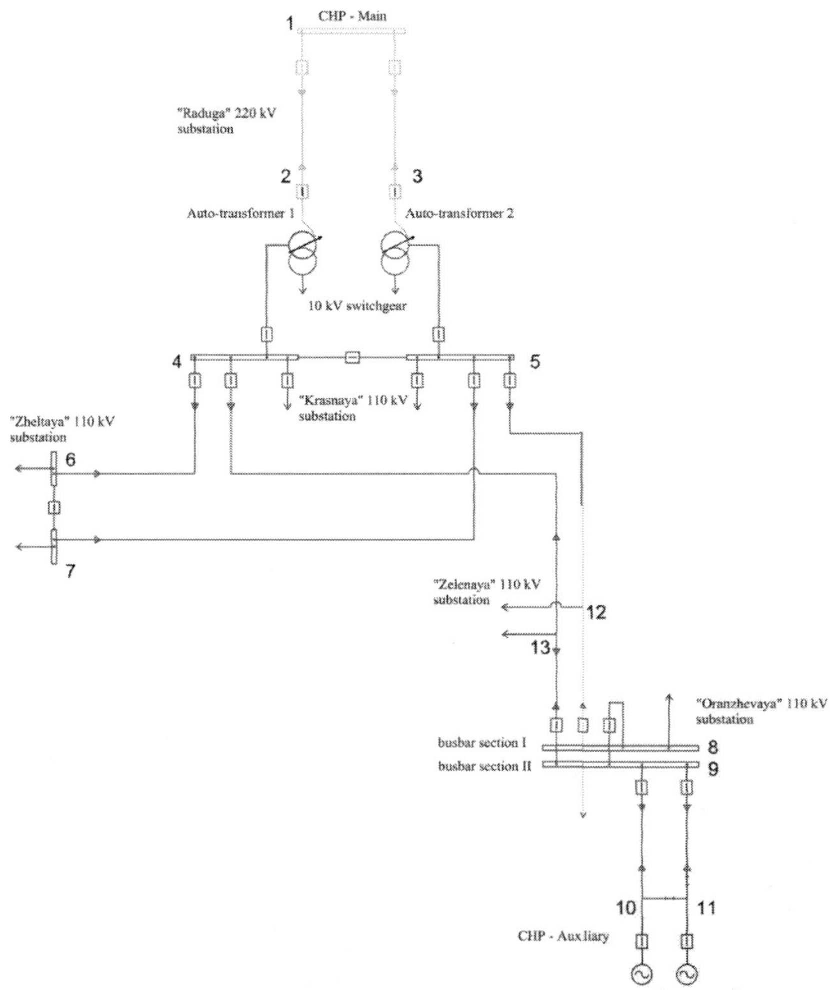

Fig. 1. Diagram of the network section

According to the results above, the following most severe modes were identified:

Mode 6: the load of the 110 kV Raduga–Oranzhevaya cable line with a tap-off to Zelenaya substation No.1 does not meet the mode parameter requirements in accordance with GOST R 58670-2019 [8]; the emergency permissible current-transmission capacity was found to have been exceeded.

Mode 2: current in CL 220 kV CHP-Main–Raduga No.2 cable line does not meet the mode parameter requirements in accordance with GOST R 58670-2019 [8]; the emergency permissible current-transmission capacity was found to have been exceeded. All solutions for the 220 kV network are considered a separate investment project and go under a different title. The solution for the 220 kV network is not

https://doi.org/10.1051/e3sconf/202338401013

considered in this work. This is primarily due to the fact that additional calculations are required for the 220 kV network in this case because the line feeds not only the load on the Raduga substation.

As a solution to the problem of current load exceeding the transmission capacity of lines in the 110 kV network, two options are considered:

1. Installation of a generator on the Zelenaya substation low-voltage (6-20 kV) or high-voltage (110 kV) busbars;

2. Reconstruction of the network section with replacement of the 110 kV cable line of the Raduga–Oranzhevaya substation, with a tap-off to Zelenaya substation No. 1, No. 2.

In both cases, an increase in maximum power consumption in the energy system of Moscow and the Moscow region according to the scheme and development program of the Unified Energy System of Russia (UES SIPR) is taken into account [9].

2.1 Installation of a small-capacity power plant on the consumer side

The power plant is installed with a 9.0 MW capacity that is enough to eliminate insufficient capacity at the level of 2032 for various emergency outages.

The generation source is installed on low-voltage busbars (10 kV) of the 110 kV Zelenaya substation. This option was chosen taking into account the following technological features of the substation design:

1. Sectional circuit breakers are installed on low-voltage busbars of the substation. In case of faults on one of the busbar sections (in the event of a voltage loss), the automatic transfer switch (ATS) triggers the operation to supply the load. If a generator is installed on high-voltage busbars in the substation and in case of faults in the line of the busbar section where the generator is installed, the network is deprived of the generator and operates in the same way as in the mode without the generator.

2. The selection took into account that a 6-20 kV cell is cheaper than a 110 kV cell.

3. Installation of the generation source on low-voltage busbars will further relieve the transformer at the substation.

2.2 Choice of power generation technology

Moscow has a centralized gas and electricity supply system. On the territory of the problematic section of the network, the place for installation of the generation source is located close to the gas pipeline. So, the power generation technology is chosen from two types: gas turbines (GTs) or gas generators (GGs).

GTs and GGs have their own advantages and disadvantages. During operation, gas engines have vibrations and low-frequency noise, they are of larger sizes relative to GTs, and are inferior to GTs in terms of emissions [10]. However, bringing noise and environmental characteristics to standard values is possible through appropriate engineering solutions [11]. Differences in the size of GGs compared to GTs when implementing small generation are not significant [12].

Application of GGs is the most profitable for priority power generation [13] due to higher efficiency than that of GTs. GGs can be repaired on site without complex diagnostic equipment once in 7-8 years [11], unlike GTs, whose repair is more complicated and performed by highly skilled specialists, usually at the manufacturer's plant. The efficiency value of GGs decreases by 3-5% when load is reduced to 50% [12]; with GTs at such load reduction, operation in some cases becomes inexpedient due to possible wear and tear of the unit. GTs also require high-pressure fuel gas, so a booster compressor must be installed.

Comparison of gas turbines and gas engines as a part of mini-CHPs in a work by Mutugullina [11] shows that gas turbines should be installed at facilities that have equal electric and thermal needs at over 30-40 MW capacity.

Based on analysis, a 9.0 MW gas power plant consisting of 4 GazEcos 2.25 MW gas generators 16GDG49 (Russian-made) can be installed to reduce the cable line load in the post-fault mode in this network.

To enable the GFPP-9.0 to work as part of an integrated power system, a combined method of regulating the rotational speed and active power of the gas turbine should be used.

For combined regulation under the mode of parallel operation with the power system, the regulation will be carried out according to the law:

$$P_{\text{r}} \rightarrow const, P_{\text{r}} \neq \varphi(f) \quad (1)$$

And in stand-alone (isolated) or island modes according to the law (grid frequency is the controlled parameter) [14]:

$$P_{\text{r}} \rightarrow P_0 \neq \varphi(f) \quad (2)$$

In stand-alone mode, the main task of the automatic control system (ACS) is to regulate the generation of active power, determined by the load magnitude.

Change of one regulation law to another in the ACS occurs automatically according to certain criteria. It is obvious that if the generating unit (GU) suddenly switches from parallel operation with the power system to island mode, then switching from one control law to another should occur quickly enough, and automation should reliably identify such a change in the operating mode of the GU and generate a command to change the algorithm in the ACS.

2.3 Reconstruction of a network section

To select a cable with the required transmission capacity, the mode where the current load exceeds the transmission capacity of the lines in the 110 kV network is calculated. The following ground characteristics are taken into account: humidity of normal soil and sand is 7-9% and specific thermal resistance of the ground is 1.2 m·K/W. After determining the cable location and selecting the screen area, a cable for 110 kV with the required core cross-sectional area is selected to solve the problem of insufficient grid transmission capacity. For a 110 kV

 https://doi.org/10.1051/e3sconf/202338401013

cable line, select the cable with cross-linked polyethylene as cable insulation, polyethylene as cable shell (XLPE PE).

3 Economic evaluation of both variants

The analyzed period for economic evaluation is 10 years, which corresponds to the practice of planning stage designing of electric networks up to 220 kV [15].

Cost of total capital investments in GPP-9.0 MW is calculated by the formula [16 - 18]:

$$C = C_{EQ} + C_{DW} + C_{CI} + C_{CM} \quad (3)$$

where C_{EQ} is the cost (in Russian rubles) of equipment of the gas-fired power plant; C_{DW} – is the cost of design work, which is 5% of the cost (in Russian rubles) of construction and installation work; C_{CI} is the cost of construction and installation work, equal to 15% of the cost (in Russian rubles) of equipment; C_{CM} – is the cost of commissioning, equal to 3% of the cost (in Russian rubles) of equipment; Shipping costs are included in the cost of equipment.

The cost of GPP equipment consisting of 4 GazEcos 2.25 MW gas generators 16GDG49 (Russian-made) will be ₽179,600,000 [19]. The cost of construction and installation works will be ₽26,940,000. Therefore, the cost of design works will be ₽1,347,000. The cost of commissioning works will be ₽5,388,000.

Capital investments:

$$C = ₽179{,}600{,}000 + ₽1{,}347{,}000 + ₽26{,}940{,}000 + ₽5{,}388{,}000 = ₽213{,}275{,}000 \quad (4)$$

GPP's own needs are taken to be 10%. Average annual time of turbines is taken to be 8,000 hours. Revenue from selling electric power per 1 kW is taken according to tariffs in Moscow and Moscow region [20]. Economic evaluation also takes into account the cost for fuel and oil purchase, costs for employee wages, including allocations for social needs. According to directive N 68 by Gosstroy (Russian State Committee for Construction) dated April 3, 2000 [21], the number of personnel working at the GPP with 3 units is taken to be 16 people. Allocations for social needs is 30%. Other costs are taken to be 10%. Depreciation charges are considered a linear method. The service life of GGs is 15 years. The discount rate is taken to be 18.5%.

For the network reconstruction project according to the organizational standards of state-owned energy company FGC UES [22], the aggregated cost of 110 kV cable line with XLPE insulation for cable laying is ₽16,920,000/km for the year 2000. This cost was converted into current prices in accordance with the letters of the Russian Ministry of Construction. The cable line is 7.8 km long. Since the network is being reconstructed, the cost of construction of collectors and switching points is not taken into account.

The results of cost-benefit analysis for the period of 10 years are given in Table 2, where C is the capital investment in the project, $C_{liquid10}$ is the liquidation value for the 10th year, NPV is net present value for 10 years, PI is the project's profitability index (ratio of NPV to capital investments).

Table 2. Results of cost-benefit analysis.

Solution method	C, mln. rub.	Costs, mln. rub./year	Revenue, mln. rub./year	$C_{liquid10}$, mln. rub.	NPV, mln. rub.	PI
Network reconstruction	1008.6	28.2	0	546.3	-950.5	0.06
Installation of distribution generator at the consumer side	213.3	185.0	270.0	71.1	175.2	1.82

The project of installing a small power plant has a better investment appeal. If this project is implemented for the estimated period of 10 years, the investor's company will make an annual profit. At the end of the estimated period, the company's value will increase by ₽175.2 million rubles, taking into account the discount rate.

4 Implications of scaling this practice for major power generating and grid companies

Apart from optimizing the cost of upgrading grid facilities, the scaling of this alternative solution to the problem of insufficient grid capacity through installation of distributed generation will allow power grid companies to reduce power flows and, consequently, cut losses in distribution grids. The following are other positive effects:

- Higher number of payments for connection to the grid;
- Higher reliability of power supply to consumers in post-fault modes and modes of high risks of power supply disruption (freezing rain, hurricane and other natural and man-made phenomena);
- Normalization of voltage levels in normal and post-fault modes (reduction of on-load tap changeovers, no need for installation of reactive power compensation equipment);
- Expanding the functions of electric grid organizations. In addition to the basic functions, such as electricity transmission and distribution, connection of new consumers and maintenance of redundancy, new functions are also possible: mutual redundancy, multi-agent response, and demand management [23-24].

The following benefits are created for generating companies:

- Risk diversification, investment optimization;
- Partial solution to the problem of regulating reactive power in the grid and, accordingly, increasing the reliability of generating equipment at major power plants [25];

- Business expansion, opening of new strategic prospects for the company's development in connection with the inevitable energy transition;

- Occupying a position in a new niche, which is developing in view of the growing number of implemented distributed generation facilities in Europe; forming a strong competitive position in comparison with other players thanks to greater industry experience;

- Introduction of distributed generation will compensate for a part of the required generating capacity due to exhaustion of a thermal power station's own resource [1].

Despite the positive effects, introduction of distributed generation to solve the problem of insufficient capacity in a distribution grid also poses new challenges to the power system as a whole. There remains the problem of dynamic stability of the grid and the need to limit short-circuit currents with significant scaling of this practice. However, implementation of this solution in particular cases, instead of reconstructing the grid due to the smallness of the generation source, will not significantly affect the network.

5 Conclusion

When considering a problematic section of Moscow's power grid, in which the current load in the transmission line does not meet the mode parameter requirements, an alternative way of solving this problem was proposed – installation of a low-capacity GPP instead of reconstructing the network.

Installation of a gas-fired mini-TPP in the example under consideration has a higher investment appeal than solving the problem in the traditional way, network reconstruction.

Integration of distributed generation to solve one of the problems of electric grid companies with the involvement of generating companies, creates a combination of advantages and opportunities for both parties. However, introduction of distributed generation also comes along with new challenges that need to be addressed.

Results obtained from this study can be used by electric grid companies as guidelines for solving the problem of insufficient grid capacity in problem areas through integration of distributed generation, in addition to traditional methods.

References

1. Khokhlov A., Melnikov Y., Veselov F., Holkin D., Datsko K. Distributed Energy in Russia: Development Potential. Moscow: Energy Center, Skolkovo Moscow School of Management, 2018. pp. 87.
2. Krylov P. A. Peculiarities of parallel operation of low-power generation sources with a powerful grid. Siberian Federal University, Krasnoyarsk, April 15-25, 2016. pp. 46.
3. Features of integration and prospects for effective use of distributed generation facilities in Russia's electric power system: Minutes of joint meeting No. 5/18 of October 18, 2018 of the Scientific and Technical Board of Unified Energy System and the Section on Reliability and Safety of Large Energy Systems, Scientific Council of the Russian Academy of Sciences on integrated energy problems. Moscow, 2018. pp. 16.
4. Boljevic, S., & Conlon, M. (2009). Fault current level issues for urban distribution network with high penetration of distributed generation. 2009 6th International Conference on the European Energy Market. doi:10.1109/eem.2009.5311425
5. Byk F.L., Myshkina L.S. Reliability of distributed power facilities. Reliability and Safety of Energy. 2021, 14 (1) pp. 45-51.
6. Inter RAO - Engineering Ltd. Scheme of power supply for Moscow (distribution networks with 6-10-20 kV voltage) for the period up to 2030, taking into account connected territories. November, SEM.6-10-20kV-PZ-v2. Online: https://www.mos.ru/upload/documents/oiv/tom_3_compressed.pdf
7. Allam Z. Enhancing renewable energy adoption in megacities through energy diversification, land fragmentation and fiscal mechanisms //Sustainable Cities and Society. 2019, V. 101841.
8. Calculations of electric power modes and determination of technical solutions for prospective development of power systems. GOST R 58670-2019
9. Scheme and Program for the Development of the Unified Energy System of Russia in 2021-2027. Approved by the Ministry of Energy of Russia, No. 88, February 26, 2021
10. Chigvintseva I.R., Burganov E.F. Monitoring of parameters of HPP power plant operation. Vestnik of Kazan Technological University, 17 (11), 2014. pp.188-190.
11. Mutugullina I. A. Comparative analysis of gas engine and gas turbine units in the context of solving energy saving problems. Vestnik of Kazan Technological University, 2014. V. 17 (12), pp. 373-375.
12. Korolenko T.V., Alekseeva O. C. Prospects for the introduction of gas engine-powered mini-CHP, 2019.
13. Belkin A. P., Dubova A. V. Evaluation of the Efficiency of Transition to Decentralized Energy Supply in Tyumen Oblast. Vestnik of Ivanovo State Energy University, 2016. No. 2, pp. 5-13.
14. Gurevich Y. E., Ilyushin P.V. Features of mode calculations in power districts with distributed generation. Nizhny Novgorod: Nizhny Novgorod Institute of Management, Russian Academy of National Economy and Public Administration under the President of the Russian Federation, 2018, pp.280. ISBN 978-5-00036-226-6, EDN YWRPSP

15. Glazunov A.A., Shvedov G.V. Designing a district electrical network. MPEI, 2010.
16. Masleyeva O.V., Agaltsova T.A., Erdili N.I. Ecological and Economic Assessment of the Life Cycle of Traditional Energy Sources and Bioenergy. Proceedings of Nizhny Novgorod State Technical University, 2019. No. 1 (124). pp. 119-129. EDN: ZANGKD.
17. Generating systems based on gas engine and gas turbines. Developer: Energy Service Center, Siberian Energy Center. [Electronic resource] https://nppsibec.ru/wp-content/uploads/2019/09/sistemy-kogeneraczii-na-osnove-gazoporshnevyh-i-gazoturbinnyh-ustanovok.pdf (accessed October 6, 2022)
18. Price list for gas engine power plants (Mini-CHPs) and diesel generators [Electronic resource] https://www.d-system.ru/money/?etext=2202.k8NPMv980pIWU7EMjHiNNrI0pRPcawUDVPWBxM12_WZ4YXFqY2tqZHZ1a3hxaHZ3.0f611771298d326202b622d508e2f4c4ef9e51d4&yclid=5873033669814323375/ (accessed October 6, 2022)
19. GazEcos 2.25 MW 16GDG49 - power supply systems [Electronic resource] https://www.gazecos.ru/calc_kvh.html (accessed October 6, 2022)
20. Full version of tariffs for 2022 [Electronic resource] https://www.mosenergosbyt.ru/individuals/tariffs-n-payments/tariffs-mo/polnaya-versiya-tarifov-mo.php (accessed: July 12, 2022)
21. Order N 68 of the State Committee for Construction of the Russian Federation dated April 3, 2000 "On approval of the Guidelines on labor rationing in the energy sector. Part 3. Standards for the number of employees of utility electric power enterprises
22. Aggregated cost indicators for power transmission lines and substations with a voltage of 35-1150 kV. Standards of FGC UES, 2012.
23. Byk F.L. et al. Functions of Regional Grid Companies in the Integration of Local Power Systems. Transmission and Distribution, 2020, No. 2. pp. 48-55.
24. Byk F.L., Myshkina L.S. Development of distributed energy and increase in electric grid efficiency. Business. Education. Law. 2019, No. 4. pp. 134-138.
25. Impact of operation mode in terms of reactive power regulations on technical condition and reliability of generating equipment operation / S. Lenev, A. Vivchar, A. Okhlopkov, V. Bitney // Proceedings of the 2022 4th International Youth Conference on Radio Electronics, Electrical and Power Engineering, REEPE 2022 : 4, Moscow, 17–19 марта 2022 года. – Moscow, 2022. – DOI 10.1109/REEPE53907.2022.9731385. – EDN JGBMNI.

 https://doi.org/10.1051/e3sconf/202338401014

Ensuring the reliability of energy systems with the application of a new method of decreasing seasonal variations of ground resistance

Ivan Pavlovich[1]*, *Siarhei* Baraishuk [1], *Alexey* Skripko[2]

[1]BSATU, Department of practical training of students, 220023 Belarus, Minsk
[2] TerraCynk ltd. technical department, 220000 Minsk, Belarus

Abstract. The possibility of improving the reliability of the grounding system is shown. The analysis of dependence of soil resistivity on humidity was carried out, on the basis of which a composition for normalization (reduction of seasonality) of soil electrical resistivity was proposed. A number of full-scale experiments were carried out to measure the resistance values of the grounding devices of experimental circuits, after backfilling the places of laying the contour elements with various compositions and introducing other types of soils and minerals.

Introduction

For the uninterrupted operation of electrical installations, reliable grounding is necessary, which can reduce the risk of electric shock to maintenance personnel, as well as protect equipment. It is well known that the electrical resistance of the ground loop is determined by many factors, such as soil porosity, soil moisture, freezing depth, mineral salt content [2]. These parameters change throughout the year, so the so-called seasonal factor is introduced to calculate the ground loop.

The purpose of this work is to develop a method for reducing the influence of seasonality on the resistance of grounding devices.

1 Ways to reduce the seasonal factor

The current spreading resistance of a grounding device depends on the types of soil (sand, clay, limestone), particle size and density, humidity and temperature, as well as the chemical composition of the soil, the presence of acids, salts, alkalis in it [1, 2]. For seasonal influences on soil resistivity, in turn, the determining parameters are humidity and temperature [3 - 4]. It can be concluded that increasing the ability of the soil to retain water with minerals and salts dissolved in it in the near-electrode space improves the properties of the grounding device.

An analysis of the previously studied dependences of soil resistance on moisture, described in particular in [9], is in good agreement with the results obtained by the authors and presented in Figure 1 and shows that stabilization of soil moisture in the amount of 12-16 mass percent is optimal, a further increase in moisture is already does not lead to any significant decrease in resistance (Fig. 1).

One of the well-known ways to reduce the effect of annual temperature changes on the GD resistance is the use of deep ground electrodes. The disadvantages of this method include the high metal consumption of work.

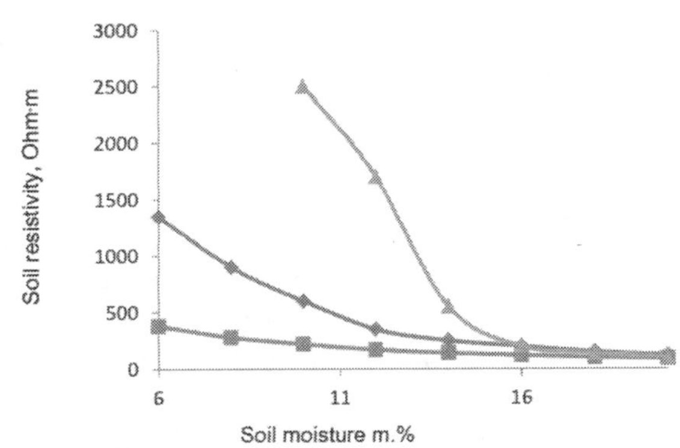

Fig. 1. Dependence of the specific resistance to current spreading of various soils on humidity, expressed in mass percent.

Another way to reduce seasonal fluctuations in resistance is to use electrolytic electrodes. Electrolytic grounding is a tubular electrode with a diameter of 50-110 mm and a length of 3 m, made of stainless steel with perforations on the walls. The electrodes are filled with an electrolyte based on mineral salts [6–7]. Known mineral activators produced in the CIS are a mixture of an ion-exchange salt modified with a halide

* Corresponding author: mrchesel20@gmail.com

activator and a surfactant. The disadvantage of such ground electrodes is the use of electrodes made of corrosion-resistant steel or non-ferrous metals as the electrode material [5]. Also, this method contributes to soil salinization and an increase in the rate of destruction of metal structures located in close proximity to the grounding device.

2 Experimental studies

In our research, a soil-substituting mixture based on hydrogel, bentonite clays, and graphite was developed. The use of hydrogels is due to its ability to bind water in the near-electrode space, helping to maintain stable humidity. Maintaining the same soil moisture throughout the year allows you to stabilize the amount of mineral salts dissolved in the soil, which has a positive effect on the resistance to the spread of electric current in the soil.

Fig. 2 shows the results of measurements of the specific resistance of a mixture of soil with hydrolyzed polyacrylonitrile in various proportions.

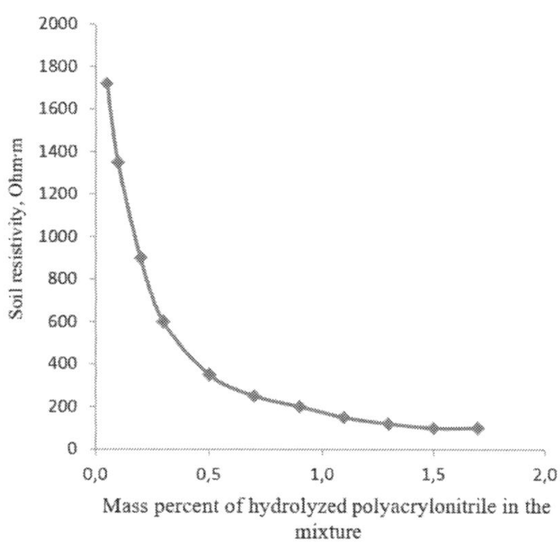

Fig. 2. Dependence of soil resistivity on the percentage of gel introduction based on hydrolyzed polyacrylonitrile.

As can be seen from the graph presented in Fig. 2, when hydrogels obtained by swelling hydrolyzed polyacrylonitrile with a dry weight of more than 1.3-1.5% of the soil mass are introduced, moisture stabilization occurs, and a further increase in concentration does not lead to a decrease in soil resistivity. , which indirectly indicates the optimal soil moisture.

The use of graphite in this mixture is necessary in order to compensate for the decrease in conductivity in mixtures when the water crystallization temperature is reached. The ongoing experimental and field studies have revealed that when using graphite powder, it makes it possible to keep the resistance of the mixture at a sufficiently low level, even at negative temperatures (Fig. 3).

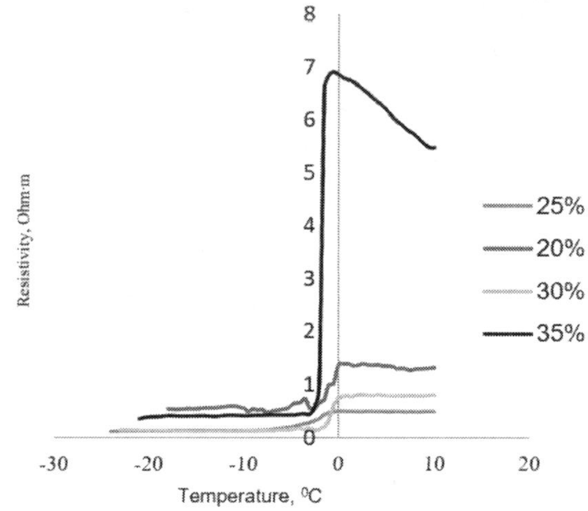

Fig. 3. Dependences of the specific resistance of the mixture on temperature at different humidity

For field experiments, grounding devices were mounted, which are two vertical electrodes made of ordinary black steel with a diameter of 12 mm and a length of 2 m. These vertical electrodes were connected by a horizontal strip 4x40 mm long 4 m. for the mixture. Systematic measurements of the control and experimental grounding devices were carried out using a four-wire method with an IS-10 meter, according to the method proposed by the manufacturer.

The results of experimental studies of the resistance of the control and experimental circuits are shown in Figure 4. From the data obtained, it can be concluded that the GD has the lowest resistance value, the near-electrode space of which is treated with an experimental mixture to reduce the spreading resistance of the GD. The results of the research show that the resistance of the ground loop changes slightly, which indicates that the influence of the seasonal factor has also decreased. It is worth noting the general decrease in the resistance of such a circuit, relative to the control one.

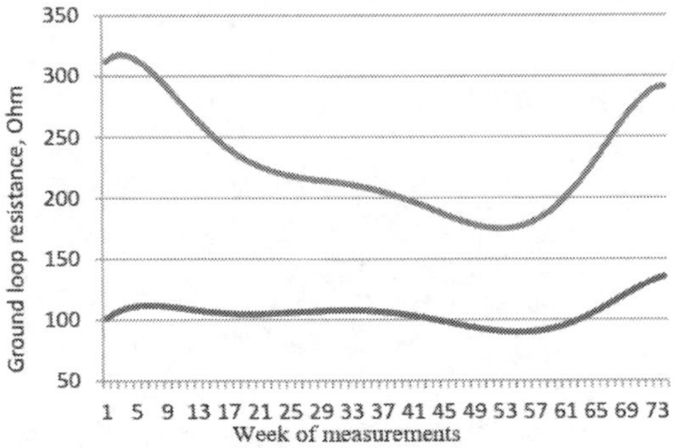

Fig. 4. Current spreading resistance values of the control (blue line) and experimental (red line) circuit, measured every week starting from September.

As can be seen from the graph, the use of hydrogel allows not only to reduce fluctuations in the resistance to spreading of the GD, but also to reduce the resistance of the GD in comparison with the control circuit. The use of the proposed additive also reduces the freezing temperature of the soil by 4-6 °C, which makes it possible to use higher average long-term low temperatures when determining the climatic zone of the site for the location of the designed electrical equipment, and to reduce the seasonality coefficients when interpreting field measurements. The conducted studies have shown that the effect on the decrease in electrical resistance can be divided into 2 factors: a change in the seasonality factor and a decrease in the specific resistance of the soil in the near-electrode space, which makes it possible to make adjustments to the calculations of the resistance of the ground electrode when using mixtures during the installation of the GD. It is obvious that the efficiency of such a replacement is the higher, the greater the difference in the electrical resistivity of the soil and the mixture, as well as the greater the amount of soil replaced in the near-electrode space. However, previous studies [6] showed that there is a relatively small optimal volume (~10 cm in the radius around the electrodes) when such a replacement is most effective and a further increase in the volume of the soil to be replaced loses its efficiency according to an exponential law.

This will ensure the longest possible prevention of soil freezing in the autumn-winter period, which is one of the factors for the high-quality operation of the grounding device.

The research was supported by the Belarusian Republican Foundation for Fundamental Research.

Conclusion

The use of a soil-substituting mixture makes it possible to reduce the effect of seasonal temperature fluctuations, which allows the ground loop to perform its functions stably, which favorably affects the stability and reliability of energy systems. Reducing the influence of external factors on the resistance of the grounding device will allow a more rational approach to the consumption of materials and reduce labor costs, as well as increase the reliability of grounding and electrical installations in general. The method developed by us to reduce the seasonality factor by using a soil-substituting mixture in the near-electrode space makes it possible to use ordinary black steel [5].

References

1. Vedeneeva L.M., Chudinov A.V. Study of the influence of the basic properties of the soil on the resistance of grounding devices // Bulletin of the Perm National Research Polytechnic University. Geology. Oil and gas and mining. - 2017. - V. 16, No. 1. - S. 89-100.
2. Trifunovic J., Kostic M. Analysis of the effect of imperfect contact between ground electrodes and the surrounding soil on the electrical properties of ground loops // Electrical engineering. - 2014. - T. 96. - No. 3. - S. 255-265.
3. IEEE Std 142-2007 IEEE Recommended Practice for Grounding Industrial and Commercial Power Systems. - Approved. June 7, 2007 - 225 p.
4. J. He, R. Zeng, Y. Gao, Y. Tu, W. Sun, J. Zou, et al. Seasonal Impact on the Safety of a Substation Earthing System IEEE Trans Power Delivery, 18(3) (July 2003)
5. Drako M.A. Corrosion of ground electrodes of electrical installations // Energy strategy, 2019, no. 6 (69), p. 44–48.
6. S.M. Baraishuk, I.A. Paulovich, M.Kh. Murodov, Kh. Abdulkhaev, A.N. Skripko Reducing the resistance of grounding devices by using soil treatment with moisture-stabilizing additives that are non-aggressive to the grounding material // Agropanorama.– 2021. – No. 5 (147).– P. 28–33/
7. Noskova S.V. Resistance of an electrolytic ground electrode. Calculation features // News of Electrotechnics. - 2020 No. 1 (121) - 2 (122). - With. 72-77.
8. Gribanov A.N. Bipron - grounding of electrical installations // Exposition Oil Gas, .– 2016 .– No. 4 .– p. 72-75.
9. P.A. Ferre, J.D. Redman, D.L. Rudolph, and R.G. Kachanoski, "The dependence of the electrical conductivity measured by time domain reflectometry on the water content of a sand", Water Resour. Res., 1998.

https://doi.org/10.1051/e3sconf/202338401015

Power consumption management and equalization of the load schedules of Azerbaijan power system

Nurali Yusifbayli [1], *Valeh* Nasibov [2]* and *Rana* Alizade [2]

[1] Azerbaijan Technical University H.Javid Avenue 25, Baku, Azerbaijan
[2] Azerbaijan Scientific-Research and Design-Prospecting Power Engineering Institute, AZ1012, Baku, Azerbaijan

Abstract. Foreign experience in applying the principles of power consumption management is analyzed in the paper. A comparative analysis of the main indicators of the load schedules of the power systems of Azerbaijan, Turkey, Ukraine, Russia and Jordan was carried out and it was found out that there is significant potential for improving such indicators of the load schedules as variation factor and the fill factor for the Azerbaijan power system. The positive effects of equalization of the load schedules of the power system of Azerbaijan are studied, based on the technical and economic indicators of the main generating capacities.

1 Introduction

Equalization of load schedules, power consumption management are one of the key trends in the development of the electric power industry in the world. The transition of electricity consumers to the status of active consumers is fully consistent with the innovative transformation of the electric power industry based on the concept of intellectualization of the power system. The basis of the concept of transformation of the electric power industry in the world is the Smart Grid concept, where power systems should be controlled in real time by a network of information and control devices and systems. At that, the management covers not only all generating sources, transmission and distribution networks, but also all types of consumers of electrical energy [1, 2].

2 Smart Grid concept

In the Smart Grid concept, the consumer acts as a full-fledged participant in the electricity market. At that, an active consumer, both by readiness to change his load, and by the change itself, can receive economic benefits, and the power system can also benefit [3, 4].

Despite the fact that the Smart Grid concept provides for the possibility of regulation of the load of all types of electricity consumers, in practice, the greatest application of load control is in the housing sector, in industry and the commercial sector [5,7].

Currently, there are two types of programs of change of the load of consumers: programs based on incentives and programs based on time.

The incentive based programs are given below:

1. Direct load control (DLC). According to preliminary agreement between the supplier and the consumer of electricity, the supplier can remotely control some of the consumer's appliances (air conditioners and water heaters).

Operation notices are usually announced in a short time. To use this method, consumers must be equipped with a remote control system so that the supplier can reassign, turn on or turn off the appliances. Direct load control is mainly used in the residential sector or in the small commercial sector. It is not suitable for the industrial sector, where breakdown in power consumption can lead to undesirable large-scale consequences.

2. Iterruptible/ curtailable (I/C) Service. In comparison with the direct load control, this method is commonly used in the industrial sector and the large-scale commercial sector. When the system is overloaded, consumers are asked to reduce some loads to a certain level. By participating in this management, consumers can receive tariff discount or bill discount. However, if consumers do not respond within a predetermined period of time, they may be fined. In this method, the frequency and duration of operations are limited.

3. Demand bidding buyback (DB). Supplier, based on the situation with production and demand, announces the total amount of electricity that needs to be reduced. Consumers place bets on the amount (on the amount of reduction in electricity consumption) based on their situation and the wholesale market. Once an offer has been accepted, consumers must provide the specified limit, otherwise or they will be penalized. This method is also suitable for major clients. Small consumers can be integrated by the system operator and involved as a single consumer.

4. Emergency Demand Response Program (EDRP). The system operator by means of this program encourages consumers to reduce power consumption when unexpected situations occur in order to increase the reliability of power supply. Under this program,

*Corresponding author: nvaleh@mail.ru

consumers are not penalized for non-fulfillment of obligations.

5. Capacity Market Program (CAP). When there is not enough reserve in the system, consumers must reduce their predetermined consumption. The announcement is usually published a day in advance. These reductions are considered as the system's capacity to replace usual resources of generating capacities. By proving their ability to reduce electricity consumption, consumers can receive a reservation payment, and by providing a discount, consumers can be encouraged. On the contrary, if they do not provide a discount, then they can be fined.

Ancillary Service Market Proqrams (A/S). Similar to demand bidding buyback, consumers are also betting on power outages. These proposals are offered to the independent system operator/regional transmission organization. These reductions are used as operational reservation. If the offer was accepted, the consumers must comply with the reservation. In this situation, the reservation is paid according to the market price. Once the reduction is announced, consumers are paid according to the spot price. In other words, the consumer is paid for both the willingness to reduce and the reduction in electricity consumption itself.

3 Emergency recovery programs

Time-based emergency recovery programs include as follows:

1) Time of Use (TOU),
2) Real Time Pricing (RTP)
3) Critical Peak Pricing (CPP).

In the TOU program, the electricity tariff is formed for the peak and for the rest of the time.

The RTP program is the same as the TOU program, with the difference that the electricity price is generated for each hour (24 different electricity prices).

The CPP program is used during limited number of days and is divided into four groups:

1. CPP with fixed time – price increase time is known, but day of application is unknown.
2. CPP with variable time of application– time and day of price increase are unknown.
3. Variable of pricing – in this program, the timing of the price increase varies.
4. Critical peak discounts. Consumers receive discounts for reducing their load during critical peak hours.

Diagram that indicates the classification of power consumption management programs is shown in Figure 1.

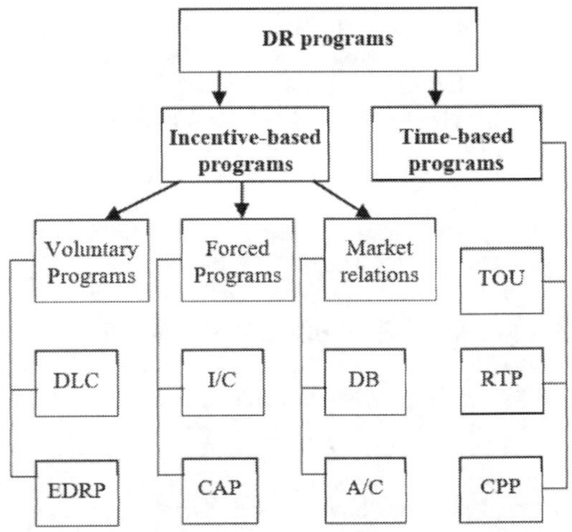

Fig. 1. Classification of power consumption management programs

The use of power consumption management programs can lead to such positive effects for the power system as:

– reduction of power losses in electrical networks,
– reduction of peak power,
– equalization of load schedules,
– reduction in transmitted power, increase in network bandwidth,
– improvement of stability,
– reduction of environmental pollution,
– improvement of voltage profiles,
– reduction or deferment of capital investments in network development,
– reduction or deferment of capital investments in development of generating capacities.

The use of power consumption management programs can lead to such positive effects for the consumer as:

– the ability to transfer power consumption to cheaper hours,
– increase in energy efficiency,
– opportunity to earn income,
– reduction of environmental pollution,
– increase of the reliability of power supply and decrease of the problems of complete disconnection from the network,
– reduction of the diktat of electricity market participants.

Here, the effects of equalization of load schedule will be considered.

It should be noted that the positive effect of equalization of load schedule includes several components:

1. Savings associated with the postponement of the construction of power plants (mainly peak),
2. Decrease of fuel burnout due to minimum and partial loads,
3. Decrease of fuel consumption for starting and stopping of power system units,

4. Reduction of costs associated with the increase in accidents, repairs, etc. due to frequent start-stops,

5. Reduction of the unloading of cost-effective power plants in order to keep the units of the Azerbaijan TPP in operation,

6. Reduction of other losses associated with the growth of expenses for own needs, losses of electricity during the transmission and distribution of electrical energy

4 Daily electrical load schedules

Despite the fact that natural gas and electricity markets have not yet been organized in Azerbaijan, in 2021 the Law "On the use of renewable energy sources in the production of electricity" was adopted in Azerbaijan, which specifies the rights and obligations of an active consumer of electricity. The active consumer can use part of the electricity he produces and sell the other part to the electricity supplier. At that, the consumer can take part in filling the load schedule for the day ahead to receive economic benefits, like other electricity producers [8].

As is known, the daily electrical load schedules, regardless of the parameters of the power system, are uneven and have several alternating characteristic sections, such as morning rise, daytime drop, evening peak, deep night dip. The nature of the change in the daily load schedule, in addition to daily rhythms, is also influenced by such parameters as weekly (working days and off-days) and seasonal (winter, spring, summer, autumn) changes in power consumption [9-12].

As can be seen from Figures 2-3, the daily load schedules differ from each other not so much in form as in the level of load and various characteristics of the load schedule.

As a rule, in working days the load is slightly higher than the load in off-days, and the load in summer and winter is higher than the load in spring and autumn.

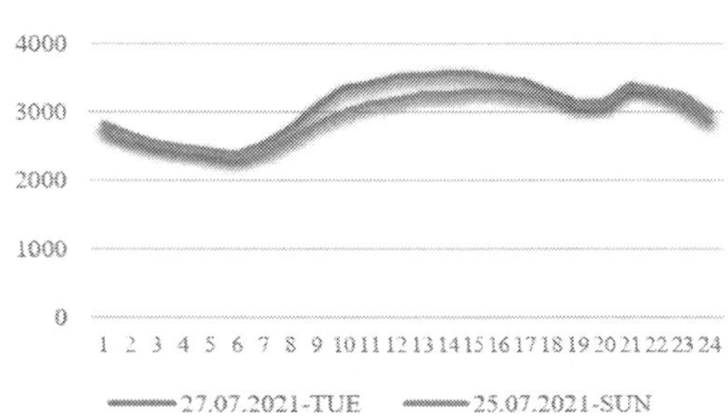

Fig. 2. Summer daily load schedule

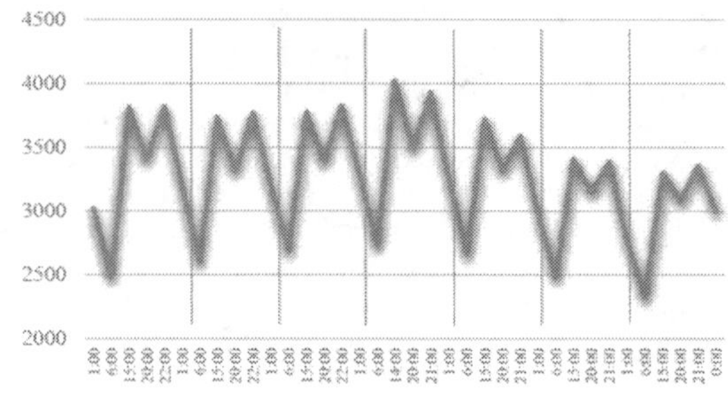

Fig. 3. Weekly load schedule

In the Azerbaijan power system, the maximum load occurs in the summer and is highly dependent on the ambient temperature. To evaluate the characteristics of the daily load schedule, parameters such as the variation factor, the fill factor, the maximum regulation coefficient, etc. are used. The lower the variation and fill factors, the worse the uniformity of the load schedule, and the worse the technical and economic indicators of coverage of the load schedule.

As can be seen from Tables 1-2, the characteristics of the daily load schedules of the Azerbaijan power system are noticeably worse compared to similar characteristics of other countries, and there is significant potential in the Azerbaijan power system to improve the characteristics of the daily load schedules.

Table 1. Characteristics of daily load schedules

Load schedule indicators, MW	Winter		Spring		Summer		Autumn	
	Work day	Off-day	Work day	Off-day	Work day	Off-day	Work day	Off-day
Maximum load (P_{max})	3252	3085	2957	2857	3566	3358	2967	2877
Minimum load (P_{min})	2042	2013	1993	1929	2390	2329	1839	1834
Moring maximum ($P\ m_{min}$)	3120	2772	2801	2532	3538	3259	2652	2517
Evening maximum ($P\ e_{max}$)	3252	3085	2957	2857	3566	3358	2967	2877
Average load (P_{av})	2727	2541	2496	2350	3077	2954	2434	2327
Variation factor (K_{var})	0,63	0,65	0,67	0,68	0,67	0,69	0,62	0,64
Fill factor (K_{fill})	0,84	0,82	0,84	0,82	0,86	0,88	0,82	0,81
Difference between maximum and average load (P_{max}-P_{av})	525	544	461	507	489	404	533	550
Load regulation range (P_{max}-P_{min})	1210	1072	964	928	1176	1029	1128	1043
Regulation coefficient (K_{reg})	0,37	0,35	0,33	0,32	0,33	0,31	0,38	0,36

Table 2. Characteristics of the daily load schedules of the countries of the world

Load schedule indicators MW	Azerbaijan		Turkey		Ukraine		Russia	
	Sum	Win	Sum	Win	Sum	Win	Sum	Win
Maximum load (P_{max})	3566	3252	39131	41590	21572	28435	110851	142381
Minimum load (P_{min})	2390	2042	28882	29521	16011	20597	89812	122162
Variation factor (K_{var})	0,67	0,63	0,73	0,70	0,74	0,72	0,81	0,85
Fill factor (K_{fill})	0,86	0,84	0,89	0,87	0,89	0,87	0,92	0,94
Regulation coefficient (K_{reg})	0,31	0,37	0,26	0,29	0,25	0,27	0,18	0,14
Relationship ($P_{sum.min}$/ $P_{win. min}$)	1,17		0,97		0,77		0,73	
Relationship ($P_{sum.max}$/ $P_{win.max}$)	1,10		0,94		0,75		0,77	

It should be noted that regardless of the day of the week or season, the basic part of the daily load schedule is mainly covered by thermal power plants with good technical and economic indicators, such as Janub PP-780 MW with specific fuel consumption (SFC) of 233 g/kWh, Shimal PP –800 MW, with specific fuel consumption of 214 g/kWh and Sumgayit PP–525 MW, with specific fuel consumption of 224 g/kWh. Half peaks and peaks of the daily load schedule of the Azerbaijan power system are mainly covered by the Azerbaijan TPP-2400 MW and relatively small modular power plants with total installed capacity of 1200 MW.

As can be seen from Figure 4, the Azerbaijan power system itself fully covers its own demand for electricity and exports a certain amount of electricity to neighboring states (mainly to Georgia and Turkey). The technical and economic indicators of coverage of the daily schedule of the Azerbaijan power system mainly depend on the operating mode of the Azerbaijan TPP with specific fuel consumption of 330-337 g/kWh.

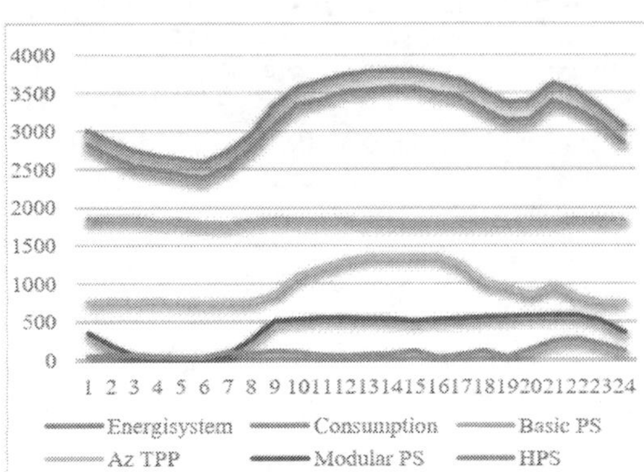

Fig. 4. Coverage of the daily load schedule

The characteristics of the load coverage modes mainly depend on the number of connected units with a capacity of 300 MW at the Azerbaijan TPP: 2-unit, 3-unit, 4-unit and 5-unit (no more than 10 days a year it is necessary to include the 6th unit in operation). At night, modular power plants usually stop, and the switched on units of the Azerbaijan TPP are unloaded to a technical minimum. In the morning, modular power plants are put into operation, and the capacity of the Azerbaijan TPP is gradually increased, if necessary. The components of the effect of equalization of the load schedule are discussed below.

The construction of 410 MW power plant (corresponding calculations and negotiations are being carried out with suppliers and civil engineers) would cost 450 million dollars, while the cost of 1 kW of power is 1,100 dollars. If we take the effect of postponement of construction to a later date (9-10 years), we will get an annual effect of $47 million.

Fuel burnout associated with the partial load of the Azerbaijan TPP is calculated based on the operating modes of this TPP. In 2021, the units of the Azerbaijan TPP were in operation: 5 units - 87 days, 4 units - 73 days, 3 units - 45 days, 2 units - 160 days.

From the typical daily load schedules, when 5 units are in operation, the fuel overspending is calculated as follows: the units carried minimal loads for 10 hours with the overspending of 21 g/kWh, while 7651 MWh of electricity was generated. For 8 hours, the units of the Azerbaijan TPP operated at power close to rated power with the overspending of 2.7 g/kWh, while 10,273 MWh were generated. During the remaining 6 hours, the units operated at average power with the overspending of 14.3 g/kWh, while 5501 MWh of electricity was produced. The total overspending for one day of operation of the Azerbaijan TPP with 5 units is approximately 21 g/kWh * 7651000 kWh + 2.7 g/kWh* 10273000 kWh + 14.3 g/kWh * 5501000 kWh = 267 t.e.f.

Considering that there were 87 such days in 2021, the annual overrun from the operation of the Azerbaijan TPP with 5 units is 267 t.e.f. * 87 days = 23229 t.e.f. in a year. Similar calculations were performed for other modes of the Azerbaijan TPP, and the results are summarized in Table 3.

Table 3. Calculation results for units of Azerbaijan TPP

Number of turn-on units	Number of days	Fuel overspending associated with partial load, t.e.f.
5	87	23229
4	73	16644
3	45	8478
2	160	22310
Total		**70661 t.e.f.**

As shown above, modular power plants stop at night and start working in the morning, in order to regulate the load schedule, the units of the Azerbaijan TPP are also sometimes stopped and then turned on (start-stop can be carried out both in weekly and seasonal regulation of the load schedule). Considering that approximately 130,000 m^3 of gas is consumed during the cold-starting of unit of the Azerbaijan TPP, 60 m^3 of gas at low-capacity modular power plants, and 130 m^3 of gas at medium-capacity modular power plants, the overspending associated with the start-stop of the units is approximately 7 million m^3 of gas per year.

In order to prevent shutdown of units of the Azerbaijan TPP, it is often necessary to operate more cost-effective power plants with partial loads, which leads to fuel overspending due to partial loads of cost-effective power plants. The table shows the values of fuel overspendings at cost-effective power plants.

Table 4. Fuel overspending values

Power Plant	Fuel overspeding in 2021, t.e.f.
Janub PP–780 MW	42000
Sumgayit PP–525 MW	10000
Shimal PP–800 MW	8800
Total	**60800**

If we bring the above values of the equivalent fuel overspendings into the volumes of natural gas (natural gas is used at the power plants of Azerbaijan), we will get 115 million m^3. Taking into account the gas overspending for the start-stop of the units, we will get 122 million m^3 of natural gas.

We are conducting special studies to determine the costs associated with the increase in accidents, repairs, etc. due to frequent start-stops, as well as losses associated with the growth of expenses for own needs, losses of electricity during the transmission and distribution of electrical energy. Without taking into account these components, the annual effect of equalization of the load schedule (at current prices for a thousand cubic meters of natural gas at $1,740), taking into account the effect of postponement of the construction of a peak station, can be estimated at $260 million per year. More than 80% of this amount falls on the share of natural gas, the price of which has recently been characterized by great instability, and therefore one can operate with the amount received conditionally.

Strategic trend for the development of the electric power industry, which provides for the widespread introduction of renewable energy sources, is adopted in Azerbaijan. Modular power plants will make it possible to cover the electrical load of the power system when introducing renewable energy sources in the amount of up to 500 MW, the coverage of large volumes will be accompanied by significant problems, primarily the need to build peak power and large capacities for the electric energy accumulation.

Thus, at current natural gas prices, the amount that can be used for the equalization of the load schedule can be estimated at $260 million. The main means of equalization of the load schedule is the use of multi-tariff systems. To do this, it is necessary to introduce modern electricity meters with appropriate functions.

Currently, the two-part tariff applied in Azerbaijan for continuous productions (production of aluminum, steel, etc.) does not contribute to equalization of the load schedule (their share in total consumption is not significant). In order to equalize the load schedule, it is necessary to stimulate the development of industry with continuous automatic production [13-15].

Conclusions

1. Power consumption management is one of the main components of the concept of modern development of the electric power industry.

2. Currently, there are two types of programs of change of the load of consumers: programs based on incentives and programs based on time.

3. The characteristics of the daily load schedules of the Azerbaijan power system are noticeably worse compared to similar characteristics of other countries, and there is significant potential in the Azerbaijan power system to improve the characteristics of the daily load schedules.

4. The technical and economic indicators of load coverage mainly depend on the number of connected units with a capacity of 300 MW of the Azerbaijan TPP.

5. The effect of equalization of the load schedule of the Azerbaijan power system at current natural gas prices can be $260 million, which can be directed to the introduction of multi-tariff systems, the development of accumulation systems and continuous automatic production with uniform power consumption.

References

1. Md Rahat Hossain, Amanullah Maung Than Oo, A B M Shawkat Ali Smart Grid, p. 24 (2013)
2. Gurtovtsev A.L., Zabello E.P. *Alignment of the graph of the electric load of the power system. Energy and fuel and energy complex*, No. 7/8., p.13-20, (2008)
3. Voropay N.I., Stennikov V.A. Active consumers in an intelligent electric power system http://energystrategy.ru/projects/Energy_21/3-6.pdf
4. Volkova I.O., Shuvalova D.G., Salnikova E.A. Active consumer in the smart energy system: opportunities and prospects,

   https://publications.hse.ru/mirror/pubs/share/folder/xnto5eu7bd/direct/71899800
5. Eid, C. Time-Based Pricing and Electricity Demand Response: Vxisting Barriers and Next Steps C.Eid, E.Koliou, M. Valles [et al.] Elsevier Utilities Policy, Volume **40**, p. 15-25 (2016)
6. Pazouki, S., Mohsenzadeh, A., Haghifam, M. R. *Optimal planning of parking lots and DLC programs of Demand Response for enhancing distribution networks reliability. IEEE Pes General Meeting Conference & Exposition,*

National Harbor, MD, USA, p. 1-5 (27-31 July, 2014)

7. Robertson, G., Harrison, G.P., Robin Wallace, A. OPF Techniques for Real-Time Active Management of Distribution Networks. IEEE Power & Energy Society General Meeting, Chicago, IL, USA, p. 3529-3537 (16-20 July, 2017)
8. Nasibov V.Kh., Alizade R.R. *Study of characteristic load curves of the Azerbaijan energy system to determine the load power reserve*, 85th meeting of the International Scientific Seminar named after. Yu.N. Rudenko "Methodological issues in the study of the reliability of large energy systems", Reliability of liberalized energy systems, Irkutsk, Russia, p. 119-124 (June 30-July 4, 2013)
9. Papkov B.V. *Active elements of consumers in the electric power system*, Methodological issues in the study of the reliability of large energy systems. Issue. 66., Minsk, Irkutsk: ISEM SO RAN, p. 218-224 (2015)
10. Hayder O. Alwan Compass Load Scheduling with Maximum Demand and Time of Use pricing for Microgrids, Virginia Commonwealth University VCU Scholars, p. 154 (2019)
11. Jun Du, Zongnan Zhang, MenghanLi ,JingGuo&Kongge Zhu Optimal scheduling of integrated energy system based on improved grey wolf optimization algorithm, Scientifc Reports (2022)
12. Jia, H. et al. *Research on some key problems related to integrated energy systems*. Autom. Electr. Power Syst. **39(7)**, p. 198–207 (2015)
13. Dunnan, L., Guangyu, Q. & Qi, Li. Research on optimization operation of integrated energy microgrid considering multi-type energy conversion and storage. South. Power Grid Technol. **12(3)**, p. 105–115 (2018)
14. Li, W. et al. Optimal dispatch model considering environmental cost based on combined heat and power with thermal energy storage and demand response. Energies **12(5)**, p. 817 (2019)
15. Zhang, X. et al. Hourly electricity demand response in the stochastic day-ahead scheduling of coordinated electricity and natural gas networks. IEEE Trans. Power Syst. **31(1)**, p.592–601 (2015)

New approaches and digital technology automation tasks processes of control and accounting of electricity in distribution networks

Omorov[1] *T.T.*, Takyrbashev[1,*] *B.K.*, Zakiryaev[2] *K.E.*, Imanakunova[3] *Zh.S.*, Koibagarov[1] *T.Zh.*, Asiev[3] *A.T.*

[1] National Academy of Sciences of the Kyrgyz Republic, Institute of Machine Science and Automation, Bishkek
[2] K. Tynystanov Issyk-Kul State University, Karakol
[3] I. Razzakov Kyrgyz State Technical University, Bishkek

Abstract. This paper studies 0.4 kV power distribution networks (PDN) and automated systems for electricity monitoring and metering (ASEMM) As is known, one of the main tasks of ASEMMs is the digitalization of PDNs, aimed at further improving the efficiency and reliability of their operation. It is advisable that new models, methods, and intelligent technologies used for automation and informatization of distribution networks should also be focused on minimizing their power losses, which are currently fairly high thus significantly compromising the technical and economic performance of automation systems employed and PDNs. Modern (conventional) ASEMMs, implemented at the facilities of utilities, do not have the appropriate technical, algorithmic, and software tools designed to reduce power losses in the PDN. This is due to the fact that conventional ASEMMs mainly solve the problem of remote data collection from the system meters and their digital processing for the purpose of revenue metering of electricity. In this regard, the paper proposes methodological, algorithmic, and digital technologies designed to solve a set of new functional tasks in the conventional ASEMMs, aimed at reducing power losses in DPNs by optimizing their operating conditions.

Introduction

In the context of the energy crisis, the most important task is the problem of saving electricity through the introduction of new scientific methods and digital technologies at power sector facilities. Analysis of the results of operation of modern automated 0.4 kV power distribution networks (PDN) shows that they lose more than 11-12% of electricity in the form of technical and commercial losses of the total amount coming to the inputs of facilities of distribution utilities. By comparison, electricity losses in developed countries average 6-7%. It follows that there is significant potential for their reduction based on the use of new modern technologies that can back further improvement of the system of monitoring and metering of electricity in PDNs.

As is known, at present, in order to automate and digitalize information processes in PDNs, new technologies in the form of automated systems for electricity monitoring and metering (ASEMM) [1, 2], which can be considered as elements of Smart Grid technology [3, 4], are actively and widely adopted. The practice of their use has shown that they fail to ensure adequately the desired level of electricity losses, since these automated systems are mainly designed for electricity revenue metering. Analysis shows that to significantly reduce electricity losses in the PDN it is necessary to additionally solve a set of problems related to the optimization of operating conditions of facilities [5-8], diagnostics of critical states of distribution grids [9-11], including the identification of places of unauthorized consumption (theft) of electricity [12-14], and the monitoring of electricity losses [15, 16] in real time. This set of tasks in the existing (conventional) ASEMMs is currently not adressed. The purpose of this study is to formulate proposals to improve conventional ASEMMs, which include in their composition new additional information subsystems, focused on significantly reducing technical and commercial losses of electricity.

Structure and tasks of conventional ASEMMs

As is known, electricity losses in low-voltage distribution networks are caused by such main factors as: current and voltage asymmetry [5, 17]; unauthorized electricity consumption (theft) in DPNs [14, 18]; non-linear properties of loads of network consumers [17, 19]; exceeding the critical levels of wear of wires of backbone network lines. These factors cause the networks to deviate from their rated operating conditions. Unbalanced PDNs is a consequence of unbalanced loads and their unequal distribution over the phases of the three-phase network, which is one of the main factors that lead to increased

*Corresponding author: b.takyrbashev@gmail.com

losses of active power in networks and transformer substations. According to published research, in lines with distributed load with a relative deviation of phase currents from their average value in the range of 0.3 - 0.5 technical losses increase on average by 35%. The results of experimental studies to determine electricity losses in power supply systems of single-family dwelling units show that technical losses from the unbalance in transmission lines and transformer substations account for more than 6% of the total amount of electricity consumed in DPNs [20]. In this case, the quality of power deteriorates, and the probability of failure of household appliances and industrial plants increases. A number of technologies [5, 21-23], which have not found wide practical application due to the complexity of their technical implementation, have been proposed to solve the problem of balancing adjustment of distribution networks. In particular, balancing adjustment devices with special transformers are practically not used in 0.4 kV networks, as they are the sources of technical electricity losses in networks, and they are complex engineering systems and have a rather high cost. As part of the process of combating the above undesirable factors, the most promising is the use of the potential capabilities of integrated hardware and software systems of ASEMM, developed on the basis of AIM, MDM, and APM technologies [24-26]. Such information systems have been developed by a number of companies abroad, such as JSC "Systems and Technologies" Group, JSC "Electrotechnical factories "Energomera", Research and Production Corporation "Lianozovo Electromechanical Plant" (JSC LEMZ R&P Corp.), Research and Production Corporation "MIR" (Russia), ADD Grup (Moldova), Yitran (Israel), Hexing Electrical Co.Ltd. (PRC), SigmaTelas (Lithuania) [27-30].

The generalized structure of conventional ASEMMs is shown in Fig.1,

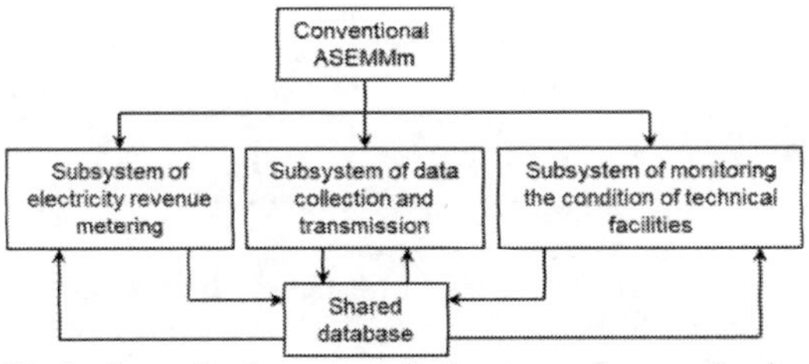

Fig.1 - Generalized structure of subsystems of conventional ASEMMs

which includes the following information subsystems:

• revenue metering of electricity consumed by DPN consumers;

• automated collection of data from the system's electricity meters and transfer of the necessary data to the upper control level;

monitoring the condition of electricity meters and other technical facilities.

To store regulatory, reference, process, measurement, and other data in the ASEMM there is a shared database set up. Through this database the necessary information is exchanged between these functional subsystems, whose main functions are:

• remotely collecting energy consumption data with a specified sampling rate (hour, week, month, etc.);

• automated metering of electricity consumed by network consumers and the provision of multi-tariff options at the same time;

• continuous monitoring of the use of energy;

• controlling power and remotely disconnecting/reconnecting network consumers;

• getting up-to-date information about meter failures and malfunctions in the system;

• calculating energy balance of the distribution network;

• setting up a shared database and preparing reports and other information materials;

• exchanging information with the upper control level.

Analysis of the functional structure of modern ASEMMs shows that the main function of these information systems is revenue metering of electricity in distribution networks. Their main advantages are:

1)automation of processes of measurement data collection, without the participation of inspectors (supervisors) in charge of power supply and electricity metering;

2)elimination of the human factor in the collection of data on energy consumption, which enables ending corrupt practices;

3)automatic disconnection of the consumer in case of late payment and exceeding the set limit on electric power consumption.

At the same time, analysis of the results of the use of conventional ASEMMs shows that the existing composition of functional subsystems is not focused on minimizing the technical and commercial losses of electricity in the PDN. In this regard, there is a need to develop new methods and digital technologies focused on improving the conventional ASEMMs used in distribution networks. One possible way in this direction is to develop a modernized information and control system on the platform (basis) of conventional ASEMMs, designed to address optimization, diagnostic, and monitoring tasks in the PDN.

Structure and tasks of the modernized ASEMM

Our analysis shows that as part of conventional ASEMMs, it is advisable to address the following functional tasks:

• automatic control of technical losses of electricity, ensuring optimization of distribution network operating conditions;

• detection and identification of the coordinates of unauthorized electricity consumption (thefts);

• diagnostics of the condition of the wires of the sections of the network backbone line that connect different consumers;

 https://doi.org/10.1051/e3sconf/202338401016

• on-line monitoring of technical and commercial electricity losses in the PDN.

Our analysis shows that addressing these tasks as part of the ASEMM system can significantly reduce technical and commercial losses of electricity, as well as improve power quality and reliability of power supply to consumers. In this regard, there is a need to develop new scientific methods and digital technologies focused on the modernization of conventional ASEMMs used in distribution networks. One of the possible ways in this direction is to create new additional information subsystems as part of conventional ASEMMs, aimed at addressing the above optimization, diagnostic, and monitoring tasks. The structure of the proposed modernized ASEMM is shown in Fig. 2.

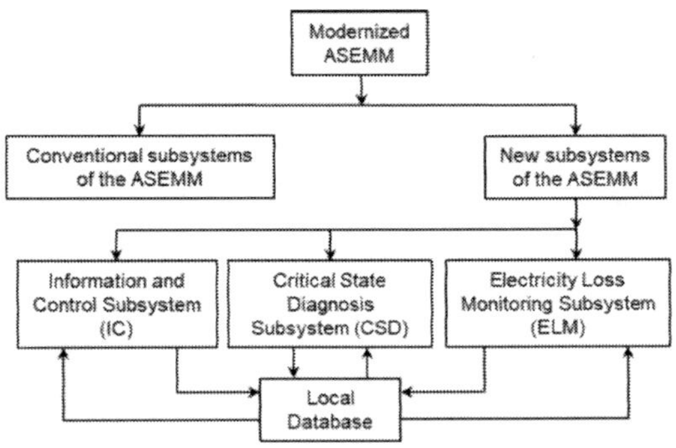

Fig.2 - Structure of additional subsystems of the ASEM

It includes conventional information subsystems of the ASEMM (Fig. 1), as well as new subsystems: IC, CSD, and ELM, designed to address the above additional functional tasks. In this case, to build an IC subsystem we use the following objective function as the performance metric of the system:

$$E=J \quad (1)$$

where J is the value of the effective current in the zero wire of the initial section of the network.

Minimization of the performance metric (1) is equivalent to the optimization of the operating conditions of the unbalanced PDN, in which the balancing adjustment of the initial section of the network and the minimum imbalance of the values of total power consumed by each of the network phases is provided. This can significantly reduce technical losses in the transformer power sources of the PDN as well as improve their reliable operation. In order to minimize the objective function E, we introduce the criterion function F(p), which defines the measure of deviation of the desired state of the distribution network in terms of power p at its input from the actual state defined by the vector p=[p_1, p_2, p_3], where p_k is the modulus of the power consumed by the k-th network phase. As a result, the problem of optimization of the distribution network operating conditions is reduced to solving the following extreme-value problem:

$$\min_{p\in P} F(p) = F(p^*), \quad (2)$$

where P - discrete admissible subset; p^* - optimal desired vector. An algorithm for solving the extreme-value problem (2) is proposed in [8].

From a technical perspective, the IC subsystem is implemented by means of a digital controller (DC) [8,31], the structure of which is shown in Fig. 3.

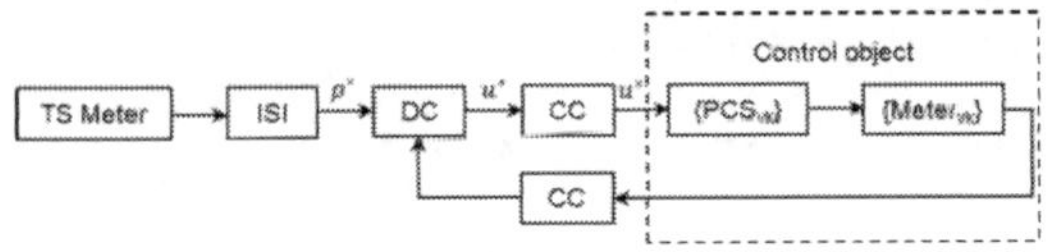

Fig. 3 - Functional structure of the IC subsystem

It is assumed that a group of loads of network consumers, to which electricity Meters$_{\nu k}$ are connected, together with the actuators of the system, is a control object. The key functional unit of the system is the digital controller (DC) based on a microcontroller unit. The DC unit generates control actions u^* applied to the object based on a special algorithm (control rule). The program governing the operation of the DC is formed by the initial state identifier (ISI) of the automatic system in the form of the setting action ρ*. The control signal u^* is a digital command code, which is formed as the vector u*=[Φ₁,Φ₂,β], where Φ₁,Φ₂ - numbers (names) of pairs of phases in which it is necessary to switch network consumers from a more loaded phase (Φ₁) to a less loaded one (Φ₂); β- vector composed of the coordinates (addresses) of consumers of the phase Φ₁ who are to be switched. This control signal u^* is transmitted to the actuators of the system through the communication channel (CC). Such actuators are phase current switches ($PCS_{\nu K}$), designed to carry out the required switching of loads of network consumers from one phase to another [7, 8]. The phase current switch ($PCS_{\nu K}$) is implemented on the basis of a separate microcontroller unit. In conventional ASEMMs, various data transmission technologies (PLC, GSM, etc.) are used as CCs.

The procedure for synthesizing an algorithm for the functioning (control) of the digital controller includes the following main steps:

1. Formation of the initial data of the control task.
2. Situational analysis of the object.
3. Control algorithm synthesis.

Input data for the task are formed by reading the information recorded in the data concentrator of the ASEMM, and by writing it to the local database of the system. As such information one uses, in particular, the active and reactive powers consumed by the phases and consumers of the network. The situational analysis is carried out to determine the structure of phase switching, i.e., to identify the names of the phases (Φ₁,Φ₂,), in which it is necessary to carry out the switching operations of the corresponding loads of the network consumers. The synthesis of the control rule u* is carried out on the basis of the found vector. Methods, algorithms, and technologies for the construction of the IC subsystem are proposed in [7, 8].

The main functions of the Electricity Loss Monitoring (ELM) subsystem are the identification and continuous monitoring of technical and commercial electricity losses in the DN. When there is unauthorized consumption (thefts) of electricity in the network, the balance of complex powers is determined by the following relationships:

$\dot{S}_k(\xi) = \dot{S}_k^a(\xi) + \dot{S}_k^T(\xi) + \dot{S}_k^x(\xi)$, $k=\overline{1,3}$,

where k- index variable denoting the number of the corresponding phase (A, B, C), $k=\overline{1,3}$; $\dot{S}_k$- complex power consumed by the k-th phase at a discrete moment of time t $=t_\xi$; $\dot{S}_k$ - total complex power consumed by all consumers of the k-th phase; $\dot{S}_k^T$ - technical power losses in the k-th phase; $\dot{S}_k^x$- uncontrolled power losses in the k-th phase of the network. In this case, the powers $\dot{S}_k(\xi)$ and are $\dot{S}_k^a(\xi)$ known values. It should be noted that in the existing ASEMMs, technical $\dot{S}_k^T$ and commercial $\dot{S}_k^x$ losses are not determined, and only the values of total power consumed $\dot{S}_k$ by the network phases and electric power consumer $\dot{S}_k^a$ are evaluated. The main task of the ELM subsystem is to identify the values $\dot{S}_k^T$ and $\dot{S}_k^x$ and $\dot{S}_k^x$ to conduct continuous monitoring of uncontrolled power losses in the distribution network on their basis. At the same time, to effectively address the tasks of the ELM subsystem, virtual DPN models are introduced for consideration, describing the desired states of real-world networks in the absence of unauthorized consumers there. Methods and algorithms for solving the problems of the subsystem in question are proposed in [12, 13].

In the "Critical State Diagnosis of the Network" subsystem two problems are solved:

• detection and identification of unauthorized power consumption (thefts) in the PDN;

• diagnosis of the condition of wires of sections of the backbone line connecting different consumers by their wear and tear level.

The solution to these problems is based on the identification of the PDN model using numerical methods [32, 33] and the estimation of complex resistances of inter-consumer sections of the three-phase network [11]. At the same time, it is possible to assess the level of wear and tear of the backbone line wires. Mathematical conditions for determining the critical states of the DPN are obtained. The results of solving the problems can be used to take appropriate organizational and engineering measures to eliminate the specified critical states of the PDN, which makes it possible to reduce electricity losses caused by unauthorized consumption of electricity and critical levels of wear and tear of wires of sections of the backbone line. Methods and technologies for building the CSD subsystem are outlined in [15, 16].

Data exchange between IC, ELM, and CSD subsystems can be performed through the local database of the information system. Input data for addressing new functional tasks come from the ASEMM data concentrator. In turn, the concentrator receives data by polling the electricity meters installed at the network consumers and in the transformer substation.

Conclusion

Despite the active and widespread adoption of integrated hardware and software systems of ASEMMs in distribution networks, technical and commercial losses of electricity remain quite high. We made several proposals on improving traditional systems of automation and digitalization of networks. They are based on the development of new functional subsystems as part of conventional ASEMMs. Such subsystems are designed to address the following major functional tasks: automatic control of electricity losses, which ensures a reduction of technical losses in distribution networks (including losses in the transformer power sources); detection and identification of places of unauthorized consumption (theft) of electricity, real-time identification and monitoring of technical and commercial electric power losses in the network; diagnosis of the condition of wires of inter-consumer sections of the backbone line by their wear and tear level.

The creation and use of software of new functional subsystems allows the modernization of the conventional ASEMM into an information and control system. This makes it possible to improve significantly its efficiency and economic performance of distribution utilities, as well as the reliability of the transformer power supply sources of networks.

References

1. Eremina M.A. Development of automatic systems of revenue energy metering // Molodoi uchyonyi. 2015. No. 3. Pp. 135-138. (in Russian)

2. Yakushev K.V. Automated system of revenue electricity metering for the retail market // Informatizaciya i upravlenie v promyshlennosti. - 2009. - No. 3(23). - Pp. 9-13. (in Russian)

3. El-Hawary, M.E. The Smart Grid—State-of-the-art and Future Trends. Electr. Power Compon. Syst. 2014, 42, pp239–250.

4. Kazmi S.A., Shahzad M.K., Khan A.Z., Shin D.R. Smart Distribution Networks: A Review of Modern // Energies, 2017.

5. Kosoukhov F.D., Vasiliev N.V., Filippov A.O. Reducing losses from current unbalances and improving the power quality in 0.38 kV networks with municipal and household loads // Elektrotekhnika 2014. No. 6. Pp. 8-12. (in Russian)

6. Kiselev M.G., Lepanov M.G. Balancing adjustment of currents in power supply networks by a power electric controller of inactive power // Elektrotekhnika. 2018. No. 11. Pp. 63-70. (in Russian)

7. Turatbek T. Omorov, Beishenaly K. Takyrbashev, Rima Ch. Osmonova. Synthesis of the managing director of the subsystem for optimization of the operating mode of the distributive electric network // Engineering Studies. 2016. No. 3. pp. 606-615

8. T.T.Omorov, K.Takyrbashev. K.E.Zakiriaev, T.Zh.Koibagarov. Digital control of electric power flows in non-symmeter distribution networks as a composition of AM-RCS // Energy Systems Research. 2021, No.1. pp.38-46.

9. Ershov A.M., Filatov O.V., Mlotok A.V., et al. System of protection of 380V electric network against overhead line wire breaks // Elektricheskie stancii. 2016, No. 5. Pp. 28-33. (in Russian)

10. Klochkov A.N. Device for detecting three-phase networks with a broken phase wire. - Bulletin of KrasSAU, No. 1, 2011, pp. 221-223. (in Russian)

11. Omorov T.T., Osmonova R.Ch., Takyrbashev B.K. Diagnosis of the condition of electric lines of distribution networks as part of the automated system for electricity monitoring and metering // Kontrol. Diagnostika. 2017. No. 5. Pp. 44-48. (in Russian)

12. Omorov T.T., Takyrbashev B.K., Osmonova R.Ch., Koibagarov T.Zh. Identification of leakage currents in distribution networks according to the data of the automated system for electricity monitoring and metering // Bulletin of the South Ural State University. Series: Energy. 2018. No. 2. Pp. 48-54. (in Russian)

13. Omorov T.T. On the issue of locating unauthorized electricity consumption in distribution networks as part of automated systems for electricity monitoring and metering // Pribory i sistemy. Upravlenie, kontrol, diagnostika. 2017. No. 7. Pp. 27-32. (in Russian)

14. Danilov M.I. On the issue of prompt detection of uncontrolled electricity consumption by data of the automated systems for electricity monitoring and metering // Pribory i sistemy. Upravlenie, kontrol, diagnostika. - 2020. No. 5. Pp. 17-22. (in Russian)

15. Omorov T.T., Osmonova R.Ch., Koibagarov T.Zh., Eralieva A.Sh. On the issue of identification of technical and commercial electricity losses in the automated information and measurement system of electricity revenue metering // Elektrichestvo. Peredacha i raspredelenie. - 2018. - No. 5 (50). - Pp. 56-60. (in Russian)

16. Omorov T.T., Takyrbashev B.K., Zhanybaev T.O., Koidagarov T.Zh. Identification and monitoring of power losses in distribution networks as part of the automated system for electricity monitoring and metering // In the collection: Methodological issues of research on the reliability of large systems in the energy industry. Proceedings of the 93rd seminar session. In 2 volumes. - 2021. - Pp. 33-42. (in Russian)

17. Zhelezko Y.S. Power losses. Reactive power. Power quality. Moscow: ENAS, 2009. 456 p. (in Russian)

18. Sapronov A.A., Kuzhekov S.L., Tyniansky V.G. Prompt detection of uncontrolled electricity consumption in electric networks with a voltage of up to 1 kV // Izvestiya vuzov. Electromekhanika. 2004. No. 1. Pp. 55-58. (in Russian)

19. Arutyunyan, A.G. On the calculation of additional power losses in three-phase four-wire networks // Elektrichestvo. - 2015. - No. 10.

20. Averbukh M.A., Zhilin E.V. On electricity losses in power supply systems of single-family dwelling units // Energetik. 2016. No. 6. - Pp. 54-56. (in Russian)

21. Patent No. 2249286 (RF). G.A. Bolshanin. Method for automated active control of voltage and current unbalance level // Bull. No. 9. 27.03.2005.

22. Patent No. 2490768 (RF). I.V. Naumov, D.A. Ivanov, S.V. Podyachikh, Gantulga Damdinsuren. Balancing adjustment device for three-phase networks with a zero wire // Bul. no. 23. 20.08.2013.

23. Patent No. 2548656 (RF). Samokish V.V. Method for balancing adjustment of phase currents of three-phase four-wire line and device for its implementation // Bul. no. 11. 27.12.2013.

24. http://www.sigmatelas.lt/

25. https://addgrup.com/

26. http://www.mirmsk.ru/stuff/career/vacancies.

27. http://www.energomera.ru/

28. https://lemz.ru/

29. http://www.yitran.com/

30. http://www.hxgroup.cn/en/

31. Omorov T.T. Takyrbashev B.K., Koibagarov T. J. Management of electricity losses in distribution networks as part of automated systems for electricity monitoring and metering // Mekhatronika, avtomatizaciya, kontrol, Vol. 22, No. 4, 2021. Pp. 192 - 199. (in Russian)

32 Omorov T.T., Takyrbashev B.K., Osmonova R.Ch. On the issue of mathematical modeling of the three-phase unbalanced distribution network // Izvestiya Vysshikh Uchebnykh Zavedenii. Problemy energetiki. 2020. Vol. 22. No.1. Pp. 93-102. (in Russian)

33. Omorov T.T., Kozhekova G.A. Synthesis of control systems of multidimensional objects by criterial constraints // Proceedings of the National Academy of Sciences of the Kyrgyz Republic, Bishkek, 2009, No. 1. Pp. 45-52. (in Russian)

Improvement of distance protection algorithms based on phasor measurements

Alexey V. Mokeev[1,2] *, *Sergey A.* Piskunov[1,2]

[1]Northern (Arctic) Federal University, Electroenergetics and Electrotechnics Department, Arkhangelsk, Russia
[2]Engineering Centre "Energoservice", Arkhangelsk, Russia

Abstract. Distance protection of overhead lines is widely used and is continually being improved. However, the distance protection cannot work stably and is blocked in several electromechanical transients. Additionally, a non-linear arc can affect the stability of the distance protection in case of a short-circuit. The authors consider ways to improve distance protection during such transients. In particular, this paper proposes the use of synchronized phasor measurement technology to improve the efficiency of distance protection.
This paper considers a new approach for the analysis of electromechanical and electromagnetic transients in a power system using process synchrophasors. As examples, cases of a short-circuit in a line with one-way and two-way powers are considered. The authors consider the influence of a non-linear arc at the fault point and consider the possibility of the coincidence of the electromechanical and electromagnetic transients.
The result is a new algorithm for operating distance protection, an improvement in the parametric method for determining the location of line damage. The calculations performed and mathematical modeling confirm the effectiveness of the proposed algorithm.

1 Introduction

The reliability of the power system is largely dependent on the technical excellence of the relay protection and automation devices. This paper is devoted to the consideration of issues related to improving the efficiency of relay protection devices with synchronized phasor measurement (SPM) technology on the example of improving distance protection (DP).

Improvement of distance protection based on synchronized phasor measurements is considered in several studies [1-3]. However, still one of the main disadvantages of distance protection is the need to block it when some electromechanical processes occur, for example, an asynchronous mode in the power system, power swings [4-6]. Additionally, the sensitivity of distance protection is reduced in the case of short-circuits with an arc.

This paper discusses new algorithm for operating overhead line distance protection, which ensure the correct functioning of the protection both in electromagnetic transients and electromechanical transients, as well as in a combination of these transients and considering the influence of the arc.

2 Theory of process synchrophasors

2.1 Short-circuit process synchrophasors

The term of a process synchrophasor was proposed in [7], where the development of the SPM theory was considered. In this paper, it is assumed that the synchrophasor of the process is the theoretical value of the synchrophasor, which can be obtained analytically or using numerical calculation methods. The concept of a process synchrophasor should be distinguished from the concept of synchrophasor estimation using various well-known methods, for example, the Fourier algorithm. Thus, the authors consider the analysis of process synchrophasors and their components to improve the overhead line distance protection algorithms.

Consider the definition of process synchrophasors using the example of a simple equivalent circuit of a power system and an RL-model of a line with a three-phase short-circuit (Fig.1).

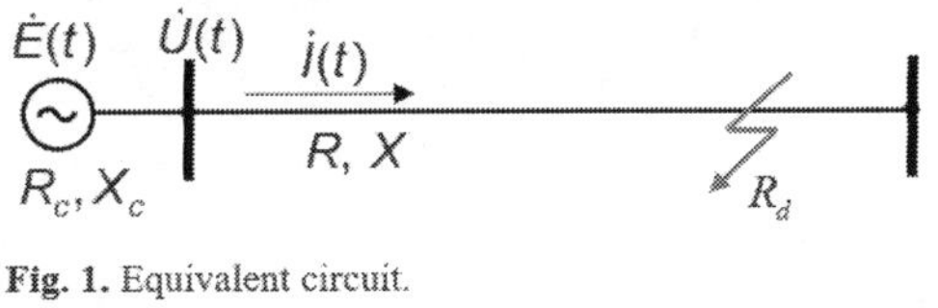

Fig. 1. Equivalent circuit.

* Corresponding author: a.mokeev@narfu.ru

The differential equation of the line in case of a short-circuit in the instantaneous values of current and voltage:

$$u(t) = i(t)R + L\frac{di(t)}{dt}, \quad (1)$$

where R, L - parameters of the short-circuited section of the line.

To obtain this differential equation in the process synchrophasors, one should replace the instantaneous values of current and voltage $u(t)$, $i(t)$ with expressions $\dot{U}(t)e^{j\omega_0 t}$, $\dot{I}(t)e^{j\omega_0 t}$ and substitute them into equation (1). After simple transformations, the following equation is obtained:

$$\dot{U}(t) = \underline{z}\dot{I}(t) + L\frac{d\dot{I}(t)}{dt}, \quad (2)$$

where $\underline{z} = R + j\omega_0 L$.

The solution of the differential equation (2) can be used to determine the current synchrophasor of an electromagnetic or electromechanical transient process for a given synchrophasor of the power system equivalent EMF:

$$\dot{I}(t) = \frac{1}{L}e^{-p_1 t}\int_0^t \dot{E}(\tau)e^{p_1\tau}d\tau, \quad (3)$$

where $p_1 = \beta + j\omega_0$, $\beta = (R_c + R)/(L_c + L)$, $\omega_0 = 2\pi 50$ рад/с, R_c, L_c - parameters of the equivalent power system.

Based on expressions (2) and (3), it is possible to determine, respectively, the total current and voltage synchrophasor of the considered electromagnetic process and its forced and free components [7]. To implement distance protection, it is sufficient to use only the first components of synchrophasors, but the use of total synchrophasors makes it possible to increase the protection speed [7, 8].

Existing distance protection algorithms [4, 5, 6, 9], methods for identifying line parameters [10], assessing the state of a distribution network [11] are based on differential equations with instantaneous values of currents and voltages. The authors propose the use of differential equations in which current and voltage synchrophasors are used instead of instantaneous values. The proposed approach has certain advantages when performing tasks such as the implementation of virtual PMU, the identification of line parameters, and the monitoring of the condition of electrical equipment.

The issue of improving the distance protection of an overhead line in the case of one-sided and two-sided measurement of current and voltage synchrophasors is considered in detail in [8]. In addition, a non-linear arc at the fault location has a negative impact on the short-circuit impedance estimate. Therefore, the compensation of this influence is important for distance protection and especially for line fault location devices.

2.2 Estimation of arc influence

The arc during a short-circuit has a complex non-linear character. Sometimes, in order to carry out simple calculations to analyse the operation of distance protection [5], the arc resistance R_d is chosen to be constant. In this case, the system of equations (1) can be represented as follows:

$$u(t) = i(t)R_\Sigma + L\frac{di(t)}{dt}, \quad (4)$$

where $R_\Sigma = R + R_d$.

Equation (4) can be represented as equation (2) by replacing R with R_Σ. For further calculations, we introduce the following variable:

$$\underline{z}_0(t) = \frac{\dot{U}(t)}{\dot{I}(t)} = R_0(t) + jX_0(t). \quad (5)$$

Thus, equation (4) will have the following form:

$$\underline{z}_0(t) = \underline{z} + L\frac{\dot{I}'(t)}{\dot{I}(t)}, \quad (6)$$

where $\dot{I}'(t) = \frac{d\dot{I}(t)}{dt}$.

Using equation (6), the real and imaginary components of the line impedance and arc resistance can be determined:

$$L = X_0(t)\left[\omega_0 + \mathrm{Im}\left(\frac{\dot{I}'(t)}{\dot{I}(t)}\right)\right]^{-1}, \quad (7)$$

$$R_\Sigma = R_0(t) - L\,\mathrm{Re}\left[\frac{\dot{I}'(t)}{\dot{I}(t)}\right], \quad (8)$$

$$R_d = R_\Sigma - \frac{L}{L_\mu}R_\mu, \quad (9)$$

$$R = R_\Sigma - R_d, \quad (10)$$

where R_μ, L_μ - unit line parameters per 1 km.

The proposed algorithm for estimating the resistance and inductance of a short-circuited section of the line expands the previously presented algorithms related to compensating for the influence of arc resistance [8].

The considered version of the algorithm for estimating the influence of the arc is simplified. For a more accurate description of the arc, various nonlinear models are used [13, 14]. To take into account the influence of one of these arc models on the operation of distance protection, it is necessary to create a new system of equations that describes the process of a short-circuit in the line.

In any non-linear arc model, its resistance becomes a time-dependent variable $R_d(t)$. Thus, the total line resistance will also change with time. On the other hand, the above expressions (7) – (10) for the line resistance do not change. However, if earlier it was sufficient to use expression (3) to simulate and find the total current synchrophasor and its components, then in

https://doi.org/10.1051/e3sconf/202338401017

the case under consideration the problem becomes more complicated, since in the differential equation one of the coefficients becomes a function of time.

In this case, the analytical solution for definition of the process current synchrophasor has the following form:

$$\dot{I}(t) = \frac{1}{L} e^{-\rho(t)} \int_0^t e^{\rho(\tau)} \dot{E}(\tau) d\tau , \qquad (11)$$

where $\rho(t) = \int_0^t p(\tau) d\tau$, $p(t) = \beta + \beta_d(t) + j\omega_0$,

$\beta = \frac{R + R_c}{L + L_c}$, $\beta_d(t) = \frac{R_d(t)}{L + L_c}$.

The difficulty in applying expression (11) for calculations lies in the fact that in most nonlinear models the arc resistance depends on the current. Therefore, equation (11) for a number of calculations is more convenient to replace with a system of differential equations, which is solved by numerical methods [14].

One of the best known non-linear arc models is the Mayr model [14]. For the case under consideration, the system of differential equations with a non-linear Mayr arc model and with the use of process synchrophasors will have the following form:

$$\begin{cases} \frac{dR_d(t)}{dt} = -\frac{R_d(t)}{\tau}\left(\frac{\left[\mathrm{Re}\left(\dot{I}(t)e^{j\omega_0 t}\right)\right]^2 R_d(t)}{P_0} - 1\right) \\ \frac{d\dot{I}(t)}{dt} = \frac{1}{L_\Sigma}\left(\dot{E}(t) - \left(\underline{z}_\Sigma + R_d(t)\right)\dot{I}(t)\right) \end{cases} , \qquad (12)$$

where $\underline{z}_\Sigma = R_\Sigma + j\omega_0 L_\Sigma$, $R_\Sigma = R + R_c$, $L_\Sigma = L + L_c$, τ - arc time constant, P_0 - arc power constant.

If it is necessary to separate the forced and free components of the current $\dot{I}(t)$, you can use the simple method considered in [7]. In this case, the free component is determined first, and then the forced components.

Further, based on the differential equation (4) with a variable coefficient, the voltage on the substation buses and the arc voltage are determined. For other arc models [14], a similar approach can be used to determine the current and voltage synchrophasors.

3 Simulation and analysis

3.1 Electromagnetic transient

This section discusses several examples of modeling a three-phase short-circuit at the end of the line, taking into account the influence of a non-linear arc. The current and voltage synchrophasors are estimated using the Fourier algorithm.

The authors propose to compare two methods for calculating the line impedance. The first method is based on the traditional approach, when complex current and voltage amplitudes are applied:

$$\underline{z} = \frac{\dot{U}(t)}{\dot{I}(t)} . \qquad (13)$$

The second method involves the use of expressions (4) - (10) to calculate the line impedance and estimate the distance to the short-circuit point. The purpose of the comparison is to show that the calculation using expressions (4) - (10) allows much more efficient use of current and voltage synchrophasors for estimating the line impedance and, accordingly, for the operation of distance protection.

Modeling is carried out according to the equivalent circuit, which is shown in Fig. 1. Initial data for the first example: frequency f_0 = 50 Hz, EMF system synchrophasor $\dot{E}(t)$ =136.8$e^{-j0.25\pi}$ kV, system impedance $\underline{z}_c$ = 1.2+j12.8 Ohm, unit line impedance $\underline{z}_\mu$ = 0.05+j0.38 Ohm, line length 30 km, arc time constant τ = 0.0001 s, arc power constant P_0 = 1 MW.

Modeling is done in MATLAB\Simulink. To calculate the process synchrophasors and instantaneous values of current, voltage and arc resistance, the system of equations (12) is used. Fig. 2 shows the instantaneous values of the short-circuit current and the total current synchrophasor module. Fig. 3 shows the change in arc resistance over time.

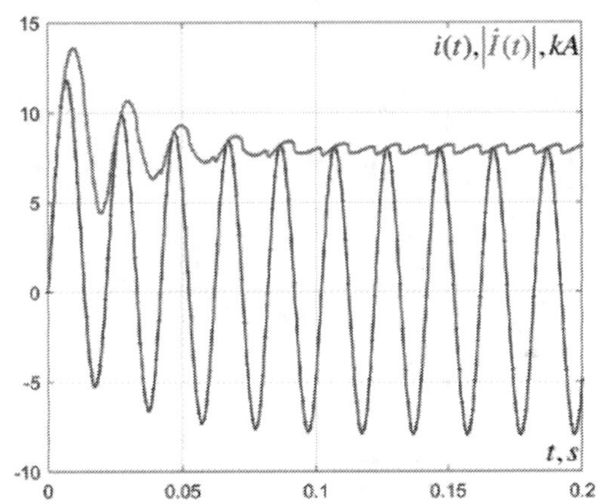

Fig. 2. Instantaneous current values and total current synchrophasor module.

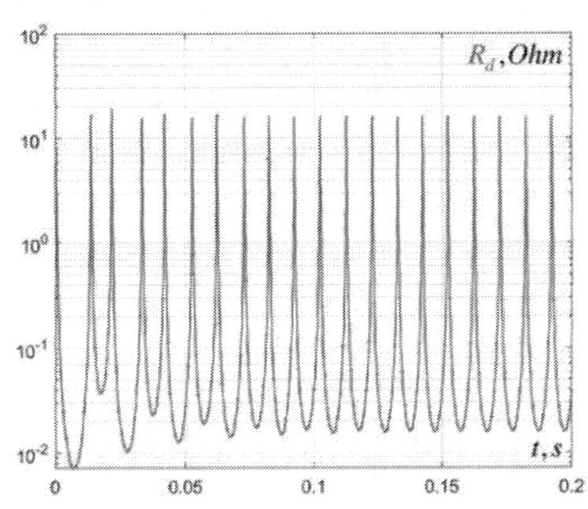

Fig. 3. Arc resistance.

We use a special form of the graph (hodograph) to present the results of calculating the line impedances by different methods. The hodograph shows the dependence of the imaginary part of the line impedance on the real part. The hodograph for the considered example is shown in Fig. 4.

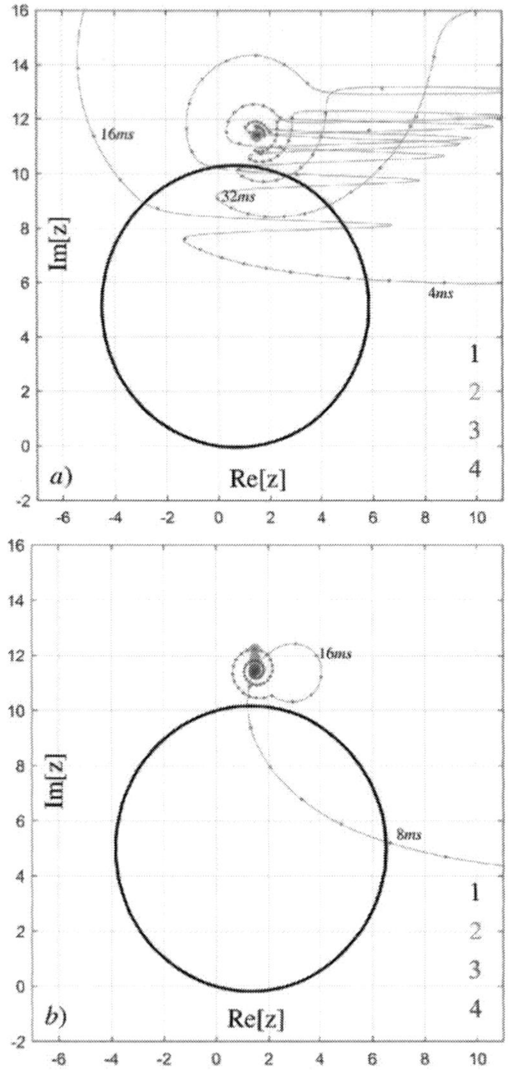

a) calculation via process synchrophasors; b) calculation through the estimation of synchrophasors using the Fourier algorithm; 1 - characteristic of the 1st stage of distance protection; 2 - calculation through expressions (4) - (10); 3 - calculation via expression (13); 4 - line impedance true value
Fig. 4. Line impedance estimation for example 1.

Fig. 4a shows the results of line impedance calculations through process synchrophasors. In this case, for the traditional calculation method (13), the arc resistance is not excluded from the line resistance, so the corresponding changes to the right area are visible on the graph. In addition, the result of the calculation approaches the true value over several periods. As a rule, with such resistance fluctuations, the distance protection is blocked.

Fig. 4b shows the results obtained by evaluating the current and voltage synchrophasors using the Fourier algorithm. The averaging principle of operation of such an algorithm makes it possible to reduce the influence of a non-linear arc. However, line resistance fluctuations continue for more than 20 ms according to the traditional calculation method (13). In the case where the proposed expressions (4) - (10) are applied, the line impedance becomes almost equal to the true value within a few milliseconds (figure 4a, 4b, line 2). The results of calculations according to expressions (4) - (10) can be made even more efficient if specially synthesized filters are used instead of the Fourier algorithm.

3.2 Combination of electromagnetic and electromechanical transient

The second example shows the results of simulating a short-circuit in a line with a simultaneous electromechanical process. The electromechanical transient process is modelled by setting an expression for changing the system EMF. Fig. 5 shows the instantaneous voltage values and voltage synchrophasor module in a substation.

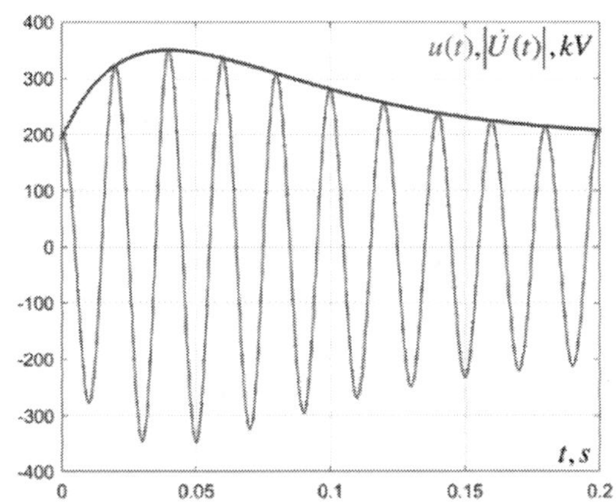

Fig. 5. Instantaneous voltage values and total voltage synchrophasor module.

Fig. 6 shows the hodographs for the example under consideration. As in the previous example, the calculations were carried out for the process synchrophasors (a) and for the synchrophasors estimated using the Fourier algorithm (b).

The simulation results allow concluding that the combination of electromechanical and electromagnetic processes leads to a decrease in the accuracy of calculations by the traditional method (13). In addition, the electromechanical transient negatively affects the operation of the Fourier algorithm, and, as a result, the estimation of current and voltage synchrophasors.

The presence of an electromechanical transient does not adversely affect the evaluation of the line impedance components when using the proposed method (4) - (10).

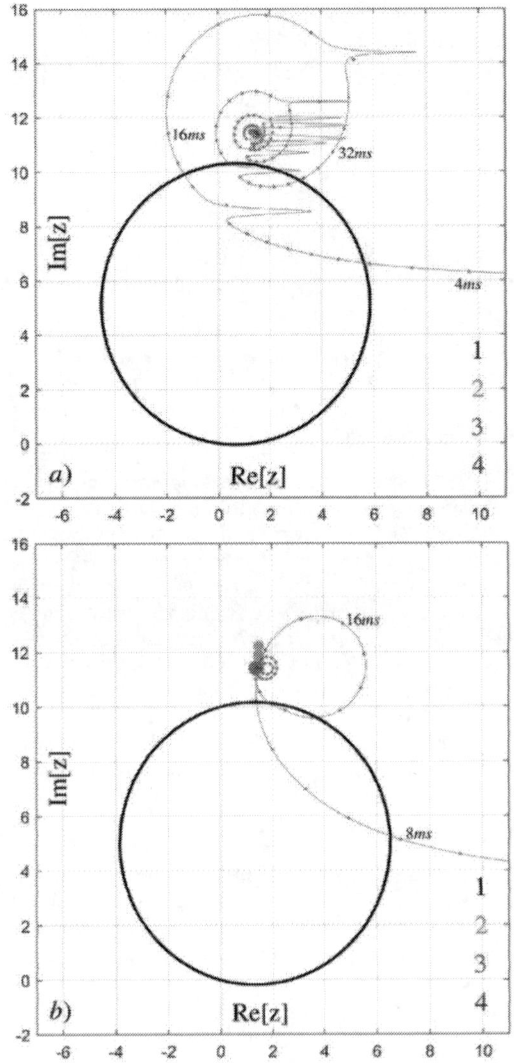

a) calculation via process synchrophasors; b) calculation through the estimation of synchrophasors using the Fourier algorithm; 1 - characteristic of the 1st stage of distance protection; 2 - calculation through expressions (4) - (10); 3 - calculation via expression (13); 4 - line impedance true value

Fig. 6. Line impedance estimation for example 2.

The authors assume that in the considered examples, the arc resistance has little effect on the line impedance estimation because the calculation was made for a short-circuit at the end of the line. Additional calculations show that with a short-circuit at the beginning of the line, for the case when the system has a large power, the influence of the arc becomes much higher. Therefore, the use of the proposed method for estimating the line impedance has advantages for other modes of power system operation.

3.3 Line with two-way power supply

Next, we consider the case when the line has power sources on both sides. In the graphs and expressions, the number 1 indicates the values related to the first power source, the number 2 to the second.

For the case under consideration, we assume that synchronized phasor measurement devices are installed at the ends of the line. Two-sided measurement of current and voltage synchrophasors has advantages over one-sided measurement, since the SPM technology is focused on distributed data processing and the creation of WAPS.

The equivalent circuit of a power system with two power supplies is shown in Fig. 7.

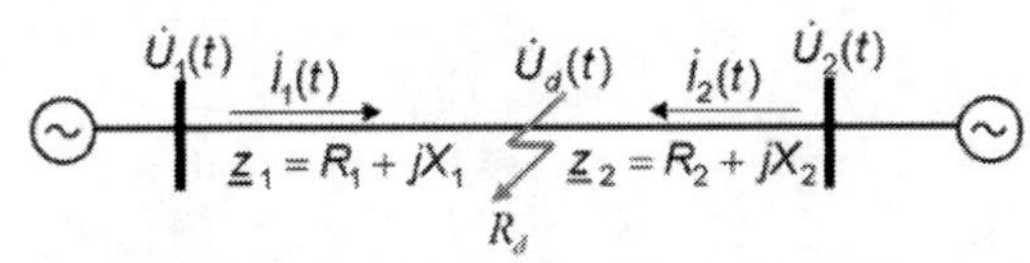

Fig. 7. The equivalent circuit of a power system with two power supplies.

Next, we consider the case when the current phasors $\dot I_1(t)$ and $\dot I_2(t)$ are not equal to each other (in modulus and phase).

System of equations for the traditional method based on complex amplitudes:

$$\begin{cases} -\dot U_1(t)+\underline{z}_1(t)\dot I_1(t)+R_d(t)\dot I(t)=0; \\ -\dot U_2(t)+\underline{z}_2(t)\dot I_2(t)+R_d(t)\dot I(t)=0, \end{cases} \quad (14)$$

where $\dot I(t)=\dot I_1(t)+\dot I_2(t)$.

This system makes it possible to eliminate the arc resistance and express the line impedance up to the short-circuit point from one of the sources:

$$\underline{z}_1(t)=\frac{\Delta\dot U(t)+\underline{z}\dot I_2(t)}{\dot I(t)}, \quad (15)$$

where $\Delta\dot U(t)=\dot U_1(t)-\dot U_2(t)$, $\underline{z}$ - total line resistance.

According to the proposed method, based on process synchrophasors for the equivalent circuit in Fig. 7, the following system of equations is compiled:

$$\begin{cases} -\dot U_1(t)+\underline{z}_1(t)\dot I_1(t)+L_1\dot I_1'(t)+R_d(t)\dot I(t)=0; \\ -\dot U_2(t)+\underline{z}_2(t)\dot I_2(t)+L_2\dot I_2'(t)+R_d(t)\dot I(t)=0, \end{cases} \quad (16)$$

where $\dot I_1'(t)=\dfrac{d\dot I_1(t)}{dt}$, $\dot I_2'(t)=\dfrac{d\dot I_2(t)}{dt}$.

We make simple transformations of system (16) and introduce the following notation:

$$\underline{z}_{10}(t)=\frac{\Delta\dot U(t)+\underline{z}\dot I_2(t)+L\dot I_2'(t)}{\dot I(t)}. \quad (17)$$

Then components of the line impedance for the proposed method can be calculated using expressions similar to (7) and (8):

$$L_1 = \mathrm{Im}\left[\underline{z}_{10}(t)\right]\left[\omega_0 + \mathrm{Im}\left(\frac{\dot{i}'(t)}{\dot{i}(t)}\right)\right]^{-1}, \quad (18)$$

$$R_1 = \mathrm{Re}\left[\underline{z}_{10}(t)\right] - L_1 \mathrm{Re}\left[\frac{\dot{i}'(t)}{\dot{i}(t)}\right]. \quad (19)$$

In addition, there is an alternative way to calculate the line impedance, when the related distance to the short-circuit point is first calculated:

$$\lambda_1(t) = \frac{\Delta\dot{U}(t) + \underline{z}\dot{i}_2(t) + L\dot{i}_2'(t)}{\underline{z}\dot{i}(t) + L\dot{i}_2(t)}, \quad (20)$$

where $\lambda_1(t) = \frac{\underline{z}_1(t)}{\underline{z}}$.

The presented expression (20) has a similar form with the well-known expressions for calculating the distance to the short-circuit point based on the method of complex current and voltage amplitudes [16-17] and current and voltage synchrophasors [18]. Nevertheless, the feature of the calculation of process synchrophasors presented in the paper makes it possible to take into account additional components necessary for accurate determination of the line impedance at the beginning of the transient process. Accordingly, this ensures the effective operation of distance protection.

As an example, Fig. 8 shows the results of calculating the distance to the short-circuit point using the proposed and traditional algorithm.

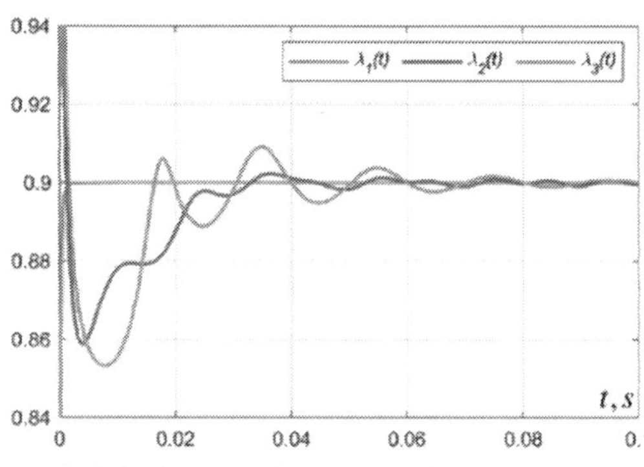

1 - calculation by expression (20), 2 - calculation by the traditional algorithm using a synthesized filter, 3 - calculation by the traditional algorithm using the Fourier algorithm
Fig. 8. Calculation of the distance to the short-circuit point.

The traditional algorithm (Fig. 8) corresponds to the expression given in [18]. Fig. 8 considers two calculation options for the traditional algorithm: using a synthesized filter (curve 2) and using the Fourier algorithm (curve 3). Curve 1 corresponds to the proposed algorithm using the synthesized filter.

Fig. 8 confirms that the proposed algorithm allows obtaining accurate values $\lambda_1(t)$ much faster than the traditional algorithm. In addition, the synthesized filter also allows you to get results that are more accurate.

In the normal operation of the power system or in the event of an external short circuit, when $\dot{i}_1(t) = \dot{i}_2(t)$, the following expressions based on current and voltage synchrophasors determine the line impedance:

$$L_1 = \mathrm{Im}\left[\frac{\Delta\dot{U}(t)}{\dot{i}_1(t)}\right]\left[\omega_0 + \mathrm{Im}\left(\frac{\dot{i}'(t)}{\dot{i}(t)}\right)\right]^{-1}, \quad (21)$$

$$R_1 = \mathrm{Re}\left[\frac{\Delta\dot{U}(t)}{\dot{i}_1(t)}\right] - L_1 \mathrm{Re}\left[\frac{\dot{i}'(t)}{\dot{i}(t)}\right]. \quad (22)$$

These expressions are derived according to the Kirchhoff law similarly to (18) and (19).

4 Discussion

In [19], voltage and current synchrophasors are used in a system of differential equations to determine the fault location in a line. However, the author [19] replaces the instantaneous values of currents and voltages with synchrophasors without additional transformations, which causes an error in estimating the distance. To reduce the error, a special algorithm based on the least squares method is used. The authors suggest that the method based on process synchrophasors would solve this problem.

In this paper, the authors present two options for calculating the distance and impedance of the line to the short-circuit point (according to expressions (17)-(19) and (20)). Despite the fact that expressions (17) and (20) are similar, there is an important difference between them. Expression (20) imposes a restriction on the X/R ratio. Taking into account the fact that X/R ratio practically does not change for an overhead line, expression (20) can be more accurate in some cases. However, further research is required to test this assumption.

5 Conclusion

The authors proposed a new algorithm for operating overhead line distance protection, which is described in this paper. This algorithm uses differential equations based on process synchrophasors instead of instantaneous values of variables. This makes it possible to ensure the correct operation of distance protection during electromechanical and electromagnetic transients.

In this paper, the authors considered several options for transients that can affect distance protection. First, this study shows how the use of process synchrophasors can reduce the effect of a non-linear arc during a short-circuit. Secondly, the authors demonstrate the advantages of process synchrophasors in the case when electromechanical and electromagnetic transients occur simultaneously.

The authors provide expressions for one-sided and two-sided measurements of currents and voltages of an overhead line. For two-sided measurement, there are two options for calculating the distance to the short-circuit point. The paper shows that the use of two-sided measurement of voltage and current synchrophasors makes it possible to simultaneously implement several signs of damage recognition.

This paper provides several options for modeling the operation of distance protection based on the proposed algorithm. The results confirm its effectiveness for various transients.

References

1. H.J.A. Ferrer et al., *Modern solutions for protection, control, and monitoring of electric power systems*, SEL (2010).
2. A.G. Phadke, J.S. Thorp, *Synchronized phasor measurements and their applications*, Technosphere (2021).
3. G. Bolandi, M. Haghifam, *Vulnerability assessment of the III stage of distance protection in real time to prevent cascading outages, based on SPM data*, Relejshchik, **1** (2019).
4. A.M. Fedoseev, *Relay protection of electric power systems*, Energoatomizdat (1992).
5. E.M. Schneerson, *Digital relay protection*, Energoatomizdat (2007).
6. A.G. Phadke, J.S. Thorp, *Computer Relaying for Power Systems*, Wiley (2009).
7. A.V. Mokeev, *Analysis of synchrophasors of transient processes in power system*, Elektrooborudovanie, **1** (2022).
8. A.V. Mokeev, S.A. Piskunov, *Application of synchronized phasor measurement technology to improve distance protection*, RPA, **3** (2022).
9. V.I. Antonov, V.K. Vanin, *Fast resistor relay for short distance power lines*, Elektr., **8** (1985).
10. D.V. Dzhumik, *Determination of the parameters of equivalent circuits of power lines by arrays of instantaneous values of currents and voltages*, TPU, **1** (2007).
11. U.G. Kononov, E.G. Zelensky, *Development of an experimental software module for state estimation based on instantaneous values of currents and voltages*, Vestnik SevKavGTU, **5** (2011).
12. A.I. Popov et al., Examples of processing low-frequency oscillations in Russia and ways to improve the analysis, SGSMA, Split (2022).
13. I.T. Titov, A.V. Bychkov, N.A. Doni, *Consideration of the influence of an open electric arc on the parameters of the operation of distance protection*, NGTU, **2** (2020).
14. L. Yuan, L. Sun, H. Wu, *Simulation of fault arc using conventional arc models, Energy and Power Engineering*, **5** (2013).
15. A.V. Mokeev, Filter synthesis for PMU, SIBCON, Astana (2017).
16. E.A. Arzhannikov, V.U. Lukoyanov, M.S. Misrihanov, *Determining the location of a short-circuit on high-voltage power lines*, Energoatomizdat (2003).
17. G. Ziegler, *Digital distance protection*, Energoizdat (2005).
18. P. Rajaraman et al., *Robust fault analysis in transmission lines using Synchrophasor measurements*, Springer, **3** (2018).
19. M. Kezunovic et al., *Application of time-synchronized measurements in power system transmission networks*, Springer (2014).

Wide area monitoring, protection, automation, and control systems for medium voltage networks

Sergey A. Piskunov[1,2]*, *Alexey V.* Mokeev[1,2], *Dmitry N.* Ulyanov[2], *Evgeny I.* Khromtsov[2]

[1]Northern (Arctic) Federal University, Electroenergetics and Electrotechnics Department, Arkhangelsk, Russia
[2]Engineering Centre "Energoservice", Arkhangelsk, Russia

Abstract. Distribution medium voltage networks have a branched structure, many power centers, long cable and overhead lines. This complicates the process of their automation, since significant capital costs for new equipment are required. New solutions based on modern technologies can help speed up this process and make it more efficient. The authors propose the use of synchronized phasor measurement technology for automating medium voltage networks. This paper considers approaches that describe the possibilities of implementing the WAMPAC principles in such networks, provides several examples where these principles apply.

1 Introduction

A gradual increase in the proportion of distributed generation and non-linear network elements reduces the reliability and stability of the operation of energy systems and increases the risk of abnormal and emergency situations, consumer shutdown [1]. Simultaneously, at first glance, some insignificant events often lead to major accidents, which can occur in the medium voltage distribution network.

In order to ensure the necessary reliability and stability of the network, a significant improvement in the infrastructure of protection, automation, monitoring, and control systems is required. The automation of medium voltage distribution networks includes two levels: the first (local), associated with the automation of transformer substations 6–10 kV. The second level (regional) involves automating distribution network power centers - step-down substations and distribution points.

According to the unified technical policy of PJSC Rosseti [2], one of the main tasks of the company in the electric grid complex is to increase the observability of the electric grid, the efficiency of managing grid assets, the introduction of modern systems for monitoring the technical condition, diagnostics and monitoring technological equipment. In this regard, it can be argued that the search for effective and comprehensive solutions for automating distribution networks is of considerable interest. Such solutions involve the use of multifunctional devices and modern technologies, which include the technology of synchronized phasor measurements (SPM) [3].

The advantage of using SPM technology to improve the reliability of energy systems is the ability to implement wide area monitoring, protection, automation, and control systems (WAMPACS). This paper discusses the propagation of WAMPACS ideas in relation to medium voltage distribution networks, including automation of their power centers.

2 Proposed approach

Currently, in the world energy industry, much attention is paid to WAMPACS in conjunction with the use of synchronized phasor measurements [4]. The scientific literature has already formed general approaches and conceptual solutions for the use of SPM technology to implement protection, automation, monitoring, and control devices [5]. However, questions remain regarding the application of WAMPACS principles to medium voltage networks.

Until recently, it was believed that the application of WAMPACS, including SPM technology, is a prerogative for high-voltage networks, since the cost of solutions based on these technologies was very high a few years ago. However, the development of these technologies recently has made it possible to significantly reduce their cost, and at the same time expand the scope of their application. Additionally, the development of communication technologies, such as cellular communication technology, makes it possible to consider the application of WAMPACS and SPM technology not only for theoretical projects, but also for practical implementation [6].

Currently, the most common SPM devices (PMUs) are developed in accordance with the IEEE C37.118 standard [7]. However, the development of other types of devices supporting SPM is also of practical interest. For example, in the concept of PJSC Rosseti [8], the main task of creating automated systems involves the introduction of inexpensive, modern, high-tech and multifunctional devices.

One of the most promising areas of application of WAMPACS and SPM technology for distribution networks is the implementation of the FLISR (fault

* Corresponding author: s.piskunov@ens.ru

location, isolation, and service restoration) system. According to statistics, about 80% of all consumer outages are associated with a disruption in the operation of the medium voltage network [9]. Therefore, given the low level of network automation, the implementation of the FLISR system is an urgent task that requires the use of efficient and inexpensive solutions.

To create a FLISR system, the use of SPM technology provides an opportunity to implement new principles of fault localization during short circuits and ground faults, and more advanced energy monitoring. Significant advantages are offered by the SPM technology for automation of power supply centers:

- distributed protection systems for busbars and a power transformer, built on the differential principle with absolute selectivity;
- automatic control of networks with distributed generation [10];
- monitoring of electrical equipment (power and measurement transformers, high-voltage breakers) [11];
- systems for data collection and transmission (reducing the time of analysis of emergency situations).

Since the automation of distribution networks is associated with large capital costs, this task requires the search for effective solutions and planning the stages of their implementation.

3 FLISR System

The FLISR system of 6–10 kV networks is an effective structure for determining and restoring emergency sections of the network. FLISR is designed to automate the supervisory control of 6–10 kV distribution networks in normal and post-emergency modes, as well as in modes with a single-phase ground fault.

Automation of distribution networks, as a rule, is carried out in several stages. Fig.1 shows the block diagram of such automation. For example, in Fig. 1, the following objects are shown: sections of the step-down substation SS-1, distribution points DP-1 and DP-2, transformer substations TS-11 - TS-5N. SS-1 supplies the overhead line with branches.

Regardless of the type of network, automation should start with power centers (stage 1). At the first stage, part of the transformer substations that supply important consumers can be automated.

In the second and third stages of network automation, equipment is installed at the remaining distribution points and transformer substations. When planning automation stages, it is important to make a reasonable choice of fault localization sections and controlled transformer substations. This is necessary so that all automation solutions do not contradict each other at different stages.

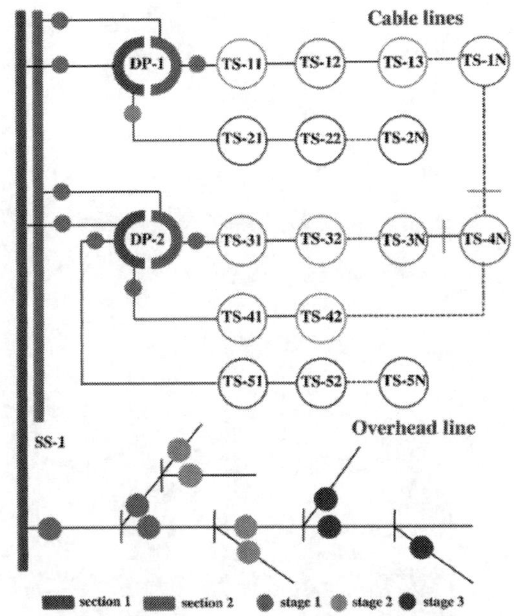

Fig. 1. Stages of network automation.

The tasks of automation include the following positions:

- installation of current measuring sensors and short-circuit current indicators at substations, distribution points and transformer substations (data collection level);
- installation of devices that provide the functions of a fault localization system, energy monitoring, remote control of circuit breakers/disconnectors, telemechanics systems (level of data transmission and processing);
- implementation of the level of dispatch control (servers with specialized software, collection, visualization and long-term storage of information, formation of remote control commands).

Automation of overhead lines has its own specifics, since overhead lines are characterized by a minimum level of automation with a significant length of lines. Fig. 1 shows one of the options for automation of overhead lines - the installation of new equipment at the branching points of the line (on supports) in the order of hierarchy from the power center of the overhead line. The equipment to be installed includes reclosers, line status indicators, and data acquisition devices.

Currently, various companies around the world are developing FLISR systems, so they have their own characteristics. Fig. 2 shows an example of the implementation of the FLISR system using the equipment of Engineering Centre Energoservice.

A feature of the presented system (Fig. 2) is the use of special fault localization devices (ENLZ), which have the support of synchronized phasor measurements. This allows for the implementation of the WAMPACS principles regarding network monitoring and control [12].

 https://doi.org/10.1051/e3sconf/202338401018

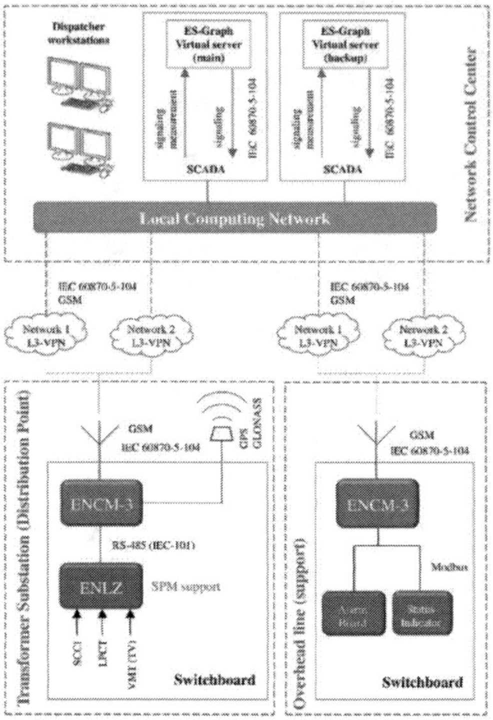

Fig. 2. Example of FLISR system.

ENLZ devices operate in fault localization systems in distribution networks. ENLZ provide synchronized phasor measurements of zero-sequence currents and voltages, processing data from short-circuit current sensors (SCCIs).

One ENLZ can simultaneously control measurements from two cable lines. The device has six optical inputs for connecting SCCIs and two analog inputs for measurements $3I_0$ and $3U_0$. Zero-sequence current transformers (LPCTs) and a voltage measuring transformer with an "open delta" winding are connected to the analog outputs using a VMT step-down transformer. The states of optical inputs and digitized analog signals are available via RS-485 in the form of telesignaling and telemetry (amplitude, phase angle).

Correct operation of the ENLZ requires accurate time synchronization, which is performed by the ENCM-3 data acquisition device with a built-in GLONASS/GPS receiver.

The specialized software tool ES-Graph ensures the operation of the FLISR system. The ground fault localization algorithm is based on the analysis of the amplitudes and angles of the zero-sequence current and voltage phasors measured in different parts of the network.

In the cable distribution network, new equipment is installed at transformer substations. ENLZ can also be installed at step-down substations and distribution points, where they measure the zero-sequence voltage.

Data is sent to the Network Control Center (NCC) via redundant dedicated cellular communication channels, which ensures relatively low equipment costs and efficient data transmission to SCADA. NCC carries out processing, transmission and storage of data on the state of the network, and generates signals for the control of switching devices.

Special overhead line status indicators (ISL) are located on supports at a certain distance from the wires. This device provides monitoring of the overhead line condition, determination and indication of the ground faults and short circuits in the network. The ISL is connected to the ENCM-3 data collection and transmission device, which is located in the switchboard on the support. All data is also transmitted to SCADA.

Thus, the considered FLISR system uses synchronized phasor measurements to perform a number of network monitoring and control tasks. In the future, the development of such systems will also make it possible to implement the functions of relay protection and automation in medium voltage networks.

4 Monitoring and protection of power transformers

The implementation of protection and monitoring systems for power transformers at step-down substations in medium voltage networks has certain features. First, the power rating of most transformers does not exceed 6.3 MVA, therefore, as a rule, the requirements for protection and monitoring of such transformers are lower than in high voltage networks. Second, a significant proportion of power transformers have reached their specified service life, so their residual value often becomes less than the necessary costs for substation automation. However, the long service life of transformers increases the probability of their failure. Therefore, planning the execution of automation tasks requires the search for rational, efficient, and integrated solutions.

Fig. 3 shows the options according to which the automation of step-down substations in medium voltage networks can be carried out, taking into account the application of the WAMPAC principles.

For small substations with the lowest level of automation, the most rational solution is to implement a simple monitoring system based on relay protection devices and a data collection and transmission system without the use of SPM. Such a system will provide control of the basic set of transformer parameters (overload capacity, short-circuit currents, overvoltages).

If, according to technical and economic indicators, the installation of SPM devices is justified, the transition to option 2 allows the implementation of a fully functional transformer monitoring system with control of a much larger number of parameters, including the equivalent parameters of the transformer circuit [11].

The most rational option is the use of multifunctional intelligent electronic devices with the SPM support, because they allow implementing several substation automation subsystems (telemechanics, energy metering, and control of electricity quality indicators).

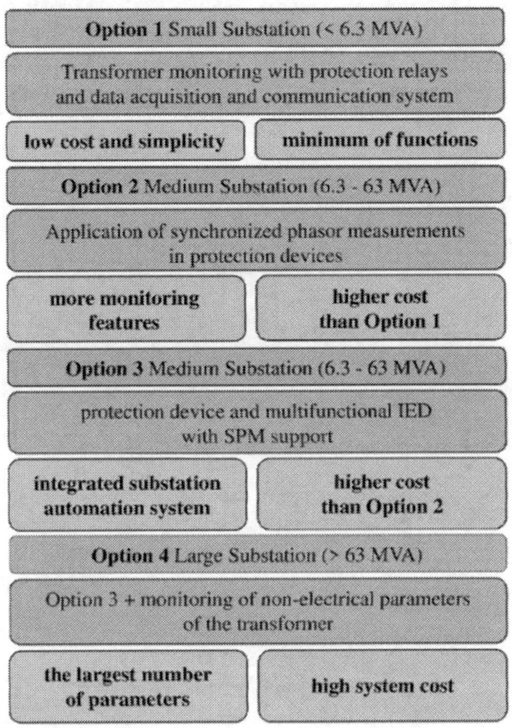

Fig. 3. Options for protection and monitoring systems for power transformers.

For larger substations, it becomes possible to implement monitoring systems with control of electrical and non-electrical parameters [13]. This approach requires significantly more costs, but at the same time, we get more effective control and performance of a greater number of tasks for monitoring and diagnosing the transformer.

5 SPM application for transformer monitoring

Most often, electromagnetic processes lead to the occurrence of defects and accidents in transformers in a large way. Therefore, it is fundamentally important that monitoring systems use a predictive way of identifying the likelihood of dangerous damage, and not identifying the consequences of damage that has already occurred.

A promising solution to this problem is the use of protection devices with transformer monitoring functions based on synchronized phasor measurements. In addition, SPM allows improving the protection functions of the transformer [11].

The paper [14] considers the issues of identification of the T-shaped equivalent circuit parameters of a power transformer based on the measurement of current and voltage synchrophasors. According to the dynamics of changes in these parameters, the monitoring system determines the defects and malfunctions of the transformer. Nevertheless, in addition to the stationary modes of transformer operation, when analyzing its state, it is important to control the parameters of the transformer equivalent circuit during transients. However, paper [14] concerns only the stationary modes of operation of the transformer, that is, this is the performance of monitoring functions. Relay protection of a transformer requires the ability to control its parameters in transient modes, so the analysis of such modes is of practical importance for developing protection devices.

Based on the current and voltage synchrophasors of the transformer, a system of differential equations can be implemented that describes its operation in non-stationary modes, similarly to a system of equations based on instantaneous values of current and voltage. There can be many such systems of equations, depending on the chosen transformer equivalent circuit and calculation conditions. As an example, we consider an L-shaped equivalent circuit of a power transformer with non-linear parameters of the magnetization branch (Fig. 4).

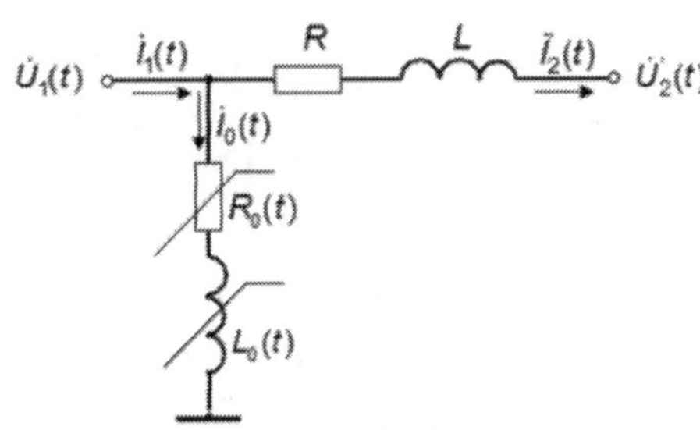

Fig. 4. L-shaped equivalent circuit.

The L-shaped equivalent circuit of a power transformer has two branches: a branch with winding impedances and a magnetization branch. A significant change in the parameters of the magnetization branch occurs during the inrush of the magnetizing current (MIC) in the transformer when it is turned on to idle or to the load, and also after the supply voltage is restored after the external short circuit is turned off.

The differential equation of the magnetization branch for the analog prototype system, which describes the mode of the magnetization inrush current, has the following form:

$$u_1(t) = i_0(t)R_0(t) + L_0(t)\frac{di_0(t)}{dt}. \quad (1)$$

Expression (1) is a non-stationary first-order differential equation because it contains two variable coefficients. Analysis of such an equation in the instantaneous values of current and voltage is not always efficient and simple. Therefore, the use of data

 https://doi.org/10.1051/e3sconf/202338401018

on current and voltage synchrophasors on the high and low sides of the transformer makes it possible to change the approach to estimating its parameters in the transient mode.

The transition from instantaneous values of current and voltage to process synchrophasors is carried out as follows:

$$u_1(t) \to \dot{U}_1(t)e^{j\omega_0 t}, i_0(t) \to \dot{I}_0(t)e^{j\omega_0 t}. \quad (2)$$

After this replacement, we transform expression (1) to the following form:

$$\dot{U}_1(t) = \underline{z}_0(t)\dot{I}_0(t) + L_0(t)\frac{d\dot{I}_0(t)}{dt}, \quad (3)$$

where $\underline{z}_0(t) = R_0(t) + j\omega_0 L_0(t)$, $\omega_0 = 2\pi 50$ рад/с.

Current synchrophasor in the magnetization branch:

$$\dot{I}_0(t) = \dot{I}_1(t) - \dot{I}'_2(t), \quad (4)$$

where $\dot{I}'_2(t)$ - current synchrophasor of the low side referred to the high side.

We introduce the following notation:

$$\underline{z}_1(t) = \frac{\dot{U}_1(t)}{\dot{I}_0(t)} = R_1(t) + jX_1(t). \quad (5)$$

Paper [15] considers the analysis of synchrophasors of various transients in the power system. We use one of the transformations [15] to estimate the inductance of the magnetization branch:

$$L_0(t) = X_1(t)\left[\omega_0 + \mathrm{Im}\left(\frac{\dot{I}'_0(t)}{\dot{I}_0(t)}\right)\right]^{-1}, \quad (6)$$

$$R_0(t) = R_1(t) - L_0(t)\mathrm{Re}\left(\frac{\dot{I}'_0(t)}{\dot{I}_0(t)}\right), \quad (7)$$

where $\dot{I}'_0(t) = \frac{d\dot{I}_0(t)}{dt}$.

The following replacement in expressions (6) and (7) allows determining the parameters of the transformer windings R and L:

$$\dot{U}_1(t) \to \dot{U}_1(t) - \dot{U}'_2(t), \dot{I}_0(t) \to \dot{I}'_2(t). \quad (8)$$

Thus, the obtained expressions (6) and (7) make it possible to determine the parameters of the transformer equivalent circuit under transient conditions. For example, in the magnetizing current inrush mode, we can determine the change in the inductance of the magnetizing branch. This change is one of the signs of identification of such a mode.

The analysis of process synchrophasors also allows the development of other new algorithms for transformer protection and monitoring [15]. For this purpose, an effective solution is the simulation of electrical circuits, for example, in MATLAB/Simulink. In particular, Simulink has a Nonlinear Transformer model that can be used for calculations based on numerical methods.

Fig. 5 shows the equivalent circuit of a no-load transformer based on the Simulink model. The peculiarity of this model is that the resistance and inductance of the magnetization branch are connected in parallel. This allows analysis of the magnetization inductance separately from other parameters.

Fig. 6 shows the Simulink transformer model developed by the authors as part of this study. Simulation of the idle mode of transformer provides data on the non-linear change in magnetizing inductance (Fig. 7). Such changes are characteristic only for this mode; therefore, the calculation of the magnetization inductance is of interest for improving the operation of the differential protection of the transformer.

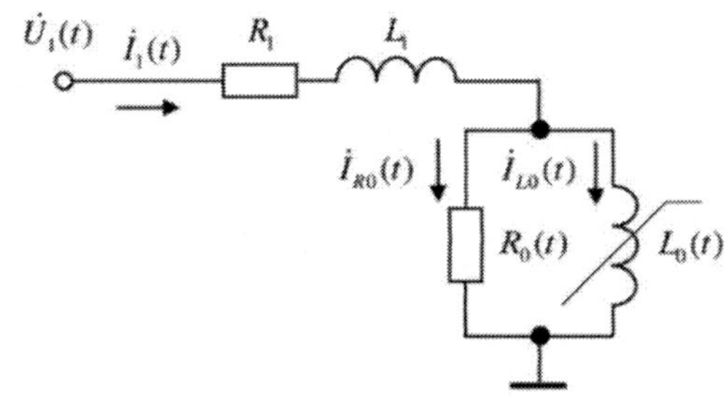

Fig. 5. Equivalent circuit of transformer model.

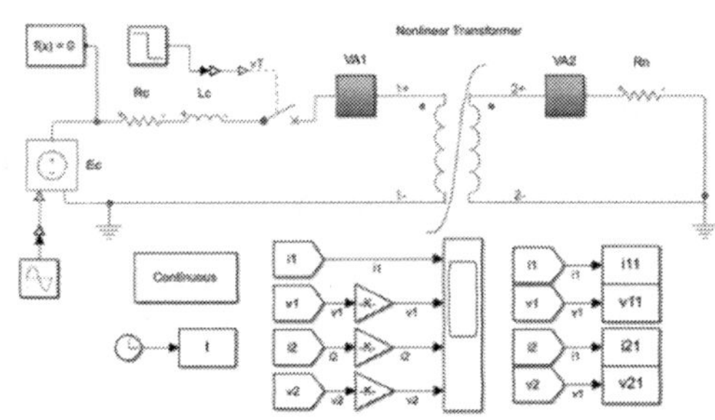

Fig. 6. Simulink model.

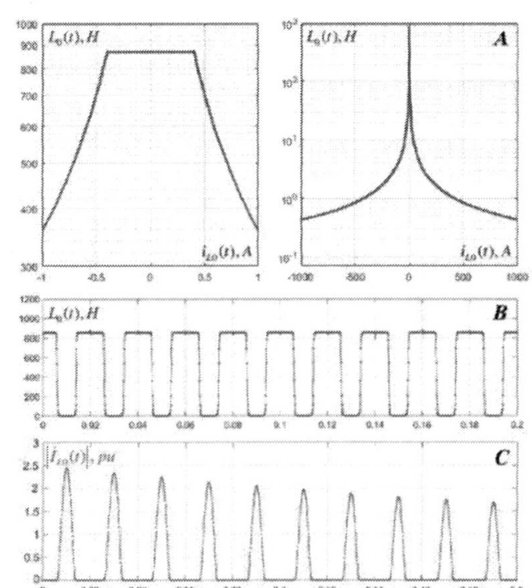

Fig. 7. Inductance (a, b) and magnetizing current (c).

The values of the magnetization inductance (Fig. 7, b) are obtained by solving a system of equations based on process synchrophasors. Significant inductance changes occur every period of fundamental frequency. Thus, the synchrophasors of the voltage and current of the transformer make it possible to determine the mode of the magnetization inrush current.

Currently, the authors are developing new algorithms for protecting and monitoring a power transformer. Testing of the proposed algorithms is already being carried out as part of a special power transformer monitoring system at substations 7 and 8 in Arkhangelsk, Russia [14].

6 Discussion

Expressions (6) and (7) contain the derivative of the current synchrophasor, the correct calculation of which in digital systems depends on the signal sampling frequency. In the general case, there are systems with a low sampling rate, which can lead to a decrease in the accuracy of calculating the current derivative and the efficiency of the proposed algorithms.

In some cases, the transmission rate of synchrophasors from PMU with a frequency of 50 Hz or less may be unacceptable. On the one hand, this problem can be solved by increasing the synchrophasor transmission rate, but this is not always acceptable. The second option is to calculate the derivative of the current synchrophasor in the PMU itself. The authors suggest that the second option may be more promising for medium voltage networks.

7 Conclusion

The development of modern technologies makes it possible to consider their application in new energy areas and directions. The technology of synchronized phasor measurements in combination with the implementation of WAMPAC principles is a promising direction for automation of medium voltage distribution networks.

The paper considers the general principles of organizing the process of automation of distribution networks based on the SPM technology, taking into account the specifics of their work.

Fault localization systems based on SPM are able to perform the functions of telemechanics, energy metering, monitoring system for the condition of electrical equipment. The control of the transformer electromagnetic parameters based on the measurement of current and voltage synchrophasors makes it possible to implement the functions of the transformer monitoring system, which ensures the detection of faults and abnormal modes of its operation.

References

1. P.V. Ilyushin, *Features of the emergence and course of emergency modes in distribution networks with distributed generation*, Vestnik KGEU, **3** (2021).
2. Unified technical policy in the electric grid complex, PJSC Rosseti, No. 450 (2022).
3. A.G. Phadke, J.S. Thorp, *Synchronized phasor measurements and their applications* (2021).
4. E. Price, Practical considerations for implementing wide area monitoring, protection and control, 59th Ann. Conf. for Protective Relay Engineers. (2006).
5. I. Ivankovic, I. Kuzle, N. Holjevac, *Multifunctional WAMPAC system concept for out-of-step protection based on synchrophasor measurements*, IJEPES, **87** (2017).
6. D.N. Ulyanov, K.V. Petrov, A.V. Mokeev *Improving the reliability of distribution networks based on the system of automatic restoration of power supply*, ISUP, **6** (2020).
7. IEEE C37.118.1-2011 Standard for synchrophasor measurements for power systems.
8. Concept for the development of relay protection, automation and automated process control systems of the electric grid complex of the group of companies PJSC Rosseti, No. 286 (2022).
9. M.R. Elkadeem, M.A. Alaam, A.M. Azmy, *Improving performance of underground MV distribution networks using distribution automation system: A case study*, ASEJ, **9** (2018).
10. P.V. Ilyushin, *Implementation features of automatic control of the power district's modes with distribution generation facilities*, RPA, **3** (2019).
11. S.A. Piskunov, A.V. Mokeev, Power transformer relay protection with its condition monitoring function, REEPE, Moscow (2021).
12. S.A. Piskunov, Mokeev A.V., E.I. Khromtsov, Application of synchronized phasor measurements in RPA devices of distribution networks, E3S WoC SUSE-2021 (2021).
13. Power transformer performance monitoring presented in SCADA (https://new.abb.com/news).
14. S.A. Piskunov, A.V. Mokeev, D.N. Ulyanov, Control, monitoring and protection systems based on synchronized phasor measurements, RSES, Volzhskij (2021).
15. A.V. Mokeev, *Analysis of synchrophasors of transient processes in the power system*, Electrical equipment, **1** (2022).

Investigation the influence of the parameters of the computational scheme on detecting the source of low-frequency oscillations

Kirill Butin[1,2*], *Aleksandr* Popov[1,2], and *Andrey* Rodionov[1]

[1]Engineering Center «Energoservice», Russia
[2]Northern (Arctic) Federal University

Abstract. The dispatching control requires the ability to detect the source of low-frequency oscillations in the power system based on the online processing of synchrophasor measurements data. This is a complex task; therefore, new approaches to its solution and variations of existing ones are constantly emerging. As a rule, a short expression of the criterion of a particular method hides a number of stages of computational data processing. The number of possible implementations of these stages increases during the investigation. The set, settings, implementation of the stages of the computational scheme, and combination of these stages affect the final result. Earlier, the authors developed a high-performance software platform that allows describing computational schemes in a general way, and then performing calculations with specified sets of algorithms. In this paper, a generalized calculation scheme for the dissipating energy method is constructed, and experiments are performed showing the influence of details of the implementation of data processing stages on the result of the online detection of the source of low-frequency oscillations.

Introduction

The monitoring of low-frequency oscillations (LFO) is one of the most relevant areas of data processing of synchrophasor measurements data and is included in the decision-making in most dispatch centers of system operators [1]. An effective solution for the problem of LFO source online localization is particularly in demand. We understand it as identifying as close as possible such a part of the power system that makes the greatest positive contribution to the development of the oscillatory process.

Paper [2] analyzes a number of LFO cases that have occurred in the Russian power system in recent years.

The interest in the problem of LFO has grown today, including in connection with the development of renewable energy sources, the introduction of which leads to the destabilization of the traditional power system, the emergence of new types of oscillations, and the complexity of control [3, 4].

For online processing, it is required to minimize the time from the beginning of oscillations mode to making a stable correct decision about the localization of their source.

The search for the LFO source can include an assessment of the role of generators in the development of the oscillatory process, comparison of the amplitude-phase characteristics of the components of measuring signals, analysis of wave propagation in the power system, comparison of measurement data with simulation results, data accumulation and machine learning, etc. [5, 3] However, none of the existing methods provides a final practical solution for real power systems [6].

Various ways of combining alternative methods are proposed [8, 9], ensembles of methods are being built [10], computing platforms are being created for comparative experiments [10, 1], test data sets are being accumulated [7], typical simulation models of power systems are being formed [12] for testing analysis methods, etc.

1 Calculation templates

In data analysis problems, the calculations often can be represented by generalized schemes or templates. The nodes of templates correspond to subtasks without indicating a specific solution of them. In the LFO analysis, such general tasks are: data pre-processing, mode detection, extraction of informative signal components, calculation of the dynamic parameters of oscillations, and the actual search for the source. On the other hand, as research develops, the variety of processing methods at each stage is constantly increasing. At the same time, the methods have different sets of parameters and software implementations. Therefore, the implementation of the whole problem is combinatorial.

Previously, the authors developed [13] and implemented in the form of a software platform [1] an approach to describe calculations in the form of generalized schemes with the possibility of varying algorithms at the user level. A computational scheme (plan) is a set of interrelated stages of calculations

* Corresponding author: k.butin@narfu.ru

(works). The plan defines the structure of the task, the work corresponds to the subtasks. To launch the plan, each work must be associated with a performer, the implementation of some data processing algorithm.

2 Dissipating energy flow method

The detection of an LFO source based on the estimation of oscillatory energy demonstrated high efficiency [7, 2]. In [14], an integral expression was derived for the energy of oscillations. In this case, the measuring signal is considered as the sum of the trend and the oscillations. In the context of a different signal model (summed modes), in [6] this expression was transformed and the so-called dissipative energy flow (DEF) method was obtained. In the DEF method, for each object, the value of W_{ij}^D is calculated in signal sections of a length of 5 to 20 LFO periods [6]:

$$W_{ij}^D \approx \int 2\pi \Delta P_{ij} \Delta f_i dt + \Delta Q_{ij} \frac{d(\Delta V_i)}{\tilde{V}_i + \Delta V_i} \quad (1)$$

The symbol Δ here denotes the operation of extracting the oscillatory component from the measuring signal. P_{ij} – active power on line $i - j$, Q_{ij} – reactive power, f_i – frequency, V_i – voltage, $\tilde{V}_i$ – average voltage value at point i. The direction of the energy dissipation flow is determined by the slope of the linear regression:

$$W_{ij}^D(t) = DE_{ij} * t + b_{ij} \quad (2)$$

Positive values of DE_{ij} indicate the flow of oscillation energy from the object to the system, and negative values indicate the opposite direction of the flow. Thus, the largest value of DE_{ij} corresponds to the LFO source. After the implementation of DEF in the ISO New England power system [11], a number of refinements (for example, node weights) were required to correct the solution. In addition, for archival records, it is allowed to manually search for the LFO source including the analysis of the power system topology. The criticism of DEF with examples of its incorrect operation is presented in [15]. Variations of the DEF method can be found, for example, in [16, 17, 18]. In other words, the potential for the development of the method has not yet been exhausted.

3 DEF template

Fig. 1 shows one of the possible calculation schemes for the DEF method. First, a set of quantities is prepared that are necessary for (1). Then the signals are packed into a special structure, the trend is removed, the modes are distinguished, the oscillation energy is calculated, and the rate of its change is calculated.

The * symbol means that the work is applied to all elements of the input data. For example, if filtering is used to extract a mode, then the same filter is applied to all values. The line thickness reflects the number of data elements in the stream.

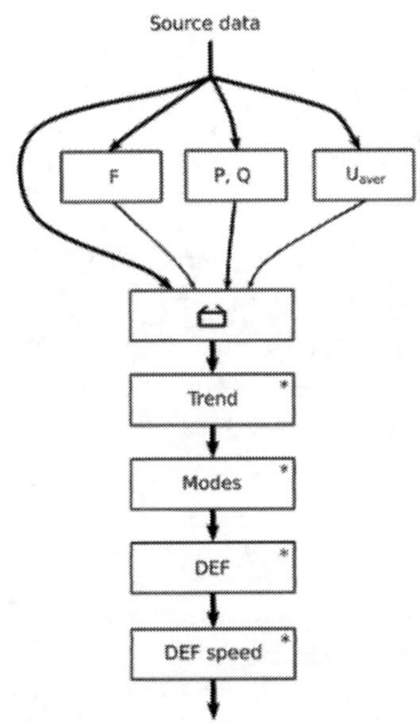

Fig. 1. DEF template.

Table 1 lists the possible performers for the selected jobs and the variable parameters.

Table 1. Calculation steps.

	Work	Workers	Parameters
1	f calculation	From data	
		Left shift	
2	P and Q calculation	From data	
		Function	
3	$\tilde{V}$ calculation		Number of averaged values
4	Removing a trend (detrending)	No	
		Averaging filter	Window function
			Win length
		Savitzky-Golay filter	Polynomial degree
			Length
5	Mode extraction	No	
		Finite impulse response (FIR) filter	Bandwidth
			Order
			Window function
		Infinite impulse response (IIR) filter	Order
			Bandwidth
6	DEF calculation	With Q	
		Without Q [17]	
7	DEF slope calculation	Linear Regression	Window length

It is supposed to be possible to add any other performers that are compatible in terms of the interaction interface.

4 Objective function

Let there be a data set of syncrophasors obtained during the LFO mode. Also, let the LFO source be known, and the data contain signals from the source.

Taking into account the application of a certain method to determine the source of an LFO, we introduce a function $s(t)$ such that $s(t) = 1$ in those cases where the source is correctly specified at time t, and $s(t) = 0$ in all other cases. Consider also the function $r(t)$ – the time until the moment t, during which $s(t) = 1$ and $s'(t) = 0$, that is, the decision was correct and did not change.

In discrete form:

$$r(n) = \begin{cases} 0, & n = 0 \\ s(n)\big(r(n-1) + s(n)\big), & n > 0 \end{cases}$$

We emphasize that the output of the function $r(t)$ is time. We will compare $r(t)$ with some setting τ_r: if $r(t) > \tau_r$, then we will say that at time t there is a stable correct solution.

The strict requirement 'the solution did not change' can be replaced by a softer 'solution almost did not change' and use, for example, the ratio of the correct answers number to the total one at a level close to 1.

The described is schematically shown in Fig. 2. For convenience of perception, r(t) is shown as a solid line.

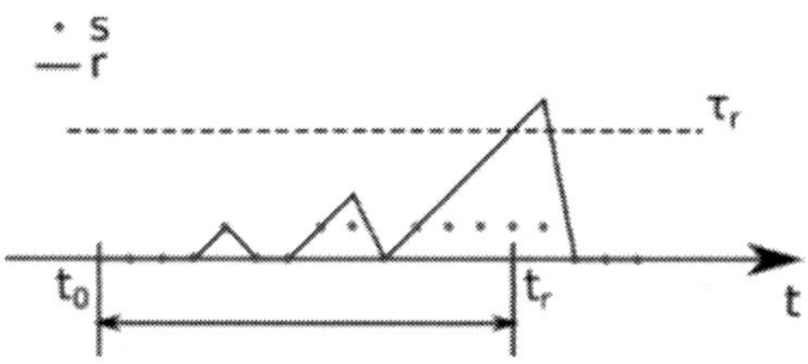

Fig. 2. Determining the time of a stable correct solution.

5 Real LFO case processing

Consider a case that occurred in January 2022 in the western part of Russia. Fig. 3 shows the time interval from the beginning of the LFO. The lines show raw frequency data for the 21 input signals. An LFO is observed in the region of 0.85 Hz.

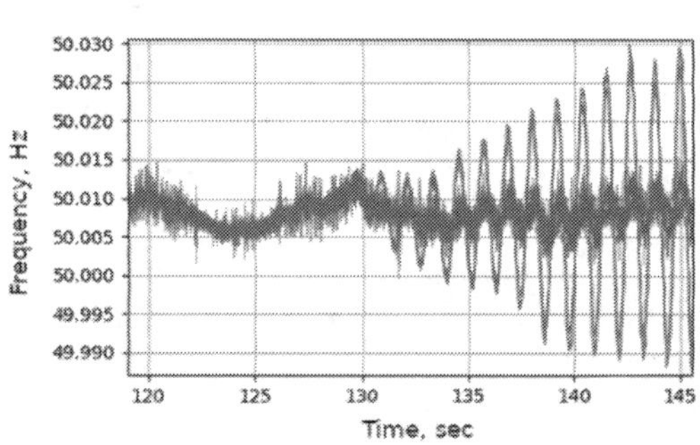

Fig. 3. LFO at frequencies.

The duration of the oscillation is about 14 minutes. Visually, the fluctuations begin at about 129 sec. Let us take the following parameters: $t_0 = 120$ sec, and τ_r=30 sec. Thus, starting from a time point of 120 seconds, the DEF method should consistently produce the same source for 30 seconds. This duration corresponds to about 25 waves of oscillations, which are enough with a margin to speak about the finding of the source [6]. The time when the DEF method began to give the correct answer (it is known to us in advance) will be considered as the moment the source was discovered.

In accordance with Table 1, we define the sets of possible performers and their parameters. The length of filters and windows varies from 5 to 1001 samples. For the IIR filter, we use the Butterworth filter of orders 2 and 3. For the Savitzky-Golay filter, the degrees of the polynomial are 5 and 6. Bandpass filter ranges: from 0.1 to 4.0 Hz, from 0.81 to 0.91 Hz, from 0.76 to 0.96 Hz, from 0.66 to 1.06 Hz, and from 0.56 to 1.16 Hz. Since the number of different combinations exceeds 10^{12}, a complete enumeration of all options is not possible, so the Monte Carlo method is used. At each step, one of the possible configurations is generated with equal probability and the DEF method is run for it. After each launch, the configuration and time of source detection are fixed.

6 Results

The calculations were carried out on the nodes of the computing cluster of the Northern (Arctic) Federal University [19]. Each node has two ten-core Intel Xeon E5-2680 processors with 64 GB of RAM.

All signals were processed in online streaming mode without creating artificial delays. Here are the main characteristics of the calculations: 127706 launches and 29.8 hours of computer time. The average processing time for one variant is 840 ms.

Fig. 4 shows a histogram for the number of correct detections (total 84002) of the LFO source depending on the time. The peak of around 140 seconds corresponds to the most frequent detection of the source. This is primarily due to the total delay of the methods. Each method can introduce a delay. Since the lengths of the filters were selected equally likely, the average total delay of the filters was about 12 seconds, which gives a peak 12 seconds after the start of the LFO.

The worst results in terms of time are given by methods either with the maximum total delay or with a too short search window for the linear regression coefficient. This is due to the fact that the DEF integral fluctuates relative to the trend and the regression coefficient constantly changes sign.

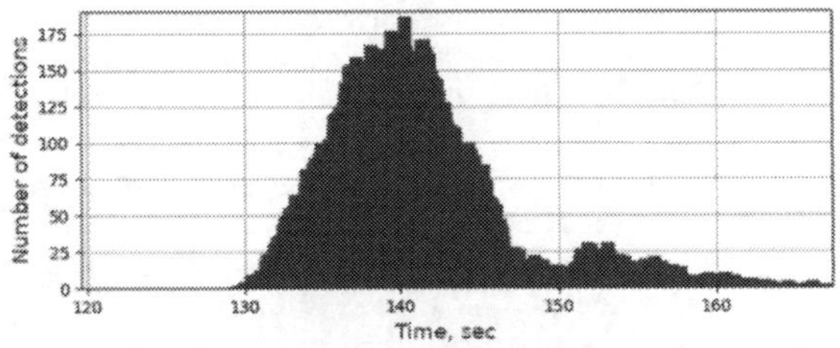

Fig. 4. Source detection frequency diagram.

The best combination of methods identified the source at 128.82 sec. Fig. 5 shows the stages of calculations.

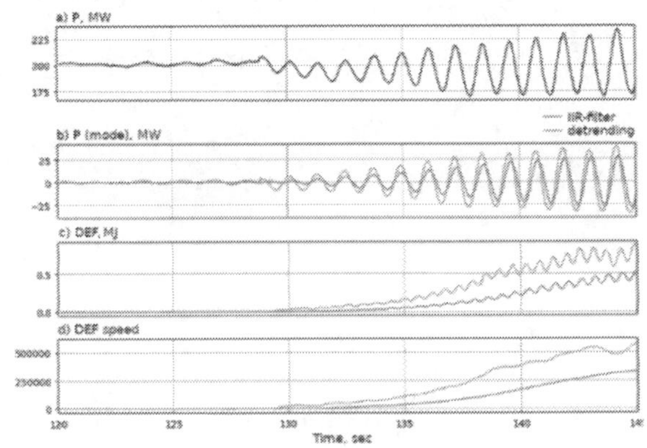

Fig. 5. Step-by-step calculation.

Fig. 5.a shows the original active power over time. Detrending was applied without further filtering using an averaging filter with a length of 35 points (Fig. 5.b). For comparison, the blue line shows the result of applying the second-order IIR filter in the band from 0.76 to 0.96 Hz. It is slightly behind and has a smaller amplitude. Fig. 5.c shows the calculation of DEF in a sliding window of 71 points. The calculation of DEF is executed without the contribution of reactive power. Fig. 5.d shows the linear regression coefficient in a sliding window. The detrending method without filtering in this case produced a positive linear regression coefficient earlier. Visually, this corresponds to the time of the beginning of the LFO.

The detrending subtask is one of the most important. Depending on the removal method, in some cases, opposite in sign values of the regression coefficient are obtained. You can remove the trend using bandpass filtering. This task is handled by an IIR filter. Its disadvantage is the accumulation of calculation errors.

Conclusions

Dispatch centers need timely LFO source detection using synchrophasor measurements. There are a variety of methods for identifying the LFO source. The DEF method is one of the promising ones. Software implementation of DEF involves some steps of calculations with variable performers and parameters. The paper presents a computational experiment on a high-performance cluster in order to select the optimal configuration that reliably determines the LFO source in a short time. The sets of possible step performers and their parameters are determined. The dependences of the final result on the calculation parameters are revealed. The possibility of using the dissipating energy flow method online has been tested. An example with a combination of calculations is given, in which the method detects the LFO source at the beginning of the oscillations. This allows us to know where the LFO source is at the time the oscillatory mode is detected. In the future, we plan to expand the test suite and use machine learning to fit the calculation parameters.

References

1. A. Popov et al. *Framework for synchrophasor measurements data processing and the case studies of the low-frequency oscillations* (https://www.naspi.org/sites/default/files/2021-04/D3S7_01_butin_energoservice_naspi_20210415.pdf)
2. A. Popov et al. *Examples of processing low-frequency oscillations in Russia and ways to improve the analysis*, in Proceedings of the 2022 International Conference on Smart Grid Synchronized Measurements and Analytics (2022)
3. L.G. Meegahapola et al. *Review on Oscillatory Stability in Power Grids with Renewable Energy Sources: Monitoring, Analysis, and Control using Synchrophasor Technology* in IEEE Transactions on Industrial Electronics, **68** (2020)
4. J. Ma et al. *Equipment-level locating of Low Frequency Oscillating Source in Power System with DFIG Integration Based on Dynamic Energy Flow* in IEEE Transactions on Power Systems (2020)
5. B. Wang, K. Sun. *Location methods of oscillation sources in power systems: a survey* in Journal of modern power systems and clean energy **5** (2017)
6. S. Maslennikov et al. *Dissipating energy flow method for locating the source of sustained oscillations* in International Journal of Electrical Power & Energy Systems, **88** (2017)
7. B. Wang, S. Maslennikov. *IEEE-NASPI Oscillation Source Location Contest-Case Development and Results* in Tech. rep. National Renewable Energy Lab (2021)
8. G. Ding et al. *Analysis of ultra-low frequency oscillation in southwest power grid based on measurement data* in Proceedings of 37th Chinese Control Conference, CCC (2018)
9. A. Popov et al. *Combined method for identifying a source of low-frequency oscillations in power system* in AIP Conference Proceedings **2552, 060004** (2023)

10. U.I.H. Banna, Solanki S.Kh., Solanki J. *Data-driven disturbance source identification for power system oscillations using credibility search ensemble learning* in IET Smart Grid 2.2, (2019)
11. S. Maslennikov, E. Litvinov. *ISO New England Experience in Locating the Source of Oscillations Online* in IEEE Transactions on Power Systems (2020)
12. C. Canizares et al. *Benchmark models for the analysis and control of small-signal oscillatory dynamics in power systems* in IEEE Transactions on Power Systems 32.1 (2016).
13. A. Rodionov et al. *Practical Issues of Processing Synchrophasor Measurement Data in the Analysis of Low-Frequency Oscillations in Power System* in Proceedings of 2nd International Youth Scientific and Technical Conference on Relay Protection and Automation, RPA (2019)
14. L. Chen, Y. Min, W. Hu. *An energy-based method for location of power system oscillation source* in IEEE Transactions on Power Systems, **28** (2013)
15. Y. Zhi, V. Venkatasubramanian. *Analysis of Energy Flow Method for Oscillation Source Location* in IEEE Transactions on Power Systems, (2020)
16. W. Hu et al. *Disturbance source location of forced power oscillation in regional power grid* in Proceedings of the 2011 IEEE Power Engineering and Automation Conference, **2** (2011)
17. S. Guo et al. *An Alternative Way to Locate Disturbance Source for Low Frequency Oscillation Considering Control Devices of Generator* in MS&E 740.1 (2020)
18. P.G. Estevez et al. *Complex Dissipating Energy Flow Method for Forced Oscillation Source Location* in IEEE Transactions on Power Systems (2022)
19. HPC NArFU (http://fujitsu-19-02.narfu.ru)

Increasing the reliability of the united power systems of central Asia in conditions of energy transition

Shamsiev[1] *Kh.A.*, Shamsiev[1] *B.Kh.* and Khamidov[2]* *Sh.V.*

[1] Coordinating dispatch center "Energy". Uzbekistan

[2] Institute of Energy Problems of the Academy of Sciences of the Republic of Uzbekistan

Abstract. The issues of increasing the reliability of the United Power System of Central Asia in the current conditions of operation and energy transition are analyzed. The issues of integrating renewable energy sources into the energy system are addressed, taking into account the forecasted volumes of their input until 2030.

Main objectives of UPS CA

Since 1960, the United Power System of Central Asia (UPS CA) and South Kazakhstan has been the basis of the electric power industry of the four Central Asian republics and adjacent regions of South Kazakhstan. The effects of interconnecting energy systems are known and are depended on the degree of their integration.

In the conditions created after 1991, the strategically justified desire of independent states for self-balance led to the loss of advantages from the unification, while until recently only the mutual exchange of electricity under intergovernmental agreements was preserved. Under the new conditions, the principles of joint operation of energy systems in the CA's UPS, the functions of coordinating bodies and the Unified Dispatch Control (UDC), have changed.

It was not possible to preserve all the advantages of parallel operation of power systems that existed under tight dispatch control, since each power system has its own criteria. The desire of each energy system to obtain maximum profit, without taking into account the interests of its partners, leads to the deterioration of its own criterion [1].

One of the main tasks of the UPS remains the reservation of energy systems in normal and emergency regimes and the implementation of contractual interstate exchanges of electricity under Intergovernmental Agreements (IGA) and the rational consumption of energy resources.

The energy systems of the states must meet the needs for base, peak and reserve power, reactive power sources, and the development of projects for the development of the electric power industry for the UPS as a whole. In this regard, alternative options for commissioning generating capacities, building new power plants with the efficient use of various energy resources should be considered, taking into account the promotion of the inclusion of Renewable energy sources (RES) and energy-saving technologies.

Integration of renewable energy sources in UPS CA

According to the Energy Industry Development Concept [3], in Uzbekistan, by 2024, it is planned to launch RES with a total capacity of 4000 MW (including SPP–2400 MW and WPP–1600 MW), and by 2030 – to increase the capacity of RES to 9000 MW.

To understand these scales, we note that the installed capacity of the Uzbekistan energy system is just over 16 GW. If the influence of RES variability can be solved by maintaining hot reserves at power units and using storage devices, then the problem of discontinuity inherent in solar stations has its own characteristics for Uzbekistan.

It is not possible to solve this problem with mobile gas stations such as gas turbines or gas-piston ones, even in such a gas-rich country as Uzbekistan. The problem with replacing the solar power that disappears in the evening is not due to a lack of generating capacities in the energy system, but with restrictions on the rate of power gain at gas stations due to existing restrictions in the country's gas transmission system (GTS) [4].

Problems with the speed of lifting the load at the stations can be solved as follows:

a) reorganize the operation of gas transmission facilities so that they operate on a flexible daily schedule of power plants, and not on an almost flat schedule as it is today – an undertaking associated

*Corresponding author: khamidov_sh@mail.ru

with high financial and time costs;

b) switch to the mode of daily mazut burning during the hours of load rise in the evening maximum, when there are restrictions on load increase due to lack of gas. To do this, it will be necessary to keep in working condition a part of the existing traditional power units that can operate on both gas and mazut. This is a very costly operation: mazut is currently burned under extreme circumstances in the winter and a transition to year-round daily fuel oil combustion is unlikely;

c) installing energy storage devices at SPP and/or in the energy system. It is also an expensive operation, which countries that massively introduce renewable energy have to take.

If the problem noted above is not addressed, a situation with large uncompensated surges of power will be created almost daily on the transit North-South of Kazakhstan loaded to the limit and, accordingly, on the interface of the UPS of Kazakhstan - UPS of Russia.

Now the allowable surge of power towards Russia is 500 MW, and from Russia towards the UPS CA – 300 MW. Taking into account the latter factor and preventing the violation of stability in the energy system of Uzbekistan, automatic shutdown of power units with a capacity of over 300 MW was introduced. Thus, when an 800 MW unit is switched off at the Talimarjan TPP, emergency automatics are activated at SALS (Special Automatic Load Shedding) in the amount of 800–300 = 500 MW, while the Control of the previous mode (CPM) is provided, which determines the required amount of SALS depending on the load, carried by the emergency shutdown unit.

Note that the problem is not only in the shortage of power reserves. Today, the UPS CA does not have enough power reserves. Thus, in the Kyrgyz energy system there are reserves of about 150–200 MW, and after the connection of the Tajik energy system, reserves of the order of 700–1000 MW will appear.

An increase in mobile reserves is possible through the joint construction of new HPPs by the countries of the region in water-rich Tajikistan and Kyrgyzstan in order to create peak capacities to compensate for the impact of renewable energy.

Taking into account the decrease in the region's water resources, attention should be focused on Pumped storage stations (PSSs), using for this purpose the existing reservoirs of daily and weekly regulation [7, 8]. These activities will help reduce imbalances from renewable energy, but this is a very long-term process.

The existing surges towards the UPS of Russia are due to the fact that with the existing centralized frequency regulation system, the regulators in the national energy systems of the UPS CA do not have time to respond to disturbances earlier than the stations of primary and secondary regulation in Russia. Large unbalances, leading to a breakdown in stability, currently occur infrequently, mainly during emergency shutdowns, in which, for one reason or another, the compensating effects of emergency automatics were not enough.

With the introduction of RES, as will be shown below, such imbalances in the UPS CA will occur every day, which casts doubt on the possibility of fully ensuring the sustainability of the parallel operation of energy systems.

On the sustainability of UPS CA with the introduction of renewable energy.

Given the high pace of RES integration in UPS CA and the duration or high cost of measures to overcome the associated problems, it should be expected that surges in power imbalances from RES will occur every day and the probability of accidents in UPS CA will, accordingly, increase, which was confirmed by systemic accidents in power systems UPS CA in 2020-2022 years.

Besides, the transit capacity of the North-South UPS of Kazakhstan is used up to the maximum load and, as a result, it will be necessary to disconnect consumers in the UPS CA from emergency automatics on a daily basis in order to maintain the stability of parallel operation.

Let's show this problem on the following example of the launching of a solar power plant with a total capacity of 3000 MW. In the current regimes in the Uzbek energy system, to cover the load during the hours of the evening maximum, it is necessary to increase the generation at thermal stations by about 1500 MW, which the existing GTS can hardly accomplishes, especially in winter (Fig. 1).

With the launch of solar power plants in Uzbekistan, with which, agreements between parties will be concluded on the basis of the "Take or pay" principle, the energy system will be forced to additionally unload up to 3,000 MW at TPPs during the day and, as a result, raise 4,500 MW in the evening, considering the replacement of the disappearing capacity at SPP.

Taking into account the regimes of "Uztransgaz" JSC do not allow changing gas consumption at thermal power plants by more than 250-300 thousand m3 per hour, it is clear that problems in the GTS will occur both during daytime unloading of thermal power plants and during evening load sets at thermal power plants, which will lead to power imbalances.

The danger of large uncompensated power imbalances was shown by the events in the UPS CA on January 25, 2022 with the full breakdown of the power interconnection [2].

The Commission of UPS CA, under the leadership of the Coordinating Dispatch Center "Energy", conducted an investigation with the involvement of independent experts, including SO UPS, identified the causes of the incident and developed a number of measures to prevent similar accidents in the future.

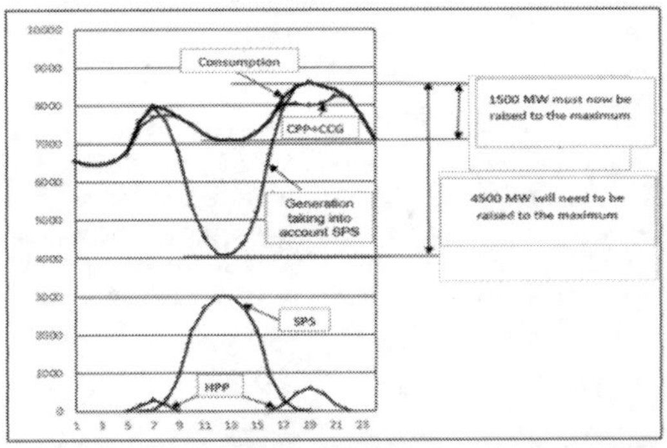

Fig. 1. Change in the daily generation of TPPs and HPPs of the Uzbek energy system before ad after the commissioning of a 3000 MW solar power plant

During the investigation of the accident, it was revealed that at the Syrdarya TPP (SDTPP) there was a long-term arc burning between the fixed and moving contacts of the lead-in disconnector of the suspended type of the 500 kV L-502 line (SDTPP – Tashkent Substation), which, due to air ionization, switched into phase-to-phase short circuit with disconnection of L-502 from High-speed protections (HSP) on both sides with successful three-phase automatic reclosing.

At the same time, the second busbar system of 500 kV was disconnected at the SDTPP from the excessive operation of the main and backup sets of differential protection of the 2nd busbar system 500 kV. After 2 minutes, because of a repeated two-phase short circuit, L-502 again disconnected from the Phase differential protection (DFZ) on both sides. The short circuit happened for the same reason. The first busbar system of 500 kV was disconnected at SDTPP from the excessive operation of the main and backup sets of differential protection of the 2nd busbar system.

A local accident with the down of the first 500 kV busbar systems at the SDTPP, with the loss of generation at it and the division of the power system created imbalance of more than 2000 MW, seven times

higher than the calculated imbalance, that fell on the North-South transit of Kazakhstan, which caused a breakdown of the dynamic stability of this transit, loaded almost to the limit.

Transit separation almost doubled the power imbalance in UPS CA, which led to the development of an emergency with a decrease in frequency to 47.1 Hz, with the operation of automatic under-frequency load hedding system at individual stations in South Kazakhstan and the shutdown of two CCGTs in the Uzbek power system and, ultimately, a frequency avalanche.

The Commission has developed a number of measures aimed at pre-venting the recurrence of such accidents in the future, the implementation of which will eliminate the occurrence of uncompensated power surges in the future. However, it seems that they are not enough to avoid the recurrence of such systemic accidents in the implementation of plans for the construction of renewable energy sources in the UPS CA.

The advanced introduction of RES in Uzbekistan and Kazakhstan can lead to:

- a sharp increase in the problems with the regulation of imbalances and the necessary power reserves for this during the entire time of the day, not only during peak hours as it is now;
- the urgent need to solve the problem with excess gas during the daytime, when solar stations will operate at full capacity, and thermal power plants - at a technological minimum;
- the problem with the increasing the speed of load at thermal power plants.

Considering previously noted acute shortage of power reserves in the UPS CA, it can be concluded that with the large-scale introduction of RES, power surges for the North-South transit of Kazakhstan will be inevitable and daily failures of the stability of this transit will become the main problem that may turn out to be the key obstacle to the integration of RES.

Increasing the transit capacity of the North-South of Kazakhstan to maintain stability during power surges, as experience shows, is an expensive and, at the same time, a temporary and quickly exhausted measure. To solve the problem of ensuring the stability of the parallel operation of the power systems of Central Asia with the UPS of Kazakhstan and the UPS of Russia, in [2] the authors proposed the idea of using back-to-back HVDC converter station (HVDC-Scheme) in the UPS CA in the problem area (Fig. 2).

Fig. 2. Suggestion for DC insertion

Conclusion

As noted above, in the accident on January 25, 2022, the shutdown of the North-South transit of Kazakhstan led to almost a doubling of the power imbalance and the development of the breakdown in the system.

If the injection of power supplied through this transit for the southern regions of Kazakhstan is maintained in an emergency, this would contribute to obtaining a regime in the UPS CA with a smaller deficit and, accordingly, at lower costs for its normalization. It is proposed to separate the UPS CA and the Southern regions of Kazakhstan from the UPS of Kazakhstan by a HVDC-Scheme installed in the L-514 (Shu - Frunze) cut-off, and by transferring the shunting 220 kV overhead lines to dead-end modes [2].

The capacity of the HVDC-Scheme must be chosen sufficient to supply the Shymkent and Dzhambyl regions from the UPS of Kazakhstan, estimated at about 1000 MW. The analysis shows that if such HVDC-Scheme would be in the UPS CA, then in the accident on January 25, the frequency would decrease after the busbar systems at the Syrdarya TPP were extinguished only to 48.5 Hz and there would be no development of the breakdown in the system with the operation of the automatic under-frequency load

shedding system at the stations and the subsequent frequency avalanche in the power system.

This approach will allow to address of several current problems:

- with centralized frequency regulation by Russian stations, all power imbalances in the UPS CA fall on cross-border communications towards the UPS of Russia and overload them, while operating through a direct current insertion, the imbalances will remain in the UPS itself;
- taking into account the preservation of supplies through the HVDC-Scheme and the frequency effect of the load, frequency regulation will be much easier than with the preservation of parallel operation with the UPS of Russia and the UPS of Kazakhstan;
- the structure of the Automatic Control of Frequency and Power flows (ACFP), which is planned to be developed and implemented in the CA UPS, will be simplified; the task of regulating deviations in the balance of flows at the border of the UPS CA with the UPS of Kazakhstan will disappear and the ACFP will focus on regulating the frequency and flows within the energy association.

1. Analysis of the accident on January 25 in UPS CA shows the presence of problems with power balancing in case of violation of the stability of the North-South transit of Kazakhstan.
2. The large-scale introduction of variable and intermittent RES, combined with the existing and expected problems in the gas transmission system of Uzbekistan, greatly increases the likelihood of uncompensated surges in the North-South transit of Kazakhstan. The RES implementation program should be linked to a large-scale reorganization of the gas transmission system with the creation of interaction between the operators of the electric and gas networks in the region to ensure the operation of the GTS in accordance with the schedule of the energy system.
3. In order to efficiently use water and create peak capacities to compensate for the impact of RES, it is proposed to increase efforts in the region for the joint construction of hydroelectric power plants, as well as pumped storage power plants, using for this purpose the existing reservoirs of daily and weekly regulation.

References

1. Nasirov T.Kh. and others. Scientific and technical problems of functioning and development of the integrated energy system of Central Asia and South Kazakhstan as an interstate association. // Publishing House "Fan" of the Academy of Sciences of the Republic of Uzbekistan. Problems of Informatics and Energy. 1993, No. 5. p. 25–29.
2. Shamsiev Kh.A., Shamsiev B.Kh. Analysis of the prospects for the development of the UPS of Central Asia, taking into account the integration of renewable energy sources. - Proceedings of the Scientific and Technical Center of the Unified Energy System, No. 86, 2022
3. The concept of providing the Republic of Uzbekistan with electric energy for 2020-2030. / Ministry of Energy of the Republic of Uzbekistan, 2020. Official website of the Ministry of Energy of the Republic of Uzbekistan.
4. Nasirov T.Kh., Nepomnyashchy V.A., Shamsiev Kh.A. Influence of wind and solar power plants on the controllability and reliability of the functioning of power systems. - Republican scientific and technical conference "New technologies - the basis for the development of energy in Uzbekistan", TSTU, December 2020.
5. Flexible Gas Markets for Variable Generation. A EURELECTRIC report, May 2014.
6. Shamsiev Kh.A. Problems in the integration of renewable energy sources into the energy system of Uzbekistan and recommendations for their solution. / Workshop "Expanding cross-border cooperation in the field of energy through the introduction of wind and solar energy in the energy systems of the CIS member states to achieve SDG 7", Almaty, December 2021.
7. Shamsiev Kh.A. Energy Security in Central Asia: Current Status and Future Directions. – Regional workshop at expert level, OECD. M.: June 8, 2021.
8. Khamidov Sh.V., Shamsiev Kh.A. Issues of RES integration in CA UPS and recommendations. Presentation at the High Level Forum "EU - Central Asia". Organizer: Energy Charter Secretariat. November 4, 2021.

E3S Web of Conferences 384, 01021 (2023) https://doi.org/10.1051/e3sconf/202338401021
RSES 2022

The innovative EHV line and its main indicators

Krasilnikova Tatyana[1], *Makhmudov* Karomatullo[1], *Samorodov* German[2] and *Novikov* Nikolay[3*]

[1]NSTU, Department of Automated Electric Power Systems, 630073 Karl Marx Avenue 20, Novosibirsk, Russia
[2]Branch of JSC «STC FGC UES» – Siberian Research Institute of Energy, 630126 Klenovaya str 10/1, Novosibirsk, Russia
[3]JSP «STC FGC UES», NRU MPEI, JIHT RAS, 115201 Kashirskoe highway 22, вuilding 3, Moscow, Russia

Abstract. Single-circuit EHV lines widely used all over the world have a significant drawback, consisting in the fact that with the most probable single-phase stable failures, the line is completely taken out of operation. In this paper, a single-circuit EHV line is considered, one phase of which is made in the form of two parallel semi-phases, any of which is used as a backup phase in emergency modes. To symmetrize the conditions in the middle of the line, the Series Capasitor (SC) are included in the two usual phases. The article presents an algorithm for calculating normal modes, the main indicators of an innovative 500 kV line, consisting in its increased capacity, reliability and economic efficiency. The main provisions are illustrated by the example of a 500 kV line with a length of 500 km.

Introduction

Single-circuit EHV lines are widely used all over the world [1-6], which have the disadvantage that with the most probable single-phase stable failures, the line fails completely. This fact gives superiority to DC lines that can operate with single-pole stable damage at one pole with a transmission of 50% of the maximum power [7]. There are a number of ways to increase the capacity of single-circuit AC lines (the use of compact lines of increased natural power [8], the use of SC [9]), in which the issue of reliability only worsens.

The reliability of a single-circuit line can be improved by using the reserve phase of the line [10], which is switched on instead of the emergency operating phase (Fig.1). The reserve phase is used only in short-term emergency modes, and the rest of the time remains disabled, which is its significant disadvantage.

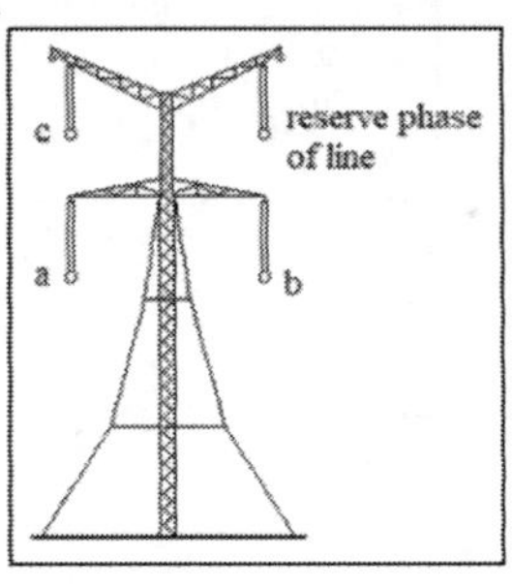

Fig. 1. A line with a reserve phase.

The purpose of this work is to develop a line that provides increased capacity with high reliability of operation, while being characterized by favorable economic indicators.

General characteristics of the innovative line

In the innovative line [11], one phase is performed in the form of two parallel operating semi-phases, one of which is used as a reserve phase in emergency modes.

Figure 2a shows the layout of the phases and semi-phases of the innovation line on the support, and Figure 2b shows its scheme.

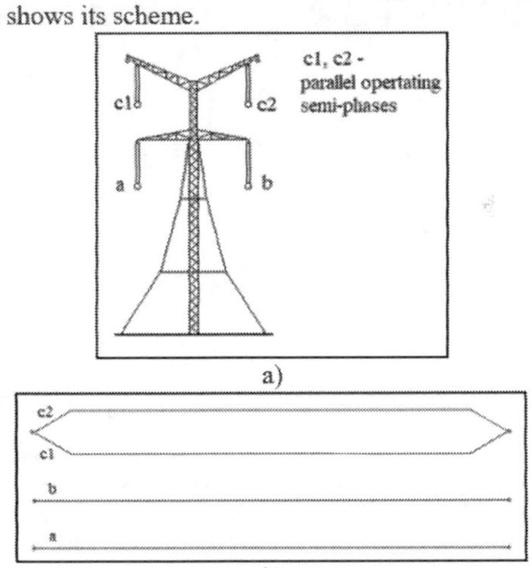

Fig. 2. A line with parallel operating semi-phases: a - the location of the phases on the support; b - the scheme of the innovative line.

Structurally, the total cross-section of the semi-phases is equal to or close to the cross-section of the individual phase. Fig.3 shows the possible designs of phases and semi-phases of the 500 kV line, with respect to which calculations are carried out in the future. The proposed phase and semi-phase designs

*Corresponding author: novikov@ntc-power.ru

differ in geometric dimensions and the number of wires in the phase: these are traditional (option A) and compact (option B) designs.

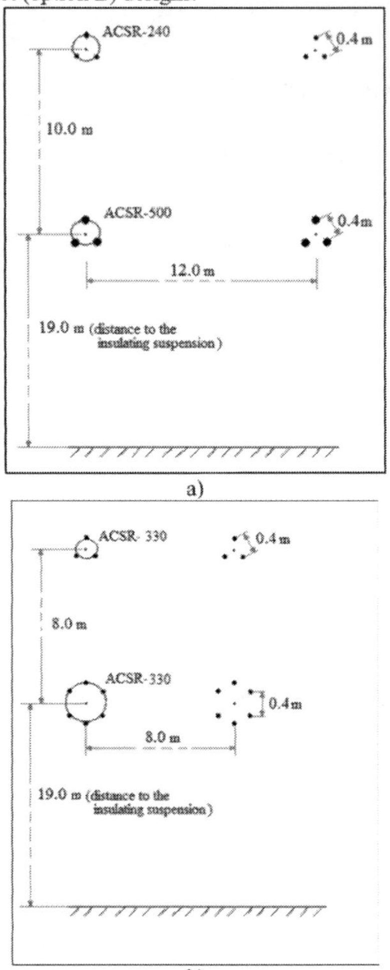

Fig. 3. Designs of phases and semi-phases of the innovative line: a – traditional design (option A); b – compact design (option B).

Mathematical model of an innovative line for calculating normal conditions

The innovative line has a phase-by-phase longitudinal and transverse asymmetry, and for the symmetry of normal conditions, it is necessary to install SC in the middle part of the normal phases, and at the ends of one of the semi-phases, shunt reactors are connected, as shown in Fig.4. Since the innovative overhead line is characterized by phase asymmetry, the universal method for calculating steady-state conditions is the matrix method, when the line and other installations are described in phase coordinates.

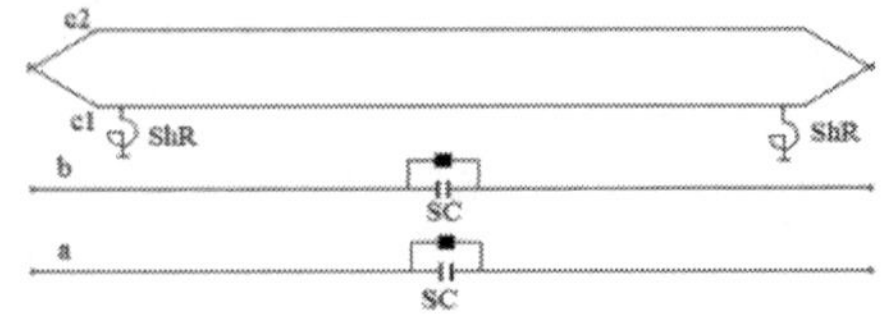

Fig. 4. Scheme of innovative overhead line with semi-phases for calculating normal conditions.

The complete phase matrix of the circuit is formed according to the above figure

$$\boldsymbol{M}_{12} = \boldsymbol{M}_{ShR} \cdot \boldsymbol{M}_{l/2} \cdot \boldsymbol{M}_{SC} \cdot \boldsymbol{M}_{l/2} \cdot \boldsymbol{M}_{ShR} \quad (1)$$

The corresponding matrices in formula (1) are defined as follows. The complete matrix of the untransposed half section of the line has the form

$$\boldsymbol{M}_{l/2} = \begin{vmatrix} \boldsymbol{A}_{l/2} & \boldsymbol{B}_{l/2} \\ \boldsymbol{C}_{l/2} & \boldsymbol{D}_{l/2} \end{vmatrix} \quad (2)$$

where

$$\boldsymbol{A}_{l/2} = \begin{vmatrix} A_{aa} & A_{ab} & A_{ac1} & A_{ac2} \\ A_{ba} & A_{bb} & A_{bc1} & A_{bc2} \\ A_{c1a} & A_{c1b} & A_{c1c1} & A_{c1c2} \\ A_{c2a} & A_{c2b} & A_{c2c1} & A_{c2c2} \end{vmatrix},$$

$$\boldsymbol{B}_{l/2} = \begin{vmatrix} B_{aa} & B_{ab} & B_{ac1} & B_{ac2} \\ B_{ba} & B_{bb} & B_{bc1} & B_{bc2} \\ B_{c1a} & B_{c1b} & B_{c1c1} & B_{c1c2} \\ B_{c2a} & B_{c2b} & B_{c2c1} & B_{c2c2} \end{vmatrix},$$

$$\boldsymbol{C}_{l/2} = \begin{vmatrix} C_{aa} & C_{ab} & C_{ac1} & C_{ac2} \\ C_{ba} & C_{bb} & C_{bc1} & C_{bc2} \\ C_{c1a} & C_{c1b} & C_{c1c1} & C_{c1c2} \\ C_{c2a} & C_{c2b} & C_{c2c1} & C_{c2c2} \end{vmatrix},$$

$$\boldsymbol{D}_{l/2} = \begin{vmatrix} D_{aa} & D_{ab} & D_{ac1} & D_{ac2} \\ D_{ba} & D_{bb} & D_{bc1} & D_{bc2} \\ D_{c1a} & D_{c1b} & D_{c1c1} & D_{c1c2} \\ D_{c2a} & D_{c2b} & D_{c2c1} & D_{c2c2} \end{vmatrix}.$$

$$\boldsymbol{M}_{SC} = \begin{vmatrix} \boldsymbol{1}_4 & \boldsymbol{Z}_{SC} \\ \boldsymbol{O}_4 & \boldsymbol{1}_4 \end{vmatrix} \quad (3)$$

where $\boldsymbol{Z}_{SC} = \begin{vmatrix} Z_{SC} & 0 & 0 & 0 \\ 0 & Z_{SC} & 0 & 0 \\ 0 & 0 & 0 & 0 \\ 0 & 0 & 0 & 0 \end{vmatrix}$, $Z_{sc} = i \cdot X_{sc}$ –phase reactance of the SC.

The zero and unit matrices of the 4th order are respectively determined by

$$\boldsymbol{0}_4 = \begin{vmatrix} 0 & 0 & 0 & 0 \\ 0 & 0 & 0 & 0 \\ 0 & 0 & 0 & 0 \\ 0 & 0 & 0 & 0 \end{vmatrix}, \qquad \boldsymbol{1}_4 = \begin{vmatrix} 1 & 0 & 0 & 0 \\ 0 & 1 & 0 & 0 \\ 0 & 0 & 1 & 0 \\ 0 & 0 & 0 & 1 \end{vmatrix}.$$

The other main element of the four-wire circuit includes shunt reactors (ShR) located at the ends of the line. In general, unvariable ShR (UShR) and variable ShR (VShR) can be used. (The ShRs at the ends of the line are not shown in Fig.4, since it is assumed that the conditions of transmission of natural power is being considered when the reactors are switched off.)

So, in particular, the complete VShR matrix for the line under consideration will be written

$$M_{VShR} = \begin{vmatrix} \mathbf{1}_4 & \mathbf{0}_4 \\ \mathbf{Y}_{VShR} & \mathbf{1}_4 \end{vmatrix} \quad (4)$$

where $\mathbf{Y}_{VShRi} = \mathbf{Z}_{VShRi}^{-1}$.

In turn $\mathbf{Z}_{VShR} = \begin{vmatrix} Z_{VShRa} & 0 & 0 & 0 \\ 0 & Z_{VShRb} & 0 & 0 \\ 0 & 0 & Z_{VShRc1} & 0 \\ 0 & 0 & 0 & Z_{VShRc2} \end{vmatrix}$,

where $\mathbf{Z}_{VShRi} = \frac{1}{Y_{VShRi}}$ –reactance of the i-th phase, i=a,b,c1,c2.

Vector-columns of specified voltages at the ends of the line are accepted

$$\mathbf{U}_1 = \frac{U_{nom}}{\sqrt{3}} e^{i\delta} |1 \quad h^2 \quad h \quad h|^t,$$

$$\mathbf{U}_2 = \frac{U_{nom}}{\sqrt{3}} e^{i\delta} |1 \quad h^2 \quad h \quad h|^t, \quad (5)$$

where $\mathbf{U}_1 -$ rated line voltage; δ –angular shift between voltage att the ends of the line; $h = e^{i2\pi/3}$ –phase operator.

Next, we find the vector column of currents at the end of the line

$$\mathbf{I}_2 = |I_{2a} \quad I_{2b} \quad I_{2c1} \quad I_{2c2}|^t = \mathbf{B}_{12}^{-1}(\mathbf{E}_1 - \mathbf{A}_{12}\mathbf{E}_2) \quad (6)$$

Given that the current in phase "c" consists of the currents of the semi-phases "c1" and "c2", we find the current at the end of the line

$$I_{2c} = I_{2c1} + I_{2c2}$$

And accordingly, the column vector of the three-phase current will be

$$I_{2abc} = |I_{2a} \quad I_{2b} \quad I_{2c}|^t.$$

To assess the level of asymmetry that occurs in the circuit, we define a vector column of symmetrical components of currents at the beginning of the line

$$\mathbf{I}_{1sym} = |I_{11} \quad I_{12} \quad I_{10}|^t = \mathbf{H}_{sym} \cdot \mathbf{I}_{1abc} \quad (7)$$

where $\mathbf{H}_{sym} = \begin{bmatrix} 1 & h & h^2 \\ 1 & h^2 & h \\ 1 & 1 & 1 \end{bmatrix}$ –transformation matrix from phase components to symmetric components; I_{11}, I_{12}, I_{10} –respectively, the currents of the positive, negative and zero sequences.

The coefficient of asymmetry in the current of the negative sequence will be determined accordingly

$$K_{12} = I_{12}/I_{11} \quad (8)$$

An important characteristic of the circuit that determines its capacity is the angular characteristic, which represents the dependence of the active power transmitted along the line on the angular shift between the voltages at the ends of the line. Thus, the total power at the end of the line is defined as the scalar product of the corresponding column vectors of voltage and current

$$S_2(\delta) = \mathbf{U}_{2abc}(\delta) \cdot \mathbf{I}_{2abc}(\delta)$$

where $\mathbf{U}_{2abc} = \frac{U_{nom}}{\sqrt{3}} e^{i\delta} |1 \quad h^2 \quad h \quad h|^t$ –vector is a column of voltages at the end of the line.

Accordingly, the angular characteristic of the scheme is found as

$$P_2(\delta) = ReS_2(\delta) \quad (9)$$

The maximum of the angular characteristic at $\delta = 90^0$determines the maximum transmitted power of the circuit

$$P_{2\max} = P_2(90^0)$$

Accordingly, capacity of the circuit is reduced taking into account the margin coefficient for static stability

$$P_{2\text{cap}} = \frac{P_{2\max}}{K_{mc}} \quad (10)$$

where $K_{mc} = 1{,}2$ –the margin coefficient for static stability.

Reduction of asymmetry in normal conditions

Based on the proposed algorithm, we will analyze the normal conditions of the 500 kV innovation line with a length of 500 km, bearing in mind the level of asymmetry that occurs in it.

In the absence of a two-phase SC in the middle of the line and a ShR at the ends of one of the semi-phases, the asymmetry coefficient reaches 14%, which significantly exceeds the permissible value for synchronous generators of no more than 6%. The decrease in the asymmetry coefficient in the negative sequence occurs not only due to the SC, but also as a result of the use of ShR.

As follows from the calculations, in the case of using UShRs, the suseptance of which is 0.65 mSm, the optimal reactance of the SC for a traditional design (option A) is 66 Ohm, and for a compact design (option B) - 27 Ohm; the corresponding asymmetry coefficients are equal for option A - 0.3%, for option B - 1.0% (Fig. 5).

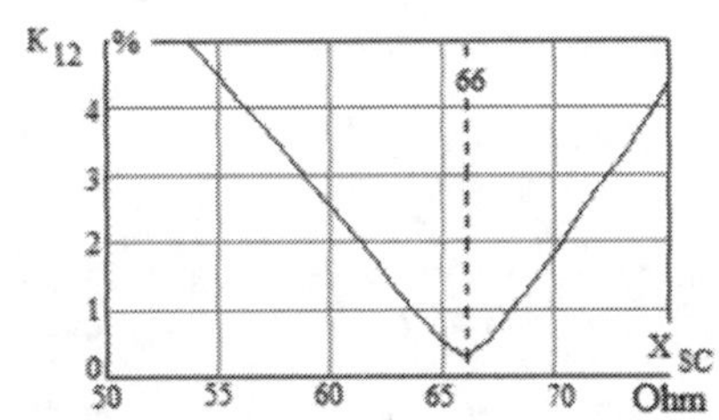

a)

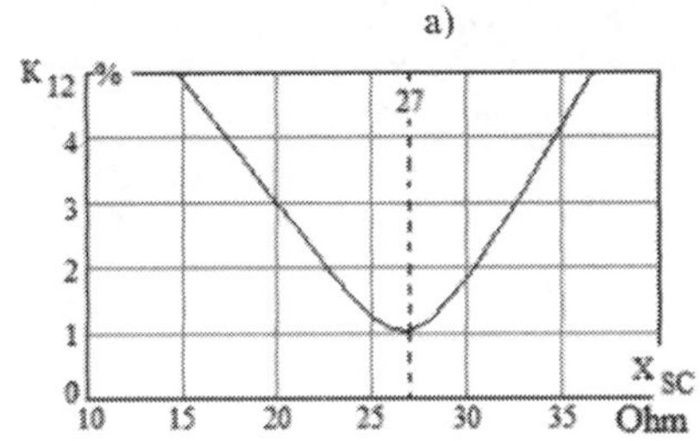

b)

Fig. 5. Dependence of the asymmetry coefficient in the negative sequence on the reactance of the SC: a – traditional design; b – compact design.

https://doi.org/10.1051/e3sconf/202338401021

Thus, it can be concluded that with the combined use of a two-phase SC and a single-phase ShR, relatively small asymmetry coefficients are provided.

The main indicators of the innovative line

Doubled capacity

The main indicator of the innovative line, first of all, is its doubled capacity compared to the traditional single-circuit line. Table 1 shows a comparison of the capacities of traditional and innovative lines calculated according to (10). The innovative line provides an almost two-fold increase in capacity compared to a single-circuit traditional line.

Table 1. Comparison of capacities of traditional and innovative 500 kV lines (500 km)

Line type	Traditional (single-circuit) line	Innovative line	
		Traditional design	Compact design
Capacity, MW	1460	2900	2880
Capacity ratio		1.99	1.97

Increased reliability

In the EHV lines, the overwhelming number of failures is single-phase. In case of unstable failures in the innovation line, as in the traditional line, the Single-Phase Automatic Reclosing is used.

A noticeable proportion of failures are sustained. The proposed scheme, in the event of sustained failures, allows you to switch to operation in post-emergency mode with the possibility of transmitting at least 50% of the power of the initial maximum conditions. So, if one of the semi-phases is permanently damaged, it is switched off by switches 2 (Fig. 6a), and the line switches to operation in emergency mode. At the same time, in order to ensure an acceptable level of asymmetry, it is necessary to shunt the SC with switches 1. In case of sustained damage to one of the phases, for example, phase "b", it is switched off by switches 3, and the corresponding semi-phase, disconnected by switches 2 from the other semi-phase, is switched on by switches 4 instead. In order to ensure an acceptable level of asymmetry, it is necessary in this case to shunt the SC with switch 1 in phase "a", as shown in Fig. 6b.

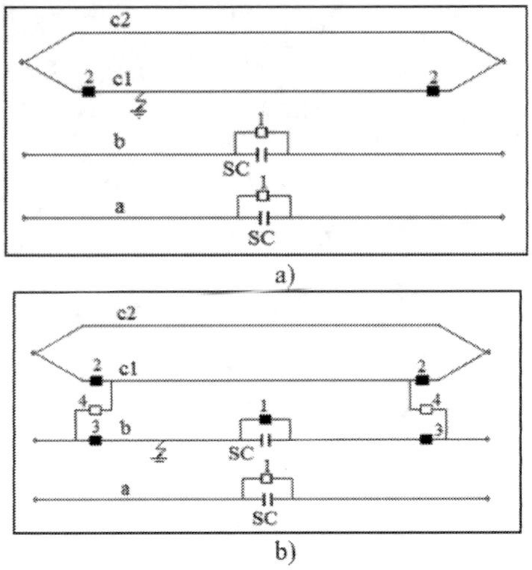

Fig. 6. Schemes of post-emergency mode with sustained damage: a – damage on the semi-phase «c1»; b – damage on the phase «b».

Increased economic efficiency

The optimal conditions of a traditional overhead line is the transmission of natural power, which for a 500 kV overhead line is rounded 900 MW and, respectively, for two circuits 1800 MW. According to the condition of heating losses, the total cross-section of phases and semi-phases in the single-chain version is assumed to be equal to the total cross-section of phases of the double-chain line.

Reliable technical and economic information is available at the level of 2000 [12], and therefore a comparative analysis is made in the prices of this period, which is quite acceptable with comparative estimates.

The cost of a single-circuit three-phase line on steel supports in the range of HV and EHV is quite accurately extrapolated by the dependence

$$K_{3ph}(U_n) = a \cdot e^{bU_n} \quad (11)$$

where the coefficients a, b are determined based on the cost of 220 and 330 kV voltage lines.

Figure 7 shows the free-standing supports of the traditional single-circuit line [13] and the innovative single-circuit line, which shows that an additional wire is added to the innovative line and its cost increases by the suspension of this wire.

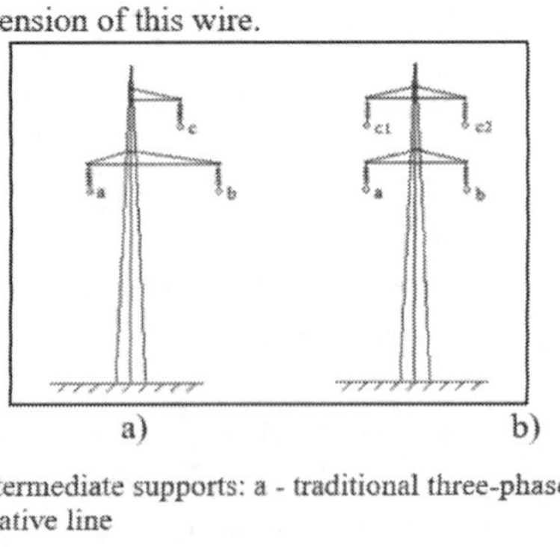

Fig.7. Intermediate supports: a - traditional three-phase line; b - innovative line

According to [14], the cost of wires and insulation is approximately 40% of the cost of the line. Then the capital investment in the innovative line, taking into account the suspension of the additional wire, will be

$$K_{in}(500) = (1 + \frac{0,4}{3})K_{3ph}(500) \quad (12)$$

The unit cost of the line depends significantly on the cross-section, and it can be estimated using the linear dependence of the cost on the total cross-section of all phases

$$K_{in} = 2,72 + 0,45 \cdot 10^{-3} S, \text{ million·rubles/km} \quad (13)$$

As a result, we obtain the unit cost of an innovative line with free-standing supports, which with its total cross-section 3*6*330 mm^2 will amount to 5.4 million rubles/km.

The unit cost of two circuits on different free-standing supports according to [12] is 8.0 million rubles/km.

For the above data, Table 2 presents a technical and economic comparison of a two-circuit traditional line with a single-circuit innovation line, which shows a noticeable economic advantage of the innovative option.

Table 2 Technical and economic indicators of two-circuit traditional and single-circuit innovative lines

Line type / Parameters		Two cicuits of traditional overhead lines on different supports	Single-circuit innovative overhead line (compact design)
Transmitted power, MW		1800	1800
The cost of the line, million rubles		4000	2700
Installed capacity of the SC and its cost	Qsc, MVar	-	250
	Ksc, million rubles	-	150
Total capital investments	million rubles	4000	2850
	%	140	100

Conclusions

In this article, a new type of single-circuit EHV line is justified, the capacity of which exceeds that for a traditional line by two times. In addition, the proposed line has increased reliability, allowing, in case of sustained failures, to switch to operation in post-emergency mode with the possibility of transmitting at least 50% of the power of the initial maximum conditions.

A technical and economic comparison of two-circuit traditional and single-chain innovative lines with a voltage of 500 kV, a length of 500 km and a capacity of 1800 MW showed that the capital costs in the two-circuit version are 40% more than in the innovative version.

References

1. I.M. Bortnik, 1200 kV Transmission Line in the USSR. The First Result of Operation, CIGRE (1988)
2. H.K. Xie, K. Mu, EHV Transmission and its Insulation: Under Planning and Construction in China, IEEE Electrical Insulation Magazine, **15**, 2 (1999)
3. G. Trudel, S. Bernard, G. Scott, Hydro-Quebec's defense plan against. extreme contingencies, IEEE Trans. Power Systems, **14**, 3 (1999)
4. A.O. Silva, H.P. Chagar, A.O. Filho, H.A Magalhaes., D.C. Barros, O.P. Santos, J.B. Silva, R.P. Guimaras, Reliability and Upgrading Studies of The 765 kV Itaipu Transmission System, CIGRE, (2000)
5. H. Zhou, Y. Su, Y. Chen, Q. Ma and W. Mo, The China Southen Power Grid, IEEE power & energy magazine, (2016)
6. A.F. Dyakov, The electric networks of super- and ultra-high voltage UES of Russia: theoretical and practical issues, Moscow: Energoprogress Corporation EEEK Publ., 2012, 696 p.
7. J. Arrillaga, Y.H. Liu, N.R. Watson, Flexible Power Transmission, The HVDC Options, John Wiley &Sons, Ltd, (2007)
8. L. Yafang, Y. Yichao, W. Huiren and others. Study of 500 kV compact transmission technology, CIGRE, (1998)
9. J. Gingras, R. Mailhot, J. Lavergne, Review of The Hydro-Quebec System operation with Series Compensation, CIGRE, (2000)
10. G. I. Samorodov, T. G. Krasilnikova, V. P. Dikoy, S. M. Zilberman and R.A. Iatsenko, *Non-Conventional Reliable AC Transmission System for Power Delivery at Long and Very Long Distances*, in Proc. 2002 IEEE/PES Transmission and Distribution Conference, Vol.2, Yokohama, Japan, (2002)
11. G. I. Samorodov, T. G. Krasilnikova, Patent No. 2765656, Single-circuit power transmission line of high or extra high voltage, Date of state registration in the State Register of Inventions of the Russian Federation 01.02.22. Bull. No. 4.
12. Handbook on the design of electrical networks, Ed. D.L. Faibisovich. - 3rd ed., revised. and additional, ENAS, (2009)
13. Handbook of high voltage electrical installations. Ed. I.A. Baumstein and M.V. Khomyakov, Energy, (1974)
14. G.N. Aleksandrov et al. Designing ultrahigh voltage transmission lines, Ed. G.N. Alexandrov and L.L. Peterson, Energoatomizdat, (1983)

The influence of power quality on the operational modes of industrial power supply systems

Alexander Sevostyanov [1,*]

[1]Nizhny Novgorod State Technical University n.a. R.E. Alekseev, Minin str., 24, Nizhny Novgorod, Russia

Abstract. About 90% of distortions involving power quality analysis value are caused by the operation of industrial power supply systems. Thus, not only do gradual changes of voltage and harmonic distortions appear, but also disturbances caused by voltage dips and interruptions, which lead to technological equipment malfunctioning as well as serious damage to power quality. However, despite the possible damage from voltage dips and short interruptions in power supply systems, there is no standardization in power quality requirements, and power quality requirements are often not taken into consideration in the design and operation of power supply systems. On the basis of the research conducted on power quality value and analysis of current regulatory documents, the author of the article feels that it is crucial to take into consideration voltage dips and interruptions in the operation of industrial power supply systems.

1 Introduction

The issue of building industrial power supply systems in which power quality is taken into consideration is intrinsic to ensure the effective and trouble-free operation of these power supply systems [1].

As a rule, the issue of increasing power quality is relevant for power grids which are connected to a large number of consumers having sharply variable, unbalanced and non-sinusoidal operational modes.

Electrical interference has a random character of voltage distortion; and when affecting the technological equipment of industrial enterprises, the interference results in breaking the technological cycle of production caused by:

- automation systems malfunction;
- the error in determining the operating parameters of power grid;
- the breakdown in operating modes of industrial equipment.

In power supply systems nearly 60% of defects and breakdowns in industrial power supply systems are related to poor power quality including failures caused by voltage dips and interruptions. The mentioned above troubles can be the reasons for failures in control and protection systems of electric consumers, technical errors in the operation of microprocessor technology, equipment damage and emergence of abnormal operating modes of the main industrial equipment.

Despite the possible damage, the regulatory documents do not contain a number of indicators concerning power quality, which seems quite important for designing power supply systems. Transition to smart power grids will contribute to an even greater number of failures in operating modes because of voltage dips and interruptions. The latter is connected with the susceptibility of digital elements of power grids to these types of interference.

2 Electromagnetic interference levels in industrial power supply systems

In connection with the development of the industrial sector of Russia's economy in current market conditions, the following issues are being addressed:

- minimizing the damage caused by a decrease in power quality;
- increasing the reliability of industrial equipment power supply;
- increasing the capacity of power sources when connecting to new consumers.

The findings in terms of power quality [2] show that for industrial facilities the indicators of power quality, in average 30% cases under their supervision, do not meet the requirements of GOST 32144-2013:

- under- and overestimated voltage 60% of measurements;
- total harmonic distortions 2% of measurements;
- n-harmonic voltage factor 20% of measurements;

* Corresponding author: sevosaa@gmail.com

https://doi.org/10.1051/e3sconf/202338401022

- unbalance factor of return voltage sequence 2% of measurements;
- unbalance of zero sequence of voltage 24% of measurements.

It should be noted that about 90% of distortions of power quality analysis value are the result of the operation of industrial electric receivers.

In most cases, technical problems related to the functioning of electrical receptors in the Russian Federation due to power quality disturbances are caused by:
- voltage harmonics 18% of disturbances;
- voltage dips or interruptions 52% of disturbances.

In the European Union countries, the main causes of technological equipment malfunctioning are:
- harmonic components - 5.4%;
- voltage sags - 23.6%;
- pulsed voltages - 29%.

In the USA, according to the statistics, equipment shutdown can be caused by:
- voltage sags and pulsed voltages - 48%;
- harmonic distortions - 22%.

Based on the data presented, it should be noted that the most frequent causes of breakdowns in industrial power supply systems are brownouts and pulsed voltages.

These disturbances occur during normal operation of industrial equipment including the following manufacturing equipment: electric arc furnaces (EAF), various rolling mills, welding machines, electrical receivers with motor load. The Figure 1 shows a voltage-dip graph based on measurements on one of the 0,4 kV feeders of the welding areas of JSC "Zavod Trud" (Nizhny Novgorod).

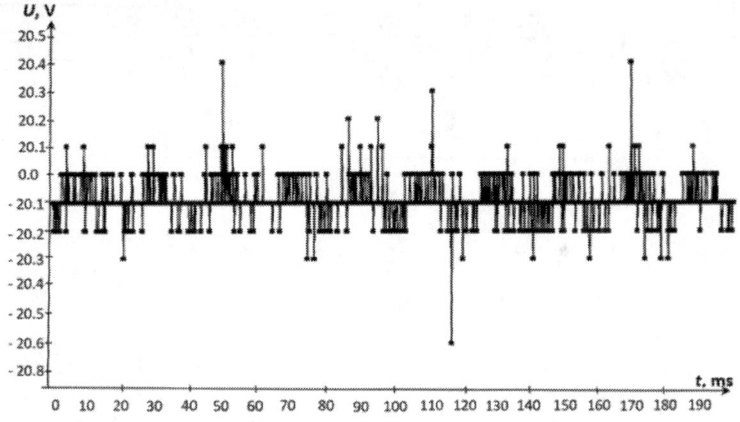

Fig. 1. Diagram of voltage dips (0.4 kV feeder of the welding area).

Let us consider the electromagnetic interference levels (EML), in particular voltage dips, introduced into power grid with industrial electrical consumers [3-5].

While operating electric arc furnaces produce irregular voltage dips with a frequency of up to 1Hz and regular voltage dips with a frequency of up to 10Hz.

When electric arc furnaces work, regular voltage dips usually have a depth not exceeding 5% of the supply voltage value, whereas irregular voltage dips have the depth up to 100%. The causes of irregular voltage dips are unfavorable arc ignition conditions during the melting of metal, arc unstable combustion, electrode short circuits with charge at furnace startup and charge collapse.

Working welding machines create a voltage sag, the value of which depends on the capacity of the machine and power supply [6]. Projection and spot welding create voltage dip up to 7% deep. Butt and multiple-spot welding create voltage dip up to 19% deep. In this case, voltage dip duration for spot, projection, multi-spot and seam welding machines is 0.02-1.0s., while for butt welding machines - 0.2-20s.

The voltage sag value created by rolling mills depends on their types. Synchronous and asynchronous direct current motors are exploited as the drivers of rolling mills. Rolling mills with drives from synchronous and asynchronous DC motors are the reason for voltage fluctuations and dips. Thus blooming and slabbing mills in 6, 10kV grid create voltage dips up to 20% in depth over 1-5s. and the average frequency of 20-40 dips per minute. Hot strip mills create voltage dips from 2 to 10% in depth over 1-5s. and an average frequency of 2 to 20 dips per minute. In the case of cold rolling mills, voltage dips are between 2 and 20% in depth for 2 to 20s. and average frequencies of 4 to 20 dips per hour.

Presses and automatic press lines during their operation create voltage dips from 1 to 5% with a frequency of occurrence of 0.17-0.2Hz.

Conveyors, fans, and compressors at equipment startup are the reason for dips not exceeding for transporters and conveyors 1.2%; fans 3-5%; compressors 4-6% in depth.

Asynchronous and synchronous DC motors create voltage dips from 1 to 20% in depth for 3-10s. The frequency of voltage sags depends on the electric motor operation modes. The highest frequency is observed in the intermittent mode of equipment operation.

3 The effect of voltage dips and interruptions on industrial electrical equipment

The influence of voltage dips and voltage interruptions on the industrial electrical consumers depends on the factors that determine the depth of voltage dips, their duration and the frequency of their occurrence. Interestingly, control and protection systems industrial electric consumers [4, 5, 7] built on the basis of microprocessor technology are the most sensitive to these types of interference. Figure 2 shows the maximum allowable voltage dips and interruptions causing the failure of microprocessor-based control systems [8].

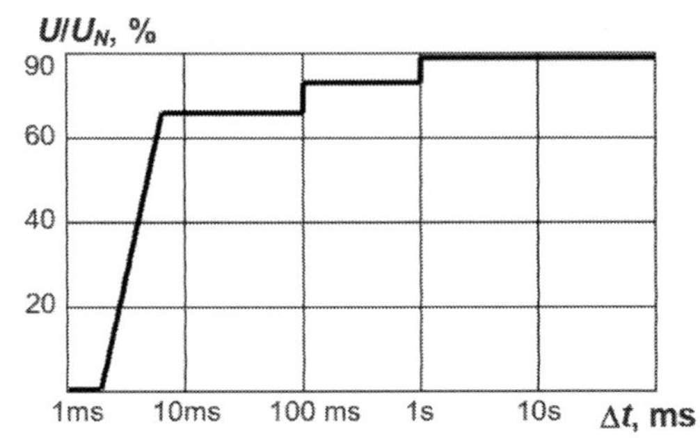

Fig. 2. The graph of allowable voltage dips for microprocessors.

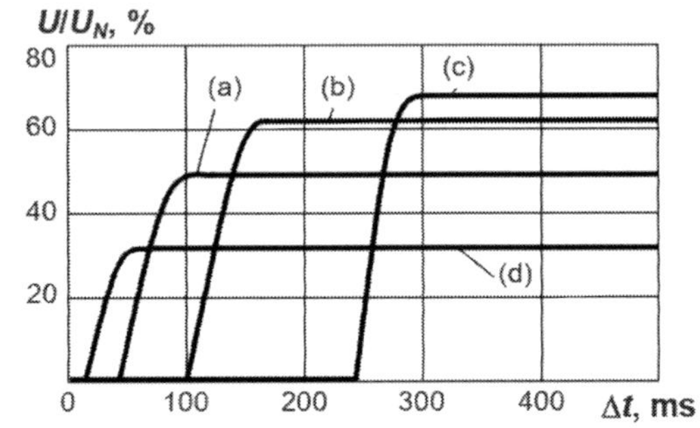

Fig. 3. The graph of allowable voltage dips: a) magnetic starters; b) variable speed drives; c) programmable controllers; d) relays.

Figure 3 shows the allowable voltage dips and interruptions which have been discovered in automatic systems including control relays, magnetic starters, programmable controllers and elements of variable speed drives.

Thus, technical failures in equipment operation will be observed at the following values of voltage dips:

- magnetic starters $\Delta U_п \geq 50\%$;
- variable speed drives $\Delta U_п \geq 60\%$;
- programmable controllers $\Delta U_п \geq 70\%$;
- microprocessors $\Delta U_п \geq 65\%$;
- relays $\Delta U_п \geq 30\%$.

Failures of these devices will be observed if the duration of voltage dips is:

$\Delta t_п \geq 40$ms for magnetic starters;

$\Delta t_п \geq 100$ms for variable speed drives;

$\Delta t_п \geq 240$ms for programmable controllers;

$\Delta t_п \geq 100$ms for microprocessors;

$\Delta t_п \geq 40$ms for relays.

For industrial power consumers with rectifier and inverter units the limits of allowable values of voltage dips and interruptions are shown in Figure 4 [9]. It should be noted that inverters have greater sensitivity to voltage dips in comparison with rectifiers. If the duration of voltage dip is more than 6-9ms and the value of sag is more than 20%, the rectifier fails to work.

If a voltage sag is not more than 15% and its duration is not more than 3ms, rectifier and inverter units shut down.

To assess the possibility of uninterrupted operation of personal computers in the context of existing voltage dips and interruptions, ITIC (Information Technology Industry Council) curves have been developed, Figure 5 [10]. According to these curves voltage interruptions with A duration not exceeding 20ms. are allowed. Voltage dips up to 10% in depth of any duration, up to 20% in depth with a duration not exceeding 10s. and up to 30% in depth with a duration of less than 0.5s are allowed too.

When assessing the reliability of industrial manufacturing equipment in the context of THE occurrence of voltage dips and interruptions, it is necessary to evaluate of damage extent caused by these dips and interruptions.

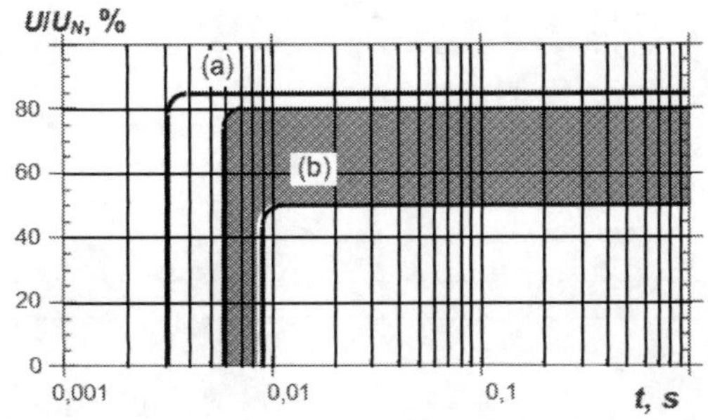

Fig. 4. Characteristics of noise immunity: a) inverter; b) rectifier.

Determination of damages from power supply interruptions should classifies into two types of damages – main and unforeseeable [11]. The main damage is associated with the damage caused by voltage dips and interruptions in the power supply of industrial enterprises, but the equipment is preserved, manufacturing process is ongoing and there is no wasted produce. In other words, the main damage is caused by the failure to implement the plan for production output. The unforeseeable damage is associated with the element of surprise, as a result of which there may be manufacturing process disturbance, equipment failure and, as a consequence, waste is produced. This damage depends on the type of consumer, the amount of electricity shortfall, the availability of the consumer's generation reserves and the possibility of changing industrial manufacturing process. If consumer's reserves is sufficient to compensate for the shortfall in production during its energy limitation, the shortfall in electricity during that time is compensated by the backup power supply system of the enterprise, thereby ensuring the continuity of the manufacturing process.

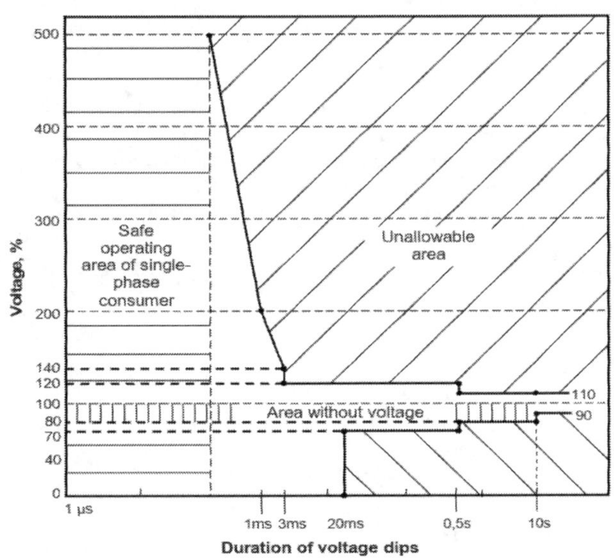

Fig. 5. Areas of allowable voltage dips and interruptions for personal computers, ITIC curves.

In the works [9, 12] the values of damages from poor quality of electric power defined in fractions from the total annual damage are given:

- voltage sags and short-term interruptions lasting less than 3 min - 57,7%;
- overvoltages - 34,4%;
- long voltage interruptions lasting more than 3min - 4,3%;
- flicker, voltage asymmetry - 2.7%;
- voltage non-sinusoidality - 0.9%.

As can be seen, the greatest damage is caused by voltage dips and interruptions. At the same time the complexity of modern manufacturing processes and high requirements to maintaining their stability lead to significant losses in various industries, which is clearly illustrated by Table 1 [5, 13]. In addition, it should be noted that GOST 32144-2014 [14] does not specify the requirements for these voltage dips and interruptions.

The presented information allows us to conclude that when power quality rationing and developing recommendations for its improvement, it is necessary to pay attention to the values that determine the depth and duration of voltage dips and interruptions.

Table 1. Damages from power outages in some industries

Industry	Damage, euro per an interruption
Semiconductor manufacture	3800000
Computer center	750000
Steel production	350000
Glass industry	250000
Pharmaceutical company	17000000

4 Recommendations for improvement of electromagnetic compatibility of industrial consumers

The methods for calculating certain types of electromagnetic disturbances are presented in the works [3-5]. Basically, the methods allow evaluating slow changes, oscillations, non-sinusoidal and asymmetric voltages. Taking into account the probabilistic nature of interference, it is better to exploit simulation modeling of power supply systems with the use of software products PSCAD and MatLab Simulink for interference calculation and prediction. Simulation modeling tools allow us to study the influence of any types of interference on industrial electrical equipment not only during normal operation, but also transient processes.

For industrial power supply systems, it is possible to ensure electromagnetic compatibility of mfanufacturing equipment by using schematic solutions as well as applying special devices in the grid. Schematic ways of ensuring electromagnetic compatibility are the most simple and economical. The following technical solutions can be recommended:

- separation of power supply to electrical consumers and manufacturing equipment, which create EMF and are sensitive to these types of interference;
- use of interlocks to limit the simultaneous work of large consumers creating voltage sags;
- increasing THE power source capacity of power supply systems;
- application of pulse-width modulation rectifiers.

In industrial power supply systems, special means can be used, including, for example, the use of [15, 16, 17]:

- uninterruptible power supplies;
- passive and active harmonic filters;
- static VAR compensators;
- transformer-thyristor voltage regulators;
- a motor soft starter;
- other devices.

Conclusions

In connection with the electric power industry transfering to intelligent technology systems in industrial power supply systems, it is possible to observe a tendency to increase the damage from voltage sags and interruptions, which must be taken into account when considering the noise immunity of industrial equipment to the effects of this kind of interference.

To increase the resistance of technological equipment to voltage sags and interruptions as well as to reduce damages, it is necessary to create the standards for voltage sags and interruptions as a part of GOST 32144-2013 and take these standards into account when considering the design and operation of industrial power supply systems.

As electrical consumers in industrial power supply systems create voltage sags and interruptions influencing other industrial consumers, it is necessary to separate consumer power supplies or to introduce uninterruptible power supply sources into the power supply system of responsible electric consumers.

References

1. L. Dobrusin, Energoekspert, 4, 30-35 (2008).
2. A. Ded, S. Sikorsky, P. Smirnov, Omsk Scientific Bulletin, **2**, 60-63 (2018).
3. G. Vagin, *Electromagnetic compatibility in the power industry (in Russian)* (Academia Publishing Center, Moscow, 2010).
4. B. Borisov, *Power supply of electrothechnological installations (in Russian)* (Naukova Dumka Publisher, Kiev, 1985).
5. I. Zhezelenko, *Electromagnetic compatibility of consumers (in Russian)* (Mashinostroenie Publisher, Moscow, 2012).
6. A. Shidlovskiy, *Electromagnetic Compatibility of industial electrical receivers (in Russian)* (Naukova Dumka Publisher, Kiev, 1992).
7. A. Ovsyannikov, R. Borisov, *Electromagnetic compatibility in the electrical power engineering (in Russian)* (NSTU, Novosibirsk, 2017).
8. IEEE 446-1995. *Recommended Practice for Emergency and Standby Power Systems for*

Industrial and Commercial Applications IEEE Orange Book (Color Book Series).

9. J.Manson, Energoexpert, 4, 49-52 (2008).
10. S. Gamazin, *Reference Book on the power supply and electrical equipment of the industrial enterprises and public buildings (in Russian)* (MPEI Publishing House, Moscow, 2010).
11. N. Savina, *Reliability of the electrical power systems (in Russian)* (Amur State University, Blagoveshchensk, 2014).
12. G. Vagin, A. Kulikov, Electric Stations, **6**, 54-59 (2019).
13. *Manual on the construction of electrical installations* (Sehneider Electric, 2009).
14. GOST 32144-2013. *Electrical energy. Electromagnetic compatibility of technical means. Quality standards of electric energy in general purpose power supply systems.* (Standardinform, Moscow, 2014).
15. I. Zhezelenko, *Electromagnetic compatibility of consumers (in Russian)* (Mashinostroenie, Moscow, 2012).
16. A. Chivenkov, *Development of methods and means of voltage and power regulation in power supply systems with autonomous power sources: Abstract of doctoral dissertation* (NSTU, N. Novgorod, 2015).
17. E. Sosnina, R. Bedretdinov, Electrical equipment: operation and repair, **4**, 24-26 (2013).

Hidden Markov Model and States Prediction of an Autonomous Wind-Diesel Complex

Yuriy E. Obzherin[1], *Stanislav M.* Sidorov[2]*, and *Sergey N.* Fedorenko[3]

[1]Sevastopol State University, Department of Higher Mathematics, 299053 Sevastopol, Russia
[2]Sevastopol State University, Department of Higher Mathematics, 299053 Sevastopol, Russia
[3]Sevastopol State University, Department of Technological Processes and Production Automation, 299053 Sevastopol, Russia

Abstract. The problem of assessing the reliability and analyzing the functioning of an autonomous wind-diesel complex, consisting of a wind power plant, working and standby diesel generators, an inverter and a storage battery, is considered. First, a semi-Markov model of an autonomous wind-diesel complex is built, which makes it possible to calculate the stationary and temporal reliability characteristics. Then, on its basis, a hidden Markov model is developed, which is used to solve the problems of predicting and evaluating its characteristics, taking into account the given parameters and the signal vector. The results of the study are obtained in a general form and are invariant with respect to the laws of distribution of random variables describing the elements of an autonomous wind-diesel complex. They allow you to simulate the functioning of the system under various distribution laws, based on statistical data, without modifying the model itself.

Introduction

Currently, renewable energy sources are being introduced into power supply system's everywhere. The combined use of wind turbines and diesel generators contributes not only to economic benefits, but also to the reduction of emissions of fuel processing products. [10, 2] However, the efficiency of wind turbines depends on many factors, in particular, on the wind load, which is stochastic. In this regard, the problem arises of constructing adequate mathematical models that take into account the stochastic nature of work and the natural resources used (for example, wind) and the presence of a diesel generator (backup energy source).

The article [3] considers an approximate model of a wind-diesel complex in Matlab\Simulink. Articles [4, 5] present the results of simulation studies. In [6, 7], the assessment of the operating modes of the power supply complex with a wind-diesel power plant is studied. In [1, 2], Markov models are used to analyze the functioning of a wind-diesel complex. It should be noted that most of the results obtained by other authors consider cases where the random variables describing the system have an exponential distribution (Markov model). The semi-Markov model proposed in this article allows the use of distributions of an arbitrary form. It also allows you to find the time characteristics of the reliability and efficiency of the system, in contrast to the Markov model.

Secondly, for territorially remote systems (or systems controlled remotely), the operator is not always able to completely obtain the information contained in the encoding of the states of the semi-Markov model when changing their states, but it is always possible to obtain some signal (information) associated with the states of the nested chain Markov, which can be considered unobservable (hidden). In such cases, the use of the theory of hidden Markov models makes it possible to solve the problems of finding and estimating characteristics based on the observed signal vector. The purpose of our work is to apply the theory of semi-Markov processes with a common phase space of states and the theory of hidden Markov models for the analysis of reliability and efficiency of an autonomous wind-diesel complex (WDC).

This article discusses the problem of assessing the reliability and analysis of the functioning of an autonomous WDC, consisting of a wind power plant, working and standby diesel generators, an inverter and a storage battery.

In this paper, first, a semi-Markov model [8-12] of an autonomous WDC is constructed, which makes it possible to calculate the stationary and temporal characteristics of reliability. Then, following the methodology proposed by the authors [18], a hidden Markov model [15–17] is developed on its basis, which is used to solve forecasting problems and evaluate its characteristics, taking into account the given parameters.

1 Semi-Markov model of an autonomous WDC

Consider system S, which is an autonomous wind-diesel complex. We will consider the inverter and the

*Corresponding author: xaevec@mail.ru

battery to be absolutely reliable in the sense that the probability of their failure is much less than the probability of failure of other elements. Therefore, we will not take them into account when constructing a semi-Markov model. Then the system S can be represented as a three-element system with a time reserve: element 1 is a wind turbine, element 2 is a diesel generator (DG), element 3 is a reserve diesel generator (RDG) (time reserve). The failure of element 1 should be understood as any event that leads to disruption of normal functioning (for example, lack of wind, failure of one of the components, etc.). The system functions as follows: if element 1 (2) fails, then element 2 (1) is switched off, element 3 is switched on, and the system operates due to the reserve of time (element 3). As soon as element 1 (2) is restored, element 3 is switched off (in this case, we assume that by the next time its characteristics are fully restored), and element 2 (1) starts working with the previous level of operating time. System failure occurs when element 1 (2) is on recovery, element 2 (1) is disabled, and the reserve time ends (failure of element 3). The reserve of time ends if either the fuel of the backup generator runs out (time of operation due to the reserve of fuel), or it fails.

Let`s assume that the uptime of elements 1 and 2 are described by random variables (RV) α_1 and α_2, respectively, having distribution functions $F_1(x) = P(\alpha_1 \le x)$, $F_2(x) = P(\alpha_2 \le x)$ and density distributions $f_1(x)$, $f_2(x)$, and the recovery time - by random variables β_1 and β_2, having distribution functions $G_1(x) = P(\beta_1 \le x)$, $G_2(x) = P(\beta_2 \le x)$ and distribution density $g_1(x)$, $g_2(x)$. A random instantly replenished reserve of time is considered, which is described by RV τ having the distribution function $R(x) = P(\tau \le x)$ and distribution density $r(x)$. Random variables α_1, α_2, β_1, β_2, τ are assumed to be independent and have finite mathematical expectations. It should be noted that SV τ is the minimum of the operating time due to the fuel reserve of the standby generator and its uptime.

On fig. 1 is shown a block diagram of the system under consideration.

Consider the discrete-continuous phase space of states of the system under consideration:

$$E = \{1112x_2, 2112x_1, 1021x_2, 2201x_1, 3020x_1x_2, 3200x_1x_2\} \quad (1).$$

Phase states are designated according to the following scheme: the first digit of the code indicates the number of the element in which the state change occurred; the rest of the numbers - what happens to each element (1 – work, 2 – disabled, 0 – recovery). The continuous components show the time remaining until the next state change.

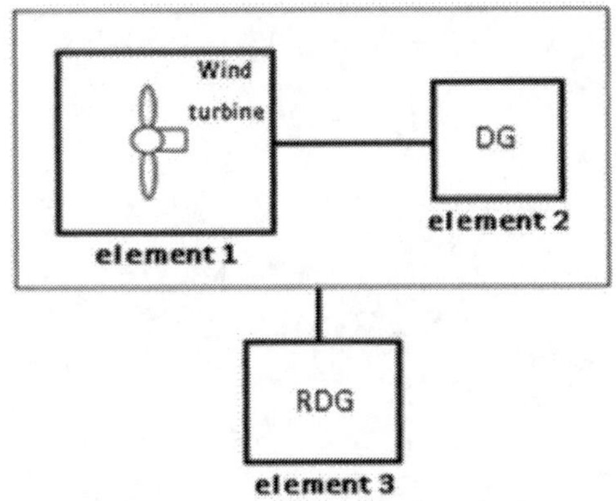

Fig. 1. Structural diagram of the considered system

Let's decipher the values of the status codes:

• $1112x_2$ ($2112x_1$) – element 1(2) has been restored, element 2(1) is enabled, element 3 is disabled, x_2 (x_1) – time until the end of element 2(1) operation.

• $1021x_2$ ($2201x_1$) – element 1(2) fails and starts recovery, element 2(1) turns off, element 3 turns on, x_2 (x_1) time until element 2(1) ends.

• $3020x_1x_2$ ($3200x_1x_2$) – system failure: x_1 (x_2) – time until the end of element 1(2) restoration, x_2 (x_1) – time until the end of element 2(1) operation.

Let`s write down the probabilities of transitions between the states of the system:

$$P_{1112x_2}^{2201x_1} = f_1(x_2 + x_1), x_1 > 0;$$

$$P_{1112x_2}^{1021y_2} = f_1(x_2 - y_2), 0 < y_2 < x_2;$$

$$P_{2112x_1}^{1021x_2} = f_2(x_1 + x_2), x_2 > 0;$$

$$P_{2112x_1}^{2201y_1} = f_2(x_1 - y_1), 0 < y_1 < x_1;$$

$$P_{1021x_2}^{1112x_2} = \int_0^\infty g_1(t)\bar{R}(t)dt = P(\tau > \beta_1);$$

$$P_{1021x_2}^{3020x_1x_2} = \int_0^\infty r(t)g_1(x_1 + t)dt;$$

$$P_{2201x_1}^{2112x_1} = \int_0^\infty g_2(t)\bar{R}(t)dt = P(\tau > \beta_2);$$

$$P_{2201x_1}^{3200x_1x_2} = \int_0^\infty r(t)g_2(x_2 + t)dt;$$

$$P_{3020x_1x_2}^{1112x_2} = 1, \quad P_{3200x_1x_2}^{2112x_1} = 1. \quad (2)$$

Let`s write down the system of equations for finding the stationary distribution of the embedded Markov chain.

$$\begin{cases} \rho(1112x_2) = \rho(1021x_2)\int_0^\infty g_1(t)\bar{R}(t)dt + \\ \quad + \int_0^\infty \rho(3020x_1x_2)dx_1, \\ \rho(2112x_1) = \rho(2201x_1)\int_0^\infty g_2(t)\bar{R}(t)dt + \\ \quad + \int_0^\infty \rho(3200x_1x_2)dx_2, \\ \rho(1021x_2) = \int_{x_2}^\infty f_1(y_2 - x_2)\,\rho(1112y_2)dy_2 + \\ \quad + \int_0^\infty f_2(x_2 + x_1)\rho(2112x_1)dx_1, \\ \rho(2201x_1) = \int_{x_1}^\infty f_2(y_1 - x_1)\,\rho(2112y_1)dy_1 + \\ \quad + \int_0^\infty f_1(x_1 + x_2)\rho(1112x_2)dx_2, \\ \rho(3020x_1x_2) = \rho(1021x_2)\int_0^\infty r(t)g_1(x_1 + t)dt, \\ \rho(3200x_1x_2) = \rho(2201x_1)\int_0^\infty r(t)g_2(x_2 + t)dt, \\ \int_E \rho(e)de = 1 \text{ (normalization condition).} \end{cases} \quad (3)$$

Let 's substitute equation 5 into equation 1 of system (3). We'll get

$$\rho(1112x_2) = \rho(1021x_2)\int_0^\infty g_1(t)\bar{R}(t)dt + $$
$$+\rho(1021x_2)\int_0^\infty dx_1\int_0^\infty r(t)g_1(x_1 + t)dt = $$
$$= \rho(1021x_2)\left[\int_0^\infty g_1(t)\bar{R}(t)dt + \int_0^\infty dx_1\int_0^\infty r(t)g_1(x_1 + t)dt\right] = $$
$$= \rho(1021x_2)\left[\int_0^\infty g_1(t)\bar{R}(t)dt + \int_0^\infty r(t)dt\int_0^\infty g_1(x_1 + t)dx_1\right] = $$
$$= \rho(1021x_2)\left[\int_0^\infty g_1(t)\bar{R}(t)dt + \int_0^\infty r(t)\overline{G_1}(t)dt\right] = $$
$$= \rho(1021x_2).$$

Similarly, substituting equation 6 into equation 2 of system (3), we obtain:

$$\rho(2112x_1) = \rho(2201x_1).$$

Consequently, system (3) is reduced to the form:

$$\begin{cases} \rho(1021x_2) = \int_{x_2}^\infty f_1(y_2 - x_2)\,\rho(1021y_2)dy_2 + \\ \quad + \int_0^\infty f_2(x_2 + x_1)\rho(2201x_1)dx_1, \\ \rho(2201x_1) = \int_{x_1}^\infty f_2(y_1 - x_1)\,\rho(2201y_1)dy_1 + \\ \quad + \int_0^\infty f_1(x_1 + x_2)\rho(1021x_2)dx_2. \end{cases} \quad (4)$$

The solution to system (4) was found in [12] and has the form:

$$\rho(1112x_2) = \rho(1021x_2) = \rho_0\overline{F_2}(x_2),$$
$$\rho(2112x_1) = \rho(2201x_1) = \rho_0\overline{F_1}(x_1),$$
$$\rho(3020x_1x_2) = \rho_0\overline{F_2}(x_2)\int_0^\infty r(t)g_1(x_1 + t)dt,$$
$$\rho(3200x_1x_2) = \rho_0\overline{F_1}(x_1)\int_0^\infty r(t)g_2(x_2 + t)dt, \quad (5)$$

the constant ρ_0 is found from the normalization condition.

Stay times in system states:

$$\theta_{1112x_2} = \alpha_1 \wedge x_2, \quad \theta_{2112x_1} = \alpha_2 \wedge x_1,$$
$$\theta_{1021x_2} = \beta_1 \wedge \tau, \theta_{2201x_1} = \beta_2 \wedge \tau,$$
$$\theta_{3020x_1x_2} = x_1,\ \theta_{3200x_1x_2} = x_2, \quad (6)$$

where $\wedge$ is the sign of the minimum.

Then the average residence times in the states are equal to:

$$E\theta_{1112x_2} = \int_0^{x_2}\overline{F_1}(t)dt,\ E\theta_{2112x_1} = \int_0^{x_1}\overline{F_2}(t)dt,$$
$$E\theta_{1021x_2} = \int_0^\infty \overline{G_1}(t)\bar{R}(t)dt,$$
$$E\theta_{2201x_1} = \int_0^\infty \overline{G_2}(t)\bar{R}(t)dt, \quad (7)$$
$$E\theta_{3020x_1x_2} = x_1, \quad E\theta_{3200x_1x_2} = x_2.$$

We divide the state space E into two non-overlapping subsets: healthy (E_+) and failure states (E_-):

$$E_+ = \{1112x_2, 2112x_1, 1021x_2, 2201x_1\},$$
$$E_- = \{3020x_1x_2, 3200x_1x_2\}$$

Let's find the stationary characteristics of the system reliability.

Using transition probabilities (2), stationary distribution (5), average residence times in states (7), we obtain:

$$\int_{E_+} m(x)\rho(dx) = \rho_0\int_0^\infty \overline{F_2}(x_2)dx_2\int_0^\infty \overline{G_1}(t)\bar{R}(t)dt + $$
$$+\rho_0\int_0^\infty \overline{F_1}(x_1)dx_1\int_0^\infty \overline{G_2}(t)\bar{R}(t)dt + $$
$$+\rho_0\int_0^\infty \overline{F_2}(x_2)dx_2\int_0^{x_2}\overline{F_1}(t)dt + $$
$$+\rho_0\int_0^\infty \overline{F_1}(x_1)dx_1\int_0^{x_1}\overline{F_2}(t)dt = $$
$$= \rho_0 E\alpha_2 E(\beta_1 \wedge \tau) + \rho_0 E\alpha_1 E(\beta_2 \wedge \tau) + $$
$$+\rho_0 E\alpha_1 E\alpha_2.$$

$$\int_{E_+} P(x, E_-)\rho(dx) = $$
$$= \rho_0\int_0^\infty dx_1\int_0^\infty \overline{F_2}(x_2)dx_2\int_0^\infty r(t)g_1(x_1 + t)dt + $$

$$+\rho_0 \int_0^\infty dx_2 \int_0^\infty \overline{F_1}(x_1)dx_1 \int_0^\infty r(t)g_2(x_2+t)dt =$$

$$= \rho_0 E\alpha_2 \int_0^\infty r(t)\overline{G_1}(t)dt + \rho_0 E\alpha_1 \int_0^\infty r(t)\overline{G_2}(t)dt =$$

$$= \rho_0 E\alpha_2 P(\beta_1 > \tau) + \rho_0 E\alpha_1 P(\beta_2 > \tau).$$

$$\int_{E_-} m(x)\rho(dx) =$$

$$= \rho_0 \int_0^\infty x_1 dx_1 \int_0^\infty \overline{F_2}(x_2)dx_2 \int_0^\infty r(t)g_1(x_1+t)dt +$$

$$+\rho_0 \int_0^\infty x_2 dx_2 \int_0^\infty \overline{F_1}(x_1)dx_1 \int_0^\infty r(t)g_2(x_2+t)dt =$$

$$= \rho_0 E\alpha_2 \int_0^\infty x_1 dx_1 \int_0^\infty r(t)g_1(x_1+t)dt +$$

$$+\rho_0 E\alpha_1 \int_0^\infty x_2 dx_2 \int_0^\infty r(t)g_2(x_2+t)dt.$$

Using the results obtained in [18, p. 61],

$$\int_{E_-} m(x)\rho(dx) = \rho_0 E\alpha_2[E\beta_1 - E(\beta_1 \wedge \tau)] +$$

$$+\rho_0 E\alpha_1[E\beta_2 - E(\beta_2 \wedge \tau)].$$

$$\int_E m(x)\rho(dx) = \int_{E_-} m(x)\rho(dx) + \int_{E_+} m(x)\rho(dx) =$$

$$= \rho_0 E\alpha_2[E\beta_1 - E(\beta_1 \wedge \tau)] +$$

$$+\rho_0 E\alpha_1[E\beta_2 - E(\beta_2 \wedge \tau)] + \rho_0 E\alpha_2 E(\beta_1 \wedge \tau) +$$

$$+\rho_0 E\alpha_1 E(\beta_2 \wedge \tau) + \rho_0 E\alpha_1 E\alpha_2 =$$

$$= \rho_0[E\alpha_1 E\alpha_2 + E\alpha_2 E\beta_1 + E\alpha_1 E\beta_2].$$

Where indicated above:
$m(x)$ – average residence times in states, $P(x, E_-)$ – probabilities of transition from operable to failure states, $E(\beta \wedge \tau)$ – mathematical expectation of the minimum of two RVs, $P(\beta > \tau)$ – mathematical expectation of that RV β is greater than RV τ,

$$P(\beta > \tau) = \int_0^\infty r(t)\bar{G}(t)dt \ ,$$

$$E(\beta \wedge \tau) = \int_0^\infty \bar{G}(t)\bar{R}(t)dt \ .$$

Using the formulas presented in [12, 13] and the expressions found above, we obtain:

- mean stationary uptime T_+:

$$T_+ = \frac{\int_{E_+} m(x)\rho(dx)}{\int_{E_+} P(x, E_-)\rho(dx)} =$$

$$= \frac{E\alpha_2 E(\beta_1 \wedge \tau) + E\alpha_1 E(\beta_2 \wedge \tau) + E\alpha_1 E\alpha_2}{E\alpha_2 P(\beta_1 > \tau) + E\alpha_1 P(\beta_2 > \tau)},$$

- average stationary recovery time T_- :

$$T_- = \frac{\int_{E_-} m(x)\rho(dx)}{\int_{E_+} P(x, E_-)\rho(dx)}$$

$$= \frac{E\alpha_2[E\beta_1 - E(\beta_1 \wedge \tau)] + E\alpha_1[E\beta_2 - E(\beta_2 \wedge \tau)]}{E\alpha_2 P(\beta_1 > \tau) + E\alpha_1 P(\beta_2 > \tau)},$$

- stationary availability factor F_a :

$$F_a = \frac{\int_{E_+} m(x)\rho(dx)}{\int_E m(x)\rho(dx)}$$

$$= \frac{E\alpha_2 E(\beta_1 \wedge \tau) + E\alpha_1 E(\beta_2 \wedge \tau) + E\alpha_1 E\alpha_2}{E\alpha_1 E\alpha_2 + E\alpha_2 E\beta_1 + E\alpha_1 E\beta_2}.$$

In the case of a non-random slack, i.e. when $R(x) = 1(x - h)$ where h is the value of the time reserve, stationary characteristics of reliability will take the form:

- mean stationary uptime T_+:

$$T_+ = \frac{E\alpha_2 \int_0^h \overline{G_1}(t)dt + E\alpha_1 \int_0^h \overline{G_2}(t)dt + E\alpha_1 E\alpha_2}{E\alpha_2 \overline{G_1}(h) + E\alpha_1 \overline{G_2}(h)}, \quad (8)$$

- average stationary recovery time T_- :

$$T_- = \frac{E\alpha_2 \int_h^\infty \overline{G_1}(t)dt + E\alpha_1 \int_h^\infty \overline{G_2}(t)dt}{E\alpha_2 \overline{G_1}(h) + E\alpha_1 \overline{G_2}(h)}, \quad (9)$$

- stationary availability factor F_a:

$$F_a = \frac{E\alpha_2 \int_0^h \overline{G_1}(t)dt + E\alpha_1 \int_0^h \overline{G_2}(t)dt + E\alpha_1 E\alpha_2}{E\alpha_1 E\alpha_2 + E\alpha_2 E\beta_1 + E\alpha_1 E\beta_2}. \quad (10)$$

As an illustrative example of the use of formulas (8)-(9), consider the system S, for which, before the start of its operation, it is assumed that the RVs $\alpha_1, \alpha_2, \beta_1, \beta_2$ have an Erlang distribution of order IV and $E\alpha_1 = 150$ h, $E\alpha_2 = 100$ h, $E\beta_1 = 24$ h, $E\beta_2 = 20$ h, and the time reserve h varies from 10 to 20 hours.

The results obtained are shown in fig. 2.

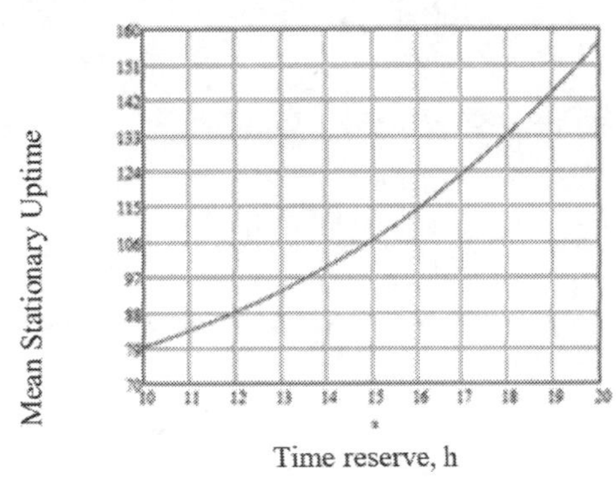

a)

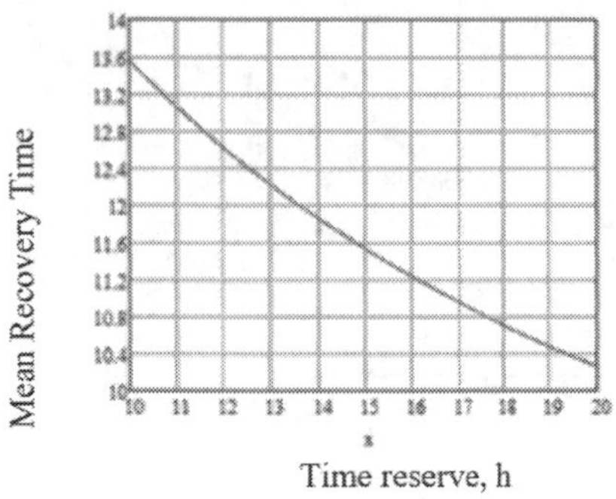

b)

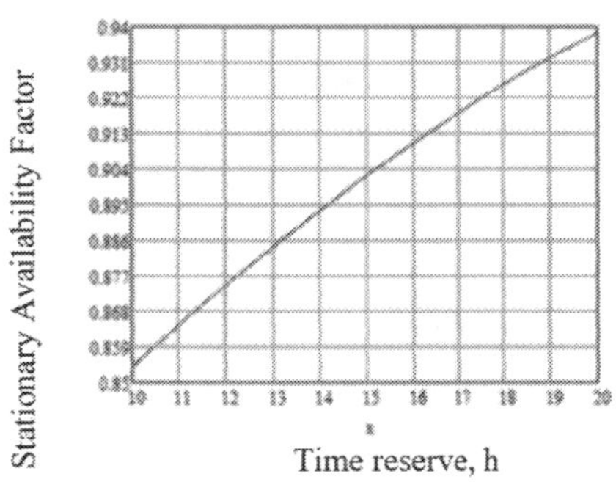

c)

Fig. 2. Graphs of stationary reliability characteristics for different values of the reserve time h: a) mean uptime, b) mean recovery time, c) stationary availability factor.

2 Finding the transition probabilities of the enlarged semi-Markov model

To construct a hidden Markov model (HMM), we enlarge the stationary distributions (5) in terms of continuous components, i.e. we apply the algorithm of stationary phase enlargement to each state. Note that the phase space of semi-Markov states is obtained by adding to the codes of physical states a set of continuous components that fix the residual times of the action of factors that change the state of the system. [13] It is desirable to enlarge these continuous components, leaving only a discrete set of physical states.

We`ll get

$$\int_0^\infty \rho(1112x_2)dx_2 = \int_0^\infty \rho(1021x_2)dx_2 =$$

$$= \rho_0 \int_0^\infty \overline{F_2}(x_2)dx_2 = \rho_0 E\alpha_2,$$

$$\int_0^\infty \rho(2112x_1)dx_1 = \int_0^\infty \rho(2201x_1)dx_1 =$$

$$= \rho_0 \int_0^\infty \overline{F_1}(x_1)dx_1 = \rho_0 E\alpha_1,$$

$$\int_0^\infty \int_0^\infty \rho(3020x_1x_2)dx_1\, dx_2 =$$

$$= \rho_0 \int_0^\infty \overline{F_2}(x_2)dx_2 \int_0^\infty x_1 dx_1 \int_0^\infty r(t)g_1(x_1+t)dt =$$

$$= \rho_0 E\alpha_2 \int_0^\infty r(t)\overline{G_1}(t)dt = \rho_0 E\alpha_2 P(\beta_1 > \tau).$$

$$\int_0^\infty \int_0^\infty \rho(3200x_1x_2)dx_1\, dx_2 =$$

$$= \rho_0 \int_0^\infty \overline{F_1}(x_1)dx_1 \int_0^\infty x_2 dx_2 \int_0^\infty r(t)g_2(x_2+t)dt =$$

$$= \rho_0 E\alpha_1 \int_0^\infty r(t)\overline{G_2}(t)dt = \rho_0 E\alpha_1 P(\beta_2 > \tau).$$

Consequently, the discrete phase space of enlarged states of the system under consideration has the form:

$$\hat{E} = \{1112,\ 2112,\ 1021,\ 2201,\ 3020,\ 3200\}.$$

Using the formulas presented in [12, p. 36],

$$\widehat{p_k^r} = \frac{\int_{E_k} \rho(de)P(e,E_r)}{\rho(E_k)}, \quad \widehat{m_k} = \frac{\int_{E_k} \rho(de)m(e)}{\rho(E_k)} \qquad (11)$$

we find the probabilities of transitions $\widehat{p_k^r}$ between the enlarged states and the average sojourn times $\widehat{m_k}$ in them.

Let`s calculate the denominators of formulas (11).

$$\rho(E_{1112}) = \rho(E_{1021}) = \rho_0 E\alpha_2,$$

$$\rho(E_{2112}) = \rho(E_{2201}) = \rho_0 E\alpha_1,$$

$$\rho(E_{3020}) = \rho_0 E\alpha_2 P(\beta_1 > \tau),$$

$$\rho(E_{3200}) = \rho_0 E\alpha_1 P(\beta_2 > \tau).$$

Average sojourn times $\widehat{m_k}$ equals:

$$\widehat{m}_{1112} = \frac{\int_0^\infty \overline{F_2}(x_2)dx_2 \int_0^{x_2} \overline{F_1}(t)dt}{E\alpha_2},$$

$$\widehat{m}_{2112} = \frac{\int_0^\infty \overline{F_1}(x_1)dx_1 \int_0^{x_1} \overline{F_2}(t)dt}{E\alpha_1},$$

$$\widehat{m}_{1021} = \frac{\int_0^\infty \overline{F_2}(x_2)dx_2 \int_0^\infty \overline{G_1}(t)\bar{R}(t)dt}{E\alpha_2} = E(\beta_1 \wedge \tau),$$

$$\widehat{m}_{2201} = E(\beta_2 \wedge \tau),$$

$$\widehat{m}_{3020} = \frac{E\beta_1 - E(\beta_1 \wedge \tau)}{P(\beta_1 > \tau)},$$

$$\widehat{m}_{3200} = \frac{E\beta_2 - E(\beta_2 \wedge \tau)}{P(\beta_2 > \tau)}.$$

Let`s find the transition probabilities $\widehat{p_k^r}$:

$$\hat{p}_{1112}^{1021} = \frac{\int_0^\infty \overline{F_2}(x_2)dx_2 \int_0^{x_2} f_1(x_2 - t)dt}{E\alpha_2} =$$

$$= \frac{E\alpha_2 - E(\alpha_1 \wedge \alpha_2)}{E\alpha_2},$$

$$\hat{p}_{1112}^{2201} = \frac{E(\alpha_1 \wedge \alpha_2)}{E\alpha_2}, \qquad \hat{p}_{2112}^{1021} = \frac{E(\alpha_1 \wedge \alpha_2)}{E\alpha_1},$$

$$\hat{p}_{2112}^{2201} = \frac{E\alpha_1 - E(\alpha_1 \wedge \alpha_2)}{E\alpha_1},$$

$$\hat{p}_{1021}^{1112} = P(\tau > \beta_1), \quad \hat{p}_{1021}^{3020} = P(\tau < \beta_1),$$

$$\hat{p}_{2201}^{2112} = P(\tau > \beta_2),$$

$$\hat{p}_{2201}^{3200} = P(\tau < \beta_2), \quad \hat{p}_{3020}^{1112} = 1, \quad \hat{p}_{3200}^{2112} = 1.$$

Then the matrix of transition probabilities $\hat{P}_i^j$ between the enlarged states of the system is equal to:

$$\hat{P}_i^j = \begin{pmatrix} 0 & 0 & \frac{E\alpha_2 - E(\alpha_1 \wedge \alpha_2)}{M\alpha_2} & \frac{E(\alpha_1 \wedge \alpha_2)}{E\alpha_2} & 0 & 0 \\ 0 & 0 & \frac{E(\alpha_1 \wedge \alpha_2)}{E\alpha_1} & \frac{E\alpha_1 - E(\alpha_1 \wedge \alpha_2)}{E\alpha_1} & 0 & 0 \\ P(\tau > \beta_1) & 0 & 0 & 0 & P(\tau < \beta_1) & 0 \\ 0 & P(\tau > \beta_2) & 0 & 0 & 0 & P(\tau < \beta_2) \\ 1 & 0 & 0 & 0 & 0 & 0 \\ 0 & 1 & 0 & 0 & 0 & 0 \end{pmatrix}$$

3 Hidden Markov model of the system under consideration

For a complete description of the HMM [16, 17], it is necessary to determine:

1. The set of model states corresponds to the set of states of the enlarged model $\hat{E}$.

2. Observable sequence alphabet (set of signals).

Let's assume that during the functioning of the system S, the states of the EMC of the enlarged model are not observed (hidden states), but only the number of operable elements is observed during the change of states of the HMM. We introduce the following set of signals:

$$J = \{0, 1, 2\},$$

where

- 0 - system failure;
- 1 - elements 1 and 2 of the system are functioning;
- 2 - the system is operational due to the backup diesel generator.

Many signals can be selected in different ways. The set of signals J is chosen in this form, because "Accurate" information about the number of operable main and redundant elements can be obtained for almost any system.

3. The matrix of transition probabilities between system states.

For our model, the transition probability matrix consists of the transition probabilities $\hat{P}_i^j$ of the enlarged semi-Markov model.

4. Connection of model states with signals.

Consider the relationship between the states of the embedded Markov chain (EMC) of the enlarged model and signals, i.e. define the connection function $R(s|\boldsymbol{x})$ [16, 18]:

$$R(s|\boldsymbol{x}) = P(S_n = s|X_n = \boldsymbol{x}), \boldsymbol{x} \in \hat{E}, s \in J,$$

$$\sum_{s \in J} R(s|\boldsymbol{x}) = 1,$$

where S_n is the n-th signal.

We assume that signals 0 and 2 are emitted correctly with a probability of 0.99 and erroneously with a probability of 0.01.

The function $R(s|\boldsymbol{x})$ of the connection between the NCM states of the integrated digital model and the signals is presented in Table 1.

Table 1. The connection function between the merged model EMC states and the signals.

Signal, s / Condition, x	s=0	s=1	s=2
1112	0	1	0
2112	0	1	0
1021	0.01	0	0.99
2201	0.01	0	0.99
3020	0.99	0	0.01
3200	0.99	0	0.01

5. The initial probability distribution of the model.

We will assume that at the initial moment of time the enlarged model can be equally likely to be in the state 1112 or 2112.

The HMM is built on the basis of the enlarged semi-Markov model.

4 Evaluation of characteristics and prediction of states

As an illustrative example, consider a system S, for which, before starting its operation, it is assumed that the CVs $\alpha_1, \alpha_2, \beta_1, \beta_2$ have an Erlang distribution of order IV and $E\alpha_1 = 150$ h, $E\alpha_2 = 100$ h, $E\beta_1 = 24$ h, $E\beta_2 = 20$ h, $E\tau = 15$ h. It is assumed that at the initial time the system is in states 1112 and 2112 with probabilities of 0.5.

Suppose that as a result of the operation of the system S, the following vector of signals is received:

$$\overline{s}_{30} = (1,2,1,2,1,2,1,2,0,1,2,1,2,1,2,0,1,2,1,2,0,1,$$
$$2,1,2,1,2,1,2,1), \quad n{=}30.$$

Consider the problem of estimating the characteristics of a hidden Markov model, taking into account the introduced parameters, following [18, 19].

1. Let's determine the probabilities of states of the hidden model at the moment of emission of the 30th signal (the last signal in the chain).

Because the 30th signal is equal to 1 (elements 1 and 2 of the system are functioning), then the task is to determine the probability of finding the enlarged system in states leading to the emission of the specified signal. As a result, we get that at the 30th step the enlarged model with a probability of 0.3553 was in state 1112, with a probability of 0.6447 in state 2112. For other states, this probability is equal to zero.

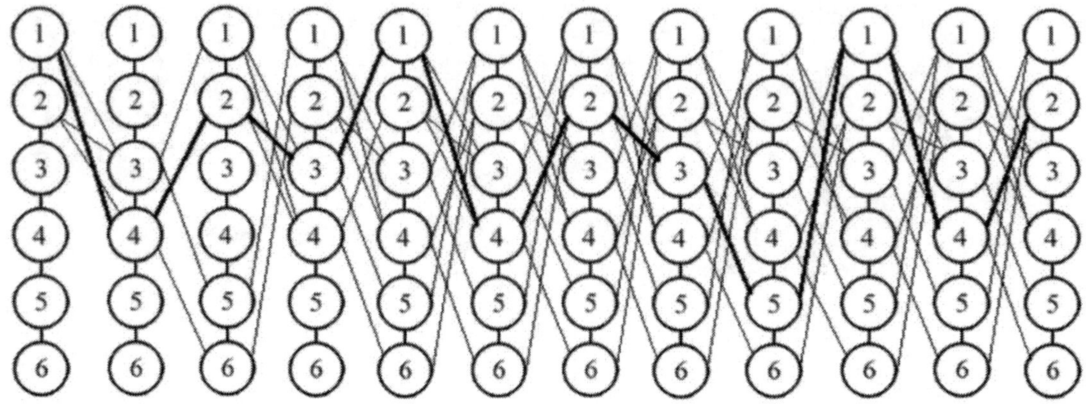

Fig. 4. The trellis of maximized transition probabilities for considered chain of signals

2. Let's find the probabilities with which the hidden model will make the transition to the states at the next 31st step.

The probabilities of the transition of the hidden model at the 31st step: to state 1021 with a probability of 0.419, to state 2201 – 0.581, to all others – with zero.

3. Let's determine the probabilities of the appearance of signals at the next 31st step.

We get that the probability of the appearance of signal 2 at the 31st step is 0.99, signal 0 – 0.01, signal 1 – 0.

4. Let's find the probability of occurrence (emission) of the received signal vector $\overline{s}_{30}$.

The probability of occurrence of the received signal vector $\overline{s}_{30}$ is equal to 0.00000243.

5. Forecasting the states of the hidden model based on the received signal vector.

Table 2 shows the most probable states of the hidden model at the transitions indicated in it and the probabilities of these states.

Table 2. Most probable states of the hidden model on transitions.

Transition number	1	7	11	17	21	26	29
Most likely state	1112	2112	2201	2112	3200	2112	2201
State probability	0.5233	0.6737	0.652	0.519	0.5136	0.6132	0.6447

Applying the Baum-Welsh algorithm [15, 16], we obtain an overestimated matrix of transition probabilities for the system under consideration. The original transition probability matrix and the transition probability matrix after reassessment are presented in Fig. 3.

Applying the Viterbi algorithm [15, 16] to the overestimated model, we determine the most probable chain of states for the received signal vector: 2201, 3200, 2112, 2201, 2112, 1021, 3020, 1112, 2201, 2112, 1021, 2112, 2201, 2112, 1021, 2112.

The trellis diagram for the first 12 received signals is shown on fig. 4. For simplicity, the following recoding of model states is used: 1112↔1, 2112↔2, 1021↔3, 2201↔4, 3020↔5, 3200↔6. The most probable states (according to the Viterbi algorithm for the overestimated model) for the vector of received signals are marked with a thick line. Thin lines connect other possible states that the system could achieve without information about the signal vector.

$$\begin{pmatrix} 0 & 0 & 0.1454 & 0.8546 & 0 & 0 \\ 0 & 0 & 0.5697 & 0.4303 & 0 & 0 \\ 0.2606 & 0 & 0 & 0 & 0.7394 & 0 \\ 0 & 0.3469 & 0 & 0 & 0 & 0.6531 \\ 1 & 0 & 0 & 0 & 0 & 0 \\ 0 & 1 & 0 & 0 & 0 & 0 \end{pmatrix}$$

a)

$$\begin{pmatrix} 0 & 0 & 0.1283 & 0.8717 & 0 & 0 \\ 0 & 0 & 0.5369 & 0.4631 & 0 & 0 \\ 0.7075 & 0 & 0 & 0 & 0.2925 & 0 \\ 0 & 0.8066 & 0 & 0 & 0 & 0.1934 \\ 1 & 0 & 0 & 0 & 0 & 0 \\ 0 & 1 & 0 & 0 & 0 & 0 \end{pmatrix}$$

b)

Fig. 3. Transition probability matrices:
a) original transition probability matrix P_i^j,
b) overestimated transition probability matrix $\overline{P}_i^j$.

Discussions

The results of this article open up great opportunities for evaluating the functioning of autonomous wind-diesel complexes. The article shows how to apply the apparatus of the theory of hidden Markov models (used for systems with a finite set of states) to semi-Markov models that have a common phase space of states, which expands the possibilities of applying the HMM theory.

The article provides only illustrative examples showing the expansion of the possibilities of modeling autonomous wind-diesel complexes using the methodology developed by the authors. However, it is obvious that it will not be difficult for energy specialists to use the parameters of real wind-diesel complexes as input data for models with subsequent calculation of characteristics.

It should be noted that the results obtained, with proper reformulation, can be used to analyze the functioning of technical systems for various purposes.

Conclusion

In this work, a semi-Markov model of an autonomous WDC is constructed, which allows calculating the stationary and temporal reliability characteristics. Then, on its basis, a hidden Markov model was developed, which is used to find estimates of the characteristics of an autonomous WDC and predict its states based on the signal vector obtained as a result of the operation. The results of the study will make it possible to predict the operating modes of an autonomous WDC. The resulting HMM allows you to re-evaluate the parameters of the constructed model (train) to increase its adequacy, according to the signals received in the process of functioning.

The results of the study are obtained in a general form and are invariant with respect to the distribution laws of random variables that describe the elements of an autonomous WDC. They allow you to simulate the functioning of the system under various distribution laws, based on statistical data, without modifying the model itself. In the future, it is planned to generalize the results for the case of a fleet (consisting of N pieces) of wind-diesel complexes.

The study was supported by the grant of the President of the Russian Federation for state support of young Russian scientists – Candidates of Sciences No. MK-329.2022.4.

References

1. A.V. Bobrov, V.A. Tremjasov, *Wind-diesel complexes in decentralized power supply: monograph* (Siberian Federal University, Krasnojarsk, 2012)
2. O.A. Grigor'eva, T.V. Krivenko, V.A. Tremjasov, Reliability analysis of an autonomous wind-diesel complex, Scientific and technical statements of the St. Petersburg State Polytechnic University, Vol. **2** (243), pp. 45–52. (2016)
3. S. Meziane, O. Feddaoui, R. Toufouti, Modeling and simulation of hybrid wind-diesel power generation system, International Journal of Renewable Energy, Vol. **8** (2), pp. 49-58. (2013)
4. R. Sebastián, A. Nevado, Study and Simulation of a Wind Hydro Isolated Microgrid. Energies, Vol. **13** (22):5937. (2020)
5. R. Sebastián, Modeling, Simulation and Control of Wind Diesel Power Systems. Energies, Vol. **15** : 1712. (2022)
6. Y.E. Shklyarskiy, D.E. Batueva. Operation mode selection algorithm development of a wind-diesel power plant supply complex. Journal of Mining Institute, Vol. **253**, pp. 115-126. (2022)
7. A.I. Baikov, A.B. Dar'enkov, E.N. Sosnina. Simulation Modeling of a Wind-Diesel Power Plant. Russ. Electr. Engin., Vol. **89**, 161–167 (2018).
8. N. Limnios, M. Nikulin. *Recent Advances in Reliability Theory: Methodology, Practice, and Inference*, eds. N. Limnios, M. Nikulin. (Springer Science+Business Media, New York, 2000)
9. Y.N. Rudenko, I.A. Ushakov. *Reliability of power systems*, 2nd edition. (Nauka, Novosibirsk, 1989)
10. Y.E. Obzherin. Semi-Markov and hidden Markov and semi-Markov models of energy systems. Proceedings of the Russian Academy of Sciences. Energy, Vol. **5**, pp. 26-32. (2019)
11. Y.E. Obzherin, S.M. Sidorov, M.M. Nikitin, Analysis of reliability and phase merging of systems with component-wise storages. Proceedings of the Russian Academy of Sciences. Energy, Vol. **6**, pp. 66–77. (2019)
12. .S. Korolyuk, A.F. Turbin, *Markovian restoration processes in the problems of system reliability* (Naukova Dumka, Kiev, 1982)
13. V.S. Korolyuk, V.V. Korolyuk, *Stochastic Models of Systems* (Springer Science+Business Media, Dordrecht, 1999)
14. V.J. Kopp, Y.E. Obzherin, A.I. Peschanskij, *Stochastic models of automated production systems with time redundancy.* (Izd-vo SevGTU,Sevastopol, 2000)
15. S. M. Ross, *Introduction to Probability Models, Ninth Edition* (Elsevier Academic Press, USA, 2006)
16. L.R. Rabiner, A Tutorial on Hidden Markov Models and Selected Applications in Speech Recognition, Proc. of the IEEE, Vol. **77** (2), pp. 257–286. (1989)
17. O. Cappe´, E. Moulines, T. Ryde´n, *Inference in Hidden Markov Models* (Springer-Verlag, New York, 2005)
18. Y.E. Obzherin, M.M. Nikitin, S.M. Sidorov, Hidden Markov model based on signals from blocks of semi-markov system's elements and its application for dynamics analysis energy systems, Smart Innovation, Systems and Technologies. Singapore, Springer, Vol. **187**, pp. 477–486. (2021)
19. Y.E. Obzherin, M.M. Nikitin, S.M. Sidorov, *Hidden Markov model of maintenance by age of system elements*, in Methodological issues of reliability research of large energy systems. 92nd meeting of the seminar established at ISEM SB RAS. In 3 books, pp. 89–98. (2020)

Methods for the Formation of Calculation Models of Electric Power Systems in the Adequacy Assessment Problem

Dmitry Krupenev[1*], *Nikolay* Belyaev[2], and *Denis* Boyarkin[1]

[1] Melentiev Energy Systems Institute of Siberian Branch of the Russian Academy of Sciences, 664033, 130, Lermontov Street, Irkutsk, Russia
[2] Scientific and Technical Center of Unified Power System, 194223, 1, lit A, Kurchatov Str., St.-Petersburg, Russia

Abstract. The article deals with the problem of generating calculation models (CM) of electric power systems (EPS) into reliability zones, designed to perform calculations of adequacy assessment indicators. The substantiation of the depth of detail of the CM is based on taking into account two properties of the EPS: reliability, namely, on the basis of calculated estimates of the limits of applicability of deterministic and probabilistic reliability and efficiency criteria, namely, on the basis of a comparison of cost estimates of measures to ensure reliability. To solve this problem, it is proposed to use a number of so-called engineering algorithms, which take into account the reliability and efficiency of the EPS. The experimental part presents the results of applying the proposed algorithms for the formation of the CM analyzed by the EPS.

1 Introduction

The formation of calculation models (CM), in the general case, is justified by the large dimension of the task of assessing and synthesizing adequacy and by the maximum reduction of the network component, which has little effect on the adequacy indicators. The formation of CM should be based on the structure of the electric power system (EPS), technical and reliability characteristics of the elements, regime and economic features of the operation of the EPS. CM are presented in the form of multi-zone models, including reliability zones (RZ) and inter-zone communications (ZC). RZ includes a part of the EPS, within which restrictions are unlikely in the transmission of active power in all possible modes for the period of the assessment of the adequacy. RZ contains a certain number of power stations and a certain load of consumers. ZC is a set of power transmission lines (TL) between RZ. Units at power plants, the load of consumers in the RZ and power lines in interzone connections when assessing the adequacy, are characterized by a certain set of reliability and technical parameters taken into account when assessing the adequacy.

In fact, the procedure for determining the CM into reliability zones and the formation of the CM is one of the stages in the adequacy assessment, since the correctness of the determination of the CM directly affects the indicators of adequacy and further problem solving based on the obtained adequacy indicators (AI), for example, justifying the redundancy level of generating capacity. The methodology for adequacy assessment, taking into account the stage of determining the CM based on the Monte Carlo method, in general, will represent the following sequence of computational steps:

1. Formation of CM (per-node) EPS into reliability zones.
2. Formation of a set of random states of the EPS for the period of adequacy assessment.
3. Minimization of the power deficit of the calculated states of the EPS.
4. Calculation of AI.

As a rule, the adequacy assessment is carried out for an annual period, taking into account many parameters that affect the reliability of the EPS, including the hourly power consumption change schedule, which for different RZs can have significant changes during the year. A change in the parameters taken into account during the analyzed period can have a significant impact on the AI values and, accordingly, form the requirements for the multiple formation of CM in the adequacy assessment cycle.

As already noted, the task of forming the CM is one of the keys in the entire process of adequacy assessment. The correctness of the obtained AI and the validity of subsequent actions depend on the correctness of its solution. Let us dwell on some aspects of the Russian practice of forming the CM. Initially, the adequacy assessment was carried out by representing the analyzed EPS of one RZ. In this case, the reliability of only generating sources and the probabilistic characteristics of the load of consumers were taken into account [1,2]. The representation of CM in this form is explained by the lack of methodological developments for the analysis of multizone models and the insufficient computing base for solving more complex problems. Most of the existing approaches to the formation of CM are of an expert nature and are based on the organizational

*Corresponding author: krupenev@isem.irk.ru

principles of EPS management. For example, one of the approaches to the formation of the CM of the Unified Energy System (UES) of Russia is its division into 7 reliability zones, where each zone corresponds to one of the unified energy systems (IPS) included in the UES of Russia. The connections between them consist of intersystem transmission lines. With the introduction of market relations in the domestic electric power industry, attempts were made to form a CM for adequacy assessment the in accordance with the free flow zones, but the division of the UES in this way, despite the deeper detail, does not have a rigorous justification for the purposes of adequacy assessment. From the standpoint of the information content of the CM, it is preferable to form the CM in such a way that each RZ represents the power system of a subject of the Russian Federation [3]. In all the above methods of forming the CM of the UES of Russia, a number of questions remain open, including the determination of the bandwidth of the ZC.

It is worth noting the foreign practice of forming the CM. For example, in the United States, the assessment of adequacy is carried out for a twenty-zone calculation model [4, 5]. The formation of reliability zones was carried out expertly using a representative sample of retrospective information about the functioning and reliability of the US energy system. The European Association for the Cooperation of Operators of Transmission Systems for Electricity (ENTSO-E) generates reports on medium-term forecasts of the adequacy of the European energy system [6]. The scope of analysis includes ENTSO-E member countries (42 transmission system operators). The calculation scheme for the analysis of the adequacy is formed according to the borders of the energy systems of the states, i.e. in practice, the energy system of each state represents the RZ. For a more detailed analysis of the situation in the field of adequacy of energy systems within states, an additional division of energy systems into reliability zones is carried out, usually based on an expert approach.

The purpose of this article is to present and analyze a number of engineering algorithms for the formation of CM.

2 The Criterion for the Formation of the CM of the Electric Power System Into Reliability Zones

Substantially, the task of determining the CM is formulated as follows: for the known structure of the EPS, the characteristics of generating capacities and power consumption schedules, the limitations of the capacity of power transmission lines and sections of the electric network, the organizational and economic conditions for the functioning of the EPS, it is necessary to determine the boundaries of the RZ for the formation of the CM and further adequacy assessment.

Mathematically, the problem of determining the CM on the RZ can be formulated as follows: given $X = \{x_1, x_2, \dots, x_N\}$ – the set of nodes of the EPS design scheme, specified in the k – dimensional space of features:

$$x^k = \{x_1^k, x_2^k, \dots, x_N^k\}, k = 1, \dots, K. \tag{1}$$

There is a distance function (metric) $\rho(x)$ between the nodes of the EPS design scheme. It should be understood that the distance functions are not a direct distance between the nodes of the calculation scheme. The distance function is formed from a set of operating characteristics of the power system, taken into account when determining the CM.

It is required to split the finite sample of nodes of the calculation scheme X into Z non-intersecting subsets:

$$S_z, z = 1, \dots, Z;\ X = \bigcup_{z=1}^{Z} S_z, \tag{2}$$

which form the RZ Z, in such a way that each RZ includes nodes whose metric ρ is less than the threshold value; otherwise, the nodes are sorted into different RZs.

In this case, each node $x_i \in X$ is assigned the number of the reliability zone $z_j \in Z$.

For the problem being solved, the most appropriate strategy would be the "divide and conquer" strategy, which consists in dividing the considered set into groups of similar objects. In the case under consideration, the number of reliability zones will tend to a minimum, i.e.:

$$Z \underset{\rho}{\rightarrow} min. \tag{3}$$

When solving real practical problems, one must understand that the division of the system into reliability zones is conditional and depends on the criteria used that form the metric ρ.

3 Formation of CM Based on Controlled Sections

When managing a power system, there is a set of controlled sections (a set of parallel connected network elements), the loading of which during operation is controlled by the system operator for not exceeding the maximum allowable flow, determined by the criteria for the allowable current load of the elements, static and dynamic stability in normal and post-emergency modes [7]. Based on taking into account the controlled sections, the algorithm for dividing the EPS into RZs can be presented in the following form:

1. Formation of the composition of the electrical network elements included in the specified controlled sections.

2. Identification of the EPS reliability zones based on the composition of the EPS elements specified in clause 1. If an object or a set of EPS objects (stations, substations) is separated by a controlled section or a set of EPS objects (stations, substations) or a set of controlled sections from the rest of the EPS, then the object or a set of objects form a reliability zone.

3. Bandwidths of interzone connections are determined in accordance with the maximum allowable active power flows over controlled sections. The generating capacity and load of the reliability zone is formed from generating units and consumers included in the reliability zone.

This method of determining the CM is acceptable for the EPS development planning horizon up to 6 years, during which the EPS structure will change insignificantly and these changes are unambiguously determined by the development scheme and program [8]. The undoubted positive side when using this approach is the certainty with the bandwidth of the ZC, which are formed on the basis of the values of the maximum allowable active power flows.

4 Formation of CM Based on Express Analysis of Adequacy

The most common in the practice of designing and operating EPS are deterministic reliability criteria of the type "*N-k* " [9]. It is obvious that for modern EPS with a large number of elements, the use of deterministic criteria for the analysis of adequacy is unacceptable, since the modes corresponding to the criteria "*N*-1" and "*N*-2" cover only a small fraction of the possible states of such EPS. Nevertheless, these criteria are quite correct to apply for express adequacy assessment. In this case, the adequacy assessment using deterministic criteria can lead to an acceptable result.

Let us estimate the boundaries of the possible application of the "*N*-1" criterion for an energy district, the power supply of which is carried out from N elements of the EPS. When the criterion "*N*-1" is met, the design states of the system under consideration are all states with a failure of no more than one element out of *N*. The probability of deficit-free operation *P* in the case under consideration can be calculated by the formula:

$$P = \prod_i (1-q_i) + \sum_i (q_i \cdot \prod_{j \neq i} (1-q_j)), i = 1, \dots, I, \quad (4)$$

where q_i is the probability of an emergency state of the i-th element. The results of calculating *P* depending on *N* at $q_i = 0.02$ are shown in Table 1.

Table 1. Probability of calculated states according to the criterion "*N*-1"

N	*P*	*N*	*P*
2	0,9996	7	0,9921
3	0,9988	8	0,9897
4	0,9977	9	0,9869
5	0,9962	10	0,9838
6	0,9943		

According to Table 1, it can be noted that with the normative level of adequacy (the probability of deficit-free work (DFP) equal to 0.996 [10]), the analysis of adequacy using the criterion "*N*-1" is possible at $N \leq 5$. In this case, the total probability of all calculated states above 0.996, and if the "*N*-1" criterion is met, that is, if there is no power shortage in case of failure of any one element, the specified standard level will be certainly provided. Accordingly, the allocation of energy districts, the power supply of which depends on 5 or less EPS elements, into separate RZs is not advisable. Deterministic criteria "*N-k*" can be correctly applied to the analysis of adequacy in relation to such energy regions.

The method based on the Monte Carlo method, but with a reduction in the number of analyzed conditions, can also serve as an express method for adequacy assessment. One of the main computational problems in applying the method for adequacy assessment based on the Monte Carlo method is the large dimension of the problem being solved, which leads to unreasonably high time costs. It should be noted that it is precisely because of the decrease in the dimension of the problem that the formation of the CM is required. When assessing adequacy based on the Monte Carlo method, they try to achieve a given accuracy criterion for each hour of the analyzed period [11,12]. For the EPS scheme, taking into account all nodes and transmission lines, for which the formation of the CM is carried out, this is an extremely time-consuming procedure, and in some cases simply impossible. In this case, one of the solution methods may be to simplify the problem, for example, to evaluate only for the hour of the maximum load of the EPS or for the monthly peaks of the load of the EPS and use the obtained indicators in determining the CM.

By definition, restrictions on power flows in the RZ are unlikely in all probable modes of operation of the EPS. From this it follows that if the DFP in adjacent nodes of the system take relatively close values, then such nodes can be combined into a reliability zone. The criterion in this case will be the value of the difference between the DFP in adjacent reliability zones:

$$\Delta = |P_i - P_j|, i = 1, \dots, n, j = 1, \dots, n, i \neq j, \quad (5)$$

where: n is the number of EPS nodes.

The algorithm for determining CM based on the proposed criterion will be as follows:

1. Adequacy assessment of the complete EPS scheme (a scheme with all nodes and power lines) using an express method.
2. Determination of the difference in the DFP between adjacent nodes of the EPS.
3. If the difference is greater than the specified criterion Δ, then such nodes are distributed over different reliability zones.
4. If the difference is less than the given criterion Δ, then such nodes are defined in one reliability zone.
5. If in ring connections the difference between adjacent nodes is inconsistent (i.e. between different nodes Δ takes values greater than or less than a given criterion), then such nodes are combined into one reliability zone.

One of the main tasks in the formation of CM is the determination of the bandwidth of the ZC. In the case of determining the CM on the basis of the controlled sections, discussed above, this problem is solved. In other approaches, this problem needs to be solved. In the general case, the throughput of the ZC is determined in accordance with [7], in the absence of the necessary information for this, based on reference data [13] in accordance with the expression:

$$P_{max} = min(P_{perm}; 0{,}8\frac{U_r^2}{x_0 l}), \quad (6)$$

where P_{perm} is the maximum transmitted power under the conditions of a long-term permissible current load, adopted according to [13] (for a 500 kV transmission line - 1700 MW, for a 220 kV transmission line - 300 MW); U_r is rated voltage of power transmission lines, kV; x_0 – linear inductive resistance of power transmission lines (Ohm/km); l is the length of the power transmission line (km).

5 Formation of CM in Accordance with the Economic Criterion

As already noted, one of the main objectives of the adequacy assessment is to substantiate the need for EPS in generating capacities and their distribution over the RZ in accordance with the restrictions on power transmission over the ZC. If an insufficient level of adequacy is identified, investment decisions are made on the construction of new generating capacities, while their volume and location are determined, among other things, by the specified network restrictions. An alternative to the construction of new generating capacities in a specific reliability zone may be to strengthen the electrical connections incident to it, provided that there is a sufficient power reserve in adjacent zones.

From this we can conclude that when forming the CM, it is sufficient to take into account the restrictions on power transmission only for those sections of the electric network, the cost of strengthening which is comparable to the cost of building new generating capacities. Taking into account other network restrictions, the elimination of which is obviously characterized by less capital intensity, will not only lead to excessive detailing of the CM and complicate the procedures for assessing and analyzing adequacy, but can also potentially lead to inefficient investment decisions for the development of generating capacities. For example, if such decisions are due to local restrictions in the distribution network, taken into account in the CM, in the presence of excess power in the EPS as a whole.

In accordance with the above, we will perform an assessment of network restrictions, the consideration of which is economically feasible when justifying the boundaries of reliability zones in CM. As initial data for such an assessment, we use aggregated price standards for standard technological solutions for the capital construction of electric power facilities in terms of electric grid facilities [14] and indexed indicators of specific capital investments in gas generation generating facilities, which, as a rule, closes the power balance, according to [15].

For Russian conditions, according to [14], specific capital investments (SCI) in an overhead TL of a voltage class of 500 kV in a typical layout, taking into account the substation part, are estimated at 59 million rubles per 1 km of length, into a single-circuit power transmission line of a voltage class of 220 kV - 28 million rubles per 1 km of length in 2021 prices. SCI in generation according to [15], taking into account indexation for 2021 and verification for specific projects, is estimated at an average of 50 million rubles per 1 MW of installed capacity. The specified SCI values are given without taking into account the cost of land acquisition.

To calculate SCI per unit of throughput of power lines, it is required to set their throughput, which can be determined by (6). SCI per unit of transmission capacity of a transmission line depends on its length and is defined as the ratio of the cost of a transmission line, determined on the basis of SCI according to the data of [14], to the transmission capacity determined by formula (6).

The results of the performed calculations are shown in Figs. 1. On the presented graph, along the abscissa axis, the length of the transmission line is plotted, along the ordinate axis - SCI per unit of transmission capacity of the transmission line. The red and blue markers show the dependencies for the 500 kV transmission line and the 220 kV transmission line, respectively. The black dotted line indicates the SCI level in the generation.

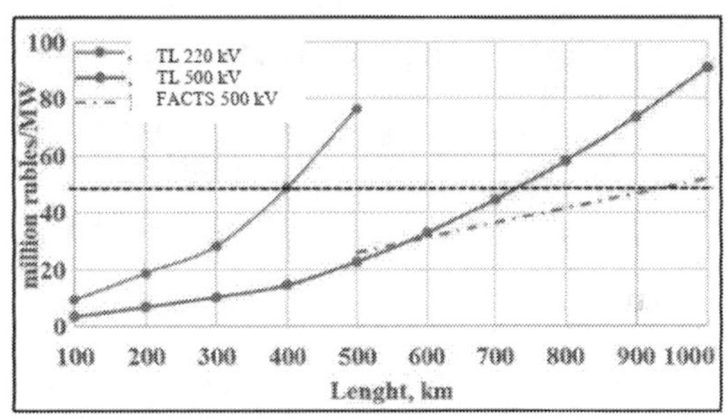

Fig. 1. Dependences of SCI per unit of transmission capacity of a transmission line on its length.

Considering the obtained dependencies, we can note the following. Dependences of SCI per unit of throughput for short transmission lines (up to 400 km for 500 kV transmission lines and up to 300 km for 220 kV transmission lines) are linear; This is due to the fact that for short transmission lines, the throughput is determined by the conditions of the permissible current load and does not depend on the length, and for long power lines, the throughput is determined by the stability conditions and decreases with increasing length, which leads to a quadratic increase in SCI. The SCI level in generation for a 500 kV transmission line is achieved with a length of about 750 km, for a 220 kV transmission line - with a length of 400 km.

Based on these results, an unambiguous conclusion can be made about the expediency of separating nodes located at a distance of more than 750 km into separate

RZ CM. In this case, if it is necessary to increase the AI, solutions for the construction of new generating capacities will definitely be more economical. The construction of a new transmission line of equivalent transmission capacity will require more costs and can only be justified taking into account additional effects, for example, optimizing the balance of electrical energy.

On the contrary, based on the proposed approach, it is advisable not to take into account the restrictions on power transmission through power lines up to 400 km long during the formation of CM, since the cost of strengthening such sections of the network is lower than the construction of new generating capacities in the corresponding volume.

It is necessary to consider in more detail the case in which the distance between the nodes is from 400 to 750 km. In this case, SCI for 500 kV transmission lines, as seen in Fig. 1, below SCI into generation. However, it should be taken into account that the cost of a 500 kV power transmission line of this length is 23.5 - 41 billion rubles, which corresponds to the cost of building a new generating facility with an installed capacity of 500 - 800 MW. Therefore, if the demand for power is below the specified range, then the construction of a new generation is more economical than a new transmission line.

Thus, based on the economic principle of detailing the CM, we can make the following recommendations for their formation:

1. EPS nodes located at a distance of up to 400 km should be combined into one RZ;

2. EPS nodes located at a distance of more than 750 km should be attributed to different RZs, restrictions on power transmission via power lines between them should be taken into account when determining the throughput of the corresponding ZC;

3. EPS nodes located at a distance of 400 to 750 km should be combined into one RZ if the potential need for the power of these nodes is more than 500 - 800 MW.

These recommendations can be adjusted taking into account the use of non-capital-intensive means of increasing the capacity of transmission lines. It is known that modern means of power flow control (phase-shifting devices, controlled reactive power compensation, etc.), collectively referred to in the literature as flexible alternating current transmission systems (FACTS), can significantly increase the capacity of long-distance transmission lines up to the permissible current load. At the same time, the coefficient of appreciation of such transmission lines due to the installation of appropriate devices, the construction of intermediate substations and switching points for them, based on the data of [12], can be estimated at 1.5. Then the dependence of SCI per unit capacity of long-range 500 kV transmission lines with FACTS will have the form indicated in Fig. 1 with a red dotted line.

Based on these results, when applying FACTS, the distance limit given in paragraph 1 can be increased to 1000 km. However, taking into account point 3 and the cost of such power transmission, it is advisable to combine nodes located at a distance of 500 - 1000 km into one reliability zone only if the potential need for the power of these nodes is more than 1000 - 1500 MW.

6 Experimental Studies

We will show the operation of each of the proposed CM generation algorithms based on the application of the IEEE RTS-96 test circuit [16].

First, we will demonstrate the division of the EPS into reliability zones in accordance with the designation of controlled sections. Since the controlled sections are not defined for the scheme under consideration, they were determined by experts. The following shows the identification of reliability zones for a part of the scheme under consideration according to the algorithm presented in the article (Fig. 2).

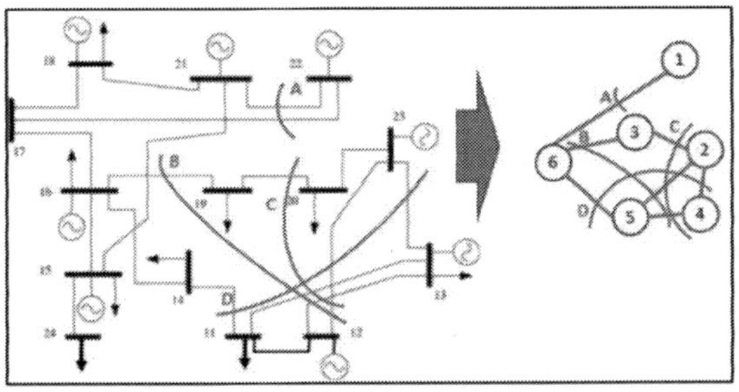

Fig. 2. Formation of CM into reliability zones along controlled sections.

Next, the formation of CM was carried out on the basis of the analysis of AI obtained by the express assessment method. In table 2 presents the probability values of a deficit-free operation of the nodes of the analyzed system.

Table 2. DFP of the considered EPS.

Node number	The probability of a deficit-free work	Node number	The probability of a deficit-free work
1	0,9661	13	0,9886
2	0,9676	14	0,9791
3	0,9656	15	0,9843
4	0,9696	16	0,9881
5	0,9674	17	0,9999
6	0,9681	18	0,9916
7	0,9917	19	0,9899
8	0,9688	20	0,9909
9	0,9729	21	0,9916
10	0,9717	22	0,9937
11	0,9999	23	0,9930
12	0,9999	24	0,9999

The value Δ=0.004 was chosen as a criterion for the formation of CM. As a result, 12 reliability zones were obtained. Table 3 shows the attribution of nodes to the RZ, and in fig. 3 shows the formed RZs in the EPS scheme.

Table 3. Assignment of nodes to reliability zones.

RZ number	Node number	RZ number	Node number
1	1,2,3,4,5, 6,8,9,10	7	15,16
2	7	8	17
3	11	9	19,20,23
4	12	10	18,21
5	13	11	22
6	14	12	24

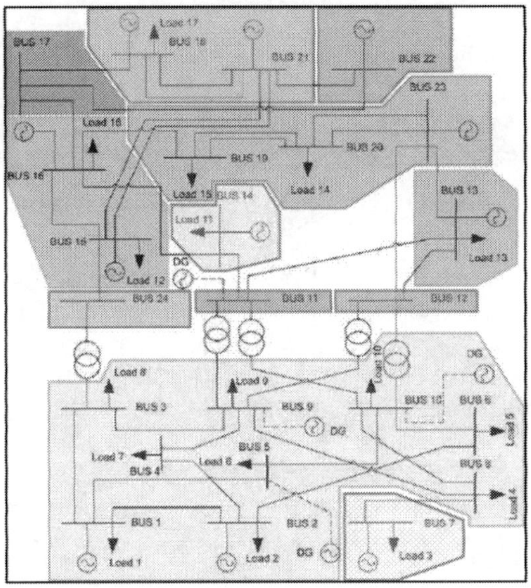

Fig. 3. The result of the formation of CM based on the express analysis of adequacy.

The next stage of experimental research is the formation of CM in accordance with the economic criterion. As a result, it turned out that the formation of CM according to this criterion for the considered EPS is impossible, since the distances between the nodes are less than 400 km, therefore, in the case of determining the optimal level of redundancy of the generating capacity, the system can be represented by one RZ.

7 Conclusion

In the present study, the problem of generating energy calculation models that are used in adequacy assessment of electric power systems is considered. The existing domestic and foreign practice of generating energy calculation models was analyzed and engineering methods for solving the problem under consideration were proposed.

The research was carried out under State Assignment Project (no. FWEU-2021-0003) of the Fundamental Research Program of Russian Federation 2021-2030 and partly supported by the Russian Foundation of Basic Research, Grant No. 21-03-12345 using the resources of the High-Temperature Circuit Multi-Access Research Center (Ministry of Science and Higher Education of the Russian Federation, project no 13.CKP.21.0038).

References

1. U. Rudenko, M. Chelzov, Reliability and redundancy in electric power systems. (Nauka, Novosibirsk, 1974) *In Russ.*
2. I. Markovich, Power systems modes (Energia, Moscow, 1969) *In Russ.*
3. G. Kovalev, D. Krupenev, L. Lebedeva. Electricheskie Stancii **2**, 44-47 (2011) *In Russ.*
4. NERC, *Probabilistic Adequacy and Measures. Technical Reference Report Final* (NERC, Atlanta, Georgia, July, 2018)
5. NERC, *2021 Long-Term Reliability Assessment* (NERC, Atlanta, Georgia, 2021)
6. ENTSO-E, *Mid-term Adequacy Forecast Executive Summary* (ENTSO-E, Brussels, Belgium, 2020)
7. Ministry of Energy, Order of the Ministry of Energy of the Russian Federation dated August 3, 2018 No. 630 "On Approval of Requirements for Ensuring the Reliability of Electric Power Systems, Reliability and Safety of Electric Power Facilities and Power Receiving Installations "Guidelines for the Stability of Energy Systems" (Ministry of Energy, Moscow, Russia, 2019)
8. Government of the Russian Federation, Decree of the Government of the Russian Federation of October 17, 2009 No. 823 (as amended on June 1, 2022) "On schemes and programs for the long-term development of the electric power industry" (Government of the Russian Federation, Moscow, Russia, 2009)
9. Ministry of Energy, Draft Order of the Ministry of Energy of Russia (modified text) "On approval of the Guidelines for the design of the development of energy systems" (Ministry of Energy, Moscow, Russia, 2021)
10. Ministry of Energy, Decree of the Ministry of Energy of the Russian Federation dated April 30, 2021 No. 321 "On the establishment of a regulatory level of balance reliability for the Unified Energy System of Russia, used in assessing the possibility of decommissioning generating equipment" (Ministry of Energy, Moscow, Russia, 2021)
11. G.F. Kovalev, L.M. Lebedeva, Reliability of Power Systems (Springer, 2019)

https://doi.org/10.1051/e3sconf/202338401024

12. D. Krupenev, D. Boyarkin, D. Iakubovskii, Reliability Engineering & System Safety, **204** (2020)
13. D. Faibisovich et al., Handbook for the design of electrical networks. (ENAS, Moscow, 2012) *In Russ.*
14. Ministry of Energy, Order of the Ministry of Energy of Russia No. 10 dated January 17, 2019 "On Approval of Aggregated Standards for the Price of Typical Technological Solutions for the Capital Construction of Electric Power Facilities in Part of Electric Grid Facilities" (Ministry of Energy, Moscow, Russia, 2019)
15. Government of the Russian Federation, Rules for determining the price of capacity sold under capacity supply agreements, approved by Decree of the Government of the Russian Federation No. 238 dated April 13, 2010 (Government of the Russian Federation, Moscow, Russia, 2019)
16. C. Grigg et al., IEEE Transactions on Power Systems, **14(3)**, 1010-1020 (1996)

Methodology for ranking critical energy facilities on the example of some emergencies in the gas industry

Sergey Senderov[1, 2]*, *Dmitry* Krupenev[1, 2], *Sergey* Vorobev[1], *Natalia* Beresneva[1], *Denis* Boyarkin[1]

[1] Melentiev Energy Systems Institute of Siberian Branch of the Russian Academy of Sciences, 130 Lermontov str., Irkutsk, Russia
[2] Federal State Budget Educational Institution of Higher Education «Irkutsk National Research Technical University», 83 Lermontov str., Irkutsk, Russia

Abstract. The paper presents the main components of a methodical approach to assessing the level of significance of critical energy facilities on the example of individual outages of the most important facilities of the Russian gas transmission network. Approaches to modelling the fuel and energy complex of the country and its constituent energy systems: gas, electricity and heat supply are presented. Algorithms for identifying critical energy facilities of the industry, regional and federal levels, as well as examples of their ranking depending on the level of expected consequences for consumers of electrical and heat energy in case of loss of performance of these critical facilities are given.

1 Introduction

In the context of the possible implementation of large-scale emergencies in the energy sector, accompanied by a decrease in the productivity of some energy facilities and related energy systems (ES), it is not always possible to ensure a deficit-free energy supply to consumers. In these cases, it is necessary to talk about minimizing the volume of possible shortages of final types of energy among consumers. Among the causes of large-scale emergencies in the energy sector, we can single out the causes caused by intentional impacts. It is logical to assume that intentional impacts can be aimed primarily at those facilities, the decrease in the efficiency of which can significantly reduce the production capabilities of some ES and the energy sector as a whole and can lead to significant shortages of the corresponding types of fuel and energy resources (FER).

It is logical to identify such critically important facilities (CF) of the ES and of the energy sector as a whole. Not all CFs of ES are CFs of the energy sector, i.e. the loss of productivity of not each of them can lead to shortages of final types of energy among consumers. The task of the study is to form an algorithm for identifying the CFs of ES and to form a methodological approach to determining the level of these CFs.

2 Algorithms for identifying CFs of ES and of the energy sector

2.1 CFs of ES

CFs of ES of the federal level can be recognized as objects, the termination of which may cause significant undersupply of the relevant fuel and energy resources throughout the country as a whole. At the same time, the shortage of supplies can be in relative terms δ_{sum} or more of the country's total need for this type of FER. As such δ_{sum} for the gas industry, earlier in [1, 2] a value of 5% was taken. Objects that are not included in the list of federal CFs of the considered ES can be defined as the CFs of the ES of the regional level, but the termination of their operation may cause undersupply of the corresponding FER to at least one of the regions in the relative value of δ_{reg} and more. As such δ_{reg} for the gas industry in [1, 2], a value of 40% was taken. The algorithm for the formation of lists of CFs at the regional and federal levels of any ES is shown in Fig. 1.

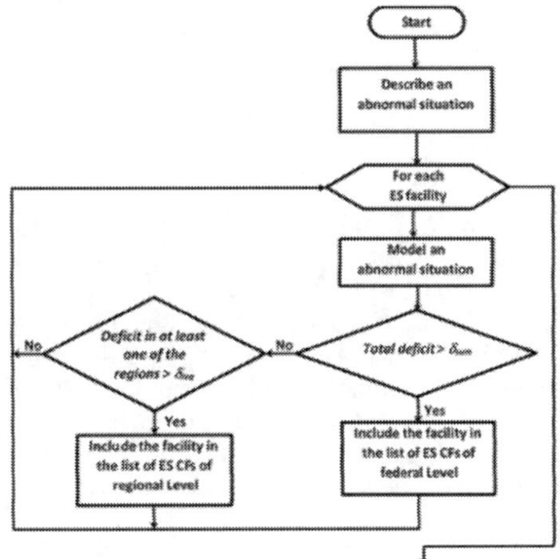

Fig. 1. Algorithm for the formation of lists of ES CFs.

Research is carried out by enumeration of various emergencies. The result of modeling studies is the

* Corresponding author: ssm@isem.irk.ru

magnitude of deficits of the corresponding FER for its consumers. Comparison of the relative value of the total deficit of this type of fuel and energy resources with the agreed value of the conditionally permissible relative deficit δ_{sum} allows us to single out some CFs of ES in their federal list. CFs of ES which were not included in this list should be checked for the formation (when they are turned off) of the deficit of this type of FER in the regions. If the relative deficit of this type of FER in any of the regions exceeds the conditionally permissible value δ_{reg}, then this CF falls into the regional list of CFs of this ES.

2.2 CFs of energy sector

The algorithm for the formation of the list of CFs at the level of the energy sector is shown in fig. 2. This algorithm begins with the analysis of the set of CFs ES of the federal level, which is formed during the operation of the algorithm presented above, in Fig. 1.

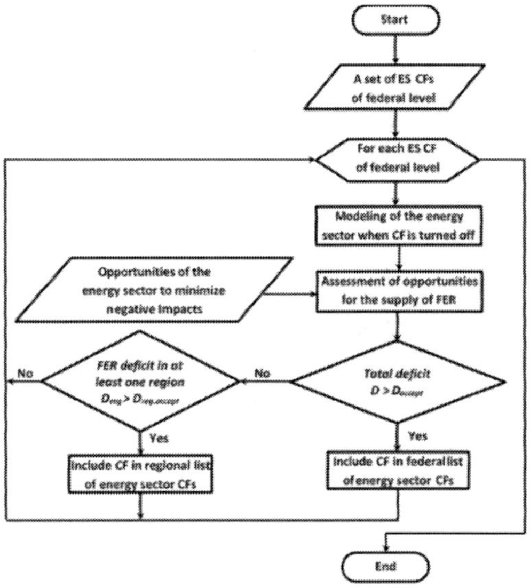

Fig. 2. Algorithm for the formation of lists of energy sector CFs.

Research is carried out using the economic-mathematical model of the energy sector [3, 4], while taking into account it's possibilities to minimize the deficit of FER. Comparison of the relative value of the total energy deficit among consumers with the value of the conditionally acceptable relative deficit (D_{accept}) makes it possible to form a federal list of energy sector CFs from some of the ES CFs of the federal level. The loss of operability of some ES CFs, which, according to the accepted rule, did not fall into the federal list of energy sector CFs, can lead to significant energy shortages among consumers in one or more regions. Facilities that are not included in the federal list of energy sector CFs, but capable of causing a shortage of energy resources greater than the conditionally acceptable ($D_{reg.accept}$) should form a regional list of energy sector CFs.

Let us dwell in more detail on the aspects of the functioning of the unified electric power system (UES) and the unified gas supply system (UGSS) of Russia that are closest to the task of research.

3 Modeling of emergency modes in the energy systems of Russia and of energy sector as a whole

3.1 Unified gas supply system

The gas industry of Russia, in the technological part, is represented by the UGSS and local gas supply systems in the east of the country. The UGSS, through a multi-line gas transmission network (GTS), connects gas fields in the European part of the country with natural gas consumers in the same European part and in the regions of Western Siberia.

The existing territorial structure of the UGSS of Russia has a number of significant drawbacks. Today, more than 85% of all Russian gas is produced in the northern regions of the Tyumen region. The main consumers of gas within the country - its European regions and gas export points - are located at a distance of 2-2.5 thousand km from the places of its production. All this gas is transported over long distances using multi-line gas transmission corridors, often with a significant concentration of gas flows in one corridor. These corridors have a large number of mutual intersections and bridges. Currently, GTS of Russia has several intersections of main gas pipelines (MG) that are potentially dangerous for the operation of the UGSS. Failure of some of them can lead to significant restrictions on gas supplies to consumers. A significant part of such intersections is UGSS CFs from the standpoint of ensuring GTS operability.

The ESI SB RAS has developed software "Oil and Gas of Russia" [1, 2]. This software includes a streaming simulation model of the UGSS of Russia. the research scheme of the model contains 382 nodes, including 22 underground gas storage facilities (UGS), 28 gas sources, 64 gas consumers (regions and points of gas export delivery), 268 nodal compressor stations (CS), 628 arcs (MG corridors and individual MG). When solving the problem of estimating the state of the system after a disturbance, the criterion for the optimal distribution of flows is the minimum gas deficit at the consumer at the minimum cost of its delivery.

Shutdown of objects leads to the solution of the problem of distribution of flows in the system in order to maximize the supply of gas to consumers. The model is formalized as a maximum flow problem [5, 6]. The calculation graph is supplemented with two fictitious nodes: O - total source, S - total sink, while additional sections are introduced that connect node O with all sources and all consumers with node S. The mathematical notation of the task has the following form:

$$\max f \tag{1}$$

under the conditions that

$$\sum_{i\in N_j^+} x_{ij} - \sum_{i\in N_j^-} x_{ji} = \begin{cases} -f, j=O \\ 0, j\neq O, S \\ f, j=S \end{cases} \quad (2)$$

$$0 \leq x_{ij} \leq d_{ij}\text{, for all (i, j)} \quad (3)$$

Here: N_j^+ a subset of arcs "incoming" to node j; N_j^- - a subset of "outgoing" arcs from node j; f - the value of the total flow through the network; x_{ij} - flow along the arc (i, j); d_{ij} - restrictions on the flow along the arc (i, j).

Problem (1)-(3) about the maximum flow has not a unique solution. The next step is to solve the problem of the maximum flow of minimum cost, i.e. the cost functional is minimized:

$$\sum_{(i,j)} C_{ij} x_{ij} \to \min \quad (4)$$

where C_{ij} is the price or unit costs for gas transporting.

The result of solving the problem is to determine the possibilities of satisfying consumers with gas, identifying the volumes of possible undersupplies of gas in one or another emergency situation. Based on these results, objects are identified, the cessation of which will lead to a potential shortage of gas in the network. By cutting off facilities, the shutdown of which will lead to a potential shortage of gas in the network less than the threshold value of 5%, a list of UGSS CFs is formed.

For example, let's choose the most indicative situations with the loss of productivity of some UGSS CFs. Using specialized models to minimize the shortage of electric power and information on the expected undersupply of gas to gas-consuming power generating facilities for the relevant unified electric power systems (UPS), we calculate the possible volumes of undersupply of electricity to consumers.

3.2 Unified electric power system

The purpose of UES modeling in this case is to determine the undersupply of electricity to consumers as a result of accidents at fuel and energy facilities. Problem to be solved: for the known structure, parameters of the elements and the schedule of power consumption of the UES, it is necessary to determine the non-admission of electricity for the period from the beginning of the emergency and the reduction of gas supply to gas-consuming power generating facilities until it is completely eliminated. The UES calculation model is a graph whose nodes are energy zones, and the arcs are interzone connections. The energy zone includes a part of the power system, as a rule, it is a regional power system, which contains a set of generating units and is characterized by power consumption in each hour of the billing period. Interzone communication includes power lines that connect energy zones. Thus, for each hour of the billing period, it is necessary to solve the following problem [7]:

need to find:

$$\sum_{i=1}^{I}(\overline{N}_{cons,i} - N_{cons,i}) \to min, \quad (5)$$

considering balance constraints

$$N_{gen,i} - N_{cons,i} + \sum_{j=1}^{J}(1 - z_{ji}\, a_{ji})z_{ji} - \sum_{j=1}^{J} z_{ij} = 0, i = 1, \ldots, I,\ i \neq j, \quad (6)$$

and linear inequality constraints on variables:

$$0 \leq N_{cons,i} \leq \overline{N}_{cons,i}, i = 1, \ldots, I, \quad (7)$$

$$0 \leq N_{gen,i} \leq \overline{N}_{gen,i}, i = 1, \ldots, I, \quad (8)$$

$$0 \leq z_{ji} \leq \overline{z}_{ji}, 0 \leq z_{ij} \leq \overline{z}_{ij}, j = 1, \ldots, J, i = 1, \ldots, I, i \neq j, \quad (9)$$

where: $\overline{N}_{cons,i}$ is the value of power consumption in energy zone i, MW; $N_{cons,i}$ – secured power consumption in energy zone i, MW; $\overline{N}_{gen,i}$ – available generating capacity in energy zone i, MW; $N_{gen,i}$ - used generating capacity in the reliability zone i, MW; $\overline{z}_{ji}$ – interzone communication capacity, MW; z_{ji}, z_{ij} – actual load of interzone communication, MW; $I=J$ is the number of energy zones.

3.3 Energy sector as a whole

The problem of modeling the energy sector as a whole is described in detail in [8] and mathematically represents a classical problem of linear programming. The statement and restrictions of this problem are written as follows:

$$AX - \sum_{t=1}^{T} Y^t = 0 \quad (10)$$

$$0 \leq X \leq D \quad (11)$$

$$0 \leq Y^t \leq R^t \quad (12)$$

where t - categories of consumers; A – matrix of technological coefficients of production (extraction, processing, transformation) and transport of certain types of fuel and energy; X - the desired vector, the components of which characterize the intensity of the use of technological methods of functioning of energy facilities (extraction, processing, transformation and transport of energy resources, fuel reserves); Y^t – the desired vector, the components of which characterize the volumes of consumption of certain types of fuel and energy by certain categories of consumers (t); D - a vector that determines the technically possible intensity of the use of individual technological and production methods; R^t is a vector with components equal to the volumes of the given consumption of certain types of fuel and energy by certain categories of consumers.

The objective function then has the following form:

$$(C, X) + \sum_{t=1}^{T}(r^t, g^t) \to \min \quad (13)$$

where, the first component reflects the costs associated with the functioning of the ESs. Here C is the vector of unit costs for individual technological methods of functioning of energy facilities; the second component is the damage from the shortage for each type of fuel and energy for each of the selected categories of consumers. The magnitude of the deficit of energy resources among consumers g^t is determined by the expression. The vector r^t

conditionally denotes "specific damage" at the consumer.

This study uses only a part of the capabilities of the presented model, implemented within the framework of the "Correctiva+" [9], concerning the production of thermal energy. As a result of the calculations, the following are determined: possible volumes of undersupply of thermal energy in the regions of the country in case of shutdown of the corresponding UGSS CFs, taking into account the interconnected work of all branches of the energy sector, reserves of energy resources, the possibility of interchangeability of various types of energy resources and their diversification.

4 Assessing the possibility of supplying energy resources in the event of the shutdown of some gas industry CFs

4.1 Natural gas

The studies were carried out on the model of the Russian gas industry described above. Baseline for calculations: average day of maximum gas consumption in the network (January 2021). On such days, the work of the network can be considered the most intense. The total gas flow through the network on that day, including exports, amounted to approximately 2,180 million m^3, while gas consumption within the country was about 1,550 million m^3/day. Studies have shown that a potential shortage of gas among consumers can be observed when the operation (one by one) of 449 UGSS facilities (242 nodes, 199 arcs and 8 intersections of the gas pipelines of the settlement graph) is stopped. Of this number of objects earlier [1; 2, 10, 11], 61 facilities were singled out, single shutdowns of which can lead to a relative gas deficit of 5% or more of the total gas demand for the UGSS. These facilities include 22 main line arcs between nodal CSs, 36 nodes (30 nodal CSs, 5 head CSs at field exits, 1 CS at UGS outlets) and 3 MG intersections between CSs. These objects represent a modern list of UGSS CFs.

Let us consider an illustrative example to assess the possibilities of reliable supply of electrical and thermal energy to end consumers and, thereby, to classify certain UGSS CFs as CFs of the energy sector. To do this, we will single out several UGSS CFs located at different points of the gas transmission network. The selected CFs include three intersections of MG corridors between CSs and five nodal CSs at MG intersections.

Based on the solution of problem (1)-(4), the possibilities of gas supply to consumers were assessed with the help of the software "Oil and Gas of Russia" in the event of shutdown of the selected UGSS CFs. Each study dealt with the failure of one CF. For example, Table 1 below provides information on the estimated possibilities of gas supply to the regions of the Russia in the event of the loss of operability of one of the UGSS CF. The real names of the UGSS CFs have been replaced with conditional ones. Table 2 provides information on the total possibilities of satisfying gas consumers in the event of a loss of operability of the UGSS CFs. These CFs in Table 2 are ranked by the value of the total gas deficit.

Table 1. Loss of operability of the UGSS CF "Intersection No. 1".

Region	Consumption	Supply	Deficit
	million m^3/day		%
Perm region	66	0,7	99
Kirov region	13,3	0	100
Rep. of Udmurtia	13,2	5,6	58
Rep. of Bashkortostan	62,4	10	84
Rep. of Tatarstan	69	50	28
Rep. of Mari El	5	0	100
Rep. of Chuvashia	7,7	0	100
Rep. of Mordovia	10	0	100
Nizhny Novgorod region	29,3	5,5	81
Kostroma region	18,2	0	100
Arkhangelsk region	14	0	100
Vologda region	33,8	0	100
Rep. of Kareliaя	3,7	0	100
Leningrad region	77,5	67,7	13
Novgorod region	15,4	12,7	18
Smolensk region	9,6	0	100
Kaluga region	6,9	4,1	41
Moscow region	140,7	98,4	30
Tula region	29,4	0	100
Bryansk region	8	0	100
Oryol region	5,5	0	100
Lipetsk region	17,5	0	100
Voronezh region	21,1	0	100
Kursk region	8	0	100
Belgorod region	28,6	0	100
Total for UGSS	**1550**	**1091**	**30**

Table 2. Total possibilities of gas supply in the event of a loss of operability of the UGSS CFs.

CF	Consumption	Deficit
	million m^3/day	%
Intersection No. 2	1217	21
CS 1	1220	21
CS 2	1351	13
CS 3	1357	12
CS 4	1389	10
CS 5	1442	7
Intersection No. 3	1454	6

The values of the calculated undersupply of gas, noted in tables 1, 2, will affect the process of fuel supply to the corresponding gas-consuming power generating sources. Due to the fact that different types of electricity generating sources operate in the UES of Russia, adequate modeling allows us to answer the question: how problems with shutting down CFS in the gas industry can affect the supply of electricity to consumers.

4.2 Electricity

The assessment of the undersupply of electricity was carried out for the conditions of the functioning of the UES of Russia in 2022, [12]. Undersupply of electricity was estimated for the period of the first two weeks of January. When assessing the undersupply of electricity, an accounting of scheduled repairs of generating equipment was carried out [7]. Table 3 shows the ratio of undersupply of electricity in the conditions of shutdown of the UGSS CF to the values of the required level of electricity consumption in the UES of Russia and in the United Power Systems (UPS) of Russia for the period under review.

Table 3. Expected relative undersupply of electricity in the UES of Russia in the event of a loss of operability of the UGSS CF.

United Power System	Required level of electricity consumption, billion kWh/day	Scenario of the UGSS CF shutdown, undersupply* (%)							
		Inter-section No. 1	Inter-section No. 2	CS 1	CS 2	CS 3	CS 4	CS 5	Inter-section No. 3
UES of Russia (without the UPS of the East)	45260	0,3	-	0,6	-	0,7	1,6	0,3	0,2
UPS of Northwest	4599	2,4	-	-	-	-	-	2,8	-
UPS of Center	9549	-	-	-	-	-	-	-	-
UPS of Middle Volga	4668	-	-	-	-	-	-	-	-
UPS of South	4722	-	-	-	-	-	-	-	-
UPS of Ural	11821	-	0,1	2,1	-	2,6	6,2	-	0,8
UPS of Siberia	9901	-	-	-	-	-	-	-	-

* - "-" means less than 0.1

Taking into account the magnitude of possible undersupply of electricity by the UPSs, it is clear that the shutdown of "Intersection No. 2", "Intersection No. 3" and "CS 2" can be largely compensated by the capabilities of the UES of Russia itself. From the standpoint of possible undersupply of electricity, these UGSS CFs cannot be attributed to the energy sector CFs. All the rest of the selected UGSS CFs, i.e. "Intersection No. 1", "CS 1", "CS 3", "CS 4", "CS 5" can be attributed to the energy sector's CFs of regional level.

4.3 Thermal energy

Undersupply of gas will also affect the generation of thermal energy in some regions. In general, in Russia, the share of thermal energy generated from natural gas in 2021 amounted to almost 76%, in the European part of the country this value exceeded 89% (calculations based on [13]). Estimated values of possible undersupply of thermal energy by regions within the UGSS coverage area in case of shutdown of the UGSS CHP are presented in Table 4.

Table 4. Relative undersupply of thermal energy in the regions in the event of a shutdown of the UGSS CFs, %.

Federal district (FD), region	Share of heating power on gas	Scenario for shutdown of the UGSS CFs							
		Intersection No. 1	Intersection No. 2	CS 1	CS 2	CS 3	CS 4	CS 5	Intersection No. 3
Russia (without Far Eastern FD	**76**	**10**	7	**5**	**4**	**1**	**3**	**2**	**2**
Central FD		*14*	*10*	*7*	*10*	*5*			
Belgorod region	98	62							
Bryansk region	99	63							
Voronezh region	96	61	61	61	61	11			
Kostroma region	79	50							
Kursk region	95	60	60	60	60	60			
Lipetsk region	80	51	51		51				
Oryol Region	99	63	63	63	63	63			
Smolensk region	93	59							
Tula region	87	56	56	56	56	56			
Northwestern FD		*11*	*5*	*5*	*5*			*16*	
Republic of Karelia	60	38						38	
Arkhangelsk region	52	33							
Vologda Region	77	49	49	49	49			49	
Leningrad region	52							23	
Privolzhsky FD		*18*	*15*	*8*	*2*				*8*
Republic of Bashkortostan	98	25	25	18					
Mari El Republic	91	58	58	58	58				
Republic of Mordovia	99	63	63	63	63				
Chuvash Republic	99	63	63	63					
Perm region	90	54	54						54
Kirov region	78	50	50	50					50
Nizhny Novgorod Region	96	29							
Ural FD							*21*		
Tyumen region	71						45		
Chelyabinsk region	84						23		
Siberian FD							*3*		
Altai region	13						8		
Omsk region	58						24		

 https://doi.org/10.1051/e3sconf/202338401025

It can be seen from the Table 4 data that in most cases consumers of the Central, Northwestern and Privolzhsky federal districts will suffer the most. First of all, this is due to the high share of gas in the production of thermal energy in these regions. According to the relative values of the possible undersupply of thermal energy, it can be seen that the shutdown of the first 3 UGSS CFs (one by one) can lead to restrictions in the production of thermal energy in Russia in the UGSS coverage area equal to or more than 5%. These CFs should be included in the federal list of energy sector's CFs. The shutdown of the remaining 5 UGSS CFs can lead to restrictions on thermal energy in some regions. These objects should be attributed to the regional level of energy sector CFs.

4.4 Energy sector as a whole

In Table 5, the results of the analysis on the attribution of the UGSS CFs to various lists of the energy sector CFs are brought together.

Table 5. Results of classifying UGSS CFs as the energy sector CFs.

UGSS CF	Energy systems CFs				Energy sector CFs	
	Power industry		Heat power industry			
	Fed.	Reg.	Fed.	Reg.	Fed.	Reg.
Intersection No. 1		+	+		+	
Intersection No. 2			+		+	
CS 1		+	+		+	
CS 2				+		+
CS 3		+		+		+
CS 4		+		+		+
CS 5		+		+		+
Intersection No. 3				+		+

Thus, the studies have shown that all analyzed UGSS CFs in terms of the degree of consequences (in case of their shutdown) can be considered as CFs of the energy sector. The first three of those CFs listed in Table 5 refer to the federal level, while the rest CFs refer to the regional level of the energy sector CFs.

5 Conclusion

The paper presents the main provisions of the methodological approach to determining the level of CF in the energy sector. Using the analysis of the interconnected operation of the UGSS, UES of Russia and regional heat supply systems as an example, it is shown that not all critical objects of one energy system can significantly affect the reliability of energy supply to consumers from another energy system directly related to the first one. An analysis of the results of model studies showed that all of the UGSS CFs selected for the example are CFs of the energy sector, but some of them should be included in the federal list, while others should be included in the regional list.

The developed approach allows us to formulate the procedure for identifying energy facilities, which are the CF of the federal and regional levels of the energy sector. First of all, attention should be paid to such facilities from the standpoint of ensuring reliable fuel and energy supply to consumers and, in general, from the standpoint of ensuring the requirements of the country's energy security at the federal and regional levels. The corresponding strategic task in the development of the energy sector should be the task of forming directions and specific ways to reduce the critical importance of such critical facilities for the potential operability of the corresponding ESs and the energy sector as a whole.

The research was funded by the Russian Science Foundation, grant no. 23-29-00460, using the resources of the High-Temperature Circuit Multi-Access Research Center (Ministry of Science and Higher Education of the Russian Federation, project no 13.CKP.21.0038).

References

1. Senderov S.M., Rabchuk V.I., Edelev A.V. Izvestiya RAN. Energetika, **1**, pp. 70-78 (2016)
2. Senderov S., Edelev A. Energy, Vol. **184**, pp. 105-112 (2019)
3. Senderov S., Krupenev D., Pyatkova N. Energy Systems Research, **Vol. 2**, **4**, pp. 32-42 (2019)
4. Pyatkova N.I., Senderov S.M. Electric energy. Transmission and distribution, **Vol. 4**, **61**, p. 50-53 (2020)
5. Khramov A.V., Enikeeva S.M., Khrustaleva N.M. et al. Methods and Models for Studying the Survivability of Energy Systems, Nauka Sib. department, pp. 86-91 (1990)
6. Ford L.R., Fulkerson D.R. Flows in networks. Princeton University Press (1962)
7. Kovalev G.F., Lebedeva L.M. Reliability of Power Systems. Springer-Verlag GmbH, (2019).
8. Senderov S.M., Rabchuk V.I., Krupenev D.S. et al. Reliability of fuel and energy supply to consumers from the standpoint of ensuring energy security. SO RAN, 132 p. (2022)
9. Edelev A.V., Senderov S.M., Sidorov I.A., Feoktistov A.G., Beresneva N.M. Control, communication and security systems, **3**, pp. 197-231 (2018)
10. Senderov S.M., Vorobev S.V. Izvestiya RAN. Energetika, **1**, pp. 59–69 (2019)
11. Senderov S., Vorobev S. Reliability Engineering and System Safety, **203**, 107046 (2020)
12. Order of the Ministry of Energy of Russia (February 26, 2021) No. 88 "On approval of the scheme and program for the development of the Unified Energy System of Russia for 2021-2027". https://minenergo.gov.ru/node/20706
13. Rosstat. Remains, supply and consumption of fuel and thermal energy, waste oil products, secondary combustible and thermal resources in 2021. Russian Federation. https://rosstat.gov.ru›mediabank › pok_76_2022

Using normalized values of indicators to determine the level of energy security on the example of the Central and Southern federal districts

Elena Smirnova[1]*, *Sergey* Senderov[1]

[1]Melentiev Energy Systems Institute ESI SB RAS, Irkutsk, Russia

Abstract. At ISEM SB RAS developed approaches to assessing the level of energy security of regions by convolution of qualitative indicators of its state of energy efficiency. At the same time, the relative weights of indicators in the overall system of their value were taken into account. The article presents a new approach to the formation of normalized values of energy security indicators, which was applied in the analysis of the main trends, scales, as well as the dynamics of changes in the state of energy security of the subjects of the Russian Federation located in the territories of the Central and Southern federal districts. Conclusions about its applicability are drawn on the example of the obtained results of indicative analysis.

Introduction

The concept of energy security (ES) is interpreted as “the state of protection of citizens, society, the state, and the economy from the threat of a shortage in meeting their energy needs with economically available energy resources of acceptable quality, from threats of disruption of uninterrupted energy supply” [1-6].

An indicative assessment of the ES level of a particular region of the country is carried out according to three, to a large extent interconnected, blocks of indicators: production and resource endowment of the fuel and energy supply system of the region; reliability of the fuel and energy supply system of the region; the state of the basic production assets (BPA) of energy systems in the region (Table 1).

Taking into account the peculiarities of the energy supply of individual regions, earlier [1,4,5, etc.], the threshold values of indicative indicators for different groups of subjects of the Russian Federation, as well as the relative shares of indicators in the general system of their values, were expertly determined. Using the method of convolution of the obtained values of the analyzed indicators, integral assessments of the state of the ES of the subjects of the Russian Federation were formed.

Table 1. The composition of the most important indicators of regional energy security.

1. The block of industrial and resource security of the fuel- and energy supply of the region
1.1. The ratio of the total available capacity of the region's power plants to the maximum electric load of consumers on its territory. 1.2. The ratio of the amount of available capacity of power plants and the capacity of inter-system connections of the region with neighboring consumers to the maximum electric load on its territory. 1.3. Opportunities to meet the needs of boiler heating oil (BHO) from the region's own sources.
2. The block of reliability of fuel and energy supply of the region
2.1. The share of the dominant resource in total consumption of BHO in the region. 2.2. Share of the largest power plant in the installed electric capacity of the region. 2.3. The level of potential supply of demand for fuel in the conditions of a sharp cooling (10% consumption of consumption) in the region.
3. Block of the state of the BPA of energy systems in the territory of the region
3.1. Degree of depreciation of the BPA in the energy sector of the region. 3.2. The ratio of the average annual input of installed capacity and reconstruction of power plants in the region over the previous 5-year period to the established capacity of the region.

Many indicators are measured in different units, and in order to obtain an integral assessment of the region's ES, the principle of normalizing the values of indicators can be applied depending on the ratio of their values to the threshold ones. For this, a modernized normalization apparatus was used, obtained on the basis of the one used earlier in [7].

*Corresponding author: smirnova.e.m@isem.irk.ru

https://doi.org/10.1051/e3sconf/202338401026

The approach proposed by the authors is universal, it differs from [7] in that it allows you to work with both increasing and decreasing values of indicators, that is, those indicators whose state improves as the value increases and with those whose state improves as they decrease. their meanings.

The conversion of indicator values expressed in different units of measurement into normalized ones is performed according to the following expression:

$$X_i^N = \frac{X_{PCi} - X_i^t}{X_{C,i} - X_{PC,i}}, \quad (1)$$

where X_i^N – is the normalized value of indicator i in the analyzed period, rel. units; X_i^t – is the actual value of the indicator in the system of initial units; $X_{PC,i}$, $X_{C,i}$ - respectively, the threshold values of the pre-crisis and crisis states of the indicator *i* in the system of initial units.

In accordance with the calculation algorithm, the normalized threshold value $X_{PC,i}^N$ is always equal to zero, since this value is the starting point for unfavorable conditions, and the value of $X_{C,i}^N$ is always equal to -1.

Below are the results of assessing the level of energy security of the constituent entities of the Russian Federation located in the territories of the Central (CFD) and Southern (SFD) federal districts, obtained using the previously applied approach. Then, in order to more conveniently compare the results and compare the dynamics of the ES state in different regions, the approach of normalizing the obtained quantitative estimates will be applied.

Results of indicative ES analysis by regions of the Southern Federal District

This section provides information on the qualitative state of energy security indicators for the subjects of the Russian Federation in the Southern Federal District, as well as a qualitative characteristic of the state of energy security of these subjects for 5 years: from 2016 to 2020, in accordance with [8-10] (Table 2 - 5). In this analysis, the Republic of Adygea is included within the Krasnodar Territory, and the city of Sevastopol is included within the Crimean peninsula.

Table 2. The status of indicators on the territory of the subjects of the Southern Federal District of the district for the block of production and resource provision of the fuel and energy supply system for 2016, 2020.

Region, area	Indicator	Dimension	The threshold values of the indicator		The meaning and status of the indicator, year			
			N	C	2016		2020	
Krasnodar Region and the Republic of Adygea	1.1	un.	0,5	0,3	0,46	PC	0,43	PC
	1.2	un.	1,5	1,2	1,78	N	1,66	N
	1.3	%	60	40	134,7	N	88,9	N
Astrakhan region	1.1	un.	0,5	0,3	1,14	N	1,62	N
	1.2	un.	1,5	1,2	3,99	N	4,58	N
	1.3	%	60	40	463,5	N	391,9	N
Volgograd region	1.1	un.	0,5	0,3	1,85	N	1,66	N
	1.2	un.	1,5	1,2	4,08	N	3,53	N
	1.3	%	40	20	24,13	PC	21,66	PC
Rostov region	1.1	un.	0,5	0,3	2,15	N	2,08	N
	1.2	un.	1,5	1,2	3,44	N	3,34	N
	1.3	%	40	20	32,45	PC	50,56	N
Republic of Kalmykia	1.1	un.	0,7	0,5	0,22	C	3,43	N
	1.2	un.	1,5	1,2	3,36	N	2,84	N
	1.3	%	60	40	18,92	C	19,29	C
Republic of Crimea and Sevastopol	1.1	un.	1,2	1,1	0,86	C	1,37	N
	1.2	un.	1,5	1,2	2,05	N	2,26	N
	1.3	%	100	80	35	C	35	C

At the end of 2020, according to the first block of indicators (Table 2), an acceptable situation is observed only in the Astrakhan and Rostov regions, due to the sufficient amount of electric power in the region and the presence of intersystem electrical connections.

In the Krasnodar Territory and the Republic of Adygea, there is a pre-crisis in terms of index 1.1, however, positive dynamics are noticeable due to an increase in the region's electricity generating capacity in 2020 due to the commissioning of a number of solar power plants. At the same time, ensuring the maximum electrical load is achieved by the presence of sufficient power of possible overflows of connections with neighboring regions.

In the Volgograd region, the pre-crisis situation is associated with insufficient opportunities to cover the required volumes of BHO from its own sources.

Heating oil production and natural gas production account for less than 30% of the required fuel consumption in the region.

In the Rostov region, there is an improvement in the situation on indicator 1.3, with an increase in the share of own sources in the balance of BHO from 30 to 50%. With a sufficient margin, the maximum electrical load is ensured.

The situation in the Republic of Kalmykia has changed significantly by 2020, and at the moment the region is sufficiently provided with its own sources of electricity to cover the required volumes. The production of BHO in the region is not enough, so the situation described by this indicator is assessed as a crisis.

The situation in the Republic of Crimea has improved, moving into the area of acceptable values, due to an increase in the installed electric capacity of the region.

According to the second block of indicators (Table 3) in the Krasnodar Territory and the Republic of Adygea, the crisis situation in the ind. 2.1 is due to the share of gas in the balance of BHO - 98%.

In the Astrakhan region indicators ind. 2.2 moved into the range of acceptable values due to a decrease in the share of Astrakhan CHPP-2 with an increase in the installed capacity of the region by 150 MW, due to the commissioning of a number of solar power plants. In the Volgograd and Rostov regions, the Republic of Crimea and Kalmykia, there is a crisis situation in terms of EB for indicator 2.1 due to too high a share of gas in the balance of consumption of BHO.

The transition from crisis values to the acceptable range was noted in the Republic of Kalmykia (ind. 2.2). Thanks to the commissioning of the Tselinskaya and Salynskaya wind farms with a total capacity of 200 MW, the share of the dominant source in the region's available capacity decreased from 95 to 28%.

Table 3. Characteristics of the state of indicators on the territory of the subjects of the Southern Federal District for the fuel and energy supply reliability block for 2016, 2020.

Region, area	Indicator	Dimension	The threshold values of the indicator		The meaning and status of the indicator, year			
			N	C	2016		2020	
Krasnodar Region and the Republic of Adygea	2.1	%	40	70	98,69	C	97,93	C
	2.2	%	50	70	45,8	N	45,54	N
Astrakhan region	2.1	%	>90		97,75	N	97,83	N
	2.2	%	50	70	51,08	PC	33,08	N
Volgograd region	2.1	%	40	70	95,41	C	86,71	C
	2.2	%	50	70	65,5	PC	58,9	PC
Rostov region	2.1	%	40	70	80,29	C	79,58	C
	2.2	%	50	70	36,07	N	30,49	N
Republic of Kalmykia	2.1	%	40	70	99,6	C	99,7	C
	2.2	%	50	70	94,7	C	28	N
Republic of Crimea and Sevastopol	2.1	%	40	70	93,54	C	96,77	C
	2.2	%	50	50	7,05	N	23,21	N

The composition of the most important indicative indicators includes another indicator - 2.3, which reflects the level of potential supply of demand for fuel and energy resources in conditions of a sharp cooling (10% surge in consumption) in the region. It is estimated based on the results of studies on the fuel and energy complex model [2,3] as the value of the provision of consumers with boiler and furnace fuel in case of possible cooling, which increases fuel consumption by 10%.

The subjects of the SFD belong to regions with a temperate (moderately cold) climate (the temperature of the coldest five-day period is from -20° to -30°C [11]). The crisis threshold for such regions is 90% [4]. The studies were carried out with a simultaneous hypothetical decrease in the average January temperature in most areas of the Southern Federal District.

For almost all subjects of the Southern Federal District, increased consumption can be fully ensured, which corresponds to the zone of acceptable states. The exception is the Krasnodar Territory and the Republic of Adygea, where the situation is assessed as pre-crisis. The Republic of Crimea belongs to regions with a relatively mild climate (the temperature of the coldest five-day period is -17°C [11]), however, due to too much dependence on natural gas for this indicator, the situation is in crisis.

According to the third block of indicators, the situation cannot be called completely acceptable in any of the subjects under consideration, Table. 4.

Table 4. Characteristics of the state of indicators on the territory of the subjects of the Southern Federal District in the block of the state of the BPF of energy systems for 2016, 2020.

Region, area	Indicator	Dimension	The threshold values of the indicator		The meaning and status of the indicator, year			
			N	C	2016		2020	
Krasnodar Region and the Republic of Adygea	3.1	%	40	60	37	N	44,7	PC
	3.2	%	2	1	13,6	N	1,4	PC
Astrakhan region	3.1	%	40	60	45,3	PC	51,7	PC
	3.2	%	2	1	7,4	N	2,7	N
Volgograd region	3.1	%	40	60	59,4	PC	62	C
	3.2	%	2	1	0,8	C	0,3	C
Rostov region	3.1	%	40	60	38	N	43,7	PC
	3.2	%	2	1	5,5	N	2,1	N
Republic of Kalmykia	3.1	%	40	60	51	PC	47	PC
	3.2	%	2	1	2,5	N	26,3	N
Republic of Crimea and Sevastopol	3.1	%	40	60	55,7	PC	42,7	PC
	3.2	%	2	1	8,3	N	0	C

In the Krasnodar Territory and the Republic of Adygeya, major overhauls and modernization were carried out at the Sochinskaya TPP and at the Krasnopolyanskaya HPP with a total increase in capacity by 10 MW, as well as the commissioning of wind turbines in 2020 at the Adygei WPP with a capacity of 150 MW and the commissioning of the Adygeiskaya SPP with a capacity of 4 MW. However, according to the results of the five-year period, the values of the indicators moved into the pre-crisis and crisis areas for both indicators of the block.

In the Astrakhan region, the values of the indicator of depreciation of the BPA of the energy industry are in the area of pre-crisis values. According to ind. 3.2 the state is acceptable, due to the development of renewable energy sources in the region, such as solar power plants: the commissioning of SPP Zavodskaya (15 MW) in 2017, in 2019 - Akhtubinskaya, Mikhailovskaya and Limanskaya, with a total capacity of 105 MW; 2020 Oktyabrskaya and Peschanaya SPP, with a total capacity of 30 MW.

In the Volgograd region, the situation in the third block of indicators is deteriorating. The lack of major overhauls, dismantling and decommissioning of equipment, insufficient modernization, all this, aggravated the general state of the BPA in the energy sector of the region. Of the latest commissioning of capacities until 2019, only the Volgograd SPP (25 MW) was put into operation in 2017.

The Rostov region has moved into a pre-crisis state from an energy sector that is acceptable in terms of depreciation of the BPA. At the same time, the commissioning of capacities over the past five years allows the region to remain within the range of acceptable values: commissioning of 324 MW of Novocherkasskaya SDPP, modernization at Rostovskaya HPP 2 (20 MW), in 2020 commissioning of four wind farms with a total capacity of 350 MW.

The Republic of Kalmykia is in a pre-crisis state in terms of the degree of depreciation of the BPA (Ind. 3.1). Thanks to the commissioning of a number of solar power plants with a total capacity of 330 MW, according to ind. 3.2 the situation is assessed as acceptable.

In the Republic of Crimea, the renewal and modernization of the BPA of the energy sector has been underway since 2016, however, by 2020, according to indicator 3.2, a crisis state has been noted in the republic. The energy system of Crimea was connected to the UES of Russia through the commissioning of an energy bridge with a capacity of 800 MW in 2015-2016. In 2019, 970 MW were commissioned (Combined Cycle Plant at Balaklavskaya, Tavricheskaya and Sakskaya TPPs). In November 2020, a 500 kV high-voltage line was put into operation to transmit power from the Rostov NPP to the south of the Taman Peninsula. As a result, the degree of depreciation of the BPA of the energy sector in the region has changed significantly, but the general condition is assessed as pre-crisis.

Integral assessment of ES by regions of the Southern Federal District

The results of the integrated assessment of the state of the ES in the regions of the SFD are presented in Table 5.

Table 5. Integrated qualitative assessment of the state of energy security in the territory of the subjects of the Southern Federal District for 2016, 2020.

Years	The order numbers of the estimated ES indicators								The sum of the specific weights by state			Quality condition ES
	1.1	1.2	1.3	2.1	2.2	2.3	3.1	3.2	Boundaries of states			
	Specific weights of indicators								C[1]	PC	N[2]	
	0,104	0,138	0,133	0,120	0,079	0,170	0,127	0,129				
1	2	3	4	5	6	7	8	9	10	11	12	13
Krasnodar Region and the Republic of Adygea												
2016	PC	N	N	C	N	PC	N	N	0,12	0,274	0,606	PC
2020	PC	N	N	C	N	PC	PC	PC	0,12	0,53	0,35	PC
Astrakhan Region												
2016	N	N	N	N	PC	N	PC	N	0	0,206	0,794	N
2020	N	N	N	N	N	N	PC	N	0	0,127	0,873	N
Volgograd region												
2016	N	N	PC	C	PC	N	PC	C	0,249	0,339	0,412	PC
2020	N	N	PC	C	PC	N	C	C	0,376	0,212	0,412	PC
Rostov region												
2016	N	N	PC	C	N	N	N	N	0,12	0,133	0,747	N
2020	N	N	N	C	N	N	PC	N	0,12	0,127	0,753	N
Republic of Kalmykia												
2016	C	N	C	C	C	N	PC	N	0,436	0,127	0,437	C
2020	N	N	C	C	N	N	PC	N	0,253	0,127	0,62	PC
Republic of Crimea and Sevastopol												
2016	C	N	C	C	N	C	PC	N	0,527	0,127	0,346	C
2020	N	N	C	C	N	C	PC	C	0,552	0,127	0,321	C

According to the results of the indicative analysis of the subjects of the Southern Federal District, an acceptable state of energy security can be noted in the Astrakhan and Rostov regions. The crisis situation in the Republic of Crimea is determined by the following factors: due to its own gas production, the republic provides only 35% of its needs; in the republic, the peak-increasing demand for fuel and energy resources cannot be met in the face of possible cold snaps; the share of gas in the BHO balance is 97%.

In the Republic of Kalmykia, the state passed into pre-crisis from the crisis due to the commissioning of a large number of solar power plants.

Normalization of EB indicator values

The next step in the work was the normalization of the obtained indicator values for all subjects of the Southern Federal District. On the basis of the obtained values for a 5-year period, graphs of the states of the ES of the regions were constructed for the corresponding blocks of indicators, fig. 1-3. The threshold values here were also taken into account as normalized; where "crisis" = - 1, "pre-crisis" = 0. Thus, all states that are in the range of values below "-1" are crisis, in the range from "-1" to "0" - pre-crisis and in the area above "0" - acceptable.

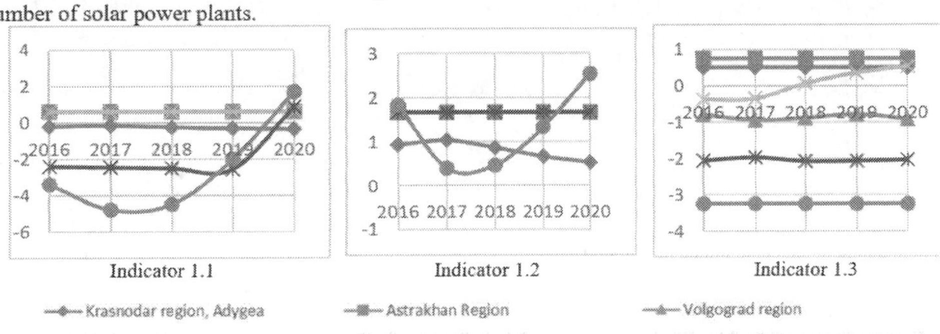

Fig. 1. Analysis of the situation with the provision of ES requirements for the first block of indicators in the regions of the Southern Federal District.

[1]The state of ES in the region is recognized as a crisis if the sum of the shares of indicators in the state "C" exceeds 0,4
[2]The state of ES in the region is recognized as normal if the sum of the specific weights of the indicators in the "N" state exceeds 0,7

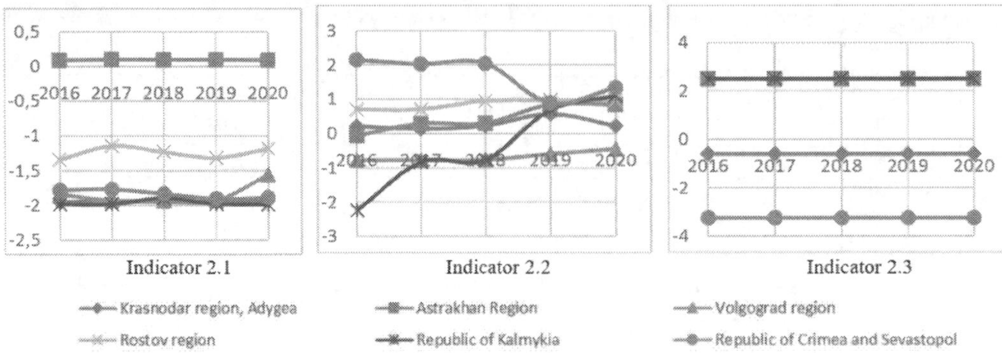

Fig. 2. Analysis of the situation with the provision of ES requirements for the second block of indicators in the regions of the Southern Federal District.

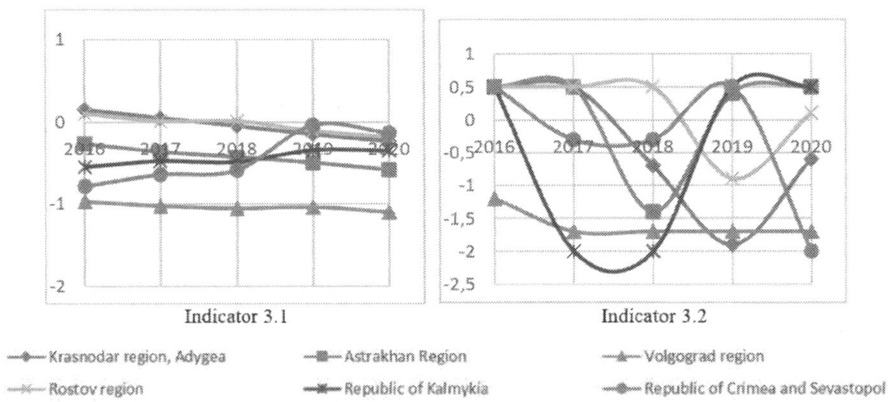

Fig. 3. Analysis of the situation with the provision of ES requirements for the third block of indicators in the regions of the Southern Federal District.

To obtain the final integral assessment of the level of ES, a convolution of the normalized values of the indicators was made, taking into account their specific weights. During the calculation, the resulting state graphs showed that too much overlap of the threshold values of individual indicators causes a bias in the overall assessment. In this regard, it was decided to reduce any significant overlap to a sufficient margin: 25% for acceptable conditions and 50% for crisis conditions. The resulting comprehensive integrated assessment of the state of the ES of the subjects of the Southern Federal District is shown in Fig. 4.

As can be seen from the graphs, in comparison with the tables, this approach allows you to track in more detail the dynamics of the state of the ES, both for each specific indicator, and in a complex way.

Results of indicative analysis of ES by regions of the CFD

The results of the indicative analysis for the CFD are presented below only in a normalized form. The state of indicators on the territory of the subjects of the Central Federal District of the district for the block of production and resource provision of the fuel and energy supply system is shown in Fig.5.

The analysis of the data obtained made it possible to identify the Ryazan and Yaroslavl regions as regions with an acceptable state of ES according to the first block of indicators. In most other regions, there is a crisis situation in terms of the level of opportunities to provide BHO from their own sources.

 https://doi.org/10.1051/e3sconf/202338401026

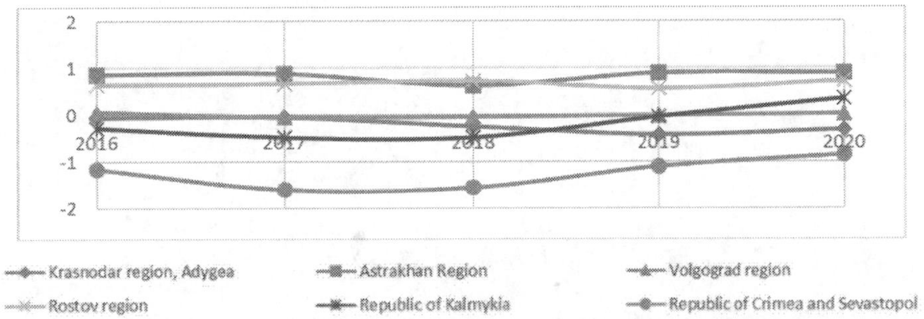

Fig. 4. Integrated qualitative assessment of the state of energy security in the regions of the Southern Federal District.

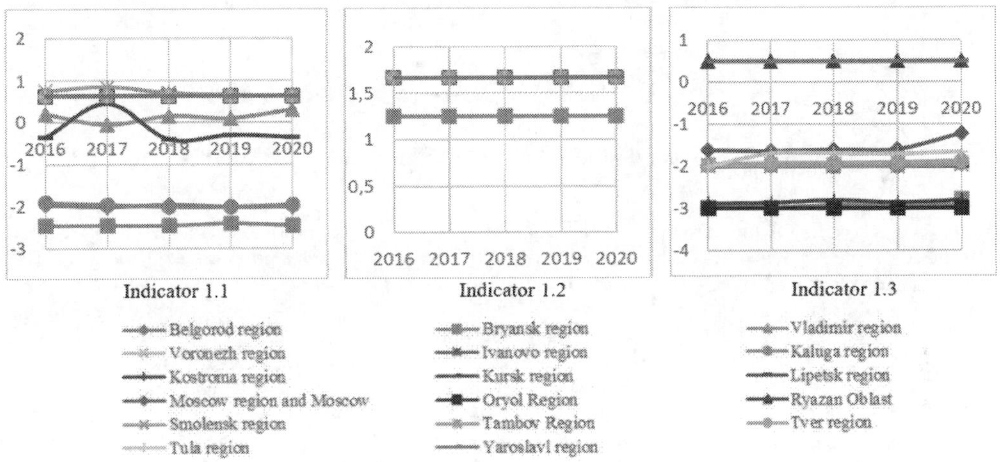

Fig. 5 Analysis of the situation with the provision of ES requirements for the first block of indicators in the regions of the Central Federal District.

According to the second block of indicators, the majority of subjects are in a crisis situation (Fig. 6). In the Bryansk and Tula regions, the situation worsened due to an increase in the share of the largest power plant in the total installed capacity of the regions. And in the Voronezh region the situation has improved and the indicators are in the area of acceptable values.

According to indicator 2.3, almost all subjects of the Central Federal District are located in the zone of crisis and pre-crisis values, since they belong to regions with a cold climate: the average temperature of the coldest 5-day period is from minus 20°C to minus 30°C [11]. At the same time, a high share of natural gas in the balance of BHO and dependence on external supplies with an insufficient amount of own fuel resources leads to insufficient coverage of the peak growing demand for fuel in conditions of sharp cooling in the European part of the country.

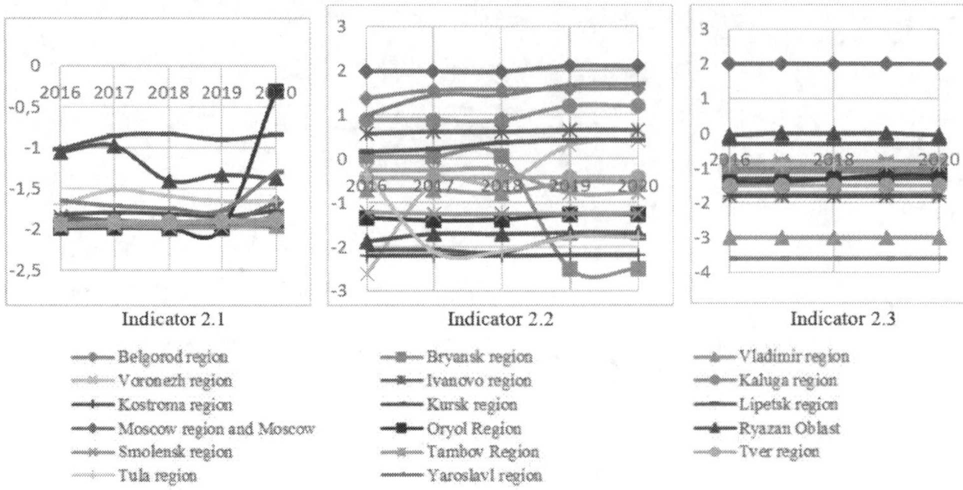

Fig. 6. Analysis of the situation with the provision of ES requirements for the second block of indicators in the regions of the Central Federal District.

According to the third block of indicators (Fig. 7), the values of indicators of the state of the BPA of the energy sector of the regions are mainly in the area of pre-crisis and crisis values.

The situation in the Voronezh and Yaroslavl regions has improved in terms of indicator 3.2, moving from the pre-crisis to the area of acceptable values. In the Voronezh region, reactor installations were launched at the Novovoronezh NPP with a total capacity of 2400 MW; in 2020, a CCPunit with a capacity of 220 MW was installed at the Voronezh CHPP-1. In the Yaroslavl region, a CCP unit was commissioned at the region's thermal power plants with a total capacity of 515 MW over a five-year period.

The results obtained are the basis for an integral assessment of the level of ES in the regions of the Central Federal District.

The final integral assessment of the level of ES in the regions of the Central District is shown in fig. 8. Based on the data, it can be concluded that, in general, the situation with energy security cannot be considered acceptable in any of the subjects.

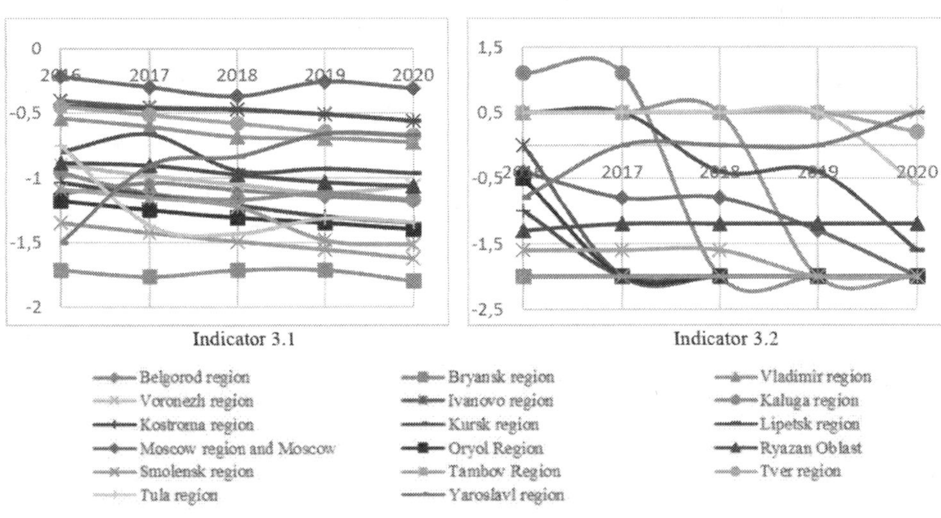

Fig. 7. Analysis of the situation with the provision of ES requirements for the third block of indicators in the regions of the Central Federal District.

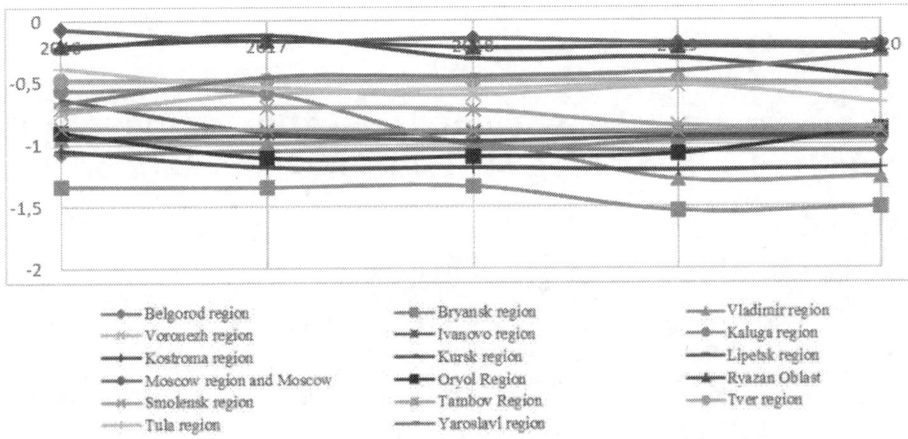

Fig. 8. Integral qualitative assessment of the state of energy security in the regions of the Central Federal District.

Conclusion

The approach considered in the article differs from the previous one in that the integral assessment takes into account not only qualitative, but also quantitative assessments of the situation with the provision of ES for individual indicators. The application of this approach makes it possible to obtain normalized values of indicators, which, in turn, correctly reflect the results of the assessment of the state of the ES in the regions, and also more clearly and in detail show the dynamics of changes in the situation with the provision of ES, both for each individual indicator and for the subjects as a whole.

The analysis showed that in the regions of the Central Federal District, which to a greater extent do not have their own sources of fuel and energy resources, the situation with the share of gas in the total consumption of boiler heating oil and with the share of the largest generating source continues to worsen. In the regions of the Southern Federal District, the situation has changed mainly in the first block of indicators, production and resource provision of the fuel and energy supply system of the regions.

In general, it should be noted that in almost all regions the situation with the aging of the basic production assets of the energy industry is deteriorating rapidly, and, consequently, with the possibility of increasing the number of emergencies with fuel and energy supply to consumers due to the failure of one or another equipment.

The research was carried out under State Assignment Project (no. FWEU-2022-0001) of the Fundamental Research Program of Russian Federation 2021-2030 using the resources of the High-Temperature Circuit Multi-Access Research Center (Ministry of Science and Higher Education of the Russian Federation, project no. 13.CKP.21.0038).

References

1. S.M. Senderov, N.I. Pyatkova, V.I. Rabchuk, G.B. Slavin, S.V. Vorobyov, E.M. Smirnova *Methodology for monitoring the state of Russia's energy security at the regional level* (Irkutsk: ISEM SO RAN, 2014)
2. V.V. Bushuev, N.I. Voropay, A.M. Mastepanov, Yu.K. Shafranik and other *Energy security of Russia.* (Novosibirsk: Science. Siberian Publishing Company RAS, 1998)
3. N.I. Pyatkova, S.M. Senderov, M.B. Cheltsov et al. *Application of two-level research technology in solving problems of energy security* (Izvestiya RAN. Energetika), **6** (2000)
4. N.I. Pyatkova, V.I. Rabchuk, S.M. Senderov, G.B. Slavin, M.B. Cheltsov *Energy security of Russia: problems and solutions.* (Novosibirsk: Publishing House of the Siberian Branch of the Russian Academy of Sciences, 2011)
5. S.M. Senderov *Assessment of the energy security level of Russian regions and the basic principles for creating an energy security monitoring system.* Safety of objects of the fuel and energy complex, **1 (1)**, 2012
6. S.M. Senderov, E.M Smirnova *Analysis of trends in the state of energy security of regions on the example of the Siberian Federal District* Energy Policy, **1**, 2019
7. N.I. Voropay, A.I. Tatarkina *Reliability of fuel and energy supply and survivability of energy systems in Russian regions* (Yekaterinburg: Ural Publishing House, 2003)
8. Statistical form of Rosstat *Information on balances, receipt and consumption of fuel and heat, collection and use of waste oil products*, 2016-2020.
9. Statistical form of Rosstat *Information on the use of fuel and energy resources*, 2016-2020
10. Statistical form of Rosstat *Technical and economic indicators of power plants, district boiler houses*, 2015-2019
11. Code of rules SP 131.13330.2018 *Construction climatology* SNIP 23-01-99 *, Moscow, 2019.

Comparative efficiency of application of the maximum clique method in searching for critically important objects of the gas industry

Sergey Vorobev[1*], *Anton* Kolosnitsyn[1], and *Ilya* Minarchenko[1]

[1]Melentiev Energy Systems Institute, 130 Lermontov str., Irkutsk, Russia

Abstract. The article proposes the use of a model of the "defender - attacker - defender" type to determine the possibilities of the gas industry to reduce gas shortages for consumers in the event of failure of its critical facilities. The calculation scheme of the gas transmission network in this study is presented in the form of a directed graph, in the nodes of which there are objects of production, consumption, storage of gas and nodal compressor stations. All found cliques are analyzed, and a list of measures aimed at reducing possible gas shortages among consumers is compiled. The article presents the results of the analysis. The results of applying the method of determining the maximum cliques to search for critical objects of the gas industry were evaluated in comparison with the results obtained earlier on this topic. Conclusions are drawn about the feasibility of using the proposed method to determine the critically important objects of the gas industry.

1 Introduction

The development of Russia's gas transmission network is associated both with a decline in production at old operating fields, and with the need to enter new areas of gas production. In addition, the network configuration is affected by changes in the volumes of consumption of the country's subjects and the reorientation of export supplies of pipeline gas. Thus, the gas transmission network is constantly changing over time both in terms of its configuration and the load of its elements.

A complex gas transmission network is a critical infrastructure and contains a significant number of facilities, the performance of which critically affects the overall performance of the network and, accordingly, its ability to reliably provide uninterrupted gas supply to consumers. The most large-scale accidents in power systems occur as a result of the failure of various critical objects of these systems [1, 2]. One of the results of such accidents is significant damage to consumers, expressed in the shortage of fuel and energy resources. Therefore, the identification of critical objects and their combinations in power systems in order to develop measures aimed at reducing the significance of these objects is an urgent task.

Currently, various studies are underway in the world to determine the critical elements of energy systems and other critical infrastructures.

A number of works are devoted to the modeling of energy systems as critical infrastructures [3, 4]. Researchers also consider the problems of vulnerability of critical energy infrastructures to terrorist attacks on them and the methodology of risk analysis for systems of interdependent critical infrastructures under various extreme weather conditions [5-10].

Previously, studies were conducted to identify critical facilities in the gas transmission network. A list of intersections of main gas pipelines in the Unified Gas Supply System of Russia has been determined, the disruption of which will lead to a relative shortage of daily gas supplies throughout the system as a whole in the amount of 5% or more [11]. Studies have been carried out to search for and determine combinations of individual sections of main gas pipelines, the simultaneous disruption of which can lead to a significant shortage of daily gas supplies through the system (5% or more) [12, 13]. Studies have been carried out to determine critically important objects in the electric power industry [14]. Taking into account the accumulated experience and on the basis of an analysis of studies currently being carried out in the world, in [11, 15] a methodology was formulated for the formation of lists of critically important objects of energy systems from the standpoint of ensuring the operability of these systems using the example of the Russian gas industry.

In these works, the definition of critically important objects and their combinations is carried out by enumeration. Multi-iteration studies were carried out, in which all elements of the computational network and their combinations were turned off in turn. Thus, those elements of the network and their combinations were determined, the failure of which would lead to the greatest shortage of gas in the system. Due to limitations in the amount of analyzed information, situations with a simultaneous disconnection of three or more network

* Corresponding author: seregavorobev@isem.irk.ru

elements were not considered. However, due to the specifics of the functioning of the gas transmission network, such emergencies are possible. Therefore, in order to more fully take into account various factors in the study of critical facilities, in this paper, the authors use the "clique" method to find the maximum number of interconnected nodes and main gas pipelines, the failure of which can cause significant damage to the system in terms of reducing gas supplies to consumers.

The problem of a clique is formulated within the framework of the methodology for modeling attacks on infrastructure systems. This technique connects two sides in a single mathematical formulation: the attack side and the defense side. Such models are based on the class of Stackelberg games [16] with prohibitions on networks, in which two actors - the leader and the follower - pursue, as a rule, opposite interests. A detailed description of mathematical models of this type of problems can be found in [17]. We also note that the mathematical description of such problems can be reduced to models of multilevel optimization of the type: attacker-defender, defender-attacker and defender-attacker-defender, which are widely used in scientific research in modeling threats and response measures at various critical infrastructure facilities [18 -20].

2 Statement of the problem of finding the maximum clique

The calculation scheme of the gas transmission network used in this study fully reflects all the technological features of a real gas transmission network and includes 388 nodes, including: 96 consumers corresponding to gas consumption regions and individual large industrial consumers, 33 gas fields, 29 underground storage facilities gas (UGS), 230 nodal compressor stations.

The scheme takes into account the peculiarity of the system, which is that the units corresponding to UGS facilities can be involved in the system both as consumers of gas and as its sources. Communication between the nodes is carried out through 755 lines, reflecting the existing main gas pipelines. The initial data on the functioning of the gas transmission network, such as the volume of gas production in the fields, the volume of gas consumption, the throughput of gas pipelines, are taken in accordance with specialized statistical information [21-23].

To solve the problem of supplying consumers with the required volumes of gas, the statement of the problem of finding the maximum flow is used. It is necessary to find the maximum possible total volume of gas that can be passed through a network with given characteristics of connections between nodes, taking into account the established throughput capacities of the lines, the available production volumes and the given consumption volume. Solving the problem of finding the maximum flow through the network allows you to determine whether the system is able to provide consumers with the required volume of gas. This statement does not take into account the gas consumption for the own needs of the gas transmission network. This gas consumption in the current study is taken into account in accordance with the results of numerous earlier feasibility studies of the functioning of the Russian gas transmission system [24].

The statement of the problem of finding the maximum flow has the following form:

$$\max f,$$

$$\sum_{j\in I}\bar{u}_{ij}x_{ij}=\begin{cases}-f, i=i_O,\\ 0, i\notin\{i_O,i_S\},\ i\in I,\\ f, i=i_S,\end{cases} \tag{1}$$

$$x_{ij}=x_{ji},\quad i,j\notin\{i_O,i_S\},$$

$$0\le x_{ij}\le d_{ij},\quad i\in I, j\in I.$$

After solving problem (1), information about the ability of the gas transmission system to close the load from consumers becomes available. Bottlenecks can also be identified, i.e. sections of the gas pipeline with maximum load, and sections with a large margin in terms of throughput. These data may contribute to making adjustments to the characteristics of the gas transmission system in the direction of increasing or decreasing the throughput of certain gas pipelines, taking into account the load from consumers.

For the Russian gas transmission system, it is proposed to define the most vulnerable objects for attack as combinations of nodes in which each node is connected to each node. In other words, we pose the problem of finding a clique of maximum size [25]. For clarification, we use the terminology from graph theory.

The gas transportation system can be represented as a directed graph, in the nodes of which there are fields, gas consumers, underground storage facilities, and compressor stations. The edges of such a graph are gas pipeline lines. The attacker's main goal is to cause maximum damage to the gas transmission system, i.e. reduction of the maximum gas flow through the gas pipeline network by attacking and incapacitating the key or linear objects of the system.

In this case, the attacking side solves the problem of finding a clique of the maximum size to disable the largest number of interconnected objects of the gas transmission system. Below is a list of the main notation for the mathematical description of the problem of finding the maximum clique.

$G = (V, E)$ - arbitrary undirected and weighted graph,
$V = \{1, 2, ..., n\}$ - set of vertices (nodes) of a graph G,
$E = V \times V$ - set of graph edges G,
$w = (w_1, w_2, ..., w_n)$ - scale vector, $w_i > 0$, $i = 1,..., n$,
$G = (V, E)$ - padded graph to graph G,
$\bar{E} = \{(i,j): i,j \in V, i \neq j, (i,j) \notin E\}$.

To find a clique of the maximum size, it is necessary to solve the following problem:

$$\max_{y\in Y} F(y),$$

$$F(y)=\sum_{i=1}^{n} w_i y_i, \tag{2}$$

$$Y=\left\{y\in\{0,1\}^n, y_i+y_j\le 1\quad \forall(i,j)\in\bar{E}\right\}.$$

The found solution of problem (2) with $w_i = 1$ determines the set of objects of the gas transmission system that form a clique of the maximum size. In the case of a non-unique solution, we will get a whole set of such cliques, which can serve as the basis for further more detailed attack planning with maximum damage on the part of the attacker.

Taking into account (1) and (2), the attacker-defender problem for the Russian gas transmission system can be represented as follows:

$$\begin{gathered} \min_{y} \max_{x} f, \\ y \in Arg \max\{C(y) : y \in Y\}, \\ r_{ij} = r_{ji} = \begin{cases} 1, & (i,j) : y_i + y_j \ldots 1,\ u_{ij} = u_{ji} = 1, \\ 0, & (i,j) : y_i + y_j < 1,\ u_{ij} = u_{ji} = 1, \end{cases} \quad i,j \in I \setminus \{i_O, i_S\}, \\ r_{i_O,j} = r_{j,i_S} = 0, \qquad j \in I, \\ \sum_{j \in I} \overline{u}_{ij} x_{ij} = \begin{cases} -f, & i = i_O, \\ 0, & i \notin \{i_O, i_S\}, \quad i \in I, \\ f, & i = i_S, \end{cases} \\ x_{ij} = x_{ji}, \quad i,j \notin \{i_O, i_S\}, \\ 0 \le x_{ij} \le d_{ij}(1 - r_{ij}), \quad i \in I, j \in I, \end{gathered} \tag{3}$$

y {0,1} – a vector that defines the nodes to which the attack is directed (the attacker's plan); *Arg max{C(y) :y ∈ Y}* set of cliques of maximum size; r_{ij} , $i,j \in I$ — parameters that define the edges that enter or exit the attacked nodes and are removed from the network along with them (the consequences of the attack). In accordance with the statement, the attacker, based on the solution of problem (3), chooses a clique whose removal from the network causes maximum damage to the throughput of the gas transmission network. The nodes included in the clique are removed from the network along with all edges adjacent to them. Thus, an attack on a node, firstly, disables the gas transmission network facilities located in this node, and, secondly, eliminates the possibility of gas transit through this node.

In case of disruption of gas supplies to consumers due to the shutdown of the click network, "bottlenecks" are formed, i.e. the most gas-loaded sections of the network. An increase in the throughput capacity of such sections, hypothetically, can reduce the shortage of gas among consumers. Such a short-term increase in the throughput capacity of a section of the main gas pipeline is possible with an increase in operating power at large main compressor stations [24]. The result of such an increase in power will be an increase in the working pressure in the gas pipeline, due to which the throughput of the section of the main gas pipeline can be increased within 10%. As a result (by using the technical capabilities of the gas transmission network), the problem of minimizing gas shortages for consumers in the event of a failure of the click network is solved.

To take into account the possibilities for increasing the throughput of sections of the gas transmission network, the original model (1)-(3) is modified into a model of the "defender-attacking-defender" type as follows.

After solving problem (3), the obtained optimal flows along the gas pipeline lines are fixed, we denote them by x^{cl}. The received maximum flow is also fixed f^{cl}.

Next, we consider new variables for flows x with given two-sided constraints. Additional restrictions are also introduced for the maximum flow of a new task. Bilateral limits for line capacities have the following form :

$$x^{cl} \le x \le 1.1x^{cl}. \tag{4}$$

These restrictions make it possible to take into account an increase in the volume of natural gas transportation through a certain line by 10%.

The minimum length flow is considered as the objective function:

$$z = \sum_{i \in I} \sum_{j \in I} L_{ij} (x_{ij} - x_{ij}^{cl}) \to \min, \tag{5}$$

L_{ij} – line length (i,j), $i \in I$, $j \in I$. In this way, the transportation distance of the additional volume of natural gas is minimized.

An additional condition for the maximum flow is introduced:

$$f \ge c_k \cdot f^{cl}. \tag{6}$$

In order to increase the maximum flow with a minimum "penalty" for its transportation, we formulate the problem according to the following principle: the minimization of the "penalty" is carried out in the objective function, and a suitable increase in the maximum flow is carried out using the introduced restriction $k*f^{cl}$.

The constant c_k indicates the possible increase in the maximum flow.

The Russian gas transportation system model (3)-(6) is described by means of the AIMMS algebraic modeling system [26]. In this system, the problem of the maximum flow and the maximum clique of nodes was also solved. The calculations were carried out on a computer equipped with an 8-core AMD FX-8350 processor (clock frequency of each core is 4 GHz) and 8 GB of RAM.

3 Calculation results

With the help of the developed mathematical models on the updated calculation scheme, as a result of a number of calculation studies, the following results were obtained.

The maximum clique for the Russian gas transmission network is 3 nodes, i.e. up to a maximum of three objects of the gas transmission system, are physically interconnected (each with each other) by gas pipelines.

Combinations of interconnected nodal compressor stations forming cliques were determined [27]. A total of 45 cliques were found, consisting of three elements. Of the received cliques, six were selected, the failure of which will lead to a gas deficit of 11–39%. It should be noted that some cliques contain critical objects found in previous studies [15].

989 pairs of cliques were received, the exclusion of which creates a shortage of gas for consumers. The failure of the largest of these pairs creates a significant synergistic effect: the first clique creates a gas deficit for consumers of 39%, the second - 11%, both cliques are unrelated. With their joint exclusion from the network, a gas deficit of 61% arises in the system as a whole (gas consumers in such a situation will not receive about 1.3

billion m^3 of gas per day). 11 pairs of cliques were found, a joint exclusion from the network, which leads to a gas deficit of 50-53% for the system as a whole. 64 cliques pairs with a possible gas deficit of 40-49%, 211 pairs of cliques with a possible gas deficit of 30-39%, 318 pairs of cliques with a possible gas deficit of 20-29%, 384 pairs of cliques with a possible gas deficit of 14-19 %.

A series of calculations was carried out to determine the possibilities for reducing gas deficits for consumers in the event of a clique failure. As the calculation scenarios, the situation with the successive failure of each of the previously received cliques was adopted. Thus, using model (3)-(6), 45 calculations were carried out with the aim of minimizing gas shortages for consumers as much as possible at the lowest cost.

As a result, in 9 cliques, it was possible to provide consumers with gas in full. These cliques are characterized by their low significance relative to the previously found cliques; the failure of each of them leads to a gas deficit of up to 5% in the system as a whole.

It was almost impossible to bypass the two cliques with the largest gas deficits (located in the areas of gas supply to the network from the largest operating fields), the decrease in the deficit in these cases reached 1% and 5%, respectively, with gas deficits from the failure of cliques of 39% and 28 %.

For 10 cliques, there was a slight decrease in deficits, up to 1%. Also, by 21 cliques, an increase in the throughput capacity of busy sections of the network by up to 10% did not lead to a decrease in gas shortages among consumers.

Only in three cliques it was possible to achieve a significant reduction in the gas deficit among consumers in the range of 7-12% for the system as a whole.

When analyzing the consequences for consumers from disruptions in the operation of critical facilities in the current configuration of the Russian gas transmission network, 27 sections were found where it is expedient with a high degree of invariance (from 25 to 48% of the scenarios under consideration) to expand capacity. In contrast to the study, which determined invariant measures to reduce the importance of critical network facilities [28], in the case of cliques, it was almost impossible to obtain invariant measures aimed at reducing gas shortages in consumers when cliques fail.

4 Conclusion

Identification of the problem of combining objects in the gas transmission network makes it possible to plan measures to ensure the reliability of gas supply to consumers. The article considers the application of methods for determining the maximum cliques to search for specific sections of the main gas pipelines of the gas transmission network.

As a result of the study, all connected sections of the network were identified and an analysis of the possible consequences for the systems in case of their occurrence was carried out.

First, it should be noted that the maximum sizes of cliques are 3 and 7 elements, respectively, for nodal and edge cliques.

Secondly, gas deficiencies arising from the occurrence of cliques found from the composition, reducing the gas deficit arising from the emergence from the totality of system compounds and their combinations [12, 15].

The results of calculations of applications of the results of previous studies, performed by various methods, models of element-by-element enumerations [12, 15], methods for determining synergistic effects [13].

Also, in the network under consideration, in order to obtain combinations consisting of three objects, it is necessary to carry out 9660036 calculations by element-by-element enumeration. With an average calculation time of one option of 6 seconds, it will take about 680 days to obtain the results of such a study. The study of the gas transmission network using the maximum clique search method was carried out much faster.

Thus, the presented study and calculation results confirmed the applicability of the maximum demand method to attract attention in the gas industry. This method can be used when looking for item captures in other network infrastructure captures.

This research was funded by the Russian Science Foundation, grant no. 23-29-00460, using the resources of the High-Temperature Circuit Multi-Access Research Center (Ministry of Science and Higher Education of the Russian Federation, project no. 13.CKP.21.0038).

References

1. Accidents on main gas pipelines in Russia in 2018-2019. URL: https://ria.ru/20190728/1556953028.html.
2. Top 5 largest and most destructive accidents on gas pipelines. URL: https://sila-sibiri-rabota.ru/avarii-na-gazoprovodax/.
3. Thompson, J. R., Frezza, D., Necioglu, B., Cohen, M. L., Hoffman, K., Rosfjord, K. (2019). Int. Journal of Critical Infrastructure Protection **Vol. 24**, p. 144-165.
4. Kai, L., Ming, W., Weihua, Z., Jinshan, W., Xiaoyong, Y. (2018). Int. Journal of Critical Infrastructure Protection **Vol. 23**, p. 79-89.
5. Tichy, L. (2019). Int. Journal of Critical Infrastructure Protection, DOI: https://doi.org/10.1016/j.ijcip.2019.01.003.
6. Tsavdaroglou, M., Al-Jibouri, S. H.S., Bles, T., Halman, J. I.M., (2018). Int. Journal of Critical Infrastructure Protection **Vol. 21**, p. 57-71.
7. Han F., Zio E. Int. Journal of Critical Infrastructure Protection, **Vol 24**, March 2019, p. 1-13. DOI: 10.1016/j.ijcip.2018.10.009.
8. Ouyang M. European Journal of Operational Research, **Vol 262**, Issue 3, 2017, p. 1072-1084. DOI: 10.1016/j.ejor.2017.04.022.
9. Aljaroudi A., Khan F., Akinturk A., Haddara M., Thodi P. Journal of Loss Prevention in the Process

Industries, **Vol. 37**, September 01, 2015, p. 101-109. DOI: 10.1016/j.jlp.2015.07.004.
10. Chen C., Li C., Reniers G., Yang F. Journal of Cleaner Production, **Vol. 279**, 10 January 2021. DOI: 10.1016/j.jclepro.2020.123583
11. Senderov S., Edelev A. Energy., 2017. DOI: 10.1016/J.ENERGY.2017.11.063.
12. Vorobev S., Smirnova E.M. RSES 2019, E3S Web Conf. **Vol. 139**, 2019. DOI: 10.1051/e3sconf/201913901016.
13. Vorobev S., Edelev A., Smirnova E. RSES 2017. E3S Web Conf. **Vol. 25**, 2017. DOI 10.1051/e3sconf/20172501004.
14. Senderov S., Krupenev D. Energy Systems Research, **Vol. 2**, No. 2 (6), 2019, pp. 41-50.
15. Senderov S., Vorobev S. Reliability Engineering & System Safety, **Vol. 203**, 107046, 2020, DOI: 10.1016/j.ress.2020.107046.
16. von Stackelberg, H.: *The Theory of the Market Economy*. William Hodge and Co., London, U.K. (1952)
17. Smith J.C., Lim C. *Algorithms for Network Interdiction and Fortification Games*. In: Chinchuluun A., Pardalos P.M., Migdalas A., Pitsoulis L. (eds) *Pareto Optimality, Game Theory And Equilibria*. Springer Optimization and Its Applications, vol 17. Springer, New York, NY. (2008) DOI: https://doi.org/10.1007/978-0-387-77247-9_24
18. Brown, G., Carlyle, M., Salmeron, J., Wood, R. Interfaces **36**(6), 530–544 (2006). DOI: 10.1287/inte.1060.0252.
19. Salmeron, J., Wood K., Baldick, R. IEEE Transactions, **19**(2), 905–912 (2004). DOI: https://doi.org/912. 10.1109/TPWRS.2004.825888.
20. Manshadi, S., Khodayar, M. IEEE Transactions on Smart Grid **6**(5), 2283–2292 (2015). DOI: https://doi.org/10.1109/TSG.2015.2397318.
21. *Export of the Russian Federation of the most important goods in 2011 - 2016 (according to the Federal Customs Service of Russia)* http://customs.ru/index.php?option=com_ newsfts&view =category&id=52&Itemid=1978&limitstart=60.
22. *InfoTEK Monthly oil and gas magazine*. **№1**, – P.154. (2021).
23. *Ministry of Energy of the Russian Federation. Statistics*. http://minenergo.gov.ru/activity/statistic.
24. Yu.P. Korotaeva, R.D. Margulova, *Extraction, preparation and transport of natural gas and condensate. Reference manual in 2 volumes. Volume II.* Nedra, 288 p., (1984).
25. Bomze, I.M., Budinich, M., Pardaloz, P.M., Pelillo, M.: *The Maximum Clique Problem.* Handbook of Combinatorial Optimization, pp. 1–74 (1999). DOI: 10.1007/978-1-4757-3023-4
26. AIMMS. URL: https://www.aimms.com.
27. Vorobev S., Kolosnitsyn A., Minarchenko I. Energies. 2022; **15**(2):501. DOI: 10.3390/en15020501.
28. Vorobev S., Senderov S., Edelev A. RSES 2020, E3S Web Conf. **Vol. 216**, 2020. DOI: 10.1051/e3sconf/202021601003.

Analysis of energy security threats based on semantic and linear optimization models of the fuel and energy complex development

Natalia Pyatkova [1*], *Alexey* Massel[1]

[1]Melentiev Energy Systems Institute, 130 Lermontov str., Irkutsk, Russia

Abstract. The article proposes the use of semantic models and linear optimization models for the analysis of energy security threats and justification of emergency situations in the event of threats, as well as the integration of mathematical and semantic (on the example of cognitive) modeling for the study of energy security problems. Also to study the energy security problems a two-level technology is proposed that integrates the stages of qualitative analysis (using semantic modeling tools) and quantitative analysis (using linear economic and mathematical models and traditional software systems).

Introduction

The main content of the tasks for studying the problem of energy security is to predict the conditions for the functioning and development of energy systems and the fuel and energy complex as a whole, taking into account possible critical and emergency situations (CS and ES), to assess the state in these conditions and identify " bottlenecks" in the systems of fuel and energy supply to consumers, in the choice of possible alternatives and specific measures to prevent emergency situations and emergencies in these systems or reduce their negative impact [1].

In modern conditions, it became necessary to improve the existing and develop new methodological, model and software tools for conducting such studies, since studies related to the concept of the risk of critical and emergency situations and their consequences begin to occupy a special place. Of particular importance is the task of analyzing possible threats and the formation on this basis of disturbance scenarios (critical and emergency situations) and the related problems of their modeling.

The article proposes the use of semantic models and linear optimization models for the analysis of energy security threats and justification of emergency situations in the event of threats, as well as the integration of mathematical and semantic (on the example of cognitive) modeling for the study of ES problems.

This work is an integral part of the energy security research conducted at the Energy Security Department of the Melentiev Energy Systems Institute Siberian Branch of the Russian Academy of Sciences [1]. Similar studies, developments and models for such studies are mainly focused on solving the problems of long-term planning of the energy industry work under normal operating conditions with a horizon of up to 15-20 years. Similar works carried out in other teams are a local or regional nature with the study of the certain problem aspects [2-5]. Comprehensive studies that allow assessing the possibilities of all energy industries interconnected work and determining the consequences for consumers of energy resources in the event of emergency situations in the operation of one industry or several industries at the same time have not been carried out before.

1. Study scheme

To study the energy security problems, a two-level technology is proposed that integrates the stages of qualitative analysis (using semantic modeling tools) and quantitative analysis (using linear economic and mathematical models and traditional software systems) (Fig. 1) .

* Corresponding author: nata@isem.irk.ru

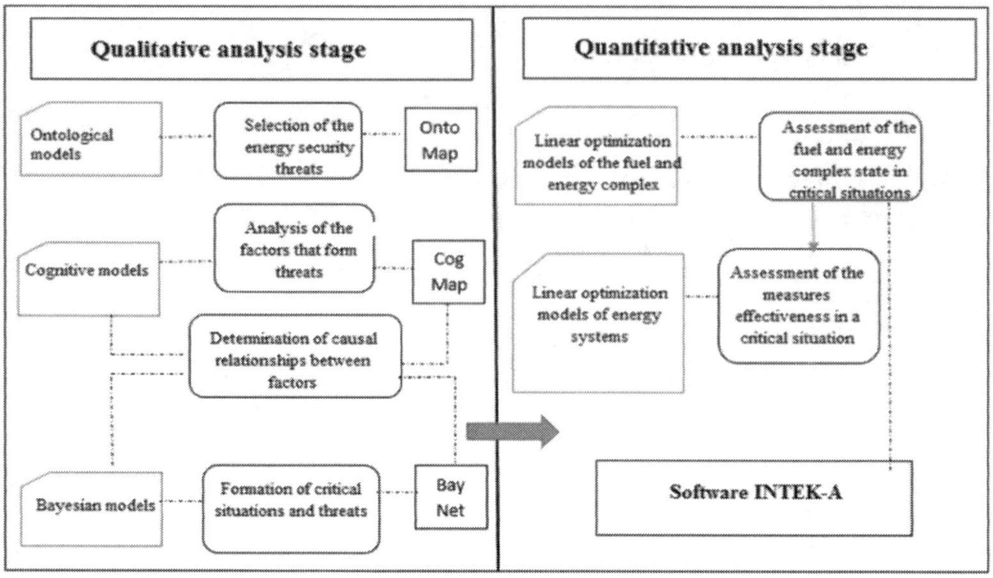

Fig 1. Tasks to be solved and modeling methods used in studies of assessing the impact of threats on the state of energy security

In modern conditions, the problem of studying threats to energy security and the factors that form these threats acquires the greatest importance. In this case, it is possible to justify the occurrence of certain critical or emergency situations.

2. Main threats to energy security

Based on the analysis of the energy sector state and the conditions for its development, the main strategic threats to energy security were identified [1]:

- insufficient level of investment in the fuel and energy sector;
- insufficient growth of explored hydrocarbon reserves;
- reduced opportunities to increase gas production;
- high depreciation and low rates of equipment renewal in the fuel and energy sector.

The analysis of these threats and the formation of CS and ES on their basis takes place at the level of qualitative analysis. At this level, it is possible to use semantic modeling methods (Fig. 1), which can be considered as methods for analyzing and substantiating threats to energy security, the formation of CS and emergency situations, and as methods for situational analysis and management.

3. Scheme of models

At the stage of qualitative analysis, using cognitive modeling methods, models of worsening/improving situations with investments and one of the important strategic threats "Reduced opportunities to increase gas production" were built (Fig. 2).

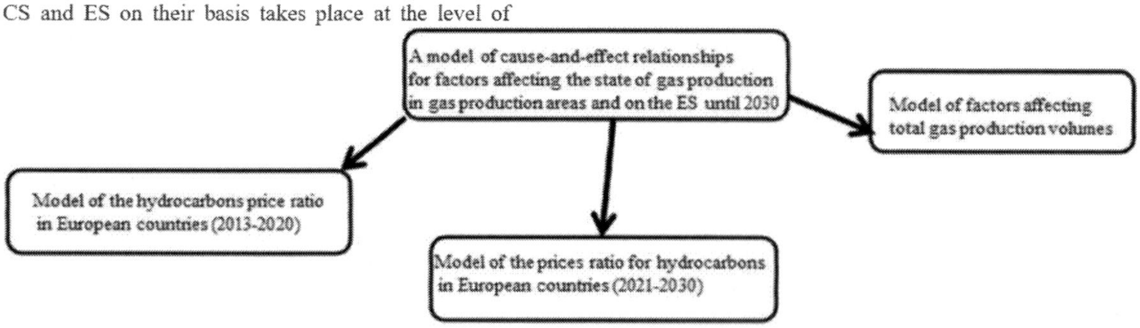

Fig. 2. The system of deterioration/improvement models for the strategic threat situation on the energy security "Reducing the opportunities to increase gas production"

Using the methods of semantic, namely cognitive, modeling for the analysis of energy security threats, it is possible to reasonably form various emergency situations and evaluate them at the stage of qualitative analysis.

At the stage of quantitative analysis, using a scenario approach, based on the knowledge of experts and of the studied threats cognitive models, critical situations are formed that arise among consumers of energy resources and in the energy industries.

Using linear optimization models of energy systems and fuel and energy complex [6], the tasks of the energy system state assessing in the conditions of threats realization to energy security in the form of critical situations are solved to determine:

- rational use of energy facilities production capacities;
- distribution of energy resources certain types by categories of consumers;
- rational use of interregional transport links capacity;
- the size of deficits in energy resources certain types for the considered categories of consumers across the country.

Conclusion

This paper proposes the use of mathematical and semantic modeling for conducting research on the analysis of threats to energy security. A two-stage scheme for conducting research and an algorithm for experimental calculations are proposed. Examples of the cognitive threat models application with the construction of cause-and-effect relationships are given.

The work was carried out within the framework of the state assignment project FWEU-2021-0003, registration number: AAAA-A21-121012090014-5.

References

1. Energy security of Russia: problems and solutions / N.I. Pyatkova, V.I. Rabchuk, S.M. Senderov, M.B. Cheltsov. - Novosibirsk: Nauka, 2011. - 211 p.

2. Malashenko Yu.E., Nazarova I.A., Novikova N.M. Management of the fuel and energy system in case of large-scale damage. Statements of optimization problems. Izv. RAN. Theory and control systems. 2018. No. 2. S. 39-51.

3. Seebregts, Ad J., Goldstein, Gary A., Smekens, Koen. Energy/Environmental Modeling with the MARKAL Family of Models. Proc. Int. Conf. on Operations Research (OR 2001), Energy and Environment Session, September 3-5, 2001. Duisburg,Germany.Access mode:http://ftp.ecn.nl/pub/www/library/report/2001/rx01039.pdf (date of the application: 01.09.2021).

4. MESSAGE (Model for Energy Supply Strategy Alternatives and their General Environmental Impact) International Institute for Applied Systems Analysis (IIASA) [Laxenburg,Austria]. Access mode: http://www.iiasa.ac.at/web/home/research/researchPrograms/Energy/MESSAGE.en.html (date of the application: 15.09.2021).

5. The National Energy Modeling System: An Overview 2009. Access mode: http://www.eia.doe.gov/oiaf/aeo/overview/index.html (date of the application: 15.09.2021).

6. Pyatkova N.I., Senderov S.M., Pyatkova E.V. Methodological features of the study of energy security problems at the present stage. Proceedings of the Russian Academy of Sciences. Energy, 2014, No. 2, pp. 81-87.

https://doi.org/10.1051/e3sconf/202338401029

In technical higher education institutions current state of the use of modern educational virtual reality laboratories in the teaching of specialized sciences

Rakhmonov *I.U.*[1,*], Ganiev *Sh.R.*[2], Alibekova *T.Sh.*[3], Nematov *L.A.*[4]

[1]Tashkent State Technical University named by Islam Karimov, Tashkent, 100095, Uzbekistan

[2]"TIIAME" NRU Bukhara Institute of Natural Resources Management, Bukhara, 105009, Uzbekistan

[3]Karakalpak State University, Nukus, 230101, Uzbekistan

[4]Bukhara Engineering Technological Institute, Bukhara, 200100, Uzbekistan

Abstract. The use of modern virtual reality laboratories in the teaching of specialized subjects in technical higher educational institutions is a rapidly developing field. These simulators provide an immersive and interactive learning experience that allows students to experiment with complex systems in a safe and controlled environment. In this article, the current state of the laboratories used in the teaching of «Basics of Power Supply» in 16 technical higher education institutions selected for research was analyzed, and the advantages and disadvantages of using modern virtual reality laboratories in this case were considered.

1 Introduction

One of the main advantages of using virtual reality simulators in technical education is the ability to provide students with hands-on experiences with equipment and systems that may be difficult or impossible in the classroom [1, 2]. For example, in fields such as energy, students can use virtual simulators to design and test complex processes and systems in industrial plants, allowing them to gain a deeper understanding of the situation and develop the skills and knowledge needed to work with these systems in real-world settings [1, 4, 6].

2 The state of the problem today

Despite these advantages, there are some challenges associated with using virtual reality simulators in technical education. One of the main challenges is the cost and complexity of developing and maintaining simulators, as well as the need for specialized hardware and software [2, 3, 5, 7-10, 12]. In addition, in order to effectively use these simulators in the classroom, it is necessary to develop the skills of using special virtual laboratories for teachers and students [11, 13, 16-19, 14].

The laboratory is performed in 4 different ways:

1. Immersive - through exposure;
2. Physical - performing while standing in front of the device;
3. Observation - research by observation of processes;
4. Virtual - execution using a specific tool.

Today there are 4 types of laboratories (fig. 1) [15, 20]:

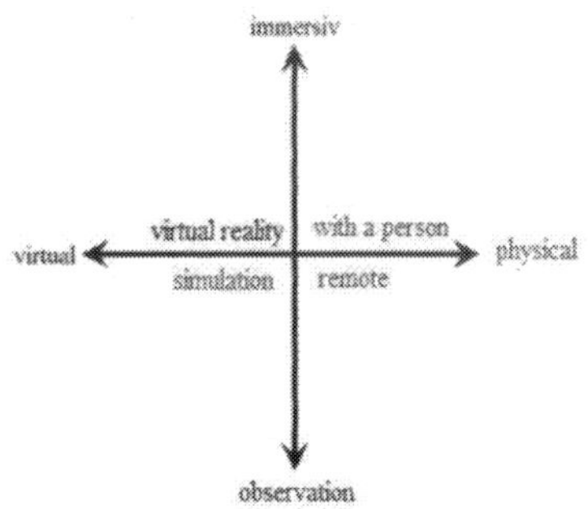

Fig. 1. Grouping of existing laboratories today

1. Traditional laboratories - they are located in physical conditions, in which the student performing the laboratory work is in a real environment [16, 11].
2. Simulations - these are interactive laboratories on a specific computer or other gadget screen, in which students perform some or all of their activities. In this, the user observes the activity through a digital lens.
3. Remote Control – In this the experiments are controlled remotely and the results are observed through digital means to see what happens as a result of the manipulation.

* Corresponding author: ilider1987@yandex.com

4. Virtual reality - when the user uses VR glasses, he performs the laboratory while standing in a real environment.

3 A modern solution to the problem

For the research, 16 higher education institutions in Uzbekistan, where the subject «Basics of electricity supply» is taught, were taken. There are 7 laboratory classes in this «Fundamentals of Power Supply» subject, and only conventional type of laboratories are used in all higher education institutions studied. The laboratories conducted in the field of «Electrical supply» require the student to simultaneously develop the ability to feel physically and work on laboratory equipment. And this demand can be met by using the laboratories that are performed only with a traditional person on a laboratory stand, or by using laboratories that perform virtual reality laboratories. Simulators do not fully cover the need, and these types of laboratories are common today, and during our research, we witnessed that these types of laboratories are available in almost all technical higher education institutions. There is no need for remote-controlled laboratories in this discipline [21, 23, 7].

The current situation in selected higher education institutions, that is, the number of existing laboratories, as well as the need for laboratories, was studied and the situation is presented in Table 1.

The analysis of the situation shown in Table 1 shows that today the need for laboratory equipment is very high, and the level of laboratory supply is on average 68% in all technical higher education institutions (Figure 2).

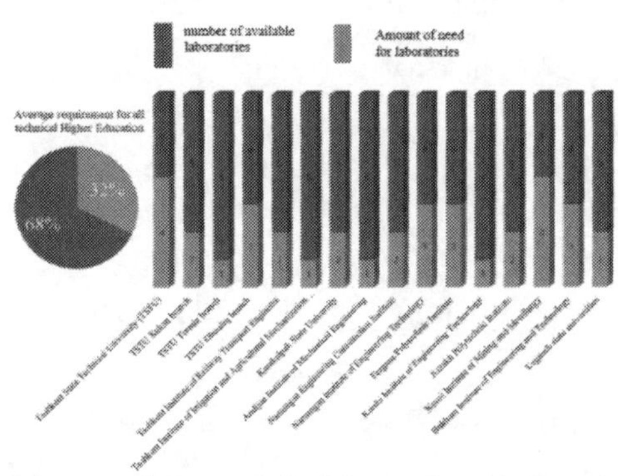

Fig. 2. Diagram of the need for laboratories «Basics of power supply» in technical higher education institutions

Table 1. The level of provision of the «Basics of Power Supply» laboratory in technical higher education institutions

№	Higher educational institution	Course of Study	Traditional laboratories available	The need for a laboratory stand
1	Tashkent State Technical University (TSTU)	5310100 5310200 5310700 5312400	4	3
2	TSTU Kukan branch	5310700	2	5
3	TSTU Termiz branch	5310200 5310700	1	6
4	TSTU Olmaliq branch		3	4
5	Tashkent Institute of Railway Transport Engineers	5310200 5310700	2	5
6	Tashkent Institute of Irrigation and Agricultural Mechanization Engineers	5430500	1	6
7	Karakalpak State University	5310100 5310200 5310700	2	5
8	Andijan Institute of Mechanical Engineering	5310200 5310700 5312400	1	6
9	Namangan Engineering Construction Institute	5310200 5312400	2	5
10	Namangan Institute of Engineering Technology	5310100 5310200	3	4
11	Fergana Polytechnic Institute	5310100 5310200	3	4
12	Karshi Institute of Engineering Technology	5310200 5310700 5312400	1	6
13	Jizzakh Polytechnic Institute	5310100 5310200 5310700	2	5
14	Navoi Institute of Mining and Metallurgy	5310100 5310200	4	3
15	Bukhara Institute of Engineering and Technology	5310100 5310700	3	4
16	Urganch state universities	5310700	2	5

The cost of having 2 different types of laboratories required to meet this need was analyzed and the results were recorded in Table 2. In order to simplify the calculations, the cost of purchasing, bringing and installing the stand of traditional laboratories was set at 100 million UZS based on the current situation. Costs to be allocated for virtual reality labs are assumed to be used by 14 students. In this case, 7 laboratories will be prepared, and two students will be allocated to each virtual laboratory to perform the laboratory (one student will be the manager, and the other student will perform the laboratory) (table 2) [22, 24, 8]:

a. The average cost of a core i5 computer to use a virtual reality laboratory: - 7 million UZS;

b. Average price of virtual reality glasses: 15 million;

c. The cost of purchasing a virtual reality laboratory: 2.5 million soums for 1 laboratory.

d. Installation costs of virtual reality laboratories: 1 mln.

When making these calculations, it is necessary to take into account that it is not necessary to allocate funds only for Virtual Reality glasses and a laboratory, since all the considered technical equipment for Virtual Reality is available in higher education institutions. In table 2, the total costs (column «General» of table 2) in the case where these conditions are taken into account, and the costs allocated in the case of using the current potential at the University (column «In the case of using the existing potential» of table 2) were calculated.

The analysis of table 2 shows that the average cost for each higher education institution is 475 million soums for the acquisition of traditional laboratory stands, 167.5 million soums for virtual laboratories, and 117.5 million soums for using the existing potential of universities. is soum. In addition, up to 4 students can work at the same time in traditional stands. Using virtual reality labs is 14.2 times more effective if two students per booth are calculated, and another advantage of using virtual reality labs is that the wear rate of devices is very small compared to other devices [9, 3, 7].

Table 2. Costs to be allocated to meet the need for the «Basics of Power Supply» laboratory in technical higher education institutions (in million UZS) [25]

№	Higher educational institution	For traditional laboratories (UZS)	For virtual reality labs	
			General	In the case where the existing potential is used
1	Tashkent State Technical University (TSTU)	300	162,5	112,5
2	TSTU Kukan branch	500	167,5	117,5
3	TSTU Termiz branch	600	170	120
4	TSTU Olmaliq branch	400	165	115
5	Tashkent Institute of Railway Transport Engineers	500	167,5	117,5
6	Tashkent Institute of Irrigation and Agricultural Mechanization Engineers	600	170	120
7	Karakalpak State University	500	167,5	117,5
8	Andijan Institute of Mechanical Engineering	600	170	120
9	Namangan Engineering Construction Institute	500	167,5	117,5
10	Namangan Institute of Engineering Technology	400	165	115
11	Fergana Polytechnic Institute	400	165	115
12	Karshi Institute of Engineering Technology	600	170	120
13	Jizzakh Polytechnic Institute	500	167,5	117,5
14	Navoi Institute of Mining and Metallurgy	300	162.5	112,5
15	Bukhara Institute of Engineering and Technology	400	165	115
16	Urganch state universities	500	167.5	117,5

4 Conclusion

In conclusion, it can be said that in the teaching of specialized subjects in technical higher education, traditional stands are used today, but there are also shortcomings in the stands. Addressing this need using modern virtual reality laboratories is both economically and socially effective. Using state-of-the-art virtual reality labs provides students with a powerful and immersive learning experience, allowing them to experiment with complex systems in a safe and controlled environment. From an economic point of view, the introduction of virtual laboratories compared to traditional laboratories is 14.2 times more profitable for the 16 higher education institutions that we examined during our research. In addition, all you need to do is pay for the virtual lab software to create virtual reality science labs from a virtual lab room.

References

1. Venikov V.A., Venikov G.V. The theory of similarity and modeling (in relation to the problems of the electric power industry). - M.: Higher school, 1984, 439 pp.
2. Nikiforov G.V., Oleinikov V.K., Zaslavets B.I. Energy saving and energy management in metallurgical production. M: Energoatomizdat, 2003.- 480 p.
3. Novikov SS Methods of operational planning and control of power consumption of large-capacity electric steel-smelting furnaces when operating on the wholesale electricity market: Avtorefer. dis. ... Cand. those. sciences. - M., 2008.- 18 p.
4. Bazara M., Shetty K. Nonlinear programming. Theory and algorithms. - Moscow: Publishing house: "Mir", 1982. - 583 p.
5. Rakhmonov I.U., Najimova A.M., and Reymov K.M. AIP Conference Proceedings **2647**. 030010. (2022). https://doi.org/10.1063/5.0104788
6. Rakhmonov I.U., Najimova A.M. AIP Conference Proceedings **2647**. 030011. (2022). https://doi.org/10.1063/5.0104791
7. Melikuziev M.V. AIP Conference Proceedings **2552**. (2023). 050021. https://doi.org/10.1063/5.0112395
8. Rasulov A.N., Usmonov E.G., Melikuziev M.V. AIP Conference Proceedings **2552**. (2023). 050018. https://doi.org/10.1063/5.0111530
9. Saidkhodjaev A.G. AIP Conference Proceedings **2552**. 050026. (2023). https://doi.org/10.1063/5.0111540
10. Rakhmonov I.U., Najimova, A.M., Esemuratova Sh.M., Koptileuov T.T. AIP Conference Proceedings **2647**. 070024. (2022). https://doi.org/10.1063/5.0104793
11. Hoshimov F.A., Rakhmonov I.U., Niyozov N.N., Omonov F.B. AIP Conference Proceedings **2647**. 030025. (2023). https://doi.org/10.1063/5.0112388
12. Rakhmonov I.U., Hoshimov F.A., Kurbonov N.N., Jalilova D.A. AIP Conference Proceedings **2647**. 050022. (2023). https://doi.org/10.1063/5.0112391
13. Rakhmonov I.U., Ushakov V.Ya., Niyozov N.N., Kurbonov N.N., Mamutov M. E3S Web of Conferences **289**. 07014. (2021). https://doi.org/10.1051/e3sconf/202128907014
14. Rakhmonov I.U., Ushakov V.Ya., Najimova A.M., Jalilova D.A., Omonov F.B. E3S Web of Conferences **289**. 07013. (2021). https://doi.org/10.1051/e3sconf/202128907013
15. A.D.Taslimov, A.S.Berdishev, F.M.Rakhimov, Melikuziev M.V. E3S Web Conf. **Vol. 139**. 2019. https://doi.org/10.1051/e3sconf/201913901081.
16. R.Karimov. AIP Conference Proceedings 2552. 030014. (2022). https://doi.org/10.1063/5.0111533
17. R.Karimov. AIP Conference Proceedings 2552. 050012. (2022). https://doi.org/10.1063/5.0111524
18. Rakhmonov,I., Berdishev,A., Khusanov, B.M, Khaliknazarov, U., Utegenov, U. IOP Conference Series: Materials Science and Engineering **883(1)**. 012104. DOI 10.1088/1757-899X/883/1/012104
19. I.U. Rakhmonov, T.H. Khakimov, I.I. Bakhadirov, B.M. Khusanov. International Journal of Advanced Science and Technology **Vol. 29**. (2020). pp. 1525-1533.
20. Erejepov, M., Novikov, A.N., Khusanov, B.M., Seytmuratov, B., Sayimbetov, Z. E3S Web of Conferences. **289**. (2021). 07018. https://doi.org/10.1051/e3sconf/202128907018
21. Usmanov E.G., Khusanov B.M. RSES 2020. E3S Web of Conferences **216**. 01161. (2020). https://doi.org/10.1051/e3sconf/202021601161
22. Stilinović, Emma. (2017). Facilitating empathy through virtual reality. Motivation and Emotion **41**. 10.1007/s11031-017-9641-7.
23. https://michaelseery.com/describing-different-types-of-virtual-labs/
24. I.U.Rakhmonov, K.M.Reymov. Statistical models of renewable energy intermittency // RSES 2020. E3S Web of Conferences **216**. 01167. (2020). https://doi.org/10.1051/e3sconf/202021601167
25. I.U.Rakhmonov, F.A.Hoshimov. Development of an algorithm for evaluating the dominant factors that have the greatest impact on the energy intensity of products // ENERGY-21. E3S Web of Conferences. 209 (2020) 07018. doi.org/10.1051/e3sconf/202020907018

Forecasting of electricity consumption by industrial enterprises with a continuous nature of production

Rakhmonov *I.U.*[1,*], Niyozov *N.N.*[1], Kurbonov *N.N.*[1], Umarov *B.S.*[2]

[1]Tashkent State Technical University named by Islam Karimov, Tashkent, 100095, Uzbekistan

[2]Karakalpak State University, Nukus, 230101, Uzbekistan

Abstract. The development of models for forecasting electricity consumption is a complex process, due to the non-linear dependence of electricity consumption on factors that affect the forecast indicators. Since the current forecasting methods do not take into account this non-linearity, the difference between the actual and forecast indicators of electricity consumption often exceeds the allowable values. To determine the required forecast indicators with high accuracy is the use of artificial intelligence methods. In this paper, when predicting electricity consumption, the method of autoregression of the integrated moving average is used. An enlarged block diagram of the algorithm for predicting power consumption using the ARIMA method has been developed.

1 Introduction

The creation and implementation in practice of forecasting models using the ARIMA method is simple in determining forecast indicators, in particular, in the process of checking the adequacy of the developed model. If the forecast error deviates from acceptable values, the automatic change of ARIMA components allows obtaining high-precision forecast indicators. Point and interval prediction can be performed using any developed model; for these types of forecasting, it is not required to develop separate forecasting models [1, 2, 4].

2 The current state of the investigated problem

It is known that the ARIMA method is part of the ARMA model, in which the autoregressive part has a differentiating component d to transform the time series into a stationary series. To do this, it is advisable to express these polynomials in the form of an operator for calculating the autoregression coefficients. To simplify the calculation and writing expressions in the form of an operator, the following definitions A are introduced [3, 5]:

A-d=0,, A-1=0. A0=-1, A1=a1, …, Ap=ap, Ap+1=0, …, Ap+d=0:

$$-(\sum_{k=0}^{p} A_k L^k)(1-L)\Delta^d y_t = -(\sum_{k=0}^{p} A_k L^k - \sum_{k=0}^{p} A_k L^{k+1})\Delta^{d-1} y_t =$$
$$= -(A_0 L^0 + \sum_{k=1}^{p} A_k L^k - \sum_{k=1}^{p} A_{k-1} L^k - A_p L^{p+1})\Delta^{d-1} y_t =$$
$$= -((A_0 - A_{-1} L^0 + \sum_{k=1}^{p} (A_k - A_{k-1}) L^k + (A_{p+1} - A_p) L^{p+1})\Delta^{d-1} y_t =$$
$$= -\sum_{k=0}^{p+1} (A_k - A_{k-1}) L^k \Delta^{d-1} y_t = -\sum_{k=0}^{p+1} \Delta A_k L^k \Delta^{d-1} y_t . \quad (1)$$

Thus, the difference operator Δ can be transferred from the time series yt on a sequence of coefficients, and the number of non-zero terms will increase by one. Repeating these calculations d times and taking into account that ΔdA0=-1, we get:

$$(1-\sum_{k=1}^{p} \Delta^d A_k L^k) y_t = (1-\sum_{k=1}^{q} b_k L^k)\varepsilon_t . \quad (2)$$

From the analysis (2) it can be concluded that in the structure of the ARMA model, MA is a process of infinite order with restrictions set by the limits. Using this method, it is possible to express a complex damage structure with a small amount of primary data. As mentioned above, this in turn is done by transforming the content of the non-stationary data into stationary data.

An enlarged block diagram of the power consumption forecasting algorithm using the integrated moving average autoregression (ARIMA) method can be represented as shown in Fig. 1. Analysis of the graph shows that power consumption is a non-stationary process [19, 6-15, 17, 20, 18].

The ARIMA prediction model is represented by three components (p, d, q).

* Corresponding author: ilider1987@yandex.com

The order of the time series difference – d is calculated as follows.

At the initial stage of creating a forecast model, a differentiation process is performed to obtain a constant time series from a changing time series. At the same time, with the help of formal tests, an assessment of the invariance of the initial time series is made. Since the processes of consumption of electrical energy are not always unchanged, then, with the help of certain transformations, the time series is reduced to a stationary mode. It includes the following parameters [16, 3, 21, 22]:

- finite differences:

$$X_t = \Delta Y_t = Y_t - Y_{t-1} \quad (3)$$

where Xt – primary difference:

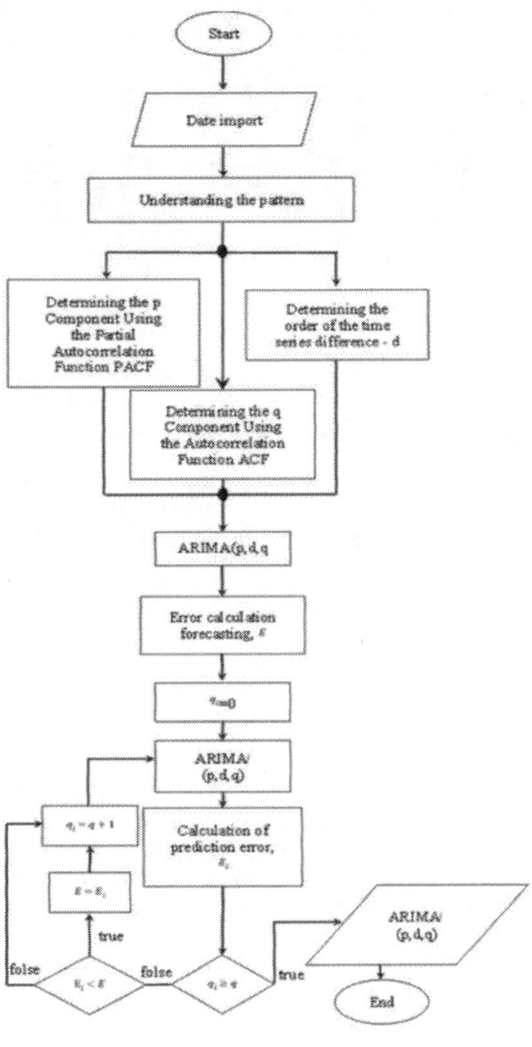

Fig.1. Enlarged block diagram of the algorithm for predicting power consumption using the autoregressive method of the integrated moving average [8]

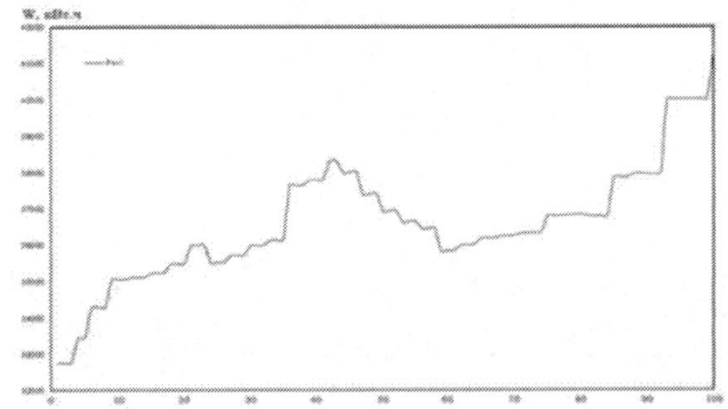

Fig.2. Electricity consumption chart of a metallurgical enterprise

$$Z_t = \Delta X_t = X_t - X_{t-1} = Y_t - 2Y_{t-1} + Y_{t-2} = \Delta^2 Y_t \,, \quad (4)$$

(Zt – secondary difference);

- logarithmic chain indices of the following form:

$$X_t = \ln(Y_t \,/\, Y_{t-1}) = \ln Y_t - \ln Y_{t-1}; \quad (5)$$

- growth rate of the next type:

$$X_t = (Y_t - Y_{t-1}) / Y_{t-1} = (Y_t \,/ Y_{t-1}) - 1; \quad (6)$$

- logarithmic series:

$$X_t = \ln Y_t \,; \quad (7)$$

- growth rate of the next type:

$$X_t = (Y_t \,/\, Y_{t-1}) \quad (8)$$

It should be noted that when determining a stationary time series, it is necessary to take into account the nature of the change in the time series graph *Xt*. In this case, the main criterion for the quality of the process is the fulfillment of the condition *Xt=f(Yt)*≈const.

For the case we are considering, the time series reaches stationarity with two orders of difference, that is, the second order is considered sufficient in our study (Fig. 3) [23, 7, 18].

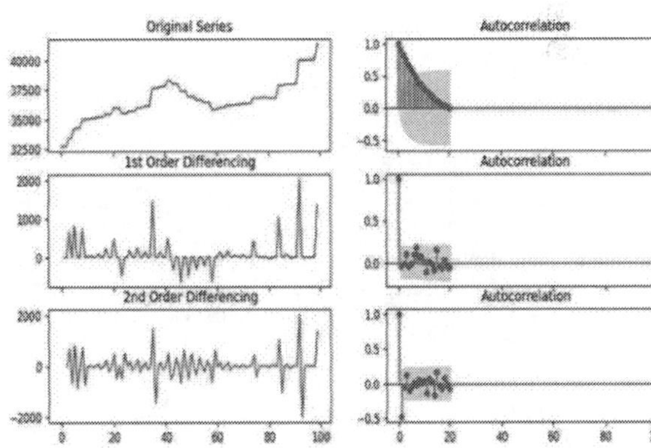

Fig.3. Determining the order of the time series difference – d

In this case, the stationarity of the series is checked first, using the extended Dickey Fuller test (adfuller) from the statsmodels package. If the series is non-stationary, then it is necessary to differentiate it. Otherwise, no difference is required, that is, d = 0. After that, differentiation is performed and the autocorrelation of the graph is obtained. As a graph line, it is enough to differentiate 1 time, so d is equal to 1.

Autoregressive process to determine autoregressive order - p.

In the autoregressive process, each time series value is the sum of a random relationship and takes previous values into account to predict future values. Any autocorrelation in the stationary series can be corrected by adding AR [7, 8, 24-28, 12]. In this case, the condition is accepted that the number of ARs is equal to the number of lags that cross the significance limit on the PACF graph (Fig. 4).

On fig. 4.4 shows a lagging of 1 PACF as it is well above the significance line. Lag 2 is also significant, but it is in the marginal zone (blue area) and p can be taken tentatively as 1.

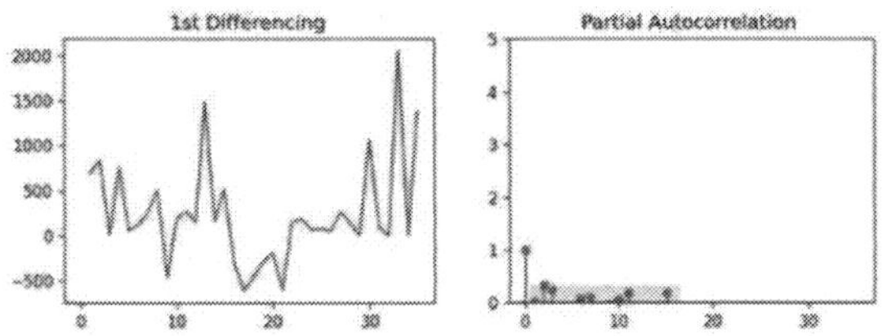

Fig. 4. Partial autocorrelation function PACF

Moving average process to determine the order of the moving average - q.

It is known that the moving average component is required to adjust the time series and the process of forming the adjustment begins with determining the observation window.

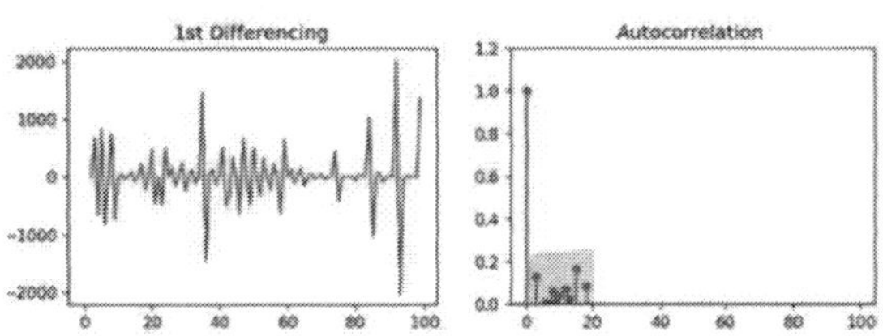

Fig. 5. Autocorrelation function ACF

The autocorrelation function (ACF) shown in Figure 5 allows you to determine the number of MA terms. Figure 4.5 shows that the pair of lags of the ACF autocorrelation function is much higher than the significance line and the dependence does not contain autocorrelation, so it is assumed that the q coefficient is equal to 1.

According to the above calculations for forecasting in the ARIMA model, the main components are determined, and they have the following form: autoregression order p=1; the order of the time series difference d=1; moving average order q=1. In general ARIMA (1, 1, 1).

The next step is to analyze the difference between the actual and forecast values, i.e. balances (imbalances). The assessment of the validity obtained using the ARIMA model is based on a comparison of the actual data with the simulation results. The evaluation results are shown in graphical form (Fig. 6) [14, 22, 15].

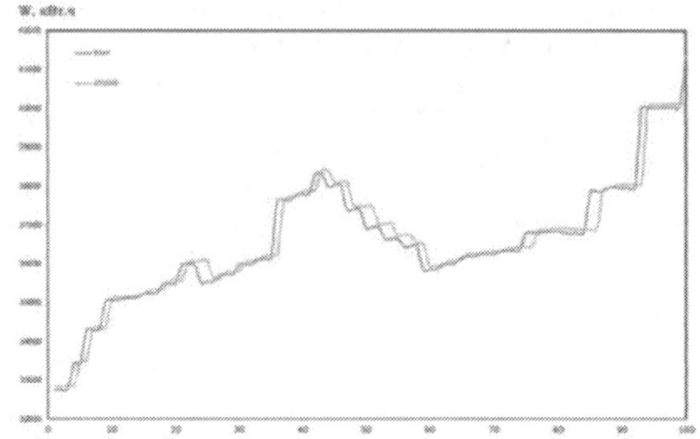

Fig. 6. Comparison of actual and forecast power consumption according to the ARIMA model (1,1,1)

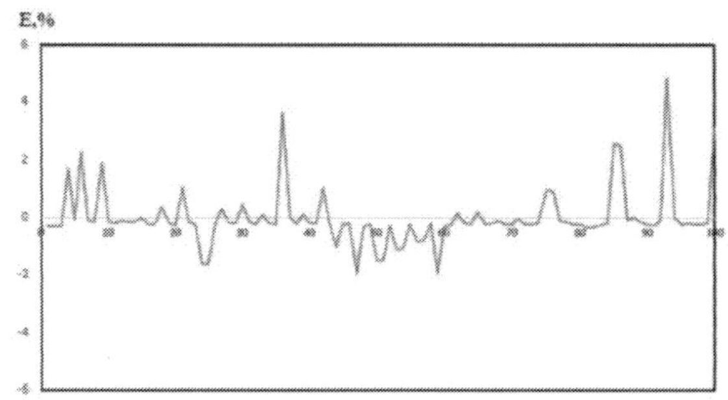

Fig. 7. Graph of the error between the actual and predicted values of power consumption according to the developed model using the ARIMA method

The degree of adequacy of the developed models is substantiated by low absolute and relative errors between actual and forecast data. Analysis of the value of forecasting errors (Fig. 7) of the error shows (6%) that the obtained mathematical models of electrical energy consumption are adequate and therefore they can be used to determine the predicted values of power consumption parameters at ferrous metallurgy enterprises [10].

3 Conclusion

Thus, it can be concluded that the use of models obtained on the basis of the ARIMA method for forecasting power consumption provides high accuracy and adequacy in determining the predicted values of power consumption parameters.

References

1. Abdullaev D.A. Operational processing of measurement information in automated control systems. - Tashkent, Sub, 1985, 128 p.

2. Venikov V.A., Venikov G.V. The theory of similarity and modeling (in relation to the problems of the electric power industry). - M.: Higher school, 1984, 439 pp.

3. Nikiforov G.V., Oleinikov V.K., Zaslavets B.I. Energy saving and energy management in metallurgical production. M: Energoatomizdat, 2003.- 480 p.

4. Novikov SS Methods of operational planning and control of power consumption of large-capacity electric steel-smelting furnaces when operating on the wholesale electricity market: Avtorefer. dis. ... Cand. those. sciences. - M., 2008.- 18 p.

5. Struchenkov V.I. Optimization methods in applied problems. - Moscow: Publishing house: "SOLON PRESS", 2009. - 320 p.

6. Bertsekas D. Conditional optimization and methods of Lagrange multipliers. - Moscow: Publishing house: "Radio and communication", 1987. - 400 p.

7. Rakhmonov I.U., Najimova A.M., and Reymov K.M. AIP Conference Proceedings **2647.** 030010. (2022). https://doi.org/10.1063/5.0104788

8. Taslimov A.D. AIP Conference Proceedings **2552.** (2023). 050023. https://doi.org/10.1063/5.0112398

9. Rakhmonov I.U., Najimova A.M. AIP Conference Proceedings **2647.** 030011. (2022). https://doi.org/10.1063/5.0104791

10. Rakhmonov I.U., Najimova, A.M., Esemuratova Sh.M., Koptileuov T.T. AIP Conference Proceedings **2647.** 070024. (2022). https://doi.org/10.1063/5.0104793

11. R.Karimov, N.Kurbanova, and ofters. Journal of Physics: Conference Series. 2094(5). **052042**. (2021). doi:10.1088/1742-6596/2094/5/052042

12. Hoshimov F.A., Rakhmonov I.U., Niyozov N.N., Omonov F.B. AIP Conference Proceedings **2647.** 030025. (2023). https://doi.org/10.1063/5.0112388

13. S.Dzhuraev, R.Karimov, and ofters. ElConRus, pp. 1166-1169. (2022). doi: 10.1109/ElConRus54750.2022.9755782

14. Rakhmonov I.U., Hoshimov F.A., Kurbonov N.N., Jalilova D.A. AIP Conference Proceedings **2647.** 050022. (2023). https://doi.org/10.1063/5.0112391

15. Rakhmonov I.U., Ushakov V.Ya., Niyozov N.N., Kurbonov N.N., Mamutov M. E3S Web of Conferences **289.** 07014. (2021). https://doi.org/10.1051/e3sconf/202128907014

16. Rakhmonov I.U., Ushakov V.Ya., Najimova A.M., Jalilova D.A., Omonov F.B. E3S Web of Conferences **289.** 07013. (2021). https://doi.org/10.1051/e3sconf/202128907013

17. Saidkhodjaev A.G. AIP Conference Proceedings **2552.** 050026. (2023). https://doi.org/10.1063/5.0111540

18. Usmonov E.G. AIP Conference Proceedings **2552.** 050020. (2023). https://doi.org/10.1063/5.0111537

19. Rasulov, A.N., Rafikova, G.R., Novikov, N.L., Ruzinazarov, M.R., Esemuratova, S. E3S Web of Conferences 289. 07006. (2021). https://doi.org/10.1051/e3sconf/202128907006

20. Melikuziev M.V., Nematov L.A., Novikov A.N., Baymuratov K.K. E3S Web of Conferences **289.** 07016 (2021). https://doi.org/10.1051/e3sconf/202128907016

21. Erejepov, M., Novikov, A.N., Khusanov, B.M., Seytmuratov, B., Sayimbetov, Z. E3S Web of Conferences. **289.** (2021). 07018. https://doi.org/10.1051/e3sconf/202128907018

22. Khakimov, H.T., Shayumova, Z.M., Kurbanbaeva, Z.K., Khusanov, B.M. E3S Web of Conferences **139.** 01076. (2019). https://doi.org/10.1051/e3sconf/201913901076

23. Rakhmonov,I., Berdishev,A., Khusanov, B.M, Khaliknazarov, U., Utegenov, U. IOP Conference Series: Materials Science and Engineering **883(1)**. 012104. DOI 10.1088/1757-899X/883/1/012104

24. I.U. Rakhmonov, T.H. Khakimov, I.I. Bakhadirov, B.M. Khusanov. International Journal of Advanced Science and Technology **Vol. 29**. (2020). pp. 1525-1533.

25. Rakhmonov I.U., Kurbonov N.N., Bijanov A.K. Types of simulators and their efficient architectures // "Bulletin of science and education" scientific and methodological journal. № 9 (129). 2022. p. 13-19.

26. Rakhmonov I.U., Niyozov N.N. The main indicators of the effectiveness of the implementation of an automated information system for the management of electricity consumption by industrial enterprises // "Universum: technical sciences" electronic scientific journal. №10 (103), 2022. - p. 52-54.

27. Rakhmonov I.U., Khoshimov F.A. Mathematical modeling of the influence of factors on the specific energy consumption // "Universum: technical sciences" electronic scientific journal. №10 (103), 2022. p. 55-58.

28. I.U. Rakhmonov, N.N. Kurbonov. Analysis of automated software for monitoring energy consumption and efficiency of industrial enterprises // E3S Web of Conferences. 216 (2020) 01178. https://doi.org/10.1051/e3sconf/202021601178

Forecasting of electricity consumption by industrial enterprises with a continuous nature of production based on the principal component method (PCA)

Niyozov *N.N.*[1,*], Rikhsitillaev *B.Kh.*[2], Jalilova *D.A.*[1], Kuatbaev *P.M.*[3]

[1]Tashkent State Technical University named by Islam Karimov, Tashkent, 100095, Uzbekistan

[2]Nuclear energy development agency under the Ministry of Energy of the Republic of Uzbekistan, Tashkent, 100025, Uzbekistan

[3]Karakalpak State University, Nukus, 230101, Uzbekistan

Abstract. Forecasting the consumption of electrical energy is important to improve its efficiency and, as a result, to improve the competitiveness of manufactured products by reducing the share of electricity costs in the cost of production. When determining the forecast indicators of electricity consumption by industrial enterprises, it is advisable to use modern high-precision forecasting methods that ensure the minimum value of the forecast error. Each enterprise must preliminarily determine, with the greatest possible accuracy, the amount and schedule of electricity consumption and then strictly adhere to them, thereby minimizing penalties and fines. The article deals with the issues of forecasting power consumption by industrial enterprises (on the example of a metallurgical enterprise) using the enlarged block diagram of the algorithm for predicting power consumption by the method of principal components (PCA) developed by the authors. Comparisons of the actual and predicted power consumption according to the developed model are made. The adequacy of the model is confirmed by small discrepancies between the actual and forecast data. This allows it to be used in determining the predicted values of power consumption parameters at ferrous metallurgy enterprises.

1 Introduction

An effective way to prevent conflicts between industrial enterprises and energy supply organizations, accompanied by additional costs of electricity and financial resources, is to increase the accuracy of forecasting the volumes and schedules of electricity consumption. The latter is ensured by the definition of these parameters for the contractual settlement period. In accordance with the Decree of the Cabinet of Ministers of the Republic of Uzbekistan dated January 12, 2018 No. 22 “On additional measures to improve the procedure for the use of electricity and natural gas”, on the basis of agreements between electricity supply companies and consumers when consuming more than 5% of electricity in excess of the limit established by the agreement, a fine is levied by introducing a multiplying coefficient to the established tariff in the amount of 1.15. It should be noted that before the entry into force of this resolution (March 2019), enterprises paid significant fines for electricity consumption in excess of the established limit. This is primarily due to the lack of a perfect methodology for forecasting electricity consumption at industrial enterprises, i.e. the presence of a large error in the forecast indicators determined using existing methods. As noted above, penalties for electricity consumed in excess of the limit lead to an increase in the price of products and reduce their competitiveness.

2 The current state of the investigated problem

Of the currently existing approximately 150 forecasting methods, 20-30 main ones are used in practice [1, 2, 4, 5, 6, 7, 8, 9, 10].

An analysis of the forecasting methods used in industrial enterprises shows that they are mainly based either on an expert assessment of the volume of electricity consumption, or on accounting for electricity consumption per unit of output [4, 5, 7, 8, 10, 12].

In order to improve the accuracy of predictive indicators, the authors of the article developed forecasting models using modern methods for predicting electricity consumption at industrial enterprises, providing minimal errors [10, 13, 15].

When developing mathematical models, information on power consumption at 100 protocol heats of a metallurgical enterprise was used as initial data

* Corresponding author: ilider1987@yandex.com

according to the criteria for sampling primary data [16, 17].

It is known that large errors in determining the predicted value of the object under study lead to additional financial costs. To reduce them, we have created a forecasting model using the principal component method [18, 5, 20].

This method is one of the methods of multivariate statistical analysis, which is used to reduce the dimension with the least loss of useful information. Reducing the number of variables in the initial data by excluding secondary ones from consideration makes it easier to operate with them, simplifies the creation of algorithms [18, 19, 8, 20, 21].

Thus, the idea of PCA is simple - to reduce the number of variables in the dataset without losing useful information. This is achieved by creating new, unrelated variables that continually increase variability. The search for such new variables is associated with the solution of a specific value / specific vector problem, and new variables are a priori determined by the dataset, not by PCA. Its solution makes the data analysis method more flexible and easily adaptable. For this, various methods and options have also been developed, adapted to data structures [22; 23].

The need to reduce the dimension of the initial data is explained by the following circumstances [24, 25, 26, 27]:

- the need for visual display of the initially selected data, which is achieved on the basis of their projection in three-dimensional space, a plane or a series of numbers;
- the desire to simplify the studied models due to the need to facilitate the calculation and interpretation of the results;
- the need to reduce the amount of stored data.

The importance of reducing the dimension of the initial information is due to the following reasons [12; 13; 14, 15, 8, 9]:

- low information content due to low variability in the transition from one object to another;
- replacement of information due to the correlation of initial features;
- the need for aggregation of initial features.

Principal component analysis has advantages over regression and Fourier analysis. It is simple and clear to use. The essence of the method is to transform a one-dimensional series into a multidimensional one using a one-parameter shift procedure. The scope of this method is the study of a multidimensional trajectory based on the analysis of the principal components and the approximation of the series with respect to the selected principal components [7, 10].

The main disadvantages of the principal component method are [5-16, 18, 23, 4]:

- the need for frequent replacement of the regulatory framework due to its rapid aging;
- the method allows you to work only with continuous data.

The principal component method was first used by Pearson in 1901 to solve problems of the best approximation of a finite set of points by straight lines and planes. For each k = 0, 1, ..., n-1, with a finite set of vectors, a task is compiled. The problem, in the formulation [8, 5, 3], is that with a minimum of the sum of squared deviations xi, from all k-dimensional linear manifolds find:

$$\sum_{i=1}^{m} dist^2(x_i, L_k) \to \min \qquad (1)$$

where $dist^2(x_i, L_k)$ – Euclidean distance from a point of a linear manifold.

In this case, any k - dimensional linear manifold is specified as a set of linear combinations. In [17-20], the problem is solved as follows:

$L_k = \{a_0 + \beta_1 a_1 + ... + \beta_k a_k \ |\beta_i \notin R\ \}$, where the parameters run over the real line R ,$a_0 \in R^n$ $a\{a_1,...,a_k\} \subset R^n$ - orthonormal set of vectors:

$$dist^2(x_i, L_k) = \left\| x_i - a_0 - \sum_{j=1}^{k} a_j (a_j, x_i - a_0) \right\|^2 \qquad (2)$$

Where $\|\cdot\|$ euclidean norm, (a_j, x_i) - Euclidean dot product, or in coordinate form:

$$dist^2(x_i, L_k) = \sum_{l=1}^{n} \left(x_{il} - a_{0l} - \sum_{j=1}^{k} a_{jl} \sum_{q=1}^{n} a_{jq}(x_{iq} - a_{0q}) \right)^2 \qquad (3)$$

Approximation problem for k=0, 1, …, n-1 in [21, 4, 25] is solved on the basis of a set of nested linear manifolds:

$$L_0 \subset L_1 \subset ... L_{n-1}$$

$$L_k = \{a_0 + \beta_1 a_1 + ... + \beta_k a_k \ |\beta_i \notin R\ \} \qquad (4)$$

To determine the reduced linear manifolds, an orthonormal set of vectors or a vector of principal components is carried out, as well as a vector:

$$a_0 = \arg\min_{a_0 \in R^n} \left(\sum_{i=1}^{m} dist^2(x_i, L_0) \right) \qquad (5)$$

That is

$$a_0 = \arg\min_{a_0 \in R^n} \left(\sum_{i=1}^{m} \|x_i - a_0\|^2 \right) \qquad (6)$$

which is the sample mean: $a_0 = \frac{1}{m}\sum_{i=1}^{m} x_i = \bar{X}$

Accordingly, the principal component vectors are defined similarly in the form of optimization problems of the same type [24, 23, 16]:

Data is centralized (by subtracting the average)

$$x_i := x_i - \bar{X}.$$

In this case, the condition $\sum_{i=1}^{m} x_i = 0$

2. The solution to the problem is the first principal component:

$$a_1 = \arg\min_{\|a_1\|=1} \left(\sum_{i=1}^{m} \|x_i - a_1(a_1, x_i)\|^2 \right) \qquad (7)$$

One of them is chosen as the final decision.

3. The projection on the first principal component is subtracted from the original data:

$$x_i := x_i - a_1(a_1, x_i) \qquad (8)$$

4. As a solution to the problem, the second main component is determined.

The second principal component is the solution to the problem:

$$a_1 = \arg\min_{\|a_2\|=1}\left(\sum_{i=1}^{m}\|x_i - a_2(a_2, x_i)\|^2\right) \qquad (9)$$

Next, at step 2k – 1, the projection onto the (k–1)-th principal component is subtracted [25-27]:

$$x_i := x_i - a_{k-1}(a_{k-1}, x_i) \qquad (10)$$

As a solution to the problem at step 2k, the k-th principal component is determined [27]:

$$a_k = \arg\min_{\|a_k\|=1}\left(\sum_{i=1}^{m}\|x_i - a_k(a_k, x_i)\|^2\right) \qquad (11)$$

One of several solutions is chosen as the final one.

The projection onto the preceding component is subtracted at each stage (2k – 1). It should be noted that, based on the solution of the optimization problem, it is easy to orthonormalize certain vectors. The calculation errors should not violate the mutual orthogonality of the vectors of the principal components, and vectors are included in the condition of the optimization problem. The presence of arbitrariness in the choice of the sign (and solve the same problem) and the most essential condition of data symmetry cause the ambiguity of the quantity.

In this case, the last principal component, the unit vector, is orthogonal to all previous vectors. When using the method of principal components, the following operations are performed [3-10, 15]:

1. The initial data are replaced by linear manifolds of lower dimension;
2. An orthogonal projection with maximum dispersion is revealed, i.e. standard deviation from the mean;
3.For the initial data, orthogonal coordinates are created with the correlation coefficient equal to zero.

To solve the problem, an enlarged block diagram of the algorithm for predicting power consumption by the principal component method was compiled, Fig.1.

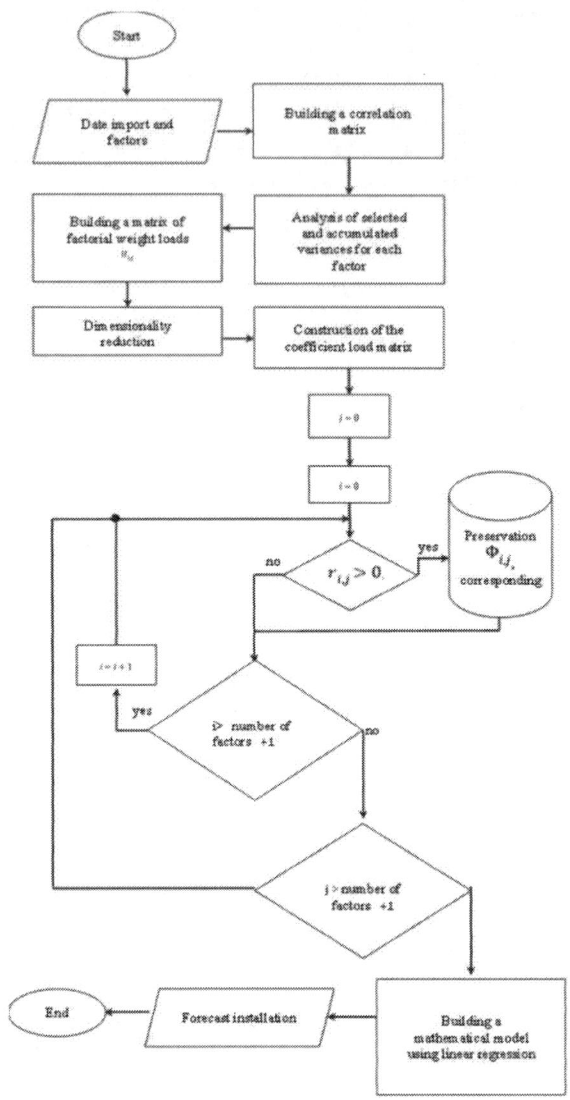

Fig.1. Enlarged block diagram of the algorithm for predicting the power consumption of the main components

The following factors were chosen as initial data: Φ_1 – furnace load, t; Φ_2 – amount of shipped metal, t; Φ_3 – weight of loaded metal charge, t; Φ_4 – duration of metal smelting, min.; Φ_5 – duration of work under current, min.

$$R=\begin{bmatrix} 1 & -0{,}31664 & -0{,}31602 & -0{,}31565 & -0{,}197091 & -0{,}21457 \\ -0{,}31664 & 1 & 0{,}999696 & 0{,}999277 & 0{,}519265 & 0{,}324651 \\ -0{,}31602 & 0{,}999696 & 1 & 0{,}999679 & 0{,}522692 & 0{,}264124 \\ -0{,}31565 & 0{,}999277 & 0{,}999679 & 1 & 0{,}521547 & 0{,}657814 \\ -0{,}19709 & 0{,}519265 & 0{,}522694 & 0{,}521547 & 1 & 0{,}321487 \\ -0{,}21457 & 0{,}324651 & 0{,}264124 & 0{,}657814 & 0{,}321487 & 1 \end{bmatrix}$$

Table 1. Contributions of principal components to the total variance of initial features

Meaning	Total variance,%	Cumulative variance, %
1	69,65	69,65
2	17,28	86,94
3	13,04	99,98
4	0,014	99,99
5	0,003	100,0

Analysis of the selected and accumulated variance for each factor shows the percentage of total variance for each factor (in this example it is 5). As you can see, the first factor (value 1) is 69,65% of the total variance, factor 2 is 17.28%, and so on. The difference due to sequence factors is shown in Table 1.

Using the method of basic components, we obtain the primary matrix of weight loads of factors. In parallel, the matrix of eigenvalues of the correlation matrix should be analyzed.
The weight loading matrix of factors looks like below:

$$U=\begin{bmatrix} -0,42577 & 0,90367 & -0,04563 & -0,29278 & -0,39021 \\ 0,97914 & 0,10209 & -0,17455 & -0,01866 & -0,00555 \\ 0,97982 & 0,10328 & -0,17077 & -0,00055 & 0,01065 \\ 0,97944 & 0,10346 & -0,17199 & 0,01924 & -0,00508 \\ 0,65063 & 0,12643 & 0,74879 & -0,88327 & -0,32493 \end{bmatrix}$$

If the size of the property field is less than the two main components, then only the first two columns of the specified matrix are taken into account.

The dependence of the principal components on the centralized normalized initial characteristics has the following form:

$$W_1 = -0,42577\Phi_1 + 0,97914\Phi_2 + 0,97982\Phi_3 + 0,97944\Phi_4 + 0,65063\Phi_5 \quad (12)$$

$$W_2 = 0,90367\Phi_1 + 0,10209\Phi_2 + 0,10209\Phi_3 + 0,10328\Phi_4 + 0,12643\Phi_5 \quad (13)$$

Next, the following matrix of factor loadings is analyzed to determine new features:

$$R=\begin{bmatrix} -0,76494 & 0,983586 & -0,07316 & -0,60971 & -0,01433 \\ 0,96225 & -0,14342 & 0,03046 & -0,01863 & -0,00559 \\ 0,961577 & -0,14261 & -0,23434 & -0,00056 & 0,01065 \\ 0,261779 & -0,14229 & 0,03310 & 0,01926 & -0,00502 \\ 0,30014 & -0,07934 & -0,25058 & 0,59948 & 0,35618 \end{bmatrix}$$

At the same time, the initial characteristics and correlation coefficients of the main components characterize the correlation matrix of factor loads and form the basis of these calculations:

$$A=\begin{bmatrix} -0,76494 & 0,983586 \\ 0,96225 & -0,14342 \\ 0,961577 & -0,14261 \\ 0,261779 & 0,84229 \\ 0,30014 & -0,07934 \end{bmatrix}$$

Those initial data for which correlation coefficients are greater than 0.7 are highly dependent on the first principal component: Φ_1 – furnace load, t; Φ_2 – weight of shipped metal, t; Φ_3 – mass of the loaded metal charge, t. The second main component is closely related to the initial data: Φ_1 – furnace load, t; Φ_4 – duration of metal smelting, min. According to the main factors determined in this way, a mathematical model is built using linear regression, which has the following form:

$$W = 36264 - 63,74\Phi_1 + 46,89\Phi_2. \quad (14)$$

According to (14), the predicted values of the object under study are determined and compared with the actual data (Fig. 2).

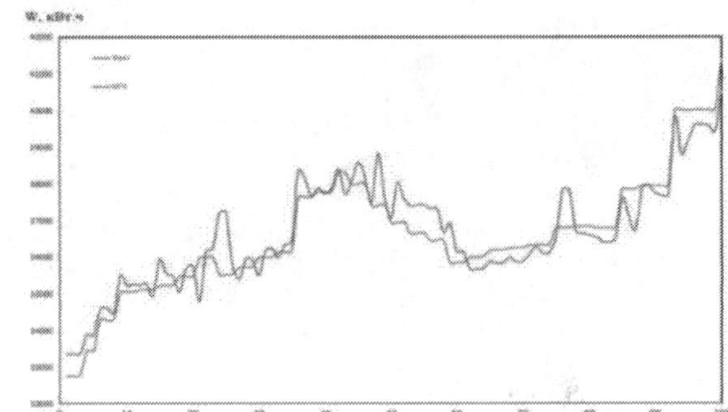

Fig.2. Comparison of the actual and forecast power consumption according to the developed model using the method of principal components

In order to assess the adequacy of the forecasting model, the results of their modeling are compared with actual data. The results are shown graphically in fig. 3.

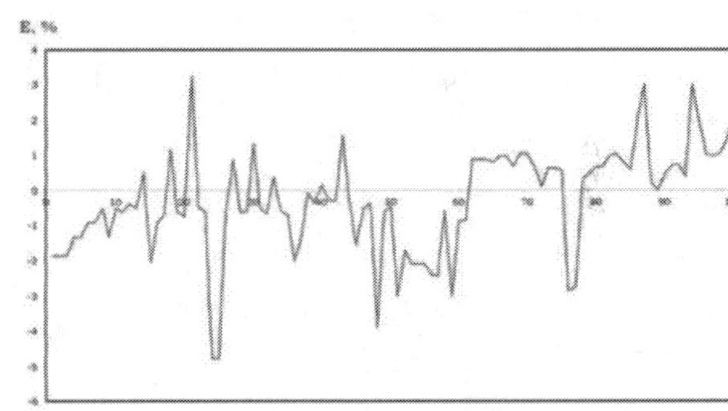

Fig.3. Graph of the error between the actual and predicted values of power consumption according to the developed model using the method of principal components

3 Conclusion

Thus, from Fig. 2 it follows that the actual and predicted values of power consumption coincide in several time intervals. For example, at points 2-15, 18-20, 26-27, 33-34 they coincide, but at other points there are significant declines that negatively affect the adequacy of the developed forecasting model. The average value of the error of the developed model lies within +3 - -5% (Fig.

3), which indicates its suitability for determining the predicted values of power consumption.

References

1. Resolution of the Cabinet of Ministers of the Republic of Uzbekistan dated January 12, 2018 N 22 "On additional measures to improve the procedure for the use of electrical energy and natural gas."

2. Bazara M., Shetty K. Nonlinear programming. Theory and algorithms. - Moscow: Publishing house: "Mir", 1982. - 583 p.

3. Struchenkov V.I. Optimization methods in applied problems. - Moscow: Publishing house: "SOLON PRESS", 2009. - 320 p.

4. Bertsekas D. Conditional optimization and methods of Lagrange multipliers. - Moscow: Publishing house: "Radio and communication", 1987. - 400 p.

5. Batsova S. V. Organization of rationing of electricity at an industrial enterprise. Economic incentives for the rational use of electricity / S. V. Batsova // Casting and metallurgy: scientific and production journal. - 2010. –№3 (56). - S. 131 - 133.

6. Chegodaev AI Mathematical methods of analysis of expert assessments // Bulletin of the Samara State University of Economics. 2010. №2 (64). S. 130-135.

7. Rakhmonov I.U., Najimova A.M., and Reymov K.M. AIP Conference Proceedings 2647. 030010. (2022). https://doi.org/10.1063/5.0104788

8. Rakhmonov I.U., Najimova A.M. AIP Conference Proceedings 2647. 030011. (2022). https://doi.org/10.1063/5.0104791

9. Rakhmonov I.U., Najimova, A.M., Esemuratova Sh.M., Koptileuov T.T. AIP Conference Proceedings 2647. 070024. (2022). https://doi.org/10.1063/5.0104793

10. Hoshimov F.A., Rakhmonov I.U., Niyozov N.N., Omonov F.B. AIP Conference Proceedings 2647. 030025. (2023). https://doi.org/10.1063/5.0112388

11. Rakhmonov I.U., Hoshimov F.A., Kurbonov N.N., Jalilova D.A. AIP Conference Proceedings 2647. 050022. (2023). https://doi.org/10.1063/5.0112391

12. Melikuziev M.V. AIP Conference Proceedings 2552. (2023). 050021. https://doi.org/10.1063/5.0112395

13. Rasulov A.N., Usmonov E.G., Melikuziev M.V. AIP Conference Proceedings 2552. (2023). 050018. https://doi.org/10.1063/5.0111530

14. Taslimov A.D. AIP Conference Proceedings 2552. (2023). 050023. https://doi.org/10.1063/5.0112398

15. Melikuziev M.V., Nematov L.A., Novikov A.N., Baymuratov K.K. E3S Web of Conferences 289. 07016 (2021). https://doi.org/10.1051/e3sconf/202128907016

16. A.D.Taslimov, A.S.Berdishev, F.M.Rakhimov, Melikuziev M.V. E3S Web Conf. Vol. 139. 2019. https://doi.org/10.1051/e3sconf/201913901081.

17. 25. Saidkhodjaev A.G., Nuriddinova Kh.R. AIP Conference Proceedings 2552. 050037. (2023). https://doi.org/10.1063/5.0133486

18. Rakhmonov I.U., Ushakov V.Ya., Niyozov N.N., Kurbonov N.N., Mamutov M. E3S Web of Conferences 289. 07014. (2021). https://doi.org/10.1051/e3sconf/202128907014

19. Rakhmonov I.U., Ushakov V.Ya., Najimova A.M., Jalilova D.A., Omonov F.B. E3S Web of Conferences 289. 07013. (2021). https://doi.org/10.1051/e3sconf/202128907013

20. R.Karimov. AIP Conference Proceedings 2552. 030014. (2022). https://doi.org/10.1063/5.0111533

21. R.Karimov. AIP Conference Proceedings 2552. 050012. (2022). https://doi.org/10.1063/5.0111524

22. S.Dzhuraev, R.Karimov, and ofters. ElConRus, pp. 1166-1169. (2022). doi: 10.1109/ElConRus54750.2022.9755782

23. Usmanov E.G., Khusanov B.M. RSES 2020. E3S Web of Conferences 216. 01161. (2020). https://doi.org/10.1051/e3sconf/202021601161

24. Erejepov, M., Novikov, A.N., Khusanov, B.M., Seytmuratov, B., Sayimbetov, Z. E3S Web of Conferences. 289. (2021). 07018. https://doi.org/10.1051/e3sconf/202128907018

25. Rakhmonov,I., Berdishev,A., Khusanov, B.M, Khaliknazarov, U., Utegenov, U. IOP Conference Series: Materials Science and Engineering 883(1). 012104. DOI 10.1088/1757-899X/883/1/012104

26. I.U. Rakhmonov, T.H. Khakimov, I.I. Bakhadirov, B.M. Khusanov. International Journal of Advanced Science and Technology Vol. 29. (2020). pp. 1525-1533.

27. Rakhmonov I.U. Automated control system for power consumption of industrial enterprises // Bulletin of the Kazan State Power Engineering University. 2022. T. 14. №4 (56). p. 30-38.

Modelling of laboratory work in the science "Fundamentals of power supply" using an educational simulator based on a programmed logic controller

Kholikhmatov *B.B.*[1,*], Samiev *Sh.S.*[1], Erejepov *M.T.*[2], Nematov *L.A.*[3]

[1]Tashkent State Technical University named by Islam Karimov, Tashkent, 100095, Uzbekistan

[2]Karakalpak State University, Nukus, 230101, Uzbekistan

[3]Bukhara Engineering Technological Institute, Bukhara, 200100, Uzbekistan

Abstract. Today, the use of programmable logic controller devices in the educational process brings great results. This study demonstrated the effectiveness of using a programmable logic controller (PLC)-based training simulator in modeling laboratory work on the subject of "Power Supply Fundamentals". The results show that the training simulator was considered effective in providing students with a flexible and safe way to learn PLC and power supply systems. Also, this article presents the working principle and algorithm of a model virtual simulator based on PLC that meets the requirements of modern education in the field of "Basics of Power Supply" developed during our research. At the same time, the methodologies used for the research are also widely covered.

1 Introduction

The field of industrial automation and control relies heavily on the use of programmed logic controllers (PLCs) to control and monitor processes in various industries. One of the key components of training and educating engineers, technicians, and operators in the use of PLCs is laboratory work. Laboratory work allows students to gain hands-on experience with PLCs in a controlled environment, but it also has its limitations, such as a lack of flexibility and the risk of equipment damage [1, 2, 10].

To address these limitations, this study aims to investigate the use of an educational simulator based on a programmed logic controller (PLC) to model laboratory work in the science "Fundamentals of power supply". The simulator will be used to provide students with a flexible and safe way to learn about PLCs and power supply systems. The objective of this study is to design and implement an educational simulator for laboratory work in the science "Fundamentals of power supply" that accurately replicates the behavior of real PLC-controlled systems.

The study will begin by reviewing the literature on educational simulators and PLCs in industrial automation and control. The methodology will then involve the collection of data from an existing PLC-controlled power supply system, the creation of a simulation model, and the integration of the simulation model with a user interface to create the educational simulator [3, 12, 6]. The effectiveness of the educational simulator will be evaluated through its use in laboratory work and through a comparison of the performance of students who have used the simulator with those who have not.

The study's goal is to provide an educational tool that will enable laboratory work in the science "Fundamentals of power supply" to be performed more efficiently, with more flexibility and safety. The results of this study will provide insight into the potential of educational simulators based on PLCs as a valuable tool for training and education in the field of industrial automation and control [4, 11-15, 5, 8].

2 The current state of the investigated problem

The methodology used in this study to model laboratory work in the science "Fundamentals of power supply" using an educational simulator based on a programmed logic controller (PLC) is divided into 6 stages [7, 16-20, 9, 12, 14]:

1. Literature review: To begin, a literature review was conducted to gather information on educational simulators and PLCs in industrial automation and control. This helped to identify the current state of the field and the challenges associated with using PLCs in laboratory work [4, 21-24, 13, 17].

2. Data collection: Data was collected from an existing PLC-controlled power supply system. This data

* Corresponding author: ilider1987@yandex.com

 https://doi.org/10.1051/e3sconf/202338401032

included the PLC program, electrical schematics, and process parameters.

3. Simulation model development: Using the data collected, a simulation model was developed that accurately replicates the behavior of the PLC-controlled power supply system. The simulation model was built using a PLC simulation software.

4. User interface design and development: The simulation model was integrated with a user interface to create the educational simulator. The user interface was designed to be user-friendly, intuitive, and interactive. It was developed using a software development tool.

5. Evaluation: The effectiveness of the educational simulator was evaluated through its use in laboratory work. The performance of students who have used the simulator was compared to the performance of students who have not. The evaluation was conducted by monitoring the students' understanding and their ability to work with the PLC-controlled power supply system.

6. Analysis and interpretation: The results of the evaluation were analyzed and interpreted to determine the effectiveness of the educational simulator in modeling laboratory work in the science "Fundamentals of power supply" and to identify areas for improvement.

The results of this study on modeling laboratory work in the science "Fundamentals of power supply" using an educational simulator based on a programmed logic controller (PLC) showed that the educational simulator was effective in providing students with a flexible and safe way to learn about PLCs and power supply systems [16, 25, 19].

During the research, the algorithm of the laboratory on "Basics of power supply" was developed as a sample. This algorithm consists of two parts (Figure 1):

1. Performing the laboratory
2. Evaluation

In the laboratory for studying the circuits of main step-down substations of industrial enterprises, the student's skills are formed in relation to this laboratory on the basis of three tasks, and his knowledge is assessed step by step. From the point of view of complexity, questions are divided into easy (green), medium difficulty (blue) and complex (red) questions. The first assignment is the question of placing the devices in series in the circuit of the main step-down substations, in which it is initially taken that c=0, the reason for the errors is to take into account the fact that the student performed the experiment without a scholarship, in this question, the device is asked in series from 1 to 6 in the range where the student will need to correctly enter the names of the devices. Entered device names are accepted and checked to see if the device name is entered correctly. If the device name is entered incorrectly, an error message will be displayed and the count will be 1 because the operation was performed incorrectly. Actions are repeated six times, because here six devices are requested, and in the next step, the command to start the step-down substations of industrial enterprises is given. In this case, the student will have to start the substation in the correct sequence. Of course, the correct sequence starts with the top source and then the consumer is connected. In the next case, it checks whether the initialization of industrial enterprises is done correctly, if it is done incorrectly, 1 error is recorded and this cycle is repeated from 1 to 10 for starting ten devices.

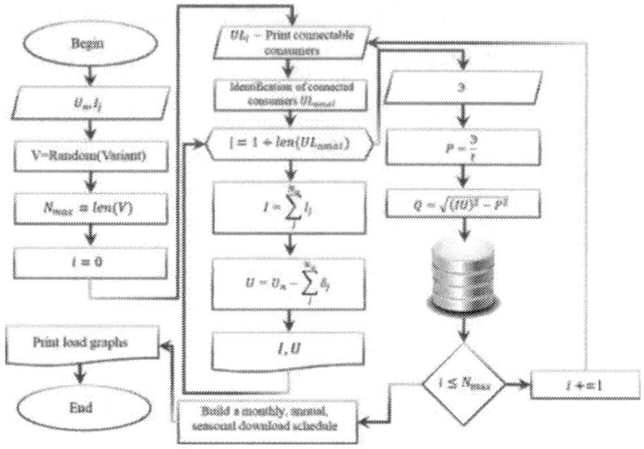

Fig. 1. Algorithm for evaluating the performance of the students who completed the training and the laboratory of the study of the schemes of the step-down substations of the industrial enterprises [21]

The next task is the commissioning of the step-down substation of industrial enterprises, in which the correct commissioning of ten devices in a row is carried out and the device is checked after disconnection, and an error is recorded if it is incorrectly started. If executed correctly, it will be repeated ten times and at the end of this assignment, the student evaluation algorithm will be activated, the maximum score and the number of repetitions will be accepted. Based on the formula $b = \frac{N_{max}-c}{N_{max}} b_{max}$ the grade of the student is derived.

Based on the above processes, one of the main tasks of the laboratory work created on the basis of virtual simulations is to provide students with theoretical and practical knowledge of the basics of electricity supply, to create skills and qualifications for applying the knowledge they have acquired in practice. Including the algorithms of the virtual simulation laboratory work, the developed software is perfectly created in all aspects. These virtual simulators serve to represent the future activity of students as specialists [25-28, 22, 24, 3].

As a result, the web page included the use of the virtual laboratory created in the framework of the research dissertation on the science of the basics of power supply, lectures, practical exercises, tests, useful tips, literature and other handouts related to the science in general.

3 Conclusion

The students who used the simulator had a better understanding of the PLC-controlled power supply system and were able to work with the system more efficiently [7]. The educational simulator was also found to be a valuable tool for training and education in the field of industrial automation and control, as it allowed students to repeat the laboratory tasks as many times as they needed to, allowing them to master the material at their own pace.

The results of this study demonstrate the potential of educational simulators based on PLCs as a valuable tool for training and education in the field of industrial automation and control [8]. However, it is important to note that the use of the educational simulator should be supplemented with practical, hands-on experience in order to provide students with a complete understanding of PLCs and power supply systems.

In summary, the use of an educational simulator based on PLC in laboratory work has been shown to be an effective method of teaching the fundamentals of power supply, providing students with a flexible and safe way to learn about PLCs and power supply systems. This approach can be effectively used in the field of industrial automation and control, providing valuable training and education.

References

1. Avila-Garzon, Cecilia & Benito, M. & Guevara, Juan & Cavanzo, Gloria. (2011). Semantic web and intelligent cooperative agents in educational simulators for mathematics: Developing a graphic simulator of conic sections to improve learning experiences. Proceedings of the International Conference on e-Learning, ICEL. 17-24.

2. Yalcin, Nedim & Vatansever, Fahri. (2020). Educational Simulator for Frequency Estimation using ANN. Academic Perspective Procedia. 3. 331-336. 10.33793/acperpro.03.01.65.

3. Sharapov, Valeriy. (2013). Programmed logic controllers in the automated control systems.

4. Bazara M., Shetty K. Nonlinear programming. Theory and algorithms. - Moscow: Publishing house: "Mir", 1982. - 583 p.

5. Struchenkov V.I. Optimization methods in applied problems. - Moscow: Publishing house: "SOLON□PRESS", 2009. - 320 p.

6. Bertsekas D. Conditional optimization and methods of Lagrange multipliers. - Moscow: Publishing house: "Radio and communication", 1987. - 400 p.

7. Batsova S. V. Organization of rationing of electricity at an industrial enterprise. Economic incentives for the rational use of electricity / S. V. Batsova // Casting and metallurgy: scientific and production journal. - 2010. –№3 (56). - S. 131 - 133.

8. Chegodaev AI Mathematical methods of analysis of expert assessments // Bulletin of the Samara State University of Economics. 2010. №2 (64). S. 130-135.

9. Rakhmonov I.U., Najimova A.M., and Reymov K.M. AIP Conference Proceedings **2647.** 030010. (2022). https://doi.org/10.1063/5.0104788

10. Rakhmonov I.U., Najimova A.M. AIP Conference Proceedings **2647.** 030011. (2022). https://doi.org/10.1063/5.0104791

11. Rakhmonov I.U., Najimova, A.M., Esemuratova Sh.M., Koptileuov T.T. AIP Conference Proceedings **2647.** 070024. (2022). https://doi.org/10.1063/5.0104793

12. R.Karimov, N.Kurbanova, and ofters. Journal of Physics: Conference Series. 2094(5). **052042**. (2021). doi:10.1088/1742-6596/2094/5/052042

13. Hoshimov F.A., Rakhmonov I.U., Niyozov N.N., Omonov F.B. AIP Conference Proceedings **2647.** 030025. (2023). https://doi.org/10.1063/5.0112388

14. Rakhmonov I.U., Hoshimov F.A., Kurbonov N.N., Jalilova D.A. AIP Conference Proceedings **2647.** 050022. (2023). https://doi.org/10.1063/5.0112391

15. Rakhmonov I.U., Ushakov V.Ya., Niyozov N.N., Kurbonov N.N., Mamutov M. E3S Web of Conferences **289.** 07014. (2021). https://doi.org/10.1051/e3sconf/202128907014

16. Rakhmonov I.U., Ushakov V.Ya., Najimova A.M., Jalilova D.A., Omonov F.B. E3S Web of Conferences **289.** 07013. (2021). https://doi.org/10.1051/e3sconf/202128907013

17. R.Karimov, A. Kuchkarov, and ofters. Journal of Physics: Conference Series 2094. **052050**. (2021). doi:10.1088/1742-6596/2094/5/052050

18. A.D.Taslimov, A.S.Berdishev, F.M.Rakhimov, Melikuziev M.V. E3S Web Conf. **Vol.139**. 2019. https://doi.org/10.1051/e3sconf/201913901082

19. Taslimov A.D., Melikuziev M.V., Najimova A.M., Alimov A.A. E3S Web of Conferences **216**. 01159 (2020). https://doi.org/10.1051/e3sconf/202021601159

20. Saidkhodjaev A.G., Nuriddinova Kh.R. AIP Conference Proceedings **2552.** 050037. (2023). https://doi.org/10.1063/5.0133486

21. Rakhmonov,I., Berdishev,A., Khusanov, B.M, Khaliknazarov, U., Utegenov, U. IOP Conference Series: Materials Science and Engineering **883(1)**. 012104. DOI 10.1088/1757-899X/883/1/012104

22. Usmonov E.G. AIP Conference Proceedings **2552.** 050020. (2023). https://doi.org/10.1063/5.0111537

23. Rasulov, A.N., Rafikova, G.R., Novikov, N.L., Ruzinazarov, M.R., Esemuratova, S. E3S Web of Conferences 289. 07006. (2021). https://doi.org/10.1051/e3sconf/202128907006

24 A.D.Taslimov, M.V.Melikuziev, A.M.Najimova, A.A.Alimov. E3S Web of Conferences **216.** 01159. (2020). https://doi.org/10.1051/e3sconf/202021601159

25 Rakhmonov I.U. Automated control system for power consumption of industrial enterprises // Bulletin of the Kazan State Power Engineering University. 2022. T. 14. №4 (56). p. 30-38.

26 Rakhmonov, I.U., Niyozov, N.N. E3S Web of Conferences, 2019. **139.** 01077. DOI: 10.1051/e3sconf/201913901077

27 Rakhmonov, I.U., Reymov, K.M. Journal of Physics: Conference Series, 2019. **1399(5).** 055038. DOI: 10.1088/1742-6596/1399/5/055038

28 Hoshimov, F.A., Rakhmonov, I.U., Niyozov, N.N. E3S Web of Conferences **209**. 07017. (2020). DOI: 10.1051/e3sconf/202020907017

Determination of the service area and location of transformer substations in the city power supply system

Melikuziev *M.V.**

Tashkent State Technical University named by Islam Karimov, Tashkent, 100095, Uzbekistan

Abstract. The article proposes a theoretically based approach to determining the service areas and location of transformer substations (TS) in the case depending on the intensity of the electric load based on the area specified in the city power supply system. Rational use of service areas and determination of the optimal area by location will help regulate the city's power supply. As a result of the scientific approach in the article, the targeted use of a transformer substation in the municipal power supply system will achieve orderly development in the city, as well as a reduction in the waste of electricity in distribution power networks.

1 Introduction

Currently, all elements of the power supply system of existing cities are components of the regional or state energy system. Such systems have multiple voltage standards scale level networks and are constantly evolving. In this case, it is important to make a decision on the basis of a timely scientific approach to determining the transformer substations and their service areas and location, which are the main element of the city power supply system. Each city power supply system has a certain hierarchy of voltages. In the substations of the municipal power supply system, as a rule, 220/35, 110/35, 110/10, 35/10, 10/0,4 kV transformers of the voltage level are used.

2 Voltage and electric load density gradation in the city electric networks

In city power supply, the total electric load is formed by the electricity of existing consumer groups. These generated electrical loads should be provided uninterrupted by transformer substations within the city power supply system. The optimal power of transformers in substations, which depends on the different voltage level in the urban area, depends on the electrical load density of the area. The electrical load density in the individual areas of the city obtained differs from the average load density in the city. The value of the electric load density depends on the specifics of the territorial construction, the number of floors of the premises located, the degree of electrification and the value of the voltage in the connection of consumers to TP. The electrical load density at each voltage level is determined by the total load at the previous network level of consumers and the additional load at the level under consideration. As a result, the electrical load density in hierarchy transition networks corresponding to electrical network voltages is determined by networks as follows [1, 10-15, 18-23, 9]:

$$\sigma_0 = \frac{S_{\sum 0.4}}{F_{city}} \quad (1)$$

$$\sigma_1 = \frac{S_{\sum 10-35}}{F_{city}} \quad (2)$$

$$\sigma_2 = \frac{S_{\sum 35-110}}{F_{city}} \quad (3)$$

$$\sigma_3 = \frac{S_{\sum 110-220}}{F_{city}} \quad (4)$$

where, $S_{\sum 0.4}$ – power load of consumers on the transformer substation 0.4 kV voltage tire, σ_0 – electrical load density;

$S_{\sum 10\text{-}35}$ – power load of consumers on a transformer substation 10 or 35 kV voltage tire, σ_1 – electrical load density;

$S_{\sum 35\text{-}110}$ – power load of consumers on a transformer substation 35 or 110 kV voltage tire, σ_2 – electrical load density;

$S_{\sum 110\text{-}220}$ – electrical load of consumers on a transformer substation 110 or 220 kV voltage tire, σ_3 – electrical load density.

The voltage by voltage rating of all transformer substations providing the city area depends on the description of consumers of electricity of individual power at each stage. Electricity consumers are considered one or three phase in terms of reliability and rated power, voltage of the power supply. On the territory of the city, a voltage of 220 kV is used, mainly in electrical networks coming from the power plant, if we see the example of the city of Tashkent. By the sequence stage, it is divided into electrical networks with a voltage of 110, 35, 10, 0.4 kV. On the territory of Uzbekistan, networks with a voltage of 330, 20, 6 kV are not used for city power supply. Transformer substations

* Corresponding author: ataslimov@mail.ru

https://doi.org/10.1051/e3sconf/202338401033

also use power transformers suitable for existing electrical network voltages [2-4].

The hierarchy of voltage and electric load density in the city distribution electrical networks is schematically shown in Fig. 1.

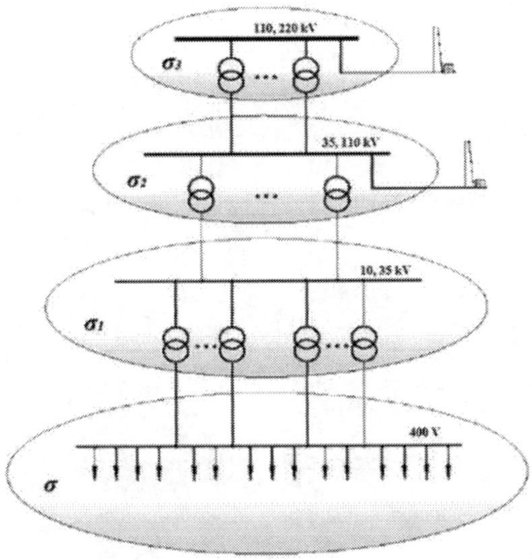

Fig. 1. Voltage and electric load density gradation in the city electric networks.

From the above expressions 1, 2, 3, 4 it follows that the actual electric load density at each voltage level varies due to different regions and factors simultaneously influencing the maximum electrical load of transformers in TS [5, 16-24].

If the electrical loads generated by consumers and the area of the city (or part of it) allocated for the building are known, the average electrical load density at the total voltage level in this city (or part of it) is determined as follows:

$$\sigma_{i.} = \frac{S_{city}}{F_{city}} \tag{5}$$

where, $\sigma_{i.}$ – i - average electrical load density for a voltage level electrical network, [MVA/km^2];

S_{city} – the total electrical load of city electrical networks (or part of it) at a certain voltage level in the city power supply system, [MVA];

F_{city} – city area (or part of it) allocated for the construction of a residential or public building, [km^2].

In the general case, the total electrical load of the city (or part) at a certain voltage level of the municipal power supply system is determined as follows:

$$S_{city} = k_{TS.i}\cdot\sum_{i=1}^{n_{TS}} S_{TS.i} = k_{TS.i}\cdot\sum_{i=1}^{n_{TS}} n_i\cdot\beta_i\cdot S_{nom.i} \tag{6}$$

where, $S_{TS.i}$ – i - electrical load of transformer substation, [MVA];

$k_{TS.i}$ – coefficient taking into account the simultaneous maximum load of transformers depending on the number of TS in the network at a given suitable voltage level;

n_i – number of transformers in substation;

β_i – load coefficient of the i - transformer in the substation;

$S_{nom.i}$ – i - rated power of substation transformers, [MVA];

n_{TS} – number of transformer substations in an urban area (or part thereof).

3 Determination of the service area of transformer substation

The service area and time of transformer substations located on the territory of the city power supply system depend on the type of consumers located in the area, the floor of the premises, the location of the electrical load depends on the geometric description. On the territory of the city, electricity consumers form an electric load in a vertical position. Taking into account this, the issue of determining the service area by the geometric location of the transformer substation is considered relevant. Determining the geometric area in the optimal position has a great influence on the development of urban planning and city power supply. By determining the limit optimal area of the transformer substation, it is possible to achieve a reduction in the length of low voltage electrical networks, as well as a reduction in the cost of waste electricity [6-10].

To provide qualitative electricity supply for all the consumers of a city they are to be rationally distributed among separate transformer substations according to their service area. The service area of each separate substation is defined by the load density in this area and the TS load, which, in turn, depends on the rated power of a corresponding transformer, the number of transformers at the substation and the load factor:

$$F_{city} = \frac{S_{TS.i}}{\sigma_{TS.i}} = \frac{n_i\cdot\beta_i\cdot S_{nom.i}}{\sigma_{TS.i}} \tag{7}$$

where, $\sigma_{TS.i}$ - is the load density in the service area of the i - substation.

In the absence of a technical solution or a single approach to the design and modelling of the elements of an city power supply system, it can have complex geometric shapes or schemes that differ for each substation by area. This makes it difficult to choose a place for new substations and purposefully develop a power supply scheme. Therefore, by the rules of city planning regulations, it is not allowed to connect additional loads to the city area or build a substation.

When determining the service area of the transformer substation can be designed and modeled with geometric shapes. For example, in geometry there are circle, square, regular six, eight corners, etc. The most convenient model is the hexagon. This makes it possible to cover almost any shaped area of the building in a regular order. This model of the transformer substation service sector can be selected as the only template in the design work. Design and modelling in such a way is considered useful in making decisions on the development of urban distribution electrical networks. It is conditionally accepted that the transformer substation should be in the center of the hexagon (in the practical case, the area should not fall beyond the boundaries of the hexagon).

Figures 2, 3 describe the template in the sample case for the service area of various rated voltage transformer substations to be placed in the city area, as well as the designable ideal view of the service. Through these presented images, a vision is generated by the service area [1, 8, 13].

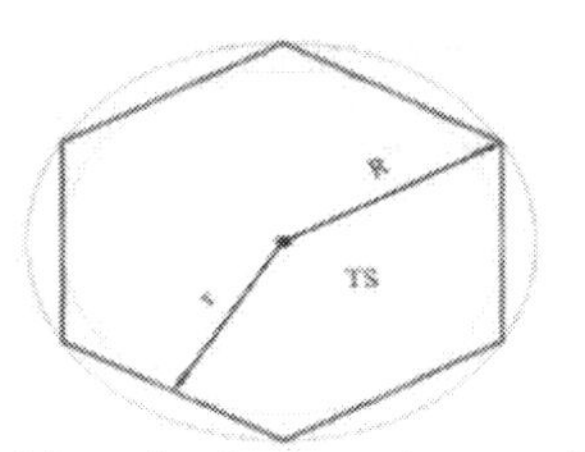

Fig. 2. The model template for the service area of a transformer substation.

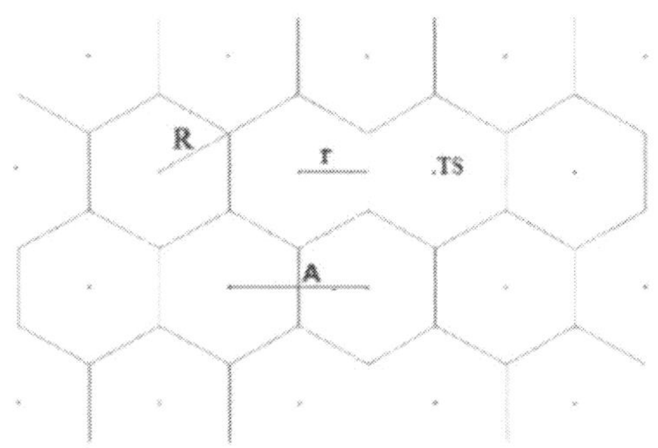

Fig. 3. An ideal model of service areas of transformer substations.

Between the basic geometrical sizes of the models shown in Figs. 2 and 3 the following relationships exist:

$$F_{TS} = 3\sin\left(\frac{\pi}{3}\right) \cdot R^2 = 2{,}6 \cdot R^2 = 3{,}46 \cdot r^2 \quad (8)$$

$$R = 0{,}62 \cdot \sqrt{F_{TS}} \quad (9)$$

$$r = 0{,}54 \cdot \sqrt{F_{TS}} \quad (10)$$

$$A = 2 \cdot r = 1{,}1 \cdot \sqrt{F_{TS}} \quad (11)$$

where, R - the radius of the circle described around the hexagon (also the side of a hexagon and the radius of TS service area);

r - the radius of the circle described inside the hexagon;

A - the theoretically minimal distance between the neighboring substations.

Considering (7) and (8-11), the following relationships are valid between the basic sizes of the geometrical model-template of TS service area, its load density and the technical parameters of the existing or future TSs [1, 9-10]:

$$F_{TS} = 2{,}6 \cdot R^2 = \frac{n_i \cdot \beta_i \cdot S_{nom.i}}{\sigma} \quad (12)$$

$$R = 0{,}62 \cdot \sqrt{\frac{n_i \cdot \beta_i \cdot S_{nom.i}}{\sigma}} \quad (13)$$

$$r = 0{,}54 \cdot \sqrt{\frac{n_i \cdot \beta_i \cdot S_{nom.i}}{\sigma}} \quad (14)$$

$$A = 2 \cdot r = 1{,}1 \cdot \sqrt{\frac{n_i \cdot \beta_i \cdot S_{nom.i}}{\sigma}} \quad (15)$$

Considering that some TS service areas under particular conditions of the city building can overlap one another, it is necessary to verify fulfilment of the condition:

$$F_{city} \le k_{rc} \cdot \sum_{i=1}^{n_{TS}} F_{TS.i} \quad (16)$$

where, $F_{TS.i}$ - the service area of the i - substation, km^2;

k_{rc} - the accepted recovering factor of service areas.

On the assumption of identical installed rated power, load factors and service area of the TS transformers, the average load density in the city (or its part) in view of (5), (6) and (12) will be:

$$\sigma_{city} = \frac{k_{TS.i} \cdot \sum_{i=1}^{n_{TS}} S_{TS.i}}{k_{pki} \cdot F_{TS.i} \cdot n_{TS}} = \frac{k_{TS.i} \cdot \sum_{i=1}^{n_{TS}} n_i \cdot \beta_i \cdot S_{nom.i}}{k_{pki} \cdot F_{TS.i} \cdot n_{TS}} \quad (17)$$

The average load density in the city (or its part) at different capacities of transformer substations can be defined as:

$$\sigma_{city} = \frac{k_{TS.i} \cdot (S_{TS.1} + S_{TS.2} + \dots + S_{TS.i})}{k_{cov} \cdot (F_{TS.1} + F_{TS.2} + \dots + F_{TS.i})} \quad (18)$$

The resulting expressions of the proposed nature of all the expressions considered transformer are determined by the value of the high voltage. The values obtained expression are quoted in Table 1 below through the expression obtained on low voltage.

Table 1. Results determined by the value of the electric charge density at the level of one voltage (on the example of Tashkent city).

S_{SS}, MVA	2x40 (skirting schemes)			
σ, MVA/km^2	2	4	8	12
S_{TS}, kVA	320	454	580	658
F_{city}, km^2	0,16	0,12	0,07	0,06
R, km	0,248	0,215	0,164	0,151
S_{SS}, MVA	2x40 (two trunk schemes)			
σ, MVA/km^2	8	16	24	32
S_{TS}, kVA	1009	1435	1763	2041
F_{city}, km^2	0,126	0,09	0,074	0,064
R, km	0,22	0,186	0169	0,157

4 Determination the location of transformer substations

In the city power supply system, the number and nominal full capacities of transformers in the transformer substation are determined by state all union standard. After the calculation is carried out, the installation of each transformer substation over a small non area of the

 https://doi.org/10.1051/e3sconf/202338401033

transformer substations is determined by the coordinates of the center for conditional loads (X_0, Y_0) through this expression:

$$X_0 = \frac{P_1 \cdot X_1 + P_2 \cdot X_2 + \dots + P_i \cdot X_i}{P_1 + P_2 + \dots + P_i} \quad (19)$$

$$Y_0 = \frac{P_1 \cdot Y_1 + P_2 \cdot Y_2 + \dots + P_i \cdot Y_i}{P_1 + P_2 + \dots + P_i} \quad (20)$$

X_i, Y_i – consumer coordinate in the area, [cm]; P_i - the calculated asset load of the consumer in the area, [kW].

The coordinates of TS determined by area may not coincide with the calculated points. In this case, the conditional loading center can fall over the building or in the direction of movement. In this case, the center of conditional loads can be moved to a point close to consumers with a large capacity [11-16, 20, 22, 24].

Let's explain using an example for a small area on determining the optimal location coordinate of TS in the city power supply system. Information about the consumer capacities of 20 consumers located in the resulting area, as well as their location coordinates, is presented in Table 2.

Table 2. An example of calculating the optimal configuration of an urban district distribution electrical network

Buildings №	Location coordinates		Computing power, kVA
	X	Y	
1	75	315	40
2	75	150	30
3	105	75	40
4	105	225	35
5	130	275	35
6	180	315	40
7	275	280	30
8	160	150	35
9	180	75	50
10	230	275	35
11	250	180	40
12	220	100	70
13	300	225	30
14	300	130	25
15	330	315	35
16	350	205	40
17	360	170	50
18	365	250	70
19	410	170	30
20	410	100	20

Based on the initial data in Table 2, from expressions (19) and (20), the transformer by area determines the location coordinate. The territory is conditionally allocated to 2 TS and is provided with lines in the form in Figure 4.

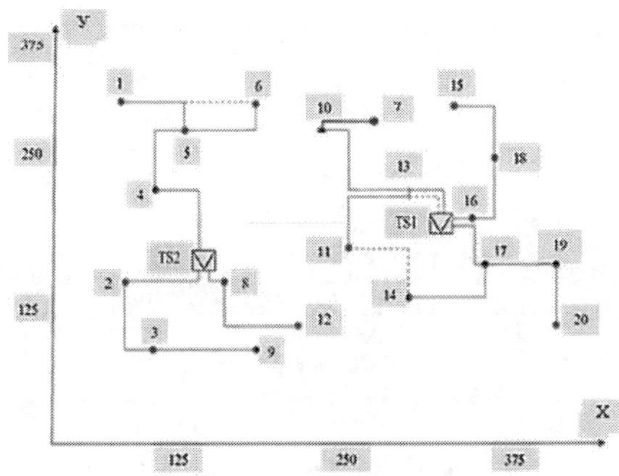

Fig. 4. Calculation of the optimal configuration of the distribution power network of the city district.

It can be seen from Figure 4 that when supplying regions, lines should be ensured that they do not intersect and do not back down. The consumers allocated for the transformer substation as well as their consumption capacities, location coordinates are shown in Table 3.

Table 3. The optimal location of each TS is the sum of the relays and capacities

TS1	X = 315	Y = 200	405
TS2	X = 150	Y = 170	375

Length of lines up to 1000 V:

- Within TS1 387 m (7, 10, 11, 13, 14, 15, 16, 17, 18, 19, 20);
- Within TS2 313 m (1, 2, 3, 4, 5, 6, 8, 9, 12);

Loading each line 135, 125, 145 kVA (TS1), 150, 120, 105 кВА (TS2); Distance between TS 167 m.

The location coordinate of TPS for consumers in networks with a voltage of up to 1000 V over the resulting area, as well as the location of power supply lines, was cited [21-26, 28, 30, 31,].

5 Conclusion

In the city power supply system, theoretical foundations for determining service areas and their radii for transformer substations of different voltage levels were developed. with the help of the developed bases, a hierarchy of electrical load density of TS service fields was proposed in accordance with the hierarchy of voltage in electrical networks. For the rational placement of TS in the territory of the city power supply, models and templates were developed based on various geometric shapes. The dependence of TS service areas and their radii on the electrical load density in the service sector for TS with two transformers 220/35-110 kV, 110/10-35, 35/10 and 10/0,4 kV was determined. Based on the consumer power and location coordinates of the consumers received by the territory, the theoretical concepts of the location of the TS and the placement of power supply lines were reasonably explained using a practical example. This method, developed, will be able to reduce the waste of resources and electricity in city electricity supply and will have a positive effect on the development and expansion trend of cities.

References

1. S. Guseva, N. Skobeleva, N. Breners y O. Borscevskis. Determination of Service Areas of Urban Transformer Substations and Distribution Using Geometrical Templates. Latvian journal of physics and technical sciences. Publicado en línea: 19 Nov 2010. Volumen & Edición: **Volumen 46.** PP: 16 – 26.
2. Yu.V.Myasoedov, N.V.Savina. Intellectualization of systems power supply of cities. Monograph. – M..: Amur State University, 2017. – 164 p.
3. Lykin A.V. Electric power systems and networks: textbook for universities // A.V., Lykin. Moscow: Yurayt Publishing House, 2019. – 360 p.
4. Dileep G. A survey on Smart Grid technologies va applications // Renewable Energy, 2020. ט 146. pp. 2589-2625.
5. Cheremisin N.M., Cherkashina V.V. A criterion method for analyzing technical and economic problems in electrical networks and systems: a tutorial. - Kharkov: Fact, 2004. - 88 p.
6. Melikuziev M.V. AIP Conference Proceedings 2552. (2023). 050021. https://doi.org/10.1063/5.0112395
7. Rasulov A.N., Usmonov E.G., Melikuziev M.V. AIP Conference Proceedings 2552. (2023). 050018. https://doi.org/10.1063/5.0111530
8. Taslimov A.D. AIP Conference Proceedings 2552. (2023). 050023. https://doi.org/10.1063/5.0112398
9. Melikuziev M.V., Nematov L.A., Novikov A.N., Baymuratov K.K. E3S Web of Conferences 289. 07016 (2021). https://doi.org/10.1051/e3sconf/202128907016
10. A.D.Taslimov, A.S.Berdishev, F.M.Rakhimov, Melikuziev M.V. E3S Web Conf. Vol. 139. 2019. https://doi.org/10.1051/e3sconf/201913901081.
11. Rakhmonov I.U., Najimova A.M., and Reymov K.M. AIP Conference Proceedings **2647.** 030010. (2022). https://doi.org/10.1063/5.0104788
12. Rakhmonov I.U., Najimova A.M. AIP Conference Proceedings **2647.** 030011. (2022). https://doi.org/10.1063/5.0104791
13. Rakhmonov I.U., Najimova, A.M., Esemuratova Sh.M., Koptileuov T.T. AIP Conference Proceedings **2647.** 070024. (2022). https://doi.org/10.1063/5.0104793
14. Hoshimov F.A., Rakhmonov I.U., Niyozov N.N., Omonov F.B. AIP Conference Proceedings **2647.** 030025. (2023). https://doi.org/10.1063/5.0112388
15. Rakhmonov I.U., Ushakov V.Ya., Niyozov N.N., Kurbonov N.N., Mamutov M. E3S Web of Conferences **289.** 07014. (2021). https://doi.org/10.1051/e3sconf/202128907014
16. Rakhmonov I.U., Ushakov V.Ya., Najimova A.M., Jalilova D.A., Omonov F.B. E3S Web of Conferences **289.** 07013. (2021). https://doi.org/10.1051/e3sconf/202128907013
17. Rakhmonov, I.U., Kurbonov, N.N. E3S Web of Conferences **216**. 01178. (2020). DOI: 10.1051/e3sconf/202021601178
18. Hoshimov, F.A., Rakhmonov, I.U., Niyozov, N.N. E3S Web of Conferences **209**. 07017. (2020). DOI: 10.1051/e3sconf/202020907017
19. A.D.Taslimov, A.S.Berdishev, F.M.Rakhimov, Melikuziev M.V. E3S Web Conf. Vol.139. 2019. https://doi.org/10.1051/e3sconf/201913901082
20. Taslimov A.D., Melikuziev M.V., Najimova A.M., Alimov A.A. E3S Web of Conferences 216. 01159 (2020). https://doi.org/10.1051/e3sconf/202021601159
21. R Karimov. AIP Conference Proceedings 2552. 030014. (2022). https://doi.org/10.1063/5.0111533
22. R Karimov. AIP Conference Proceedings 2552. 050012. (2022). https://doi.org/10.1063/5.0111524
23. Hoshimov F.A., Rakhmonov I.U., Niyozov N.N., Omonov F.B. AIP Conference Proceedings 2552. (2023). 030025. https://doi.org/10.1063/5.0112388
24. Rasulov A.N. AIP Conference Proceedings 2552. (2023). 05017. https://doi.org/10.1063/5.0111530
25. Rakhmonov I.U., Hoshimov F.A., Kurbonov N.N., Jalilova D.A. AIP Conference Proceedings 2552. (2023). 050022. https://doi.org/10.1063/5.0112391
26. Usmonov E.G. AIP Conference Proceedings 2552. (2023). 050020. https://doi.org/10.1063/5.0111537
27. Saidkhodjaev A.G. AIP Conference Proceedings 2552. (2023). 050026.https://doi.org/10.1063/5.0111540
28. Bobojanov M.K., Mahmutkhonov S.K., Aytbaev S. AIP Conference Proceedings 2552. (2023). 05011. https://doi.org/10.1063/5.0113890.
29. Rakhmonov, I., Berdishev, A., Niyozov, N., Muratov, A., Khaliknazarov, U. IOP Conference Series: Materials Science and Engineering, 2020. 883(1). 012103. DOI: 10.1088/1757-899X/883/1/012103
30. Rakhmonov, I.U., Niyozov, N.N. E3S Web of Conferences, 2019. 139. 01077. DOI: 10.1051/e3sconf/201913901077
31. Rakhmonov, I.U., Reymov, K.M. Journal of Physics: Conference Series, 2019. 1399(5). 055038. DOI: 10.1088/1742-6596/1399/5/055038

 https://doi.org/10.1051/e3sconf/202338401034

Regulation system by voltage booster transformer

Usmanov *E.G.* *

Tashkent State Technical University, Tashkent, 100095, Uzbekistan

Abstract. In this article, the features of the oscillatory circuit are investigated in order to determine the possibility of their application for controlling thrusters when building new, simple and reliable circuits for regulating booster voltage stabilizers from the value of the supply voltage. In Ferro resonant circuits connected to a voltage source with low internal resistance, with a certain combination of parameters, oscillations are excited at the fundamental frequency, the initial phase of which has a shift relative to the initial phase of the applied voltage.

1 Introduction

Electricity, as a special type of product, has certain characteristics that make it possible to judge its suitability in various production processes.

The set of characteristics under which power receivers are able to perform their functions are united by the general concept of power quality. The quality of electricity is assessed by the damage caused to the national economy [1-5, 8, 10].

The quality of electricity, along with reliability and safety, is one of the mandatory requirements for power supply systems. The quality of electricity is characterized by a combination of properties and indicators of energy quality.

Ensuring the required quality of electrical energy is a problem that is present at all stages of the existence of electrical energy, including generation, transmission, distribution and consumption. The main parameter and indicator of the quality of electrical energy is considered to be voltage and its quality.

2 The current state of the investigated problem

One of the reasons for the deterioration of the quality of electricity is the so-called "dips" of voltage, which are observed during the switching of powerful steel-smelting electric furnaces, semiconductor converters, etc. Voltage dips can be both short-term and long-term, one-time and reusable per shift in enterprises with a continuous nature of production. Voltage dips lead not only to a deterioration in the operation of power receivers in these enterprises, but also to a complete stop of the entire technological process. For example, in the spinning shops of textile mills, when the spinning machine stops, it takes up to three hours to fully restore their work, which leads to great economic damage [6-12, 14, 15].

Devices that could compensate for voltage dips in the supply network must meet the following requirements: greater sensitivity to changes in the voltage value at the input of this device and the speed of regulation of the voltage value of the power source [13, 14, 20-26].

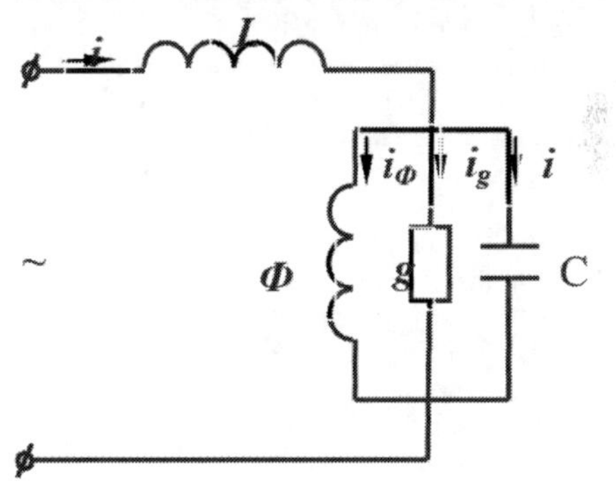

Fig. 1. Parallel oscillating contour

There are a large number of works devoted to voltage regulation using a booster transformer (VDT), the control system of which is made on the basis of semiconductor elements [8, 15, 17-19, 27-30].

It should be noted that these systems have a relatively complex semiconductor circuit base, and in regions where the average ambient temperature in summer can reach 50^0 C and above, small-sized semiconductor devices begin to work with a large error or completely fail.

Despite the recent rapid growth of semiconductor technology, the variety of physical properties and numerous possibilities of electroferromagnetic circuits still attract the attention of researchers to them.

This article discusses the issue of creating simple and reliable control systems for VDTs based on a parallel oscillatory circuit connected in series with a linear inductance (Fig. 1) [28, 9, 15].

* Corresponding author: husanovbaxtiyor7@gmail.com

 https://doi.org/10.1051/e3sconf/202338401034

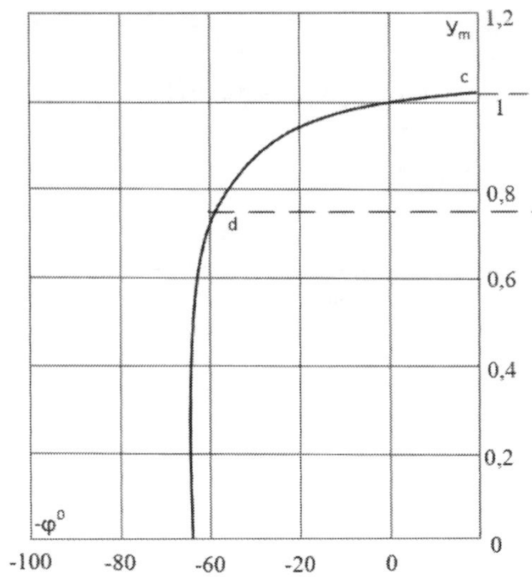

Fig. 2. Dependence φ=f(Um)

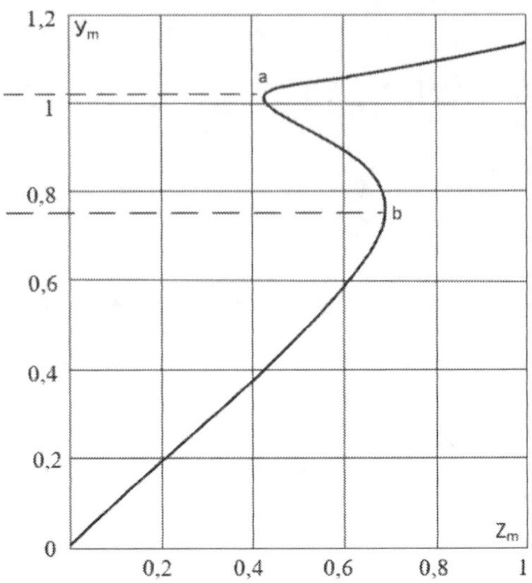

Fig. 3. Dependence Y_m=f(Z_m)

To conduct a theoretical analysis, we will take the following assumptions [1, 2, 16]:

1. Hysteresis losses, eddy currents and active resistance of the winding of a non-linear PV are considered constant and are taken into account by the conductivity g connected in parallel to the PV.
2. We neglect losses in capacitance C and leakage inductance of PV windings.
3. Only the fundamental harmonic of the harmonically changing magnitude of the magnetic flux is taken into account.
4. PV magnetic flux curve belongs to the category of symmetrical characteristics, so it must be approximated by a power function of an odd degree.

In addition, we neglect the losses in the core of the linear inductor, in view of their smallness.

The study is carried out by the method of slowly changing amplitudes. The circuit under study is described by the following differential equation:

$$u = w\frac{d\Phi}{dt} + L_0\frac{di}{dt} \tag{1}$$

Where

$$i = i_C + i_g + i_{\Phi Э} \tag{2}$$

Here, the amplitude value of the first harmonic of the magnetic flux, determined from the condition of equality of the amplitude values of the currents in the winding of the PV and the capacitor, is taken as the basic value of the magnetic flux C connected in parallel, which corresponds to the point of intersection PV and capacitor C.

$$Y_m = \sqrt{[\beta(1 - X_m^6]^2 - \gamma^2} \tag{3}$$

$$tg\psi = -\frac{\beta(1 - X_m^6) - 1}{\gamma} \tag{4}$$

To build the CVC of the circuit and determine its amplitude-phase characteristics, it is necessary to determine the dependencies I=f(Φ) and ϕ=f(U)

$$Z_m = X_m\sqrt{(1 - X_m^6)^2 - \left(\frac{\gamma}{\beta}\right)^2} \tag{5}$$

$$tg\varphi = -\frac{(X_m^6 - 1)tg\psi - \frac{\gamma}{\beta}}{(X_m^6 - 1) - \frac{\gamma}{\beta}tg\psi} \tag{6}$$

The results obtained by formula (6) have both positive and negative values:

- the negative value of the angle φ corresponds to the phase advance of the load current, from the mains voltage;
- the positive value of the angle φ corresponds to the lag of the phase of the load current from the voltage of the power source.

Fig. 2 and Fig. 3 show the dependences φ=f(Ym) and Ym= f(Zm) characteristics of the circuit under consideration, constructed using equations 1, 2, 3 and 4.

As can be seen from these Fig. 3, the dependence Y_m = f(Z_m) has a falling section (ab) where the magnitude of the current varies inversely with the magnitude of the applied voltage, the width and slope of which can be changed by changing the circuit parameters. On the dependence curve *φ=(Ym)* (Fig. 2), there is also a falling section (cd) here the initial phase of the voltage and

 https://doi.org/10.1051/e3sconf/202338401034

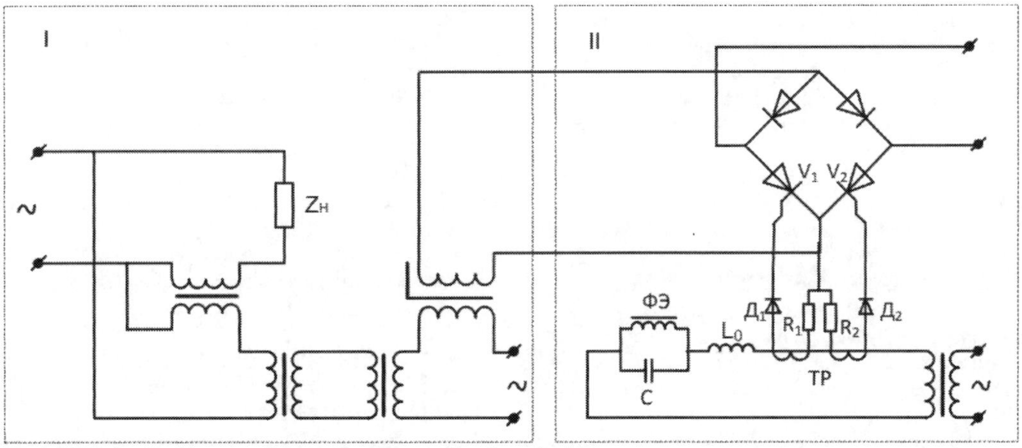

Fig.4. VDT circuit with a smooth control system as a function of voltage

current phase shift in the unbranched section of the circuit under consideration varies inversely with the magnitude of the applied voltage.

So, based on the above, this circuit can be used in the VDT control system for smooth voltage regulation at the load as a function of the power supply voltage [3, 18, 23, 25, 27] (see Fig. 4).

This device consists of two parts: I - power part i.e. booster transformer and II - VDT control system, which consists of a step-down transformer, a parallel oscillatory circuit connected in series with a linear inductance, a trans reactor with two secondary windings included in an unbranched section of a parallel oscillatory circuit.

This device consists of two parts: I - power part i.e. booster transformer and II - VDT control system, which consists of a step-down transformer, a parallel oscillatory circuit connected in series with a linear inductance, a trans reactor with two secondary windings included in an unbranched section of a parallel oscillatory circuit. The secondary windings of the trans reactor are connected through diodes and resistances to the control electrodes of the thyristors located in the arms of the bridge circuit. It should be noted that the power of the VDT control system is not more than 10 W.

3 Conclusion

1. The amplitude value of the load current in relative units and the phase angle between the load current and the mains voltage are revealed.
2. The analysis of the phase relations of the resonant circuit, connected in series with the inductance, showed that the initial phase of the current on the unbranched part of the circuit changes inversely with the applied voltage and has a wide range of dependence on the change in the input voltage.
3. Possibilities are established for obtaining amplitude -phase relationships of a given type when using a circuit for smooth control of the thyristor states as a function of the mains voltage and creating a continuously adjustable booster transformer based on them.

References

1. Bessonov L.A. Nonlinear electrical circuits. - M.: Higher School, 1997, p.343.
2. Novikov SS Methods of operational planning and control of power consumption of large-capacity electric steel-smelting furnaces when operating on the wholesale electricity market: Avtorefer. dis. Cand. those. sciences. - M., 2008.- 18 p.
3. Rasulov A.N., R in zinazars M.R. Methodology for calculating electroferromagnetic circuits. Scientific journals: "GLOBUS". (42-48bet) St. Petersburg-2016.
4. Kadyrov T.M., Alimov Kh.A., Usmanov E.G. Analysis of the regimes of electroferromagnetic circuits by reducing the equations of state to relative units. Uzbek Journal of Problems of Informatics and Energy, No. 6, 2000, pp. 48 -51.
5. Usmanov E.G. RSES 2020. E3S Web of Conferences **216.** 01160. (2020). DOI: 10.1051/e3sconf/202021601160
6. Usmonov E.G. AIP Conference Proceedings 2552. 050020. (2023). https://doi.org/10.1063/5.0111537
7. Erejepov, M., Novikov, A.N., Khusanov, B.M., Seytmuratov, B., Sayimbetov, Z. E3S Web of Conferences. **289.** (2021). 07018. https://doi.org/10.1051/e3sconf/202128907018
8. Rakhmonov I.U., Najimova A.M., and Reymov K.M. AIP Conference Proceedings **2647.** 030010. (2022). https://doi.org/10.1063/5.0104788
9. Rakhmonov I.U., Najimova A.M. AIP Conference Proceedings **2647.** 030011. (2022). https://doi.org/10.1063/5.0104791
10. Rakhmonov I.U., Najimova, A.M., Esemuratova Sh.M., Koptileuov T.T. AIP Conference Proceedings **2647.** 070024. (2022). https://doi.org/10.1063/5.0104793
11. Hoshimov F.A., Rakhmonov I.U., Niyozov N.N., Omonov F.B. AIP Conference Proceedings **2647.** 030025. (2023). https://doi.org/10.1063/5.0112388

12. Rakhmonov I.U., Hoshimov F.A., Kurbonov N.N., Jalilova D.A. AIP Conference Proceedings **2647.** 050022. (2023). https://doi.org/10.1063/5.0112391

13. Rakhmonov I.U., Ushakov V.Ya., Niyozov N.N., Kurbonov N.N., Mamutov M. E3S Web of Conferences **289.** 07014. (2021). https://doi.org/10.1051/e3sconf/202128907014

14. Rakhmonov I.U., Ushakov V.Ya., Najimova A.M., Jalilova D.A., Omonov F.B. E3S Web of Conferences **289.** 07013. (2021). https://doi.org/10.1051/e3sconf/202128907013

15. Khakimov, HT, Shayumova, ZM, Kurbanbaeva, ZK, Khusanov, BM Development of optimal modes and mathematical models of energy performance of electric steelmaking production//E3S Web of Conferences, 2019, 139, 01076.

16. Rakhmonov,I., Berdishev,A., Khusanov, B.M, Khaliknazarov, U., Utegenov, U. IOP Conference Series: Materials Science and Engineering **883(1)**. 012104. (2020). DOI 10.1088/1757-899X/883/1/012104

17. I.U. Rakhmonov, T.H. Khakimov, I.I. Bakhadirov, B.M. Khusanov. International Journal of Advanced Science and Technology **Vol. 29**. (2020). pp. 1525-1533.

18. IU Rakhmonov, TH Khakimov, II Bakhadirov, BM Khusanov. Differentiated block method of analysis and forecasting indicators of power consumption of the industrial complex // International Journal of Advanced Science and Technology Vol. 29, no. 11s, (2020), pp. 1525-1533.

19. Rakhmonov, I.U., Kurbonov, N.N. E3S Web of Conferences **216**. 01178. (2020). DOI: 10.1051/e3sconf/202021601178

20. Hoshimov, F.A., Rakhmonov, I.U., Niyozov, N.N. E3S Web of Conferences **209**. 07017. (2020). DOI: 10.1051/e3sconf/202020907017

21. Rakhmonov, I.U., Reymov, K.M. E3S Web of Conferences **216**. 01167. (2020). DOI: 10.1051/e3sconf/202021601167

22. Rakhmonov, I.U., Hoshimov, F. E3S Web of Conferences **209**. 07018. (2020). DOI: 10.1051/e3sconf/202020907018

23. Usmanov EG, Khusanov BM Phase relations in resonant circuits with a wide falling section on the amplitude characteristic // RSES 2020. E3S Web of Conferences. 216 (2020) 01161. https://doi.org/10.1051/e3sconf/202021601161

24. Taslimov A.D. AIP Conference Proceedings **2552.** (2023). 050023. https://doi.org/10.1063/5.0112398

25. Melikuziev M.V., Nematov L.A., Novikov A.N., Baymuratov K.K. E3S Web of Conferences **289.** 07016 (2021). https://doi.org/10.1051/e3sconf/202128907016

26. A.D.Taslimov, A.S.Berdishev, F.M.Rakhimov, Melikuziev M.V. E3S Web Conf. **Vol. 139**. 2019. https://doi.org/10.1051/e3sconf/201913901081.

27. A.D.Taslimov, A.S.Berdishev, F.M.Rakhimov, Melikuziev M.V. E3S Web Conf. **Vol.139**. 2019. https://doi.org/10.1051/e3sconf/201913901082

28. Taslimov A.D., Melikuziev M.V., Najimova A.M., Alimov A.A. E3S Web of Conferences **216**. 01159 (2020). https://doi.org/10.1051/e3sconf/202021601159

29. Saidkhodjaev A.G., Nuriddinova Kh.R. AIP Conference Proceedings **2552.** 050037. (2023). https://doi.org/10.1063/5.0133486

30. Saidkhodjaev A.G. AIP Conference Proceedings **2552**. 050026. (2023). https://doi.org/10.1063/5.0111540

Modes and technological features of electrolysis consumers of electricity

Rasulov *A.N.*[1]*, Paxratdinov *A.Dj.*[2]

[1]Tashkent State Technical University named by Islam Karimov, Tashkent, 100095, Uzbekistan

[2]Nukus Mining Institute, 230100, Uzbekistan, Nukus, Turtkul highway 181

Abstract. The article analyzes the use of direct voltage and current for electrolysis consumers powered by valve controlled converters, generating higher harmonics into the network, consuming reactive power fr0om the network and significantly affecting the quality of electricity and the efficiency of the entire power supply system.

1 Introduction

At present, the issue of rationalizing production, improving product quality, saving material and labor resources is of great importance. In the field of energy, this is characterized by specific figures for saving fuel, electricity and other material resources in economic development plans [1-5, 10, 8, 6].

Significant reserves of fuel savings - energy resources are available in electrical networks where electricity is transmitted according to complex schemes, undergoes multiple transformations, and in order to ensure its required parameters, it is necessary to systematically correct the scheme, equipment composition and mode by available controls using various characteristics (statistical, economic) of the system power supply and consumers of load nodes.

Therefore, the determination of the optimal level of voltage in the supply network is important and should be carried out on the basis of studying the patterns of changes in the productivity of electrolysis plants, the consumption of raw materials and materials, and the total loss of power and energy [6-12, 16, 17, 20].

All these factors are jointly displayed in the economic characteristics of the object. The economic effect caused by the improvement in voltage quality for electrolysis plants with a continuous technological process can be determined analytically.

An analysis of the individual components of the economic characteristics shows that the components of the output and the specific consumption of electrical energy have a very significant impact [18, 19, 13-15, 5].

2 The current state of the investigated problem

Due to the positive features of direct current consumers (electrolysis baths, electrified transport, etc.), they are increasingly being used in various areas of the national economy. Constant voltage and current for supplying such consumers from controlled valve converters with good technical and economic characteristics. These converters have two existing disadvantages, which are the generation of higher harmonics into the network and the consumption of reactive power from the network. Although appropriate measures are taken to eliminate these shortcomings, the actual values of these harmonics and reactive power vary widely and significantly affect the quality of power and the efficiency of the entire power supply system. Therefore, the analysis of these phenomena on the example of specific powerful consumers is of great importance. Given the wide variety of such consumers, we will consider the issues of power quality in rather energy-intensive installations for the production of chlorine and naustic soda [21-25, 7, 11, 14].

The block diagram of the main processes for the production of products is shown in Figure 1. In the electrolysis shop, two series of vank are installed, equipped with 125-135 baths in each series with a rated current of 25 kA and a voltage of 425 V. Each series of baths is powered by two converter units installed at the substation. In normal mode, each unit individually provides the current load of the series. However, to improve the reliability of the power supply of the electrolysis plant, both units are in operation. Under the action of a direct current passing through an aqueous solution of $NaCl_2$, the movement of Na^+ *and* Cl ions occurs, and chlorine is released at the anode, and caustic sodium and hydrogen are released at the cathode:

$$2NaCl + 2H_2O = Cl_2 + 2NaOH + H_2.$$

With such a clean electrolysis process, the current efficiency of the whole products will be 100%.

Electrolysis is accompanied by complex processes on the surface of the electrodes - a double electric layer that

* Corresponding author: ruzinzarov88@bk.ru

appears at the electrode-solution interface and electrokinetic processes in them, depending on the operating mode of the installation. The double electric layer is an elementary capacitor with a sufficiently large active conductivity and capacitance. The higher harmonics of the currents in these capacitors significantly affect the electrode processes and, with their optimal value, can improve the weight of the complex processes and increase the current output of the installation. The electrical parameters for electrode capacitors vary depending on the value and frequency of the current of higher harmonics. The reactive component of these resistances can be either capacitive or inductive. Thus, the presence of an alternating component in the anode current leads to a violation of the stationarity of electrolysis processes and an improvement in the performance of the installation. Therefore, the determination (synthesis) of the parameters of the equivalent circuit of the electrolysis bath and the analysis on their basis of the optimal values and frequencies of the higher harmonic currents are essential [21, 26-31, 14, 10].

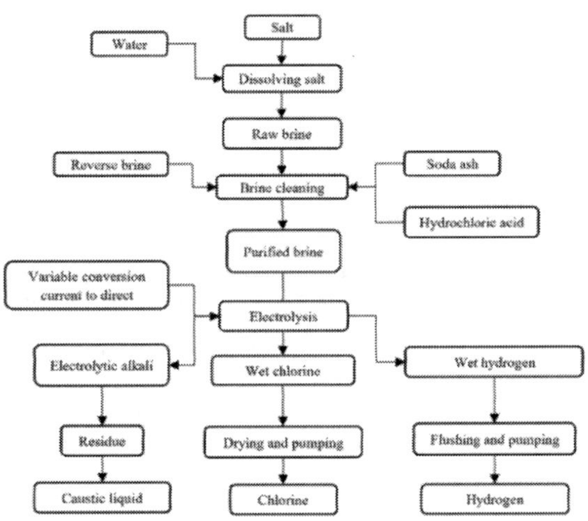

Fig 1. Structural diagram of the main processes of production of products [32]

The electrolysis process is characterized by indicators:

a) current output coefficient, %,

$$\eta = \frac{P_a}{P_t}, \tag{1}$$

where P_a and P_t are the theoretically possible and actually obtained amount of the *NaOH* product:

$$P_a = VC_{NaOH}, \tag{2}$$
$$P_t = Itk; \tag{3}$$

here V is the volume of the obtained electrolytic alkali, l; C_{NaOH} alkali concentration, g/l; I is the average current load on the bath, kA; t is the duration of the bath, h; k electrochemical equivalent (for *NaOH* k=1.492 g, for Cl_2 k=1.323g);

b) The consumption of electrical energy per one received product

$$\beta = \frac{E_a}{P_a}; \tag{4}$$

c) Coefficient of use of electrical energy

$$\mu = \frac{W_u}{W_a}; \tag{5}$$

where W_a and W_u are actually and useful breakfast electric energy:

$$W_u = ItU_T\frac{\eta}{100}, \tag{6}$$
$$W_a = ItU_B; \tag{7}$$

here $U_t = 2.18$ V is the theoretical *NaCl* decomposition voltage; U_b is the actual voltage at the bath terminals, V.

Expression (5), taking into account (6) and (7), can be written as follows:

$$\mu = \frac{2.17\eta}{U_b 100}; \tag{8}$$

Consequently, the main technical and economic indicators of cold production (P_a, β, μ) are determined by the current output, voltage and load current of electrolysis plants.

The voltage balance at the cell terminals is characterized by the following equation:

$$U_H = E_a + E_c + \Delta U_a + \Delta U_e + \Delta U_d + \Delta U_c \tag{9}$$

where E_a, E_c are the anode and cathode potentials, $\Delta U_a, \Delta U_e, \Delta U_d$ are the voltage drop in the anode, electrolyte and diaphragm, respectively, V. ΔU_c is the voltage drop in the cathode, conductive elements and contacts, V.

Theoretically, the anodic chlorine evolution potential is 1.3595 V (at 18°C). The overvoltage of chlorine evolution under operating conditions varies within 0.11-0.16 V.

The cathode potential under production conditions at a temperature of 80 - 85°C is 1.0-1.06 V. The cathode overvoltage, regardless of the material, varies within 0.25-0.39 V.

The increase in voltage in the electrolyte ΔU_3 depends on the interelectrode distance and the electrical conductivity of the electrolyte, which is determined by its composition and gas content temperature [13, 16].

To ensure a minimum voltage drop in the electrolyte, electrolysis should be carried out at a higher temperature (92±3) °C and a brine concentration of 310-315 g/l.

In the process of electrolysis, the voltage of the electrolyzer depends on the parameters of the installation mode and on the voltage level of the supply substation.

 https://doi.org/10.1051/e3sconf/202338401035

$$U_b = f(I, U, C_{NaCl}, C_{NaON}, t); \quad (10)$$

Therefore, when optimizing the operating modes of the installation, along with technological parameters, it is necessary to take into account a significant stress factor.

The importance of such accounting is determined by the high energy intensity of the installations - up to 40% of the cost of production falls on the share of electricity and therefore the effect of voltage changes is significant.

Specific electricity consumption is 2685-2813 kWt·h/t. A change in the voltage and internal resistance of the baths leads to the fact that the current of the installation varies within a fairly wide range (about 15% of the rated current – 25 kA). Such a change has a bad effect on the productivity and technical and economic performance of the installations. To the greatest extent, current changes are caused by voltage changes in the 35 kV supply network. Influenced by the ripple of the rectified voltage and current. Therefore, when optimizing the installation mode, it is important to investigate the quality of AC and DC voltage [9, 26].

The converter units use a twelve-phase rectification scheme.

For this, the secondary winding of the supply transformer is split into four parts. Two secondary windings are connected in a star, and the other two in a delta. As a result, the phase vectors of the EMF of the secondary windings are shifted by 30°. The valves are connected in a three-phase bridge and give a rectified current of 6250 A. The rectified voltage is regulated in steps by changing the transformation ratio under load. A smooth change in voltage within the control stage is performed using a saturation choke with adjustable inductance [28, 21].

The joint operation of the two units provides compensation for the fifth and seventh harmonic currents in the supply network, provided that they are evenly loaded. Separate work creates the distribution of higher harmonics of this order along the supply lines and transformer, causing significant numbers in them.

The average value of the rectified voltage of the converter.

$$U_{do} = \frac{m}{\pi} U_m \sin\frac{\pi}{m}, \quad (11)$$

Amplitude value of *k*-y harmonic U_{mk}

$$U_{mk} = \frac{m}{\pi} \int_{-\frac{\pi}{m}}^{\frac{\pi}{m}} U_m \cos \omega t \cos km\omega t d\omega l = \\ = \frac{m}{\pi} \sin\frac{\pi}{m} U_m \frac{2}{(km)^2 - 1} = \\ = U_{do} \frac{2}{(km)^2 - 1}; \quad (12)$$

Frequency spectrum of the harmonic series

$$f_k = kmf, \quad (13)$$

where f is the frequency of the mains voltage; m is the number of secondary phases; k is the multiplicity factor *(k=1,2,3……n)*.

Effective value ripple voltage

$$U_d = \sqrt{U_{do}^2 + U_{dV}^2}, \quad (14)$$

where U_{dV} is the effective value of the variable component of the rectified voltage V.

Relationship

$$\left.\begin{aligned} k_n &= \frac{U_{dV}}{U_d} = \frac{\sqrt{\sum_{k=1}^{n} U}}{U_d} 100\% \\ k_{nk} &= \frac{U_{mk}}{U_{do}} = \frac{2}{(km)^2 - 1} \end{aligned}\right\} \quad (15)$$

The values of the coefficients k_n and k_{nk} characterize the quality of the rectified voltage for 6 and 12 phase rectification circuits and the load mode of the operating installation are given below [9, 14]:

Table 1. These harmonics in the rectified voltage give rise to their corresponding current harmonics in the purpose of the electrolyzers.

Phases, m	6	12
	1,6	0,8
k_n, to at frequency, Hz		
100	-	0,003
300	0,0135	0,0135
600	0,0156	0,0079
900	0,0095	-

The voltage on the buses of 35 kV station varies within 36.5 - 38 kV. The voltage level changes 2-4 times a day. The duration of one or another voltage level is from 1 to 8 hours. Most of the time, the voltage on the RU35 kV buses is 38 and 37 kV.

With this voltage mode on the 35 kV buses of the station, the voltage level at other substations of the power center satisfies the established standards.

However, for electrolysis energy consumers located near the station, such a voltage regime does not meet the requirements of GOST and it is necessary to have automatic control devices in the power supply circuit of the plant [11].

To regulate and control the voltage on the AC and DC buses, a set of control devices is provided in the power supply circuit.

3 Conclusion

1. As it was shown, a change in the load current of a series of electrolyzers with a voltage deviation in the supply network significantly affects the technical and economic performance of the installations. Therefore, the determination of the optimal level of voltage in the supply network is important and should be carried out on the basis of studying the patterns of changes in the productivity of electrolysis plants, the consumption of

raw materials and materials, and the total loss of power and energy.

2. The actual voltage level on the 35 kV buses of the converting substation varies within 37–38 kV, which is close to the optimal value. The load current of the series varies within 21 - 29 kA, its average value for the controlled period is 24.2 kA and does not correspond to the optimal value. Such a change in the load mode of electrolyzers is caused both by voltage deviations in the network and by changes in the electrochemical and technological parameters of the electrolysis plant. Therefore, the determining factor in

optimizing the operating mode of the electrolysis plant is the load current. The implementation of automatic response of the load current and maintaining it within the limits close to optimal, in the existing mode, a voltage of 35 kV provides the best technical and economic indicators of the technological process for the production of chlorine and naustic soda.

References

1. Damasi B.B. Introduction to electrochemical kinetics - M: Chemistry, 1975. – 416p.
2. I.V. Zhezhelenko, Yu.L. Saenko, Indicators of the quality of electricity and their control at industrial enterprises, Moscow: energoavtomisat 2000
3. Yu.S. Zhelezko. Loss of electricity. reactive power. Electricity quality. Energo avtomizdat. 2010, 465p.
4. Aberson M.L. Voltage regulation optimization - m: energy 1975 – 160p.
5. N.G. Volkov. The quality of electricity in power supply systems. Tomsk: Publishing House of Tomsk Polytechnic University, 2010. - 152p.
6. Zhezhelenko I. V. Higher harmonics in power supply systems of industrial enterprises - M.: Energoatomizdat, 2010. - 375 p.
7. Savina N.V. Electricity quality: textbook Blagoveshchensk: Amur State. un-t, 2014. - 182 p.
8. Rasulov A.N., Usmonov E.G., Melikuziev M.V. AIP Conference Proceedings **2552.** (2023). 050018. https://doi.org/10.1063/5.0111530
9. Rasulov, A.N., Rafikova, G.R., Novikov, N.L., Ruzinazarov, M.R., Esemuratova, S. E3S Web of Conferences 289. 07006. (2021). https://doi.org/10.1051/e3sconf/202128907006
10. A Rasulov, R Karimov, K Shamsiyev, A Bekishev, N Kurbanova, N Musashayxova. IOP Conf. Series: Materials Science and Eng. **883(1).** 012142. (2020). doi:10.1088/1757-899X/883/1/012142
11. Rakhmonov I.U., Najimova A.M., and Reymov K.M. AIP Conference Proceedings **2647.** 030010. (2022). https://doi.org/10.1063/5.0104788
12. R.Karimov. AIP Conference Proceedings 2552. **030014**. (2022). https://doi.org/10.1063/5.0111533
13. R.Karimov. AIP Conference Proceedings 2552. **050012**. (2022). https://doi.org/10.1063/5.0111524
14. S.Dzhuraev, R.Karimov, and ofters. ElConRus, pp. 1166-1169. (2022). doi: 10.1109/ElConRus54750.2022.9755782
15. Rakhmonov I.U., Najimova A.M. AIP Conference Proceedings **2647.** 030011. (2022). https://doi.org/10.1063/5.0104791
16. Rakhmonov I.U., Najimova, A.M., Esemuratova Sh.M., Koptileuov T.T. AIP Conference Proceedings **2647.** 070024. (2022). https://doi.org/10.1063/5.0104793
17. Hoshimov F.A., Rakhmonov I.U., Niyozov N.N., Omonov F.B. AIP Conference Proceedings **2647.** 030025. (2023). https://doi.org/10.1063/5.0112388
18. Rakhmonov I.U., Hoshimov F.A., Kurbonov N.N., Jalilova D.A. AIP Conference Proceedings **2647.** 050022. (2023). https://doi.org/10.1063/5.0112391
19. Rakhmonov I.U., Ushakov V.Ya., Niyozov N.N., Kurbonov N.N., Mamutov M. E3S Web of Conferences **289.** 07014. (2021). https://doi.org/10.1051/e3sconf/202128907014
20. Rakhmonov I.U., Ushakov V.Ya., Najimova A.M., Jalilova D.A., Omonov F.B. E3S Web of Conferences **289.** 07013. (2021). https://doi.org/10.1051/e3sconf/202128907013
21. Rakhmonov, I.U., Hoshimov, F. E3S Web of Conferences **209**. 07018. (2020). DOI: 10.1051/e3sconf/202020907018
22. Rakhmonov, I., Berdishev, A., Niyozov, N., Muratov, A., Khaliknazarov, U. IOP Conference Series: Materials Science and Engineering, 2020. 883(1). 012103. DOI: 10.1088/1757-899X/883/1/012103
23. Rakhmonov, I.U., Niyozov, N.N. E3S Web of Conferences, 2019. **139.** 01077. DOI: 10.1051/e3sconf/201913901077
24. Rakhmonov, I.U., Reymov, K.M. Journal of Physics: Conference Series, 2019. 1399(5). 055038. DOI: 10.1088/1742-6596/1399/5/055038
25. Taslimov A.D. AIP Conference Proceedings **2552.** (2023). 050023. https://doi.org/10.1063/5.0112398
26. Melikuziev M.V., Nematov L.A., Novikov A.N., Baymuratov K.K. E3S Web of Conferences **289.** 07016 (2021). https://doi.org/10.1051/e3sconf/202128907016
27. A.D.Taslimov, A.S.Berdishev, F.M.Rakhimov, Melikuziev M.V. E3S Web Conf. **Vol. 139**. 2019. https://doi.org/10.1051/e3sconf/201913901081.
28. A.D.Taslimov, A.S.Berdishev, F.M.Rakhimov, Melikuziev M.V. E3S Web Conf. **Vol.139**. 2019. https://doi.org/10.1051/e3sconf/201913901082
29. Taslimov A.D., Melikuziev M.V., Najimova A.M., Alimov A.A. E3S Web of Conferences **216**. 01159 (2020). https://doi.org/10.1051/e3sconf/202021601159
30. Saidkhodjaev A.G., Nuriddinova Kh.R. AIP Conference Proceedings **2552**. 050037. (2023). https://doi.org/10.1063/5.0133486
31. Saidkhodjaev A.G. AIP Conference Proceedings **2552**. 050026. (2023). https://doi.org/10.1063/5.0111540
32. Usmonov E.G. AIP Conference Proceedings **2552.** 050020. (2023). https://doi.org/10.1063/5.0111537

E3S Web of Conferences **384**, 01036 (2023)
RSES 2022
https://doi.org/10.1051/e3sconf/202338401036

Impact of climate change on electrical energy losses in electrical networks

A.D. Taslimov*

Tashkent State Technical University, 2 University Street, Tashkent, Uzbekistan

Abstract. One of the global problems in the world is greenhouse gas emission. There are several reasons for this problem, one of which is electricity. This article provides a brief summary of the current state of electricity waste and analyzes the impact of electricity waste on greenhouse gas emissions, as well as ways to reduce greenhouse gas emissions from the point of view of electricity generation. The situation of greenhouse gas emission in the territory of Uzbekistan and how much of it corresponds to the share of electricity losses is shown.

1 Introduction

The growth of electricity losses in electric networks is an objective process for countries with a crisis economy and reformed energy sectors, and is a sign of the existing gap between consumers' ability to pay and electricity tariffs. Inadequate investments in electric grid infrastructure and electricity accounting system, quality supply of electric energy, structure of electric currents according to voltage levels, lack of full-scale automated information systems for collecting and transmitting information about electric energy balances in electric networks.

2 The current state of the investigated problem

Analytical results of current research shows that the main part of electricity losses in power grids falls on low-voltage distribution power grids, as you can see from Figure 1. [1-3, 8, 16-20].

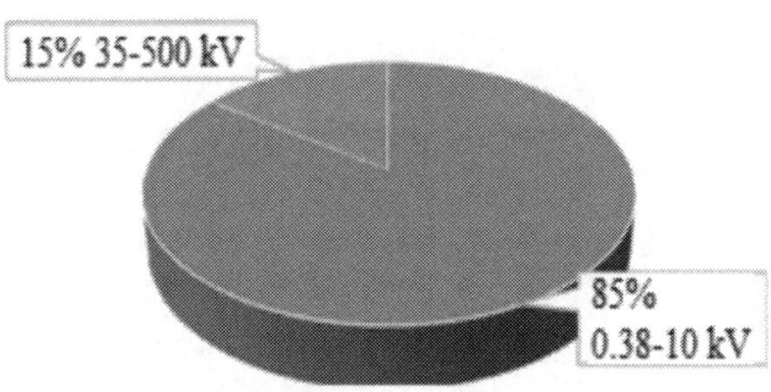

Fig. 1. Percentage of electrical energy losses

In addition, when strategies to reduce climate change events are considered, reducing electricity waste is rarely taken into account. However, ongoing scientific research shows that by reducing electrical energy consumption, greenhouse gas emissions reduction is an important tool [2, 14-15].

Electricity in transmission and distribution systems requires additional electricity generation to compensate for losses, which necessarily increases the demand for additional fossil energy resources. Renewable energy resources are directly related to greenhouse gas emissions.

The calculations show that the amount of global emission offsets is about one billion metric tons of carbon dioxide equivalent per year, which is in the same range as the annual emissions of heavy trucks or the entire chemical industry (in 2021 from the global chemical industry greenhouse gas emissions were 925 MT CO2).

Most of the electricity is generated in central power plants and sent over long distances through high voltage transmission lines and delivered to consumers through the distribution network. As power moves through the electrical grid, resistance in the metal wires creates heat in the power lines. This results in a part of the energy obtained from fuel energy resources used to generate electricity being lost in transit.

We can see the proof of this through Joule's lens law:

$$Q = I^2 \cdot R \cdot t \ [J]$$

Here, Q is the amount of heat released when an electric current flows through the conductor. I is the value of the electric current, R is the active resistance of the conductor.

Here, Q is the amount of heat released when an electric current flows through the conductor. I is the value of the electric current, R is the active resistance of the conductor.

Annual greenhouse gas emissions due to electrical energy waste in the transmission of electricity in the

* Corresponding author: ataslimov@mail.ru

power grid are more than the emissions of some industrial enterprises Figure 2 [4, 11-13, 21-25, 9, 10].

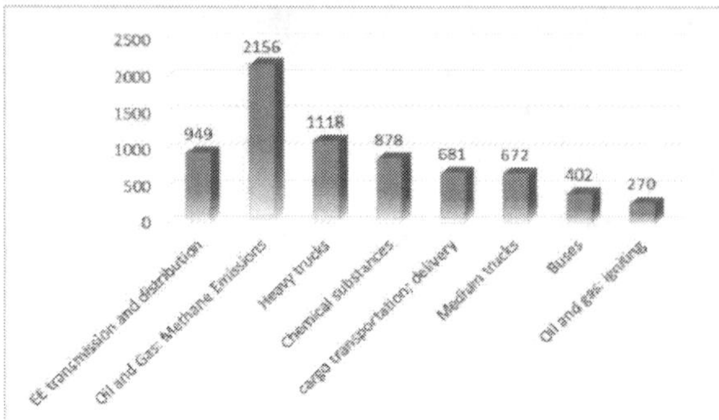

Fig. 2. Causes of greenhouse gas emissions and their amount. It is measured in million metric tons of carbon dioxide equivalent

Also, according to the data of 2020, the share of greenhouse gas emissions in the year section is shown in 3 Figure. electricity generation accounts for the second largest share of greenhouse gas emissions (25% of 2020 greenhouse gas emissions). However, about 60% of electricity comes from burning fossil fuels, mainly coal and natural gas.

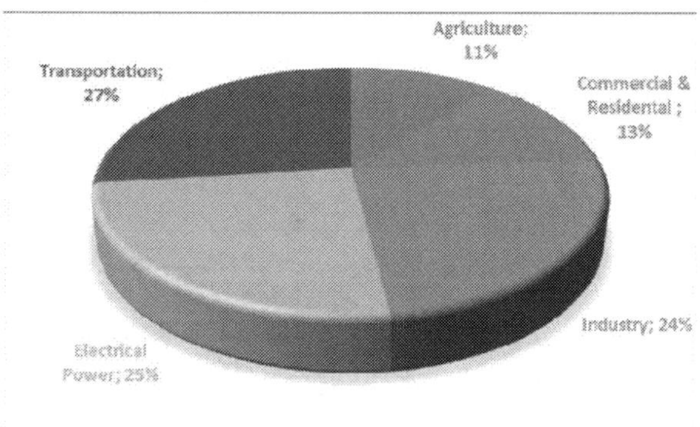

Fig. 3. Share of greenhouse gas emissions in 2020

The electricity sector includes the generation, transmission and distribution of electricity. Carbon dioxide (CO2) accounts for the majority of greenhouse gas emissions from the sector, but smaller amounts of methane (CH4) and nitrous oxide (N2O) are also emitted. These gases are released during the burning of fossil fuels such as coal, oil and natural gas to generate electricity. Less than 1% of the sector's greenhouse gas emissions come from the insulating chemical sulfur hexafluoride (SF6) used in electricity transmission and distribution equipment [26-31, 9, 12, 14].

Burning coal requires more carbon to produce electricity than burning natural gas or oil. Although coal use accounted for about 54% of CO2 emissions from the sector, it accounted for only 20% of the electricity generated in the United States in 2020. In 2020, natural gas use accounted for 39% of electricity, and oil use increased by more than 1%. In 2020, the remaining generation came from non-fossil fuel sources, i.e. nuclear (21%) and renewables (20%), including hydroelectricity, biomass, wind and solar. Most of these renewable energy sources, such as nuclear, hydroelectric, wind and solar power plants, emit almost no radiation [5, 18, 14, 16].

Studies have shown that electricity consumption varies greatly from country to country. In 2016, total transmission and distribution electricity losses reached 19% in India and 16% in Brazil. But they were more than 50% in Haiti, Iraq and the Republic of Congo. This means that only half of the electricity produced reaches consumers or is counted as usable power, and the other half is wasted on the road.

In developed countries, electricity waste was lower: the United States accounted for 6% of electricity waste in 2016, Germany 5% and Singapore 2%. These figures show that it is more efficient to transmit electricity over short distances to large population centers than to send electricity over long distances to consumers located in many rural areas.

The easiest way to solve the technical losses is to use the most advanced technologies and to modernize the existing infrastructure for long-distance transmission and local distribution of electricity. Transmission improvements can be made, for example, by replacing inefficient (low-efficiency) power transmission lines (EUL), reducing the resistance in the EUL and thus reducing electrical energy waste, as well as by controlling the high-voltage power flow [9, 18, 4, 5].

Similarly, improvements in distribution can be achieved through proper management of load and power distribution, as well as optimal layout of distribution lines. Innovations such as the adoption of digital technologies to direct power flows can also play a role.

Solutions for non-technical losses are more difficult and can only partially reduce the associated emissions. The causes of high electricity wastage are varied, such as hurricanes or wars in Haiti and Puerto Rico in recent years, or extreme events such as a combination of poor governance, corruption and poverty, as seen in India. For both types of losses, countries with a large share of fossil fuel generation and inefficient grid infrastructure can reap higher environmental benefits by reducing emissions and reducing electricity waste in transmission and distribution [16, 17, 19].

There are various options for reducing greenhouse gas emissions associated with electricity generation, transmission and distribution. The following table lists these possibilities and gives examples, Table 1 [6, 7-10].

Table 1

Types	Ways to reduce greenhouse gases	Examples
Increasing the efficiency of power plants operating with fossil energy resources	Improving the efficiency of existing fossil fuel power plants through the use of advanced technologies, and replacing high-emission power plants with low-emission ones.	Converting a coal-fired boiler to natural gas or co-firing with natural gas. Conversion of a single-cycle gas turbine to a combined cycle turbine.
Renewable energy	Using renewable energy sources rather than fossil fuels to generate electricity.	By connecting power plants powered by wind, solar, hydro, and geothermal sources, as well as some biofuel sources to the grid
End-use energy efficiency improvement	Reducing electricity consumption and peak demand by improving energy efficiency and savings in homes, businesses and industry.	By classifying the payment of the main electricity consumers for the consumed energy
Nuclear energy	Applying electricity from nuclear energy instead of energy from burning fossil fuels.	Extending the service life of existing nuclear power plants and creating new nuclear power plants.
Carbon sequestration	Popularization of electricity generation from CO2	

3 Situation of greenhouse gas emissions in uzbekistan

The share of greenhouse gas emissions is also increasing significantly in the territory of Uzbekistan, for example:

- Greenhouse gas emissions (CO2) in 2016 amounted to 10,523 (kt CO2 equivalent), which is 6.11% more than in 2015.
- In 2017, greenhouse gas emissions (CO2) amounted to 10,953 (kt CO2 equivalent), which is 4.09% more than in 2016.
- In 2018, greenhouse gas emissions (CO2) amounted to 11,247 (kt CO2 equivalent), which is 2.68% more than in 2017.
- In 2019, greenhouse gas emissions (CO2) amounted to 116.71 (kt CO2 equivalent), which is 3.77% more than in 2018.

Using the above information, we can find out how much electricity is used to generate greenhouse gases, Table 2 [7].

Table 2

years	Greenhouse gas emission value (kt CO2 equivalent)	Share of electricity %	The cost of electricity in the production of greenhouse gas emissions (kt CO2 equivalent)	The cost of electricity in the production of greenhouse gas emissions (MWh)
2016	170,89	25	42,7225	51,47289
2017	176,31		44,0775	53,10542
2018	183,81		45,9525	55,36446
2019	188,759		47,18975	56,85512

So, according to the section of 2019, about 56.8 MWh of electricity, that is, a certain part of the waste of electricity, was spent on the creation of greenhouse gas emissions.

4 Conclusion

Through research, we found out that one of the main causes of greenhouse gas emissions is the losses of electricity that occurs through the transmission and distribution of electricity. Also, the situation of greenhouse gas emissions in the territory of Uzbekistan and how much greenhouse gas emissions were released during the years 2016-2019 and what is the share of electricity losses calculated.

References

1. Decision of the President of the Republic of Uzbekistan dated November 23, 2016 No. PQ-2661 "On the program of further modernization and renewal of low-voltage power networks in 2017-2021".
2. Handbook on the design of electrical networks / ed. D.L. Faibisovich. - 4th ed., Rev. and add. - M.: ENAS, 2012 .-- 376 p.
3. Bereznev Yu. I. Choice of wire cross-sections of 0.38 and 10 kV lines in conditions of uncertainty and the problem of unification of their scale // educational institutions. Energy problems: scientific, technical and production journal. - 2009. - No. 9/10. S. 38-40.
4. Efentiev S.N., Zuev E.N. Economic current intervals of wire cross-sections of overhead lines - yesterday, today, tomorrow. // ELECTRO. Electrical engineering, power engineering, electrical industry. í2005. íɒ 3. - P. 43-48.
5. Adel Aljwary, Ziyodulla Yusupov, Olimjon Toirov, Rustam Shokirov, Mitigation of load side harmonic distortion in standalone photovoltaic based microgrid, E3S Web of Conferences, ICECAE 2021, 304, 01010 (2021) https://doi.org/10.1051/e3sconf/202130401010
6. Taslimov A.D. Snizhenie poter elektroenergii v svyazi s deep unifikatsiey sechenii kabeley raspredelitelnyx setey // "Problemy energo- i

resursosberezheniya". Tashkent, 2009. - No. 3-4. - S. 83-88.

7. M. I. Fursanov. A. A. Zolotoy. V. V. Makarevich. Raschety tekhnologicheskogo raskhoda (poter) elektroenergii na ee transport v elektricheskikh netsyakh energosistem. UDK 621.311.1.017(075.8) BBK 31.280.7ya7 F95

8. A.D.Taslimov AIP Conference Proceedings **2552.** (2023). 050023. https://doi.org/10.1063/5.0112398

9. Melikuziev M.V., Nematov L.A., Novikov A.N., Baymuratov K.K. E3S Web of Conferences **289.** 07016 (2021). https://doi.org/10.1051/e3sconf/202128907016

10. A.D.Taslimov, M.V.Melikuziev, A.M.Najimova, A.A.Alimov. E3S Web of Conferences **216.** 01159. (2020). https://doi.org/10.1051/e3sconf/202021601159

11. A.D.Taslimov, A.S.Berdishev, F.M.Rakhimov, M.V.Melikuziev. E3S Web Conf. **Vol.139.** (2019). https://doi.org/10.1051/e3sconf/201913901082

12. Rakhmonov I.U., Najimova A.M., and Reymov K.M. AIP Conference Proceedings **2647.** 030010. (2022). https://doi.org/10.1063/5.0104788

13. Rakhmonov I.U., Najimova A.M. AIP Conference Proceedings **2647.** 030011. (2022). https://doi.org/10.1063/5.0104791

14. Rakhmonov I.U., Najimova, A.M., Esemuratova Sh.M., Koptileuov T.T. AIP Conference Proceedings **2647.** 070024. (2022). https://doi.org/10.1063/5.0104793

15. Hoshimov F.A., Rakhmonov I.U., Niyozov N.N., Omonov F.B. AIP Conference Proceedings **2647.** 030025. (2023). https://doi.org/10.1063/5.0112388

16. Rakhmonov I.U., Hoshimov F.A., Kurbonov N.N., Jalilova D.A. AIP Conference Proceedings **2647.** 050022. (2023). https://doi.org/10.1063/5.0112391

17. Rakhmonov I.U., Ushakov V.Ya., Niyozov N.N., Kurbonov N.N., Mamutov M. E3S Web of Conferences **289.** 07014. (2021). https://doi.org/10.1051/e3sconf/202128907014

18. Rakhmonov I.U., Ushakov V.Ya., Najimova A.M., Jalilova D.A., Omonov F.B. E3S Web of Conferences **289.** 07013. (2021). https://doi.org/10.1051/e3sconf/202128907013

19. R Karimov. AIP Conference Proceedings **2552.** 030014. (2022). https://doi.org/10.1063/5.0111533

20. R Karimov. AIP Conference Proceedings **2552.** 050012. (2022). https://doi.org/10.1063/5.0111524

21. S Dzhuraev, R Karimov and others, Study and Analysis of Power Quality in the Electrical Networks of the Outdoor Lighting of the Dushanbe City (2022 Conference of Russian Young Researchers in Electrical and Electronic Engineering (ElConRus), 2022), pp. 1166-1169, doi: 10.1109/ElConRus54750.2022.9755782

22. R Karimov, A. Kuchkarov and others, Analysis and study of energy efficiency by the operation of a voltage stabilizer (Journal of Physics: Conference Series 2094, 052050, 2021). https://doi:10.1088/1742-6596/2094/5/052050

23. Melikuziev M.V. AIP Conference Proceedings **2552.** (2023). 050021. https://doi.org/10.1063/5.0112395

24. Rasulov A.N., Usmonov E.G., Melikuziev M.V. AIP Conference Proceedings **2552.** (2023). 050018. https://doi.org/10.1063/5.0111530

25. Usmanov E.G., Khusanov B.M. RSES 2020. E3S Web of Conferences **216.** 01161. (2020). https://doi.org/10.1051/e3sconf/202021601161

26. Erejepov, M., Novikov, A.N., Khusanov, B.M., Seytmuratov, B., Sayimbetov, Z. E3S Web of Conferences. **289.** (2021). 07018. https://doi.org/10.1051/e3sconf/202128907018

27. Rakhmonov, I., Berdishev, A., Niyozov, N., Muratov, A., Khaliknazarov, U. IOP Conference Series: Materials Science and Engineering, 2020. 883(1). 012103. DOI: 10.1088/1757-899X/883/1/012103

28. Rakhmonov, I.U., Niyozov, N.N. E3S Web of Conferences, 2019. **139.** 01077. DOI: 10.1051/e3sconf/201913901077

29. Rakhmonov, I.U., Reymov, K.M. Journal of Physics: Conference Series, 2019. 1399(5). 055038. DOI: 10.1088/1742-6596/1399/5/055038

30. R.Karimov, N.Kurbanova and others. Conference Series 2094(5), 052042, 2021). https://doi:10.1088/1742-6596/2094/5/052042

31. A.D.Taslimov, A.S.Berdishev, F.M.Rakhimov, Melikuziev M.V. E3S Web Conf. **Vol.139.** (2019). https://doi.org/10.1051/e3sconf/201913901081

Economic interval analysis of loads for selection of cross-section surfaces of electrical transmission lines

Abdurakhim Taslimov[1]*, *Farrukh* Rakhimov [2], *Feruz* Rakhimov [2]

[1]Tashkent State Technical University named Islam Karimov, 2 Universitet str., Tashkent, Uzbekistan.

[2]Navoi State Mining and Technological University, 76 Galaba shox str., Navoi, Uzbekistan.

Abstract. In this article, the implementation of comprehensive measures for the use of modern approaches to the selection of the main parameters of the system in the development of the electric power industry of our Republic, including the development of methods for optimizing, updating and increasing the reliability of electric transmission networks, as well as the selection of the most optimal cross-sectional surface of the conductors of power transmission lines written about the problem and the widespread use of the method of economic intervals to solve this problem. In addition, the action strategy for the further development of the Republic of Uzbekistan in 2017-2021 defines the tasks of "... to reduce the energy and resource capacities of the economy, to widely introduce energy-saving technologies into production, which are implemented in accordance with the target parameters of reducing energy consumption in economic sectors,..." and these tasks. It has been reflected that determining the optimal parameters of electrical networks, carrying out theoretical and scientific work on increasing their reliability, and receiving appropriate recommendations for the design, construction, modernization and operation of rural electrical networks are considered to be one of the most important issues for implementation.

1 Introduction

The problem of choosing the optimal cross-sectional surface of power line conductors has always been a problem, and now a lot of attention is paid to this problem. To solve this problem, the method of economic intervals is widely used. Academician Budzko I.A., Venikov V.A., Levin M.S., Kholmsky V.G., Blok V.N., Efentev S.N., Zuev E.N., Suvorova I. developed the theoretical aspects of this method. A., Herkusov A.A., Freishteter V.P., Martyanov A.C., Varygina A.O., Fedotov A.I. and a number of other research scientists [1-8]. Subsequent studies have substantiated and refined this method both theoretically and practically.

2 The current state of the investigated problem

The availability and adequacy of economic intervals for wire and cable loads are shown in [9, 26]. If initially the economic range of the load was determined for static load, then in the process of determining them, the dynamics of load growth [10, 29] and the possibilities of reconstruction of power lines were taken into account [11-18].

However, it should be noted that mainly the economic ranges of loads are widely used in the calculation and selection of wires of overhead transmission lines. To date, economic current density is widely used to select cross-sectional areas of cable lines, but it does not meet the condition of minimum total costs. In this regard, it was necessary to determine the economic ranges of loads and the tasks associated with them in order to choose the cross-sectional surfaces of the cables [19-25].

3 Study of changes in economic load intervals

The limits of the economic range of the load are determined from the following condition [12]:

$$З_i = З_{i+1} \tag{1}$$

where $З_i$ is the total cost of the cable line with a cross-sectional area F_i, $З_{i+1}$ is the total cost of a cable line with a cross-sectional area F_{i+1}.

As for determining the economic intervals of loads, the total costs for cable lines are determined as follows. Capital investments in the construction of cable lines are usually made within one year. It is difficult to reconstruct the cable line to increase its carrying capacity, and because it is not useful during its service life, new capital costs are not included in it. Taking into

* Corresponding author: ataslimov@mail.ru

account these characteristics of cable lines, the capital cost shown in the expression is written as follows:

$$E_M \sum_{t=1}^{T} K_t (1+E_{MK})^{\tau-t} = E_M K \qquad (2)$$

In this case, we assume that the costs are included in the first year of operation τ=1.

Due to the fact that capital funds do not change, depreciation expenses remain constant, in which we can write the following expression:

$$\sum_{t=1}^{T} (U_{at} - U_{a(t-1)})(1+E_{MK})^{1-t} = U_{a1} = p_a K \qquad (3)$$

Operating costs do not depend on the cross-sectional area of the cables and do not change over time. Therefore, they may not be taken into account when determining the economic intervals of loading. When determining the economic intervals of loads, the length of power lines is not important and is usually taken as 1 km.

Taking into account the above, as well as expressions (2) and (3), the expression for determining the total costs of the transmission line can be written as follows:

$$З = (E_M + p_a)K \frac{U_И 10^{-5}}{U_H^2 \gamma F}\left[S_1^2 + \sum_{t=2}^{T}(S_t^2 - S_{t-1}^2)(1+E_{MK})^{1-t}\right] \qquad (4)$$

The analysis of technical and economic indicators of cable lines in rural areas showed that all cross-sectional surfaces of cables have economic intervals in accordance with the conditions of availability of economic intervals of loads.

If the load during the considered period is constant and does not change over time

$$S_t = S_{t-1} = S$$

then according to (4) $З_i$ and $З_{i+1}$ in expression (1) take the following form:

$$З_i = (E_M + p_a)K_i + \frac{U_И S^2 10^{-5}}{U_H^2 \gamma F_i} \qquad (5)$$

$$З_{i+1} = (E_M + p_a)K_{i+1} + \frac{U_И S^2 10^{-5}}{U_H^2 \gamma F_{i+1}} \qquad (6)$$

Equating expressions (5) and (6) in accordance with condition (1) and solving the obtained equation with respect to the load, we find the expression for determining the limits of the economic intervals for the interconnected cross-sectional surfaces for the case where the load growth dynamics is not taken into account.

$$S_{чег} = \sqrt{\frac{(E_M + p_a)(K_{i+1} - K_i)U_H^2 \gamma F_{i+1} F_i 10^5}{U_И (F_{i+1} - F_i)}} \qquad (7)$$

As can be seen in expression (7), the sectors that determine the limits of the economic intervals of the load can be divided into two groups: fixed (constant) and variable. The first includes specific permeability and standard cross-sectional areas F_i and F_{i+1}. The group of constant factors can be conditionally included E_M - standard coefficient of efficiency, p_a - standard coefficient of amortization output and U_H - nominal voltage.

The economic intervals of the load have a completely different effect on the time variation of the load. If the growth of the load is expressed in relation to the value at the end of the accounting period, then the limit values of the economic load intervals can be determined by the following expression [12].

$$S_{чег} = \sqrt{\frac{(E_M + p_a)(K_{i+1} - K_i)U_H^2 \gamma F_{i+1} F_i 10^5}{U_H (F_{i+1} - F_i) A}} \qquad (8)$$

where A is the coefficient that determines the law of growth of the load.

The value of the coefficient A is determined by expressions (10) - (15), respectively, in relation to the exponential, linear, simple modified exponential laws of load growth and the saw-shaped description of the load.

The obtained expression (8) differs from the similar expression (7) only by the part $A^{-\frac{1}{2}}$ in order to take into account the load growth dynamics. If the expression (7) is taken as a basis, then the expressed part of $A^{-\frac{1}{2}}$ is actually the relative values of the limit of economic load intervals. We call them displacement coefficients and denote them by k_c . During the study, they are determined according to the corresponding value of A as follows:

$$k_c = A^{-\frac{1}{2}} \qquad (10)$$

$$A^{Э} = (1+k_{ўc})^{2(1-T)} + \left[1-(1-k_{ўc})^{-2}\right] \times \sum_{t=2}^{T} (1+k_{ўc})^{2(t-T)}(1+E_{MK})^{1-t} \qquad (11)$$

$$A^{Ч} = (1+k_{ўc})^{-2} + (1+k_{ўc})^2 + \sum_{t=2}^{T}(1+k_{ўc})^{2(t-T)}(1+E_{MK})^{1-t} \qquad (12)$$

$$A^{MЭ} = \left[1-(1-\frac{1}{k_T})m\right]^2 + \sum_{t=2}^{T}\left\{\left[1-(1-\frac{1}{k_T})m^t\right]^2 - \left[1-(1-\frac{1}{k_T})m^{t-1}\right]^2\right\} \times (1+E_{MK})^{1-t} \qquad (13)$$

$$\begin{aligned}A^{АЭ} &= (1+k_{ўc})^{2(t-t_M)} + \left[1-(1+k_{ўc})^{-2}\right]\times \\ &\times \sum_{t=2}^{t_{p1}} (1+k_{ўc})^{2(t-t_{p1})}(1+E_{MK})^{1-t} + \\ &+\left[k_{p1}^2(1+k_{ўc2})^{2(t_{p1}-t_{p2}+1)}-1\right]\times(1+E_{MK})^{1-t_{p1}} + \\ &+k_{p1}^2\left[1-(1+k_{ўc})^{-2}\right]\sum_{t=t_{p1}+2}^{t_{p2}}(1+k_{ўc2})^{2(t-t_{p2})}(1+E_{MK})^{1-t} + \\ &+\left[k_{p1}^2(1+k_{ўc3})^{2(t_{p2}-T+1)}-k_{p1}^2\right](1+E_{MK})^{1-t_{p2}} + \\ &+k_{p2}^2\left[1-(1+k_{ўc3})^{-2}\right]\sum_{t=t_{p2}+2}^{T}(1+k_{ўc3})^{2(t-T)}(1+E_{MK})^{1-t}\end{aligned} \quad (14)$$

$$\begin{aligned}A^{АЧ} &= (1+k_{ўc}t_{p1})^{-2} + \\ &\left[(1+k_{ўc1})^2 + k_{ўc}\sum_{t=2}^{t_{p1}} \begin{matrix}(2+2k_{ўc1}t-k_{ўc1})\times \\ \times(1+E_{MK})^{1-t}\end{matrix}\right] + \\ &k_{p1}^2(1+k_{ўc2}t_{p2})^{-2}\times \\ &\times\left\{\begin{matrix}\left[1+k_{ўc2}(t_{p2}+1)\right]^2 + \\ +(1+E_{MK})^{-t_{p1}} + \\ +k_{ўc2}\sum_{t=t_{p1}+2}^{t_{p2}}(2+2k_{ўc2}t-k_{ўc2})(1+E_{MK})^{1-t}\end{matrix}\right\} + \\ &+k_{p1}^2(1+k_{ўc3}T)^{-2}\times \\ &\times\left\{\begin{matrix}\left[1+k_{ўc3}(t_{p2}+1)\right]^2 + \\ +(1+E_{MK})^{-t_{p2}} + k_{ўc3}\times \\ \times\sum_{t=t_{p2}+2}^{T}(2+2k_{ўc3}t-k_{ўc3})(1+E_{MK})^{1-t}\end{matrix}\right\} - \\ &-(1+E_{MK})^{-t_{p1}}\left[1+k_{p1}^2(1+E_{MK})^{t_{p1}-t_{p2}}\right]\end{aligned} \quad (15)$$

The analysis of the expressions for determining the displacement coefficient shows that they do not depend on the cross-sectional area of the cables and are actually determined by the relative load growth factor (multiplier of the load growth) and the calculation period. Using the coefficient A, the expression of the total costs of the power line for any law of load growth can be written as follows.

$$З_i = (E_M + p_a)(K_0 + kF_i) + \frac{U_H S_T^2 A 10^{-5}}{U_H^2 \gamma F_i} \quad (16)$$

To determine the effect of different laws of load growth on displacement coefficients, the compared options should be equal. If the compared growth laws give the same multiple of load growth, then the comparison conditions are satisfied. Therefore, it is necessary to express the previously mentioned laws of load growth with the growth factor.

For the law of exponential growth, the load growth factor at t=T is determined as follows:

$$k_T^Э = \frac{S_T}{S_0}(1+k_{ўc}^э)^2 \quad (17)$$

Similarly for the linear law of load growth

$$k_T^Ч = \frac{S_T}{S_0}(1+k_{ўc}^ч)^2 \quad (18)$$

If the load growth factor is determined from expressions (17) and (18), then the comparison conditions are fulfilled.

A given load factor:

$$k_{ўc}^э = \exp(\frac{\ln k_T}{T}) - 1 \quad (19)$$

$$k_{ўc}^ч = \frac{k_T - 1}{T} \quad (20)$$

For the load to grow according to a simple modified exponential law

$$\ln m = \frac{\ln 0{,}05 S_T - \ln(1{,}05 S_T - S_0)}{T}$$

or considering $k_T = S_T / S_0$,

$$\ln m = \frac{\ln 0{,}05 - \ln(1{,}05 - \frac{1}{k_T})}{T} \quad (21)$$

In order to study the influence of different laws of load growth on the displacement coefficient, appropriate calculations were carried out. The calculation results are presented in graphs 1-2.

The value of the displacement coefficient depends significantly on the law of growth of the load, the value of the multiple growths of the load and the calculation period T.

The length of the calculation period has a different effect on the value of the drift coefficients. In the linear law of load growth and any duration of the calculation period, the displacement coefficient is less than the exponential one, and the difference between them increases as the duration T increases.

This difference increases as the load increases.

According to the simple modified exponential law, the coefficient of displacement is larger with the growth of the load and the short duration of the calculation period, decreases with the increase of the calculation period, and becomes smaller compared to the exponential growth of the load. The characteristic of the

 https://doi.org/10.1051/e3sconf/202338401037

change of the displacement coefficient with respect to the change of the multiple growth of the load is approximately the same as the change of the calculation period [27, 28].

4 Influence of different laws of load growth on the shear coefficient

Analysis of shift coefficients in the change of the load in the saw-shaped description: their value shows the dependence of the laws of growth of loads between reconstructions, the coefficient of growth of the load, the value of the coefficient of distribution of loads, the time of reconstruction and the duration of the accounting period. In comparable conditions, the coefficient $k_c^{э.ўс}$ is always greater than $k_c^{ўс.ч}$. The large number of variable parameters does not allow to build appropriate relationships.

In all possible cases, the shift factor is always greater than one. Thus, when choosing cross-sectional areas of cables for lines whose load does not change over time, the limits of economic intervals of loads are always greater than those of lines whose load changes.

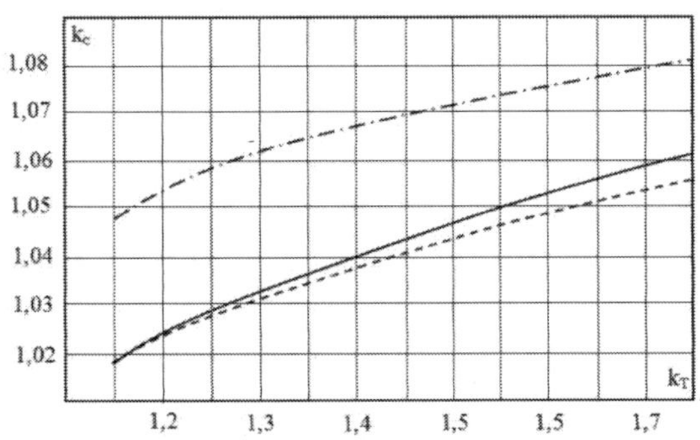

a)

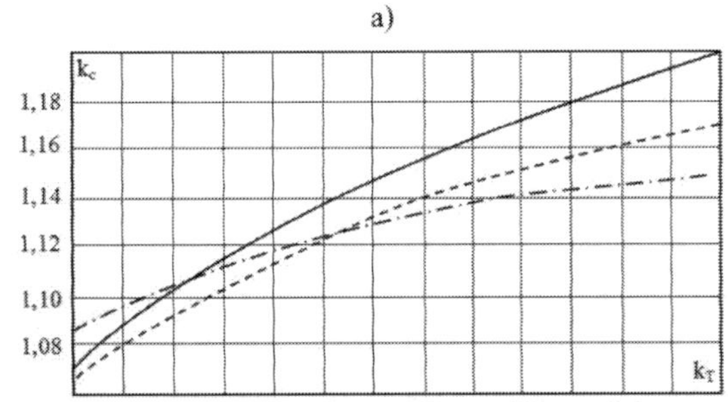

b)

—— $k_c^э$; – – – $k_c^ч$; – · – $k_c^{м.э}$;

Fig.1. a) T=5 years; b) T=10 years; Dependence of the change of displacement coefficients on k_T

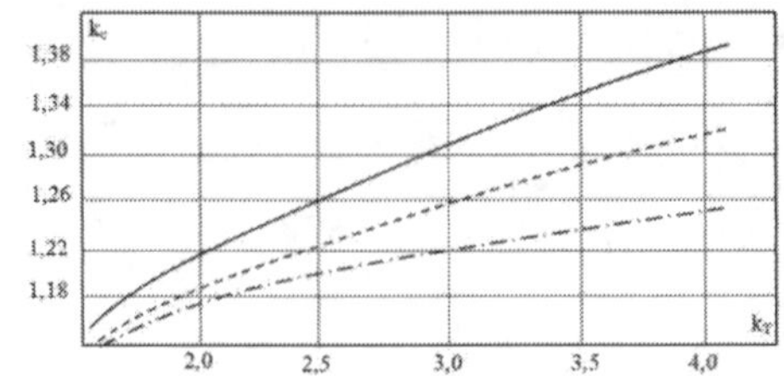

a)

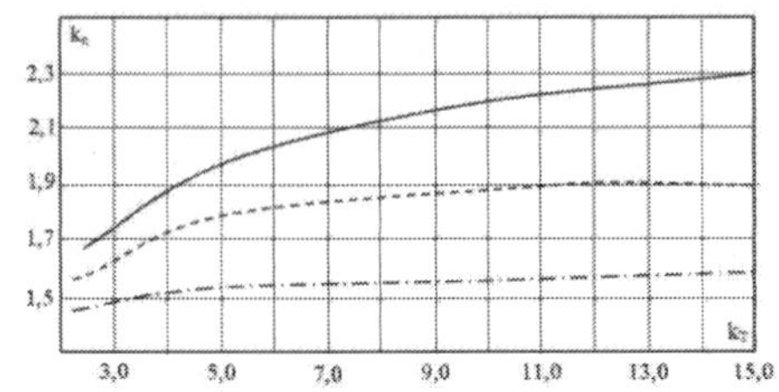

b)

—— $k_c^э$; – – – $k_c^ч$; – · – $k_c^{м.э}$;

Fig.2. a) T=15 years; b) T=30 years; Dependence of the change of displacement coefficients on k_T

The upper limit of the displacement coefficient is limited by the permissible load currents in the cable. If we take the value of the long-term permissible load current [13, 28] of the cables with the appropriate cross-sectional area, the upper limit of the economic range established for the case of constant loads, we get the values of the technically maximum permissible displacement coefficients. Studies have shown that their value is from 1.76 to 4.47, and with the increase of the cross-sectional area of the cables, the allowable displacement coefficient decreases [30-33, 14, 20].

5 Conclusion

The analysis of the values of the coefficient of displacement k_c shows that its value is 1.735 for the accounting period of 20 years with an increase in the load by 12% per year according to the exponential law. Thus, taking into account the dynamics of restrictions on long-term permissible loading, the choice of cross-sectional areas of cables in the economic intervals of loading is practically not affected.

The case is more complicated considering the limit on the allowable voltage drop. Economic loading intervals do not take this limitation into account. Therefore, after choosing the cross-sectional area of the cables for economic load intervals, it is necessary to check the selected cross-sectional area according to the permissible voltage loss.

Thus, it is possible to determine the limits of the economic intervals of the load, without taking into account the change of the load over time, and using the shift coefficients of the limits of the economic intervals, the limits of the economic intervals can be determined for any law of growth of the loads.

References

1. Suvorova I. A. Choice of cross-sections of conductors and rational voltages of distribution electrical networks in modern conditions. Dissertations for the degree of candidate of technical sciences. - Novocherkassk, 2015. - **pp**. 91-121.

2. Gerkusov A.A. Application of the method of economic intervals in the selection of cross-sections of wires with a split phase. St. Petersburg State Polytechnical University Journal. 2017. - **Volume 23**, No. 1. - S. 157-167. DOI: 10.18721/JEST.230115.

3. Freishteter V.P., Martyanov A.S. Choice of an economically justified cross-section of wires and cable cores of power transmission lines during design // Neftyanoye Khozyaistvo. - 2011. - **No. 4.** - pp. 117-121.

4. Varygina A.O., Savina N.V. Development of methods for selecting conductor cross-sections and their adaptation to modern conditions. // Bulletin of AmSU. - 2018. - Issue 81. - **P.** 50-54.

5. Fedotov A.I., Gerkusov A.A. Modernization of the method of economic intervals in the selection of cross-sections of wires of overhead power lines. // Energy problems. - 2003. - No. 1-2. - **pp** 136-140.

6. Gerkusov A. A. Analysis of methods for selecting the cross-sections of wires of overhead power lines // Scientific and technical statements of the St. Petersburg State Polytechnic University. - 2014. - No. 3 (202). - **pp.** 131-137.

7. Zuev E.N. Determining the boundaries of economic current intervals based on the minimum of discounted costs. Vestnik MPEI. - 2000. - No. 4. - **pp.** 75-77.

8. Efentiev S.N., Zuev E.N. Economic current intervals of wire cross-sections of overhead lines - yesterday, today, tomorrow // Electrical engineering, electric power industry, electrical industry. - 2005. - No. 3. - **pp.** 43-48.

9. Ananicheva S.S., Kotova E.N. Design of electrical networks: a tutorial. - Yekaterinburg: Ural Publishing House. un-ta, 2017. - 164 **pp.**

10. Efentev S. N. Development of the technique of technical and economic analysis when choosing the main parameters of electrical networks, taking into account the uncertainty of the initial information: Dis... cand. tech. Sciences. - M., 2004.-135 **pp.**

11.Zuev E.N., Efentev S.N. Problems of choosing economic expedient sections of wires and cable cores. Moscow: MEI Publishing House, 2005. - 83 **pp.**

12.Taslimov A, Rakhimov F, Nematov L, Markaev N, Bijanov A and Yunusov R. Economic load intervals for selecting 10 kV cable cross sections for agricultural consumers// IOP Conf. Series: Materials Science and Engineering **883** (2020) 012102 IOP Publishing (https://doi:10.1088/1757-899X/883/1/012102).

13.Yanukovych G.I Calculation of power lines for agricultural purposes. Textbook allowance for students with. X. universities. 2nd ed., add. - Minsk: BGATU, 2004. - 105 **p.**

14. A.D. Taslimov, A.S. Berdishev, A.A. Yuldashev, F.M. Rakhimov. Optimal tendency of selecting cable cross-sections for agricultural electrical networks // Published 1 April 2020 Published under licence by IOP Publishing LtdJournal of Physics: Conference Series, Volume **1515**, Applied Physics and Cyber-Physical Systems (doi:10.1088/17426596/1515/2/022056).

15. A.D.Taslimov, A.S.Berdishev, F.M.Rakhimov, Melikuziev M.V. E3S Web Conf. **Vol.139**. 2019. https://doi.org/10.1051/e3sconf/201913901082

16. A.D.Taslimov, A.S.Berdishev, F.M.Rakhimov, Melikuziev M.V. E3S Web Conf. **Vol. 139**. 2019. https://doi.org/10.1051/e3sconf/201913901081.

17. Koptsev L A, Koptsev A L 2011 Rationing and forecasting of electricity consumption in an industrial enterprise *J Industrial power engineering*. **1** pp 18–23

18. Hoshimov FA, Rahmonov I U 2014 Rationing of electricity production in the rolling of ferrous metallurgy *J European Science review* **11-12** pp 56-59

19. Saidkhodjaev A G, Najimova A M and Bijanov A K 2019 Method for determining the maximum load of consumers in city power supply systems E3S Web Conf **139** https://doi:10.1051/e3sconf/201913901078

20. Taslimov A D, Rakhmonov I.U. Conference Series **1399.** doi:10.1088/1742-6596/1399/5/055046

21. Rakhmonov I.U., Najimova A.M., and Reymov K.M. AIP Conference Proceedings **2647.** 030010. (2022). https://doi.org/10.1063/5.0104788

22. Rakhmonov, I.U., Niyozov, N.N. E3S Web of Conferences, 2019. **139.** 01077. DOI: 10.1051/e3sconf/201913901077

23. Rakhmonov I.U., Najimova, A.M., Esemuratova Sh.M., Koptileuov T.T. AIP Conference Proceedings **2647.** 070024. (2022). https://doi.org/10.1063/5.0104793

24. Rasulov A N, Usmanov E G, Bobojanov M K and Karimov R Ch. E3S Web Conf **139** https://doi:10.1051/e3sconf/201913901079

25. Rakhmonov I.U., Najimova A.M. AIP Conference Proceedings **2647**. 030011. (2022). https://doi.org/10.1063/5.0104791

26. Melikuziev M.V. AIP Conference Proceedings **2552.** (2023). 050021. https://doi.org/10.1063/5.0112395

27. Rasulov A.N., Usmonov E.G., Melikuziev M.V. AIP Conference Proceedings **2552.** 2023. PP, 050018, 1-4. https://doi.org/10.1063/5.0111530.

28. Rakhmonov I.U., Ushakov V.Ya., Najimova A.M., Jalilova D.A., Omonov F.B. E3S Web of Conferences **289.** 07013. (2021). https://doi.org/10.1051/e3sconf/202128907013

29. Melikuziev M.V., Nematov L.A., Novikov A.N., Baymuratov K.K. E3S Web of Conferences **289.** 07016 (2021). https://doi.org/10.1051/e3sconf/202128907016

30. Hoshimov F.A., Rakhmonov I.U., Niyozov N.N., Omonov F.B. AIP Conference Proceedings **2647.** 030025. (2023). https://doi.org/10.1063/5.0112388

31. Rakhmonov I.U., Hoshimov F.A., Kurbonov N.N., Jalilova D.A. AIP Conference Proceedings **2647.** 050022. (2023). https://doi.org/10.1063/5.0112391

32. Rakhmonov I.U., Ushakov V.Ya., Niyozov N.N., Kurbonov N.N., Mamutov M. E3S Web of Conferences **289.** 07014. (2021). https://doi.org/10.1051/e3sconf/202128907014

33. Taslimov A.D. AIP Conference Proceedings **2552.** (2023). 050023. https://doi.org/10.1063/5.0112398

Development and research of non-contact starting devices for electric consumers and motors

Makhmanazar Fayziyev[1], *Furkat* Tuychiev[2], *Ruslan* Mustayev[1]*, *Yunus* Ochilov[1]

[1]Karshi Institute of Engineering Economics, 225 Independence Avenue, Karshi, Uzbekistan

[2]Tashkent State Technical University named by Islam Karimov, Tashkent, 100095, Uzbekistan

Abstract. This article presents the analysed information about power consumers and electric motor starters in order to ensure the reliable operation of the power supply system, which is relevant today. Disadvantages of contact starters used for small power electric motors in the power supply system and advantages of non-contact starters are given. In addition, it is based on the creation of additional opportunities when using this device, that is, it creates opportunities for protection and automatic control. Contactless starters are recommended for some heavy-duty electric motors. The article describes these cases in details.

1 Introduction

Nowadays, in addition to the increase in the type of electric consumers, there is also an increasing demand for starting devices that can reliably connect them to the network. In order to ensure the reliable operation of the power supply system and to eliminate defects in contact devices, it is advisable to use starters that can eliminate new types of defects.

These starters allow reliable connection and disconnection of small power consumers and electric motors. It is advisable to use single-phase and three-phase consumers in places where there is a risk of fire, as a device for starting electric motors in the oil and gas processing industry, mining industry and residential buildings. These starters eliminate excessive noise and sparking (arcing) that occurs in contact systems.

Below are the advantages, disadvantages and special features of using this starter instead of a contact starter.

2 The current state of the investigated problem

Today, contact starters are used to start electrical devices and motors. In particular, these starting devices are used in the oil and gas industry. Examples of contact starting devices: automatic circuit breakers, magnetic starters, contactors, and electromagnetic relays [1]. In the contact part of these starting devices, during overloading and many times when reconnecting, the contacts are destroyed by the electric arc and are welded to each other. In turn, it can cause a fire hazard. Raw materials processed in the oil and gas industry and mining industry are often considered to be highly flammable products. Their processing technologies and electric drives used in the transportation of rocks cannot be imagined without launching devices. Contact parts of contact starters also have some requirements, but in the process of disconnecting many times, the contacts begin to lose their characteristics slowly [2]. In the experiment, we can see that during the operation of the magnetic starters, due to defects in their contacts, the circuit is not completely disconnected, which causes the electrical device to be under voltage. This can endanger the lives of service personnel. During the operation of contact starters, especially during start-up and cut-off, the moving mechanisms cause a certain amount of noise. As an example, we can consider the operation of elevators installed in high-rise residential buildings. Many contact actuation and control systems installed in control cabinets generate noise during elevator movement. This noise causes inconvenience to the apartments located near the control cabinet. One-speed or two-speed types of elevator electric drives are also very common. Two-speed elevator electric drives allow people in the elevator cabin to move without feeling overloaded and reduce the time it takes to get to their floor [3]. But a two-speed elevator requires more contact starters than a single-speed start when starting electric drives. In this

* Corresponding author: ruslanmustayev89@gmail.com

 https://doi.org/10.1051/e3sconf/202338401038

case, their working sequence increases and increases the amount of noise. Reducing the amount of these noises is one of the current problems [4, 5].

Contact starters consume significantly more power through their electromagnet during start-up than the power consumption in normal operation. We can also see this in Table 1 below.

The amount of noise in the enterprise increases due to the fact that electrical devices used in automated production enterprises work in different operating modes and are connected and disconnected many times in different situations. Some electrical consumers need to be disconnected from the network many times. It is impossible to perform these actions with contact starters, because they are limited to a certain number of disconnections [6].

For frequent disconnection, connection and disconnection times should be very short.

Table 1. Operation of magnetic starters and power consumption during start-up and start-up time.

Current rating of magnetic starters.	1	2	3	4	5	6
Power consumption of magnetic starter coils after starting W	68	87	200	280	350	530
Power consumption of magnetic starter coils after starting W	8	8,6	20	40	45	60
Start-up time of magnetic starters, ms	17	22	25	20	20	25

In the case of contact starters, they are made in 4 different versions.

1. Without inertia (0.001 s);
2. Fast-acting (0.001 ÷ 0.05 s);
3. Normal (0.05 ÷ 0.15 s);
4. Slow down (0.15 ÷ 9 s). [7,8].

To overcome the above shortcomings, it requires the development of single-phase and three-phase non-contact starting devices. Internet materials and educational literature were analyzed in preparation of these launchers. When analyzing the working principles and circuits of single-phase and three-phase starting devices, some shortcomings were noticed. For example, single-phase starters use only one non-contact starter unit. Such a starting device always requires the detection of phase and neutral conductors and the connection of a contactless starting device to a fixed phase conductor[9,10]. Always complying with this requirement will cause additional inconvenience to users. If the demand is not met, the consumer disconnects the neutral conductor from the network, and the phase pole remains connected to the consumer. In the starter that we offer, a non-contact starter unit is installed on both the phase and neutral conductors. Polarity reversal does not cause any defects. In any connection, the starting device can perform the operation of connecting and disconnecting the phase and neutral poles. [11, 12].

3. Single-phase and three-phase contactless starting device.

Currently, devices based on semiconductor elements (thyristor and triac) and various hybrid modules based on these elements are used to start and stop uncontrolled electric motors operating in alternating current. Their use shows a steady trend of transition from the usual contact switching of the stator circuits of asynchronous electric motors to non-contact starting devices in a number of drives. Non-contact starting devices are mainly made on the basis of semi-conductor elements and used in various circuits [13, 14]. Thyristors and triacs are mainly used to adjust the load current. When thyristors are used, an opposite-parallel circuit is used to connect and disconnect single-phase circuits. Two thyristors are used for each phase [15, 16].

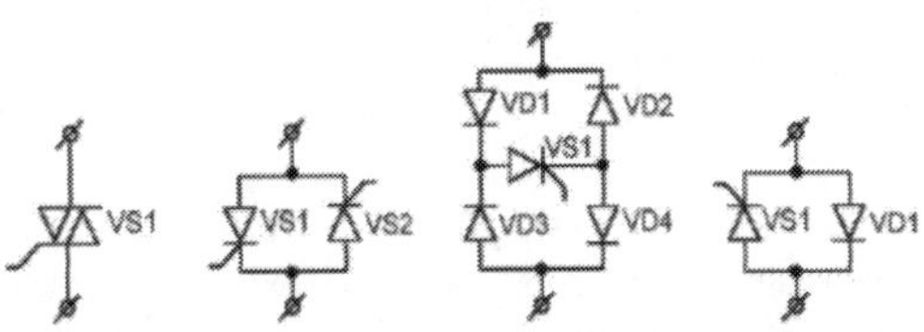

Fig. 1. Non-contact starters are semi-conductor elements of the power circuit

Three-phase starters use six thyristors. When using a circuit breaker, the elements are reduced by half, one for each phase. Basically, thyristor and triac control electrodes are provided by the phase currents to which they are connected. Only the voltage of this supply circuit and the current flowing through the circuit should be adjusted to the required value of the thyristor control electrode. For this purpose, circuits use resistance and semiconductor diodes that correct the type of current [17, 18]. In the scheme we propose, the current of the control circuit is taken from the phase, and only it is controlled by the microcontroller whether it is supplied to the control electrode or not. At startup, the microcontroller provides the required voltage to the optocoupler. In this case, the optocoupler opens and connects the necessary current circuit to the thyristor control electrode. The thyristor opens and disconnects from the consumer network. If the consumer needs to be disconnected from the network, a shutdown signal is given to the microcontroller. Then the microcontroller stops the voltage supplied to the optocoupler. The optocoupler stops working, which in turn cuts off the control circuit. The control circuit to the thyristor is cut off using an optocoupler, and the current supplied to the thyristor control circuit is stopped. The thyristor also separates the consumer from the network [19, 20].

Three-phase non-contact starters work similarly to single-phase non-contact starters. In this case, it is necessary to install three single-phase starting blocks for each phase. The block includes a thyristor or triac, an optocoupler, a voltage-reducing resistor, and a rectifying diode [21, 22]. The microcontroller device is common for all three phases. In a microcontroller device, the number of input and output ports differs from each other. The microcontroller we offer has 14 digital 6 analog input-output ports, with the help of which we can assemble a starting circuit that allows starting and reversing single-phase and three-phase electric motors, starting and reversing two-speed elevator electric motors [23].

4 Conclusion

In conclusion, we can say that contactless starting devices are required to be installed for electric motors and consumers used in a certain difficult operating mode. They eliminate dangerous situations and excessive noise. In addition, it can be used instead of a contactor in the set of reactive power compensators used in industrial enterprises. For smooth adjustment of reactive power, capacitor banks of different values of reactive power are connected and disconnected from the network in a certain logical sequence. In this case, the amount of reactive power can increase and decrease smoothly. Several contactors are used for this. Connecting and disconnecting them many times increases this risk in environments with high noise and fire sources. Therefore, single-phase and three-phase starters can overcome these shortcomings. It allows for easy automation of modern power system devices. It is also possible to create the possibility of starting by a remote control system with a microcontroller contactless starting device.

References

1. Mukhanova P. P. Actual questions of practical physics and energy SUMGAIT, Nov. 12-13, 2020.108-133 p.

2. ZAO "Uralelektro K" Cposob izgotovleniya electromagnitnogo kommutatsionnogo apparatus. Patent 2170978 C1, H01H 49/00, 20.07.2001 https://patents.s3.yandex.net/RU2170978C1_20010720

3. Bobojanov M. K. Abstract dissertation (1994).

4 *Usmanov E.G.* Departments and E. Faculty, "The noncontact switch for switching of the big capacities" European research **№ 4 (27)**, 2017 11-13 p.

5 V. Ts. P. Toka and O. Kamoliddin, "Udk 621.316.5 principle of constructing thyristor switches in AC circuits," pp. 131-133, 2016.

6 Klevtsov A.V. *Beskontaktnye stroystva puska i tormogeniya elektrodvigateley*. Uchebnoe posobie M: Infra-Engineering, 2018 -188 p.

7 Yu.K. Rozanov, M.V. Ryabchitsky, A.A. Kvasnyuk, *Silovaya elektronika* 2-e izdanie, stereotyopnoe Moskva Izdatelskiy dom MEI 2009.

8 V.T. Devices and A. Motors, "Contactless forcing circuits in braking devices of induction motors," pp. 31-36. https://rep.bntu.by/bitstream/handle/data/2638/31-36.pdf?sequence=1

9 M.K. Bobajanov, M.M. Fayziev, A. Abdurasulov, R.A. Mustaev, and S.E. Bobojanov, "Technical sciences," **vol. 14**, no. 92, P- 5–8, 2020.

10 M.K. Bobajanov, M.M. Fayziev, R.A. Mustaev, Babayev.O.E. et al. Applying the non-contact devices for starting a single-phase asynchronous electric motor // Bulletin of science and education. 2021. no. 11-2 (114). WITH.31-35p. https://scientificjournal.ru/images/PDF/2021/114/VNO-11-114-II-.pdf

11 M.K. Bobajanov, M.M. Fayziev, R.A. Mustaev, and Bozorov I.R. Application of non-contact starting devices three-phase asynchronous electric motor Science, technology and education 2021. No. 2 (77). Part 2 65-67 p. https://3minut.ru/images/PDF/2021/77/NTO-2-77-II-.pdf

12 M.K. Bobajanov, M.M. Fayziev, R.A. Mustaev. Non-contact devices for starting electric motors. Innovative technologies. Special issue, 2022, ISSN 2181-4732 11-13 p. https://ojs.qmii.uz/index.php/it/article/view/96

13. M.K.Bobojanov, S.Mahmutkhonov, and S.Aytbaev. Investigation of the Problems Non-Sinusoidal of the Voltage Form. AIP Conference Proceedings 2552, 050011, (2023), https://doi.org/10.1063/5.0113890

14. M.Bobojanov. Development and Research of Two Speed Motor with Pole-Changing Winding. AIP Conference Proceedings 2552, 050034, (2023), https://doi.org/10.1063/5.0114077

15. M.K.Bobojanov, R.Ch.Karimov, T.H.Qosimov, S.D.Zh.Dzhuraev. Development and experimental study of circuits of contactless device for automation of compensation of reactive power of capacitor batteries. E3S Web of Conferences, 289, 07012, (2021), https://doi.org/10.1051/e3sconf/202128907012

16. D.Rismukhamedov, M.Bobojanov, F.Tuychiev, K.Shamsutdinov. Development and research of pole-changing winding for a close pole ratio. E3S Web of Conferences, 264, 03057, (2021), https://doi.org/10.1051/e3sconf/202126403057

17. M.Bobojanov, D.Rismuxamedov, F.Tuychiev, K.Shamsutdinov, K.Magdiev, Pole-changing motor for lift installation. E3S Web of Conferences, 216, 01164, (2020), https://doi.org/10.1051/e3sconf/202021601164

18. R.Karimov, M.Bobojanov. Analysis of voltage stabilizers and non-contact relays in power supply systems. E3S Web of Conferences, 216, 01162, (2020), https://doi.org/10.1051/e3sconf/202021601162

19. R.Karimov, M.Bobojanov, N.Tairova, ... A.Egamov, N.Shamsiyeva. Non-contact controlled voltage stabilizer for power supply of household consumers. IOP Conference Series: Materials Science and Engineering, 883(1), 012120, (2020), DOI 10.1088/1757-899X/883/1/012120

20. M.K.Bobojanov, O.E.Ziyodulla, M.T.U.Ismoilov, E.I.U.Arziev, G.Z.Togaeva. Study of the efficiency of conveyors of mining transport systems of mining complexes. E3S Web of Conferences, 177, 03023, (2020), https://doi.org/10.1051/e3sconf/202017703023

21. R.C.Karimov, M.K.Bobojanov, A.N.Rasulov, E.G.Usmanov. Controlled switching circuits based on non-linear resistive elements. E3S Web of Conferences, 139, 01039, (2019), https://doi.org/10.1051/e3sconf/201913901039

22. E.G.Usmanov, A.N.Rasulov, M.K.Bobojanov, R.C.Karimov. Non-contact voltage relay for switching windings of a boost transformer. E3S Web of Conferences, 139, 01079, (2019), https://doi.org/10.1051/e3sconf/201913901079

23. Kh.G.Karimov, M.K.Bobozhanov. New pole-changing windings of asynchronous motors. Elektrichestvo, 1, pp. 27-32, (1996).

Analysis of payment priority for electricity consumed in industrial enterprises on the base of classified tariffs

Makhmanazar Fayziyev[1], *Yunus* Ochilov[1], *Komoliddin* Nimatov[1], *Ruslan* Mustayev[1] *

[1]Karshi Institute of Engineering Economics, 225 Independence Avenue, Karshi, Uzbekistan.

Abstract. The following issues are considered and resolved in the article: the advantage of using tariffs stratified by time to increase the efficiency of electricity consumption by industrial enterprises; Analysis of the tariff system in the electric power system of the Republic of Uzbekistan and the essence of the effectiveness of the application of differentiated tariffs; In order to justify the application of time-differentiated tariffs, the electricity consumption of the industrial enterprise is analyzed; parameters of electricity consumption are optimized according to different tariffs. In order to regulate the load schedule of the energy system using time-graded tariffs in the morning and evening "peak" periods, a method of reducing the load in "peak" periods is proposed in exchange for increasing energy consumption in the night periods.

1 Introduction

The population of Uzbekistan is increasing year by year, and today we are among the countries with a rapidly growing economy. Therefore, the demand for electricity and gas will continue to increase. In the new tariff reform of the energy network, every citizen of Uzbekistan will switch to a fair payment system for electricity and gas. The draft decision of the Cabinet of Ministers of the Republic of Uzbekistan "On changing the prices of fuel and energy resources" was submitted for public discussion. Based on the international experience and the interests of the poor and low-income society of the population, it is decided to implement the "social norm" system throughout the republic in order to increase the effectiveness of social protection. With the draft decision, starting from July 1, 2022, the prices for the republic were set based on the social norm, that is, two or more payments will be made for gas and electricity used in excess of the specified amount.

2 Current status and relevance of the problem being covered

The tariffs system directly regulates the relations between electricity producers and suppliers and electricity consumers. The 1, 2 and 3-shift work mode of many production enterprises, the electric energy system of the enterprise is in the operating mode at the peak of the electric energy consumption, which leads to stress [1]. The average electricity consumption during the night shift is 20-25% of the electricity consumption during the day shift. This, in turn, shows that electricity consumption is low at night and sharply increases during the day. One of the main problems of enterprises is to adjust load schedules for normal and cost-effective energy consumption [1]. The introduction of time-differentiated tariffs into the energy consumption regimes of production enterprises makes it possible to reduce the excess of energy consumption in "peak" periods. Load periods are introduced into the energy system of time-graded tariffs to reduce consumption in "peak" periods. Day and night tariff system with two time periods (zones) is used for two periods of the day [1, 2].

Table 1. Seasonal periods of electricity consumption and time-differentiated tariffs

The seasons	Three-period residents need electricity			
	Morning peak	Daytime period	Evening peak	Night period
Winter time	From 06:00 until 10:00	From 10:00 until 18:00	From 18:00 until 22:00	From 22:00 until 06:00
	675 soums	520 soums	675 soums	250 soums
Summer time	From 05:00 until 9:00	From 9:00 until 17:00	From 17:00 until 21:00	From 21:00 until 05:00
	675 soums (0,614 $)	520 soums (0,472 $)	675 soums (0,614 $)	250 soums (0,227 $)

Taking into account the above tariff periods, the calculation of tariff periods stratified by time for electricity for three periods (zones) of the day is expressed on the basis of the following equation [1]:

$$ES_{orr} = S^p E^p + C^{yap} E^{yap} + C^T E^T \quad (1)$$

* Corresponding author: ruslanmustayev89@gmail.com

https://doi.org/10.1051/e3sconf/202338401039

in which, So'r is the approved single-rate or average fixed tariff price for consumers; E – total power consumption of consumers; S^p, C^{yap}, C^T – A- "peak", "semi-peak" and night prices of daily electricity consumption; E^p, E^{yap}, E^T – A - peak, semi-peak and night time electricity consumption [1, 2].

It can be seen that the consumer's demand for electricity is in two different orders on working days and on weekends, and based on this situation, it is recommended to use a two- and three-period tariff system [3, 4]. A reduction of the payment for the electricity consumption at night by 1.5 times compared to the amount of payment for the electricity consumed during the day in residential premises will make it possible to smoothen the schedule of loads in the electric power system .

All wastes and energy consumption related to energy consumption in the classified electric balance of production enterprises belong to the main technological process. But the organizational structure of the electrical balance is much deeper and originates from the period of energy spent by each consumer on production [2].

The main part of the electrical balance of production enterprises is that most of the waste of electricity is in the network and transformers. Waste in the network cannot be determined by measurement, so the amount of waste is determined by calculation.

Daily electricity consumption modes of production enterprises are observed depending on the periods of the day with the help of consumption graphs. Daily load graphs provide the ability to regulate the operation of the enterprise's electrical equipment and analyze it for subsequent periods. It will also help to develop measures to clarify electricity consumption regimes in the near future.

The time-graded tariff load schedule is calculated based on the maximum and minimum hours of electricity consumption during the day and is used to determine the capacity. The graph of electricity consumption of production enterprises is characterized by a certain change of electricity consumption during the day [2].

The most complete monitoring period of electrical energy-consuming equipment in enterprises helps to work with accurate calculations, work at a uniform level of energy consumption, and improve the rhythm of production. Regulation of electricity consumption in production enterprises is one of the most important tasks for the electric power system: first of all, it is responsible for the rational use of fuel in the production of electricity, and it increases the reliability of the equipment at power plants.

A number of new measures on the rational use of energy are being accepted in the republic's electricity market. These measures are important for the renewal and modernization of the energy sector, as well as for encouraging the rational use of electricity by energy consumers. As a result of the use of time-graded tariffs depending on the volume of consumed electricity in the republic's electricity market, the possibility of technically and economically efficient use of electricity was created.

In the energy system of the republic, the information about the tariff differentiated by time is being explained to the consumers. As a result, consumers will be aware of the established tariff system. As a result, consumers who have a comprehension of tariffs will have the understanding that depending on the tariff periods, they will benefit greatly from switching to tariffs that are convenient for them [1, 2].

In order for consumers to use energy resources wisely, it will be necessary to conduct a lot of explanatory work on such tariffs [1, 2]. Otherwise, consumers will have a negative concept of the energy system. The demand for electricity consumption is formed on a daily, weekly, monthly, seasonal and annual basis. In addition, during the period of maximum loads ("peak"), the level of electricity consumption is much higher than the average load in the corresponding period [3, 6]. The capacity of the power system is determined by the period of maximum loads, the constant consumption of electricity to cover the period of maximum loads and to ensure the necessary power reserve [2, 7]. This, in turn, leads to the application of a system of time-graded tariffs [3]. The purpose of introducing time-differentiated tariffs is to move electricity consumption to low-load, night hours and encourage consumption during this time. In practice, the experience of foreign countries shows that several tariff systems are used to encourage the reduction of electricity demand during "peak" hours.

3 Application of 2 topics at the scale of industrial enterprises

The introduction of time-differentiated tariffs into the energy consumption modes of production enterprises makes it possible to reduce the excess of energy consumption in "peak" periods. Load periods are introduced into the energy system of time-graded tariffs to reduce consumption in "peak" periods. Day and night tariff system with two time periods (zones) is used for two periods of the day [4, 5, 6].

The consumer is ready to change electricity tariffs at any time, only when it is economically beneficial for the supply organizations and consumers. Nevertheless, the transition to time-graded tariffs requires solving serious legal issues. Determining and approving the norms of electricity consumption periods in production enterprises should be socially based[1, 11].

Consumers need to install electronic meters suitable for tariff periods (zones) that carry out accurate calculations of electricity consumption. Electronic meters that make accurate calculations will further strengthen the relationship between the power plant company and consumers [2, 12, 14]. Consumers who have switched to the differentiated tariff system, where the indicators of one-, two- and three-period (zone) tariff electricity meters are indicated in Appendix 8, can calculate their electricity consumption using two- or three-period (zone) tariff meters. they can make a book [1, 11, 13].

 https://doi.org/10.1051/e3sconf/202338401039

The electricity consumed by "NISHAN TEXTL" LLC, Karshi city, Kashkadarya region, was calculated and compared according to the same energy tariff and differentiated electricity tariffs, and the following result was obtained.

The annual maximum and minimum electricity consumption during the period of time until March 2021 in the process of electricity production of "NISHAN TEXTL" LLC was analyzed.

The máximum daily electricity consumption in the organization is $P_{his} = 178\,kW$, normal tariff (450 soums).During the day, the máximum amount of electricity $t_{day} = 16\,hours$ is observed, and during, the working day $n_{year} = 215$ the day

$$W = P_{his}t_{day}n_{year} = 178 * 16 * 30 = 85440\,kW \quad (2)$$

When the máximum monthly electricity of the Enterprise is found and calculated according to the normal tariff rate (450 soums),it is as follows

$N = W * \alpha = 85440 * 450 = 38448000$ soums (3495 $

Table 2. Power demand and prices of consumers by periods

Tariff times	Monthly concumtion in kw*h	Consumption share, %	Tariff prices,soums/kW*s	Costs are soums
From 06 Up to 10	10253	0,12	675	6920775
From 10 until 18	43575	0,51	520	22659000
From 18 until 22	6835	0,08	675	4613625
from22 until 06	28195	0,39	250	7048800
General:	85440			41242200 (3749 $)

Currently, the enerprise is considered a class II consumer and pays 450 soums for 1 kW*h of electricity it receives from the system.

85440*450=38448000 soums (3495 $)

If the tariff system is not used correctly,the Enterprise Will be charged monthly for the day of electricity consumption:

41242200-38448000=2794200 soums (254 $),Will suffer.

The capacity of the power system is determined by the period of maximum loads, the constant consumption of electricity to cover the period of maximum loads and to ensure the necessary power reserve. This, in turn, leads to the application of a system of time-graded tariffs. The purpose of introducing time-differentiated tariffs is to move electricity consumption to low-load, night hours and encourage consumption during this time.

After the enterprise switches to a differentiated tariff, it is necessary to distribute according to the provided capacities [2, 3].

Table 3. The state of energy consumption after the transition to the system of differentiated tariffs

Tariff times	Monthly consumption in kW*h	Consumption share, %	Tariff price, soum/kW*s	Costs are soums
From 06 until 10	9398	0,11	675	6343650
From 10 until 18	23923	0,28	520	12439960
From 18 until 22	7690	0,09	675	5190750
From 22 until 06	44429	0,52	250	11107200
General :	85440			35081560 (3190 $)

In the process of determining the electricity consumption of "NISHAN TEXTL" LLC, the maximum and minimum electricity consumption in one month of 2022 was analyzed. The following result was obtained when the maximum daily electricity consumption at the enterprise (June 1, 2848 kW·h) was calculated at the sum of the normal tariff (450 soums, as of 2022) in the 16-hour work mode:

85440*450=38448000 soums (3495 $)

During the day of electricity consumption, if the enterprise uses the tariff system correctly, for one month: *38448000-35081560*=3366440 (306 $) soums, it will benefit.

4 Conclusion

Electricity consumption costs of the enterprise were calculated using current tariffs and compared with differentiated tariffs.

- It has been proven that 3366440 soums (around 306$) can be profited for the benefit of the enterprise every month in "NISHAN TEXTL" LLC with the correct use of differentiated tariffs.

If the differentiated tariff system is used continuously, this indicator will increase.

- Electricity consumption by consumers was carried out according to the indicators of the installed metering devices according to the normal tariff.

The dependence of the time-differentiated tariff system on EENHAT for calculating electricity consumption is studied. As a result, it was possible to increase the accuracy of the calculation of the cost of electricity consumption for each period.

References

1. Kadirov.K.Sh scientific dissertation "Application of time-differentiated tariffs to increase energy efficiency of consumers" Tashkent 2019 Pages 21-22.
2. M.K.Bobojanov, Y.O.Ochilov conference "PROBLEMS OF ENERGY AND SOURCES SAVING" "Using Of Differentiated Electricity Tariffs For Residential Buildings"
3. M.M.Fayziyev, M.K.Bobojanov, Y.O.Ochilov conference «INNOVATIVE TECHNOLOGIES» «For electricity efficiency of payment of payments based on differentiated tariffs analysis»
4. Kadyrov K.Sh. Academy of Sciences of the Republic of Uzbekistan, To the Communications and Informatization Agency of Uzbekistan, Informatics and Journals of energy problems of Uzbekistan. 2014, **No. 6**, pp.93-97.
5. Markovich I.M. Modes of energy systems. - M.: Energy.1969.S-351
6. M.K.Bobojanov, S.Mahmutkhonov, and S.Aytbaev. AIP Conference Proceedings **2552**, 050011, (2023), https://doi.org/10.1063/5.0113890.
7. M.Bobojanov. AIP Conference Proceedings **2552**, 050034, (2023), https://doi.org/10.1063/5.0114077.
8. M.K.Bobojanov, R.Ch.Karimov, T.H.Qosimov, S.D.Zh.Dzhuraev. E3S Web of Conferences, **289**, 07012, (2021), https://doi.org/10.1051/e3sconf/202128907012.
9. D.Rismukhamedov, M.Bobojanov, F.Tuychiev, K.Shamsutdinov. E3S Web of Conferences, **264**, 03057, (2021), https://doi.org/10.1051/e3sconf/202126403057.
10. M.Bobojanov, D.Rismuxamedov, F.Tuychiev, K.Shamsutdinov, K.Magdiev. E3S Web of Conferences, **216**, 01164, (2020), https://doi.org/10.1051/e3sconf/202021601164.
11. R.Karimov, M.Bobojanov. E3S Web of Conferences, **216**, 01162, (2020), https://doi.org/10.1051/e3sconf/202021601162.
12. M.K.Bobojanov, O.E.Ziyodulla, M.T.U.Ismoilov, E.I.U.Arziev, G.Z.Togaeva. E3S Web of Conferences, **177**, 03023, (2020), https://doi.org/10.1051/e3sconf/202017703023.
13. R.C.Karimov, M.K.Bobojanov, A.N.Rasulov, E.G.Usmanov. E3S Web of Conferences, **139**, 01039, (2019), https://doi.org/10.1051/e3sconf/201913901039.
14. E.G.Usmanov, A.N.Rasulov, M.K.Bobojanov, R.C.Karimov. E3S Web of Conferences, **139**, 01079, (2019), https://doi.org/10.1051/e3sconf/201913901079.

Reactive power compensation for sustainable development of power grids in cities of Uzbekistan

Saidkhodjaev *A.G.*[1]*, Bobojonov *Y.M.*[2]

[1]Tashkent State Technical University named by Islam Karimov, Tashkent, 100095, Uzbekistan

[2]Karakalpak State University, Nukus, 230101, Uzbekistan

Abstract. The article deals with relevant and much needed at this stage problems of reconstruction and modernization of urban power supply systems and issues of energy saving in urban power grids. For the first time, it is proposed to carry out energy saving in the residential and public sectors of the residential zone of cities on the basis of reactive power compensation. There are given an analysis of experimental studies and measurements of power consumption and based on them recommendations and ways of energy saving in urban power grids.

1 Introduction

Currently, special attention is paid to improving energy efficiency, energy saving and rational power supply to consumers in cities. If earlier electricity consumption in cities for household needs was 11%, then due to the rapid construction of housing and communal facilities in cities, electricity consumption in 2022 increased to 40% of all usefully supplied electricity. One of the important problems of urban power supply is the use of reactive power compensation in urban networks. Due to the fact that the small-motor load has increased and the power factor $cos\varphi$ has decreased. As a result, the consumption of reactive power increased, and the losses of electricity increased. To solve this problem, it is necessary to install multifunctional meters at each transformer station (TP) and, depending on the amount of reactive power consumed, calculate the necessary automated installations of capacitor banks [1-4, 6].

2 The current state of the investigated problem

The task of calculating the mode of urban electrical networks can be formulated as follows: it is necessary, as accurately as possible, to determine the maximum (calculated) load of the same type, each individual consumer and the total total load of consumers included in this electrical network, to reduce to a possible minimum the loss of electrical energy (which depends on many factors and therefore have different meanings) in the transmission and distribution of given flows of active and reactive power, provided that the established quality of electricity is ensured and reliability requirements are met, while using a new patented method for determining the total design load and choosing the optimal number and power of the district heating substation [5, 7-13, 16].

Consider the energy saving and energy efficiency of low- voltage urban electrical networks based on reactive power compensation (using the example of power supply to consumers in a microdistrict) [14, 15].

At present, due to the growth of the well-being of the population, an increase in the number of and power of household electrical appliances, the consumption of active electricity and the consumption of reactive energy increases, since the small- motor load of household electrical appliances and fluorescent LED lamps have a low Cos φ, this leads to additional losses on reactive power. In this regard, the governmental inspection "Uzenergoinspection" adopted $cos\varphi$ =0,85. Therefore, the time has come to apply reactive power compensation in urban electrical networks and reduce electricity losses in low-voltage networks by installing compensating devices at urban transformer substations and large utility power consumers.

Based on the measurements of daily power consumption regimes and technical and economic calculations, the following results were obtained:

* Corresponding author: ilider1987@yandex.ru

Table 1. Comparative table of loads determined by the new method and after reactive power compensation [17]

TP	Load determined by the new method			Load determined after compensation			Difference		
	P, MVt	Q MVar	S MVA	P MVt	Q MVar	S MVA	P MVt	Q MVar	S MVA
Total	3,03	1,96	3,72	3,03	0,93	3,18	0.0	-1,03	-0,54
TP1	0,58	0,36	0,68	0,58	0,21	0,61	0.0	-0,15	-0,07
TP2	0,61	0,38	0,7	0,61	0,18	0,63	0.0	-0,2	-0,07
TP3	0,76	0,47	0,89	0,76	0,2	0,79	0.0	-0,27	-0,1
TP4	0,4	0,25	0,47	0,4	0,098	0,41	0.0	-0,15	-0,06
TP5	0,39	0,24	0,46	0,39	0,14	0,41	0.0	-0,1	-0,05
TP6	0,43	0,27	0,51	0,43	0,09	0,44	0.0	-0,18	-0,07

In the general case, for a microdistrict, a decrease in active energy consumption due to an increase in cosφ [2].

$$\Delta W_{hc} = \left\{\left[{}^{1}/_{cos^2\varphi_1} - {}^{1}/_{cos^2\varphi_2}\right] / {}^{1}/_{cos^2\varphi_1}\right\} \cdot K_p \quad (1)$$

Where with $cos\varphi_2$ – before compensation (0.85) ;
$cos\varphi_2$ – after compensation (0,95) ;

The relative active component of the current (coinciding in phase with voltage) will be taken equal to unity.

Relative full current is before implementations:

$$I_1=1/0,85=1,176\ A$$

Relative full current is after implementations:

$$I_2=1/0,95=1,052\ A.$$

Energy savings from reactive power compensation at one transformer substation amounted to 21040,2 kWh, current reduction by 10-13 %, line capacity increases and TP by 3-6 %, there is an increase $cos\varphi$ up to standard, reduction of electricity losses, saving of non-ferrous metal and transformer power.

Table 2. Results of measurement and calculation at a transformer substation

Time	Before compensation				After compensation			
	RU- 10kV f-r Bakhor1 RP Dagestanskaya 10kV Assets	RU- 10kV f-r Bakhor -1 RP Dagestan Skye 10 kV reagent	RU- 0,4 kV input T-2 TP – 2280 Assets	RU- 0,4 kV T-2 input TP- 2280 Reactive	RU- 10kV f-r Bakhor-2 RP Dagestanskay a10 kV Assets	RU- 10kV f-r Bahor RP Dagestan 10kV Reagent	RU- 0,4 kV input T- 2 TP-2280 Active	RU- 0,4 kV T-2 input TP- 2280 Reactive
00:00	115,03	56,57	108,80	41,43	108,73	15,13	103,42	4,71
00:30	106,20	56,83	100,00	40,17	100,20	15,00	95,56	4,44
01:00	99,86	56,23	94,00	39,89	94,27	15,53	89,56	4,76
01:30	93,86	55,89	87,94	39,83	89,20	15,80	84,67	3,69
02:00	88,54	55,63	83,03	39,66	83,13	14,80	79,02	3,51
02:30	85,89	54,43	80,51	38,97	79,93	14,27	75,96	3,38
03:00	84,26	54,69	78,80	39,49	78,00	13,87	73,56	3,20
03:30	83,40	54,51	77,89	39,71	76,67	14,13	72,89	3,64
04:00	82,03	55,03	76,63	39,71	76,13	14,13	72,04	3,29
04:30	81,51	54,43	76,06	39,26	75,07	14,40	71,24	3,16
05:00	79,89	54,09	74,57	38,91	74,60	13,53	71,02	3,47
05:30	80,14	53,66	74,80	39,09	74,93	14,60	71,29	3,38
06:00	81,51	53,83	76,00	38,74	74,87	13,60	71,06	3,47
06:30	72,86	45,51	68,17	29,94	75,47	12,40	70,97	4,09
07:00	69,77	41,49	64,74	22,29	54,67	9,60	51,52	5,23
07:30	71,06	40,54	66,34	20,86	59,93	28,00	57,08	15,33
08:00	74,14	34,89	69,37	22,23	64,40	27,40	61,25	15,87
08:30	68,83	35,74	64,29	23,49	65,47	28,87	62,28	15,96
09:00	66,69	34,63	62,06	23,09	67,27	26,47	63,94	16,93
09:30	66,26	33,60	61,77	21,31	68,73	27,20	65,20	15,07
10:00	69,94	32,91	65,49	22,06	70,47	25,73	66,62	14,84
10:30	70,03	33,17	65,71	23,20	74,20	22,60	70,67	12,97
11:00	71,57	33,86	67,00	24,51	78,67	22,87	74,98	15,33
11:30	66,60	35,91	61,89	25,37	78,60	23,20	75,16	13,32
12:00	70,71	38,06	63,37	26,29	79,53	22,13	75,87	13,33
02:30	69,51	39,00	63,77	27,37	78,87	21,53	75,07	14,38
13:00	72,26	40,11	64,29	26,57	79,07	23,73	75,60	15,56
13:30	71,31	37,37	63,09	25,60	80,67	23,00	76,76	15,02
14:00	70,97	38,06	65,71	25,89	79,80	20,27	76,22	12,84
14:30	70,89	36,86	65,49	26,29	77,93	20,33	74,09	14,76
15:00	73,97	36,77	67,26	25,77	79,53	20,93	76,09	12,52
15:30	74,49	37,54	69,43	26,11	80,93	21,93	76,93	13,82
16:00	76,46	36,69	71,54	25,89	80,93	20,47	77,16	13,60
16:30	78,86	37,46	72,29	26,51	83,33	20,67	79,47	14,02
17:00	81,86	37,71	74,97	27,09	85,80	22,33	81,87	15,42
17:30	88,03	40,29	81,37	29,71	90,40	22,20	86,58	14,67
18:00	96,34	41,23	88,11	29,14	99,33	24,33	95,07	17,24
18:30	107,57	41,74	99,66	30,46	107,93	23,60	103,64	18,1
19:00	135,34	52,89	127,09	41,20	134,53	24,27	129,02	12,53
19:30	148,71	63,26	140,46	48,17	139,47	20,93	133,64	9,82
20:00	148,54	61,54	140,23	46,40	137,73	18,60	132,22	8,13
20:30	147,00	58,54	139,54	44,51	135,40	17,73	129,60	5,73
21:00	146,66	58,11	139,54	44,00	135,20	15,07	129,51	6,18
21:30	144,94	57,43	137,09	43,94	132,80	14,73	127,38	6,18
22:00	141,34	57,94	134,11	42,91	130,87	14,93	125,38	5,73
22:30	137,49	56,57	130,40	42,69	125,60	14,67	120,09	5,29
23:00	131,06	57,43	124,29	42,80	119,73	14,53	114,76	4,89
23:30	121,54	56,57	114,69	41,31	113,67	14,13	108,44	4,67

The widespread use of multifunctional "smart" meters ASKUE allows you to automate the whole process of calculating and consuming electricity. The use of capacitor banks at the TP of the microdistrict to compensate for reactive power, as studies have shown, has saved 52,73 million sums, while the payback period for invested money was less than 2 years. Due to the increase in the consumption of reactive energy in the residential and public sectors, it is currently necessary to installation in all transformer paragraphs or large consumers of automated capacitor banks for reactive power compensation [5. 16].

The economic effect from the introduction of an automatic capacitor plant consists of the following components: savings on losses and due to the rational control of reactive power.

Consider economic component work, compensating installation on the example of a microdistrict of the city.

Before implementation automatic condenser installations $cos\varphi$=0,85.

After implementation automatic condenser installations $cos\varphi$=0,95.

In Uzbekistan and in Tashkent city, with the support of IFIs were implemented Automatic metering systems – AIM, with the extensions. On the basis of an investment projects, the reconstruction and modernization of 0,38,6-10 kV electrical networks, including in the city of Tashkent, began in the cities of Uzbekistan. When processing experimental studies, the Fisher criterion and its special cases - Student's criterion, as well as statistical and regression analyzes of natural measured data were applied [7]. A comparison of the options for the proposed method for calculating electrical loads and the standard one (according to KMK 2017 - 2019) showed that the difference in electrical loads is 21,6%, the proposed new method is more accurate, this saves electricity, transformer power, reduces electricity losses, reduces losses voltage and power, and meets the energy-saving program.

3 Conclusion

Measurements and analysis of electricity consumption in the residential and public sector showed that the calculation of the load using the new methodology saves 21,6%.

Savings due to the use of automated condenser units for our experimental microdistrict in monetary terms amounted to 52,25 million sums

The use of AIM in urban electrical networks made it possible to reduce the balance difference and saves 8÷12%.

Carrying out a number of organizational and technical measures in urban low-voltage networks will reduce electricity losses and solve the issue of energy saving.

References

1. Saidkhodjaev A.G., Saidkhodjaeva M, Modernization of Energy Sector Myths or Reality web publication, IEAA / International Association for Energy Economics, Ostober 2011. Washington. DC USA.
2. A.G. Saidkhodjaev 1, Yu Bobojonov 2, UM Turdiev 3. Perspectives of electric supply of municipal housing consumers in cities of Uzbekistan. International Journal of Advanced Science and Technology. (Scopus). Vol. 29, no. 11s, (2020), pp. 1560-1564 ISSN: 2005-4238 IJAST Copyright 2020 SERSC 1560.
3. A.G. Saidkhodjaev, B. Kh. Ametova, M.M. Mamutov. E3S Web of Conferences **139.** 01072 (2019).
4. A.G. Saidkhodjaev, A.Najimova, A.Bijanov. Method for determining the maximum load of consumers in city power supply systems. Published by EDP Sciences (Scopus) E3S Web of Conferences 22 01072 (2019).
5. Rakhmonov I.U., Najimova A.M., and Reymov K.M. AIP Conference Proceedings 2647. 030010. (2022). https://doi.org/10.1063/5.0104788
6. Rakhmonov I.U., Najimova A.M. AIP Conference Proceedings 2647. 030011. (2022). https://doi.org/10.1063/5.0104791
7. Rakhmonov I.U., Najimova, A.M., Esemuratova Sh.M., Koptileuov T.T. AIP Conference Proceedings 2647. 070024. (2022). https://doi.org/10.1063/5.0104793
8. Hoshimov F.A., Rakhmonov I.U., Niyozov N.N., Omonov F.B. AIP Conference Proceedings 2647. 030025. (2023). https://doi.org/10.1063/5.0112388
9. Rakhmonov I.U., Hoshimov F.A., Kurbonov N.N., Jalilova D.A. AIP Conference Proceedings 2647. 050022. (2023). https://doi.org/10.1063/5.0112391
10. Rakhmonov I.U., Ushakov V.Ya., Niyozov N.N., Kurbonov N.N., Mamutov M. E3S Web of Conferences 289. 07014. (2021). https://doi.org/10.1051/e3sconf/202128907014
11. Rakhmonov I.U., Ushakov V.Ya., Najimova A.M., Jalilova D.A., Omonov F.B. E3S Web of Conferences 289. 07013. (2021). https://doi.org/10.1051/e3sconf/202128907013
12. Saidkhodjaev A.G. Shaharlarning electr taminoti muammolari va rivozhlanish istiqbollari. Monograph. - T.: Oz RFA "Fan" 2012. -256 b.
13. Saidkhodjaev A.G., Eshov A.T. A software package for developing methods for calculating and optimizing the modes of electrical loads in urban power supply systems. Certificate of official registration of the computer program DGU №02278 11.12. 2019.

14. Saidkhodjaev A.G., Khamrokulov U.Sh. Computer program "Methods for calculating energy saving in industrial and urban electrical networks." Certificate of official registration of the computer program DGU №. 05994 23.10. 2019.

15. Saidkhodjaev A.G., Nuriddinova Kh.R. DGU №20517 12.16.2022.

16. Saidkhodjaev A.G., Melikuziev M.V., Kadraliev B.B. DGU №19698 11/29/2022.

17. Saidkhodjaev A.G. A new method for determining electrical loads based on interval analogues of interpolation splines. Magnitogorsk. FGBOU VPO MSTU. Journal. "Automated technologies and production" 2015. № 2 (8) 28-30 p.

 https://doi.org/10.1051/e3sconf/202338401041

Influence of the consumer to power quality at the point of connection

Makhsud Bobojanov[1*], *Sultonkhuja* Mahmutkhonov[1]

[1]Tashkent State Technical University named by Islam Karimov, Tashkent, 100095, Uzbekistan

Abstract. This article is devoted to the development of a method for assessing the impact of consumers on the deterioration of the quality of electricity in the power supply system by using equivalent circuits. When developing a method for assessing the impact of the consumer on the quality of electricity, the main factors from the electromagnetic compatibility criteria were taken into account. A method has been developed for assessing the work of a consumer in terms of voltage quality at a common connection point, which was verified using the equivalent circuit method, using the developed method, it is possible qualitatively and quantitatively assess the impact of each consumer on the emergence unacceptable voltage higher harmonic.

1 Introduction

The development of new technologies has led to increased requirements for the electromagnetic environment. The requirements of (standards) GOST 13109-67 from 1967 for power quality were revised in 1987 due to the development of semiconductor technology and revised again in 1997 due to the development of microprocessor technology [1,3].

Quality indicators of electric energy are determined according to "Electromagnetic compatibility of technical equipment of electric energy, quality norms of electric energy in general electric supply systems" and International Standard 32144-2013 [1- 5].

Consumers of electric energy can fully perform the tasks assigned to them only under certain conditions. The parameters determining such conditions are called power quality. Deviation of quality marks in any direction causes inefficient use of electricity. It also causes less effective use of electrical devices and equipment and less output.

Non-compliance with the requirements of the PQ leads to significant material damage associated with a decrease in the service life and failure of electrical equipment, an increase in electricity losses and the appearance of defective products, improper operation of relay protection devices [3-15].

2 Method of assessing the impact on the disorder to power quality from consumers at the point of connection

A common feature of the methods of assessing the impact of consumers on power quality disturbances is that the voltage quality disturbances at the node of the electric network are considered a priori to include the share contribution of individual consumers. In other words, the violation of the quality of voltage is addictive. For such systems, the principle of superposition is valid, i.e. the reaction of the system to any combination of external influences is equal to the sum of the reactions to each of these influences separately.

Additivity is the value of a quantity that corresponds to the whole object is a property of mathematical or physical quantities consisting of the sum of the values of the quantities corresponding to its parts. It can be seen that the concept of the share contribution of the consumer can be called additive only if it is proved that the power quality violation has an additional property.

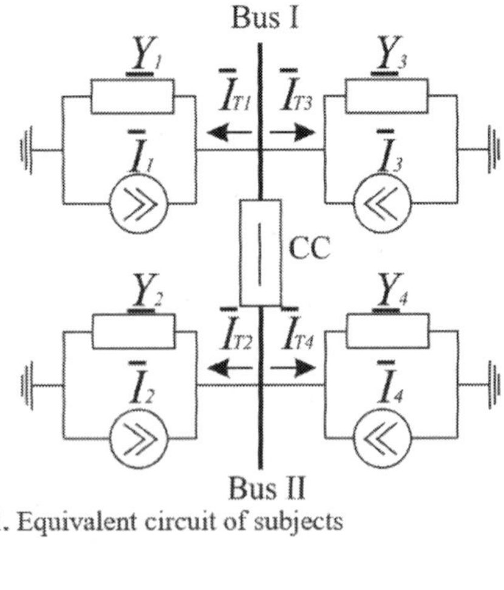

Fig. 1. Equivalent circuit of subjects

* Corresponding author: Sultonmk@mail.ru

As mentioned above, in order to determine the influence of non-sinusoidal and non-symmetry of the voltages at the point of connection of consumers in the power supply system, the switching scheme of each object is used in the studied harmonic or in the active bipolar form for reverse sequence currents with the main frequency. In the case shown in Fig.1, 4 consumers are connected to a point of connection. Here, all parameters are assumed to be related to the studied higher harmonic or the reverse sequence of the fundamental frequency.

I_1, I_2, I_3, I_4- sources of current quality disturbance (harmonic current or reverse sequence current) set for the average interval set in the switching scheme; Y_1, Y_2, Y_3, Y_4- power supply system conductivities. In the power supply system, linear (symmetric) consumers are represented only by conductance in the switching scheme, and a non-linear or symmetrical consumer is represented as a current source. Using the example of the equivalent circuit in Fig. 1, to prove the additivity or non-additivity of the voltage quality violation, it is necessary to compare the voltage quality violation UPOC when all consumers work together with the sum of the voltage quality violations U_{1SS} and U_{2SS} created by consumers. If the section breaker (SB) is off, a comparison is made on each of the two bus sections.

This voltage is found from the following formula:

$$U_{POC} = \frac{I_1 + I_2 + I_3 + I_4}{Y_1 + Y_2 + Y_3 + Y_4};$$
$$U_{1SB} = \frac{I_1 + I_3}{Y_1 + Y_3}; \qquad (1)$$
$$U_{2SB} = \frac{I_2 + I_4}{Y_2 + Y_4};$$

The sum of U_{1SS} and U_{2SS} does not give the voltage value U_{POC} at the point of connection when a sectional breaker is added.

The obtained result shows that the disturbance of voltage quality is not addictive. Therefore, it is wrong to use concepts about the contribution or share of the consumer in the violation of voltage quality. The method of determining the share contribution is imperfect because it does not correspond to the processes that occur in the network at high harmonics and cannot be used to assess the effect of consumers on voltage disturbances. In this regard, the assessment of the consumer's impact should be done in a different way, which does not use the concept of "share contribution" and takes into account the non-addictiveness of voltage quality violations.

There are two important factors to consider when developing an impact assessment method. On the one hand, the impact assessment should be autonomous according to the principles of electromagnetic compatibility. This means that each consumer should only be responsible for the quality-degrading current and conductivity of the controlled switching circuit parameters. On the other hand, the influence of the consumer on the level of high harmonics and voltage asymmetry depends not only on its parameters, but also on the parameters of the switching circuits of other consumers at a certain common connection point, as well as on the parameters and scheme of the external power network. Therefore, it is necessary to develop a method that clearly ensures the above two factors and the responsibility for improving the quality of electricity between consumers and electricity supply companies [2-3,5].

A method has been developed that does not use the concept of proportional contribution and takes into account the non-addictiveness of the voltage at the common connection point to assess the effect on voltage quality degradation by consumers.

$$U_{poc} = \frac{I_1+I_2+I_3+\cdots+I_N}{Y_1+Y_2+Y_3+\cdots+Y_N} = \frac{\sum_{m=1}^{N} I_m}{\sum_{m=1}^{N} Y_m} \qquad (2)$$

A scheme similar to that shown in Fig.1 was adopted as the initial switching scheme. In general, the voltage quality impairment value at the common connection point for N power consumers is determined by formula (2). Based on this formula, all power supply consumers can change both the sign and the denominator of this expression, so they affect the magnitude of the voltage quality violation.

It should be noted that the modules and arguments of quality-destroying currents are not the same for each of the conductivities I_m and Y_m, harmonics of consumers of power supply enterprises.

Therefore, expression (2) is written separately for each of the analyzed higher harmonics and the inverse sequence constituents of the fundamental frequency and its analysis is carried out.

$$I_E^{\Gamma} = I_1 + I_2 + I_3 + \cdots + I_N = \sum_{m=1}^{N} I_m;$$
$$Y_E^{\Gamma} = Y_1 + Y_2 + Y_3 + \cdots + Y_N = \sum_{m=1}^{N} Y_m; \qquad (3)$$
$$I_E^{A} = I_1 + I_2 + I_3 + \cdots + I_N = \sum_{m=1}^{N} I_m;$$
$$Y_E^{A} = Y_1 + Y_2 + Y_3 + \cdots + Y_N = \sum_{m=1}^{N} Y_m;$$

here I_{Σ}^{Γ} and Y_{Σ}^{Γ} is a geometric (vector) sum of quality-destroying currents and conductivities of power supply system consumers;

I_{Σ}^{A} and Y_{Σ}^{A} – the arithmetical sum of quality-disrupting currents and conductivities of power supply system consumers.

From this can be form the following expression:

$$K_{CI} = \frac{|I_{\Sigma}^{\Gamma}|}{I_{\Sigma}^{A}}; \; K_{CY} = \frac{|I_{\Sigma}^{\Gamma}|}{I_{\Sigma}^{A}} \qquad (4)$$

Here K_{CI}, K_{CY} are the current and conductivity coefficients in the phase, respectively.

https://doi.org/10.1051/e3sconf/202338401041

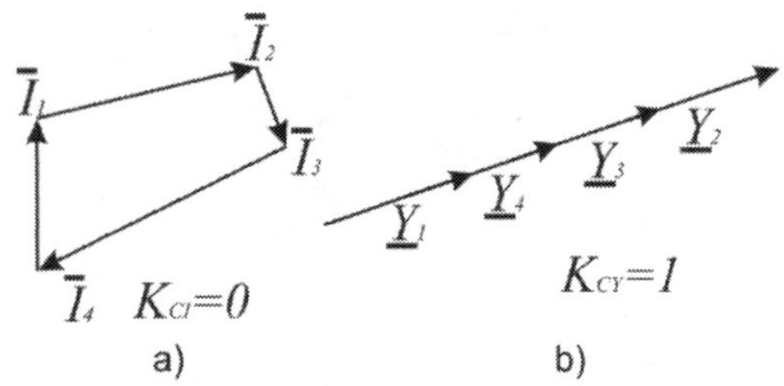

Fig.2. Geometric representation of phase matching coefficients

a) $K_{CI} = 0$ b) $K_{CY} = 1$

The phase matching coefficients describe the mutual location of currents and conductance vectors that violate the power quality in the complex plane and can take values from 0 to 1. Thus, Figure 2.- a shows one of the cases where the vector sum of the currents is zero, and the coincidence coefficient in the phase of the quality-disrupting currents is also zero. Figure 2-b shows an example where all conductances of consumers of the power supply system match in phase and $K_{CY} = 1$.

From the expression (4), the following can be obtained:

$$|I_\Sigma^\Gamma| = K_{CI} \cdot I_\Sigma^A ; \quad |Y_\Sigma^\Gamma| = K_{CI} \cdot Y_\Sigma^A \tag{5}$$

Taking into account the expressions (2) and (5), the expression of the breakdown voltage modulus at the common connection point is as follows:

$$U_{poc} = \frac{K_{CI} \cdot I_\Sigma^A}{K_{CY} \cdot Y_\Sigma^A} \tag{6}$$

Introduce the concept of the profit factor:

$$K_{profit} = \frac{K_{CY}}{K_{CI}} \tag{7}$$

From (6) can be form the following expression:

$$U_{poc} = \frac{1}{K_{profit}} \cdot \frac{I_\Sigma^A}{Y_\Sigma^A} \tag{8}$$

The profit factor can take values from 0 to $+\infty$. This shows how beneficial it is for consumers in the electricity supply system to work together in this best equivalent scheme.

3 Verification of the method using an equivalent circuit

To explain how useful is it to work together in an equivalent circuit using Fig. 2. for two cases. Figure 2, (a) shows the case where profit factor $K_{profit}=0$, that is, the utility of joint work is zero. Expressions (7) and (8) show that this parameter corresponds to $K_{CY} = 0$, and the geometric sum of conductivities is zero.

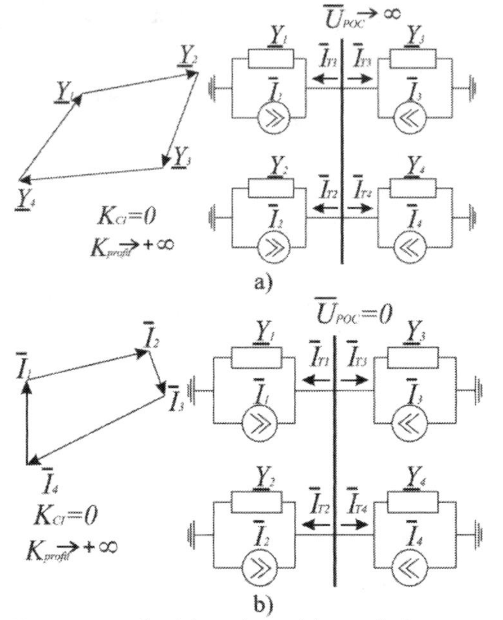

Fig. 3. A graph of the values of the profit factor

a) $K_{profit} = 0$ b) $K_{profit} \to +\infty$

This means that there is a resonance of the currents, and when $K_{profit} = 0$, the breakdown of the quality of the voltage at the point of connection U_{poc} tends to infinity.

At the same time, all connected to the point of connection suffer from non-compliant power quality indicators, which confirms that it is inappropriate for consumers to work together in this mode. Figure 2-b shows the case where K_{profit} tends to $\to +\infty$.

According to (8), this means that K_{cI} is zero (Fig. 2, a), and therefore violation of the voltage quality (10) is also zero.

In general, the admissible condition for voltage quality failure mode is the inequality:

$$U_{poc} \leq U_{permv} \tag{9}$$

here U_{prsv} is the permissible value of the n- th harmonic component of voltage or the reverse sequence of voltage determined by the K_{Un} and K_{2U} coefficients.

Thus, with a certain normal permissible value of the K_{UN} N_D coefficient and the nominal voltage U_{nom} at the point of connection, the value of U_{permv} is equal to:

$$U_{n.per} = \frac{K_{un\,ND} \cdot U_{nom}}{100\%} \tag{10}$$

Similar to the expression (10), using K_{2U}, it is possible to find the permissible value of the reverse sequence of the voltage at the fundamental frequency Umf.

Taking into account (3) and (8), we can write:

$$\frac{1}{K_{profit}} \cdot \frac{I_1+I_2+I_3+\cdots+I_n}{Y_1+Y_2+Y_3+\cdots+Y_n} \leq U_{prsv} \quad (11)$$

From (11) we get:

$$K_{eff}U_{prsv}Y_1 - I_1 + K_{eff}U_{prsv}Y_2 - I_2 + \cdots$$
$$+K_{eff}U_{prsv}Y_N - I_N \geq 0 \quad (12)$$

A necessary condition for the admissibility of the breakdown voltage regime is the observance of inequalities (9) - (12). A sufficient condition for the fulfillment of the inequality (12) is that the expression on the left side is non-negative at each value of the sum. This condition is reasonable because it provides the same requirements for all consumers of the power supply system, regardless of their capabilities and other parameters. All N power consumers connected to a point of connection are responsible for keeping U_{poc} within acceptable limits. Thus, for the k-th consumer of the power supply system taken separately, we can write the following expression:

$$K_{eff}U_{prsv}Y_k - I_k \geq 0 \quad (13)$$

From this:

$$K_{eff}U_{prsv} \geq \frac{I_k}{Y_k} \quad (14)$$

Could define $\frac{I_k}{Y_k} = U_k^{avt}$, Uk is the autonomous breakdown voltage generated by the k - participant of the power supply system.

Autonomous breakdown voltage is a breakdown voltage in which the studied consumer can work separately from other participants of the power supply system.

Condition for fulfillment of inequality (9):

$$U_k^{avt} \leq K_{profit}U_{prsv} \quad (15)$$

The expression (15) makes it possible to clearly find out whether the influence of the k-th subject on the voltage disturbance at the common connection point is allowed. Let's consider two limiting cases:

If I_k=0 with non-zero transmittance (linear or symmetrical consumer), then U_{kavt}=0.

In this case, the condition (15) is always satisfied, and such a consumer will never cause an unacceptable voltage quality violation at a power grid node [3-6].

If I_k=0, with degraded current (disruptive consumer), then U_{kavt} tends to →+∞. Thus, according to (12), only a consumer with a quality-destroying load will always cause an unacceptable voltage quality violation at the common connection point. To fulfill the condition (15), such a consumer should increase its throughput Y_k by installing filters or compensating devices, thereby reducing U_k^{avt}.

In order to quantify the impact of the K-th consumer of power supply system on the voltage disturbance at the point of connection, it is proposed to introduce the concept of the influence coefficient:

$$K_{ta.k} = \frac{K_{eff} \cdot U_{prsv} - U_k^{avt}}{K_{eff} \cdot U_{prsv}} = 1 - \frac{U_k^{avt}}{K_{eff} \cdot U_{prsv}} \quad (16)$$

- The influence coefficient can take values from -∞ to 1, including:
- if $K_{vl.k}$ is negative, then the k-th consumer has an unacceptable effect on the breakdown voltage at Upoc;
- if $K_{vl.k}$ =0, the k-th consumer in the power supply system $U_k^{avt} = K_{eff}U_{prsv}$;
- if $K_{vl.k}$ is positive, then the k-member of the power supply system has a positive effect on U_{poc}.

4 Conclusion

Summarizing the above research work on assessing the impact of consumers on the deterioration of the power quality through the use of equivalent circuits in the power supply system, can conclude the following:
- it is shown that the distortion of voltage cannot be unambiguously divided between the participants of the energy supply system;
- proportional to their influence, which confirms the incorrectness of the existing method noted in other works.
- the proposed method takes into account the features of existing approaches and regulatory documents, according to which each consumer must ensure the electromagnetic compatibility of their own electrical installations in terms of emissions of distortion currents into the external network.
- using the developed method, it is possible qualitatively and quantitatively assess the impact of each consumer on the emergence unacceptable voltage higher harmonic.

References

1. GOST 32144-2013. Electric Energy. Standards for the quality of electrical energy in general-purpose power supply systems. http://docs.cntd.ru/document/1200104301
2. Fedosov D.S. Razrabotka metoda ozenki vliyaniya potrebiteley na nesinusoidalnost b nesimmetriyu napryajeniy v elektricheskoy seti. Dissertation, Irkutsk, 2014.
3. Bobojanov, M.K., Mahmutkhonov S.K and others. AIP Conference Proceedings **2552.** (2023). 05011. https://doi.org/10.1063/5.0113890.
4. M.K.Bobojanov, S.Mahmutkhonov, and S.Aytbaev. AIP Conference Proceedings **2552.** 050011. (2023). https://doi.org/10.1063/5.0113890
5. Mahmutkhonov S.K, Kuuatbaev P.K. Analysis of load curves for DSP -100 furnace. Scientific and Theoretical Journal ISSN 2541-7789 (Online) **№4** (73), 2022. PP. 15-19.
6. M.K.Bobojanov, O.E.Ziyodulla, and others. E3S Web of Conferences **177**. 03023. (2020). https://doi.org/10.1051/e3sconf/202017703023
7. Mahmutkhonov S.K. Effect of non-linear consumers on the quality of electricity. Actual

problems of the power supply system // Proceedings of the international scientific and technical conference. - Tashkent. Tashkent State Technical University, 2021. PP. 267.
8. R Karimov. AIP Conference Proceedings **2552.** 030014. (2022). https://doi.org/10.1063/5.0111533
9. R Karimov, AIP Conference Proceedings **2552.** 050012. (2022). https://doi.org/10.1063/5.0111524
10. S Dzhuraev, R Karimov and ofters, Study and Analysis of Power Quality in the Electrical Networks of the Outdoor Lighting of the Dushanbe City (2022 Conference of Russian Young Researchers in Electrical and Electronic Engineering (ElConRus), 2022), pp. 1166-1169, https://doi:10.1109/ElConRus54750.2022.9755782
11. R Karimov, A. Kuchkarov and ofters, Analysis and study of energy efficiency by the operation of a voltage stabilizer (Journal of Physics: Conference Series 2094, 052050, 2021). https://doi:10.1088/1742-6596/2094/5/052050
12. M.Bobojanov. AIP Conference Proceedings **2552.** 050034. (2023). https://doi.org/10.1063/5.0114077
13. M.K.Bobojanov, R.Ch.Karimov, T.H.Qosimov, S.D.Zh.Dzhuraev. E3S Web of Conferences **289.** 07012. (2021). https://doi.org/10.1051/e3sconf/202128907012
14. M.Bobojanov, D.Rismuxamedov, F.Tuychiev, K.Shamsutdinov, K.Magdiev. E3S Web of Conferences **216.** 01164. (2020). https://doi.org/10.1051/e3sconf/202021601164
15. R.Karimov, M.Bobojanov. E3S Web of Conferences **216.** 01162. (2020). https://doi.org/10.1051/e3sconf/202021601162

E3S Web of Conferences **384**, 01042 (2023) https://doi.org/10.1051/e3sconf/202338401042
RSES 2022

Electrical load graphs and indicators

Rasulov *A.N.*[1], Ruzinazarov *M.R.*[1,*]

Tashkent State Technical University named by Islam Karimov, Tashkent, 100095, Uzbekistan

Abstract. The article provides methods for analyzing power consumption modes, discusses the main directions for improving the efficiency of economic incentives for the rational use of electricity with a tense energy balance during hours of maximum load of the power system and insufficient filling of the electrical load schedule at night, while improving the methods of economic regulation of accounting for electricity consumption, regulating electrical load schedules.

1 Introduction

It is advisable to plan and manage power consumption modes using and analyzing electrical load graphs. When developing the operating modes of the power system, the maximum load and active power balance, the total demand for electricity for the period under review, the expected change in load during the day, and a number of other factors are taken into account. Daily calculated load schedules allow to analyze the operating mode of electrical equipment of industrial enterprises for the past and subsequent periods of mode regulation, as well as to develop measures to optimize power consumption modes in the near future. Load schedules are used to determine the declared power during the hours of daily maximum load of the power system for the estimated period of planning and limiting power consumption [1, 2].

2 The current state of the investigated problem

In world practice, the study of the most important objects Graphs of electrical loads of industrial enterprises are characterized by a certain change in power consumption during the day. In this case, said change in load is determined by a base value, limited to those usually related to night time and zones of semi-peak and peak loads. These zones, respectively, refer to periods of low power consumption and to the hours of morning and evening load peaks.

The schedules of an enterprise with continuous production are determined by a fairly uniform power consumption during the day. Enterprises operating in one or two shifts have electrical load schedules that vary in their power consumption during the day. An analysis of the electrical load graphs of electrical industry enterprises shows their significant uneven power consumption during the day, exceeding in most cases the morning maximum load in relation to the evening one. At the same time, the time intervals of the morning and evening load maxima of the analyzed enterprises coincide with the maximum of the energy system [5-14].

As the main indicators of the daily schedules of the electrical load of industrial enterprises, which most fully characterize the dynamics of changes in the mode of power consumption, the following can be taken:

Load curve fill factor

$$K_з = \frac{P_{ср}}{P_{мак.с}}$$

Coefficient of non-uniformity of the load curve

$$K_H = \frac{P_{\min.c}}{P_{мак.с}}$$

Maximum factor

$$K_M = \frac{P_{мак.с}}{P_{ср}}$$

Number of hours of use of the maximum active load

$$T_{мах} = \frac{W_г}{P_{мак}}$$

Where P_{cp}- the average load per day, kW; $P_{\min}$ f_0 c, P(мах.c)- respectively, the minimum and maximum load per day, kW; W_r- the amount of electricity consumed per year, kWh; P_{max}– annual maximum load, kW.

The problem of compacting the schedules of the electrical load of enterprises is of great economic

* Corresponding author: ruzinzarov88@bk.ru

importance for the consumers themselves, especially for the power system. Leveling the level of power consumption at the enterprise helps to improve the rhythm of production, the fullest load of technological and power equipment, greater consistency in the work of the main, procurement and auxiliary industries, improve planning [15-21].

Of exceptional importance is the regulation of power consumption in industrial enterprises contributes to the rational consumption of fuel for the generation of electricity, increase the level of operation and reliability of equipment at power plants.

In order to interest industrial enterprises in the greatest reduction in load during peak hours of the power system, it is envisaged, by mutual agreement between the power system and the enterprise, to set a power limit not for the entire period of passage of the maximum load, but for individual agreed hours of the specified period. In this case, consumers are used - regulators with a significant, more than 30%, power reduction [22-26].

Analysis of power consumption modes is carried out in a certain sequence.

The construction and analysis of load graphs and the determination of indicators of power consumption modes are carried out. In this case, the actual daily winter and summer load curves are used, obtained for the most characteristic periods of time, determined by various power consumption modes.

After constructing the load schedules, an analysis of the changes in these schedules is carried out both for individual production areas and for the enterprise as a whole.

The analysis of changes in the electrical load curves of enterprises in the electrical industry shows their significant density in all the main parameters that determine their formation. The load curves were characterized by pronounced daytime and evening load maxima for all the enterprises under consideration, both for the summer and winter periods. So, for example, for the summer schedule, the filling factor kz for the analyzed period varied from 0.39 to 0.75, and the number of hours of using the daily maximum load T max changed from 9.3 to 17.3%. For the winter period, k z in the range of 0.48 to 0.85 m T max from 11.6 to 16.84. At the same time, the main consumers that determine the formation of load schedules for the analyzed enterprises were energy-intensive installations: foundry installations, electrothermal furnaces and installations; installations of contact cooking; electroplating installations; compressor rooms for generating compressed air, ventilation units; aquifers [5, 27-29]

To fill the night period of reducing the load in the future, the issues of intensive introduction of robotics and automated production complexes are of great importance. The widespread introduction of "unmanned technology" will greatly contribute to increasing labor productivity, and the most effective regulation of power consumption modes, a significant reduction in the cost of generating electricity in power systems.

Variable calculations to determine the effectiveness of measures to regulate power consumption modes at industrial enterprises [7, 30-35] should be carried out taking into account that the understatement by the consumer of the declared capacity and the corresponding payment for the declared capacity during the peak hours of the power system can lead to a decrease in output during the specified period and the need to produce a large amount of products at other hours, i.e. equalities must be fulfilled:

$$nT = n_1(T - T_1) + n_2 T_1 = na_1(T - T_1) + na_2 T_1 \quad (1)$$

where nis the hourly production before changing the power consumption mode; n_1and n_2- respectively, the hourly output of products outside the zone of maximum load and during hours of maximum; T - the planned annual number of hours of operation of the enterprise; T_1- the duration of the enterprise during the period of maximum load of the power system; h, a_1- coefficient taking into account the increase in hourly output outside the zone of maximum load of the power system; a_2is a coefficient that takes into account the decrease in the hourly output of products by the consumer in the zone of the maximum load of the power system [4].

The annual economic effect of the consumer, taking into account the change in output and the specific consumption of electricity when operating in the zone of maximum load - the power system and the rest of the period is determined from the expression [30]

$$\begin{aligned} Э_T &= a\Delta P + n\gamma[T(1 - a_1 b_1) + T(a_1 b_1 - a_2 b_2)] + n\omega[T(1 - a_1 b_3) + \\ &+ T(a_1 b_3 - a_2 b_4)]\beta - \Delta K(E_4 + E_a) - \Delta\Phi \end{aligned} \quad (2)$$

Where ΔP_{is} the decrease in the enterprise's capacity during peak hours of the power system kW, γis the cost per unit of production (without the constant part and cost of electricity before changing the mode of power consumption; b_1,b_2are coefficients that take into account changes in the consumption of raw materials, materials per unit of output outside the zone of maximum load of the power system and in the zone T_1;b_3 b_4- coefficients taking into account the change in the specific consumption of electricity outside the zone of maximum load of the power system and in the zone T_1; ω- specific power consumption per unit of production before the implementation of adjustment measures; a - the main rate of the two-part tariff for 1 kW of the declared capacity sum / (kW * year); β–

additional payment for 1 kWh of electricity according to the established tariff, UZS/(kWh); ΔK- additional capital investments associated with a change in the mode of power consumption, soums, E_4- the standard coefficient of efficiency of capital allocations; E_a- coefficient of depreciation; ΔΦ- change in the wage fund in connection with the regulation of electricity consumption, soum.

After the transformation of expression (2), the effectiveness of the adjustment measures is determined by the inequality:

$$a > \frac{\Delta K(E_H + E_a) + \Delta\Phi}{\Delta P} + \frac{ny}{\Delta P}[T(a_1 b_1 - 1) + T_1 a_2 b_2 - a_1 b_1)] + \frac{n\omega}{\Delta P}[T(a_1 b_3 - 1) + T_1(a_2 b_4 - a_1 b_3)]\beta \quad (3)$$

When the load decreases during peak hours of the power system due to an increase in output during other hours, the total amount of output remains unchanged. In accordance with expression (1), the change in hourly output will be determined by the relationship between the coefficients a 1 and a 2 :

$$a_1 = \frac{T - a_1 T_1}{T - T_1}; \quad a_2 = \frac{T}{T_1} - a_2 \frac{T - T_1}{T - T_1} \quad (4)$$

3 Conclusion

Reducing the load during peak hours causes additional capital investments ΔK>0and an increase in the payroll of production personnel ΔΦ>0. But at the same time, the output is provided without increasing raw materials, materials and additional electricity, so the coefficients b 1 , b 2 , b 3 , b 4 will be equal to one and the annual economic effect will be determined by the expression [4 12]

$$Э_Г = a\Delta P$$

When the load is reduced during peak hours of the power system and the planned output is maintained, $Э_г$=0. the following condition will be met:

$$a = \frac{\Delta K(E_H + E_a) + \Delta\Phi}{\Delta P} \quad (5)$$

In this case, the enterprise does not receive the economic benefit from the load control enterprises. At the same time, their implementation is expedient for the energy system and the national economy as a whole.

References

1. Salnikov A.Kh., Shevchenko L.A. Rationing of consumption and economy of fuel and energy resources. M. Energoatomizdat, 1986.
2. Gordeev V.I. Regulation of the maximum load of industrial electrical networks. Moscow: Energoatomizdat, 1986.
3. Zykin F.A., Kakhanovich V.S. Measurements and accounting of electric energy. Moscow: Energoatomizdat, 1982.
4. Pospelov G.E. Electrical systems and circuits. G.E. Pospelov, V.T. Fedik, P.V. Pychev. Minsk: Technoprint, 2004. 720s.
5. Fursanov M.I. Determination and analysis of electricity losses in electrical networks of energy systems / M.I. Fursanov. - Minsk.: Belenergosberezhenie 2005. - 207p.
6. Golovkin P.I. Energy system and consumers of electrical energy. Moscow: Energoatomizdat, 1986.
7. Mikhailov V.V. Tariffs and modes of power consumption. Moscow: Energoatomizdat, 1986.
8. R Karimov, Study of Graphs of Reactive Power Consumption at Enterprises of the Republic of Uzbekistan (AIP Conference Proceedings, 2552, 050012, 2022). https://doi.org/10.1063/5.0111524.
9. S Dzhuraev, R Karimov and ofters, Study and Analysis of Power Quality in the Electrical Networks of the Outdoor Lighting of the Dushanbe City (2022 Conference of Russian Young Researchers in Electrical and Electronic Engineering (ElConRus), 2022), pp. 1166-1169,doi: 10.1109/ElConRus54750.2022.9755782.
10. A.N.Rasulov, M.R.Ruzinazarov, N.Toirova, T.Sh.Alibekova. Graphical-analytical method for constructing load characteristics // RSES 2020. E3S Web of Conferences. 216 (2020) 01171.https://doi.org/10.1051/e3sconf/202021601171
11. Rasulov, A.N., Ruzinazarov, M.R., Novikov, N.L., Aslanova, G.N., Esemuratova, S. Loading mode of a ferromagnetic current stabilizer. // RSES 2021. E3S Web of Conferences 289,07007. https://doi.org/10.1051/e3sconf/202128907007
12. Rasulov, A.N., Rafikova, G.R., Novikov, N.L., Ruzinazarov, M.R., Esemuratova, S. Graph-analytical method of load mode analysis of ferromagnetic current stabilizer. E3S Web of Conferences 289,07006.//RSES2021. https://doi.org/10.1051/e3sconf/202128907006
13. Rafikova, G.R., Ruzinazarov, M.R., Makhmutkhonov, S.K. Features stabilization of the consumer current using electro-ferromagnetic circuits. DOI: https://doi.org/10.1051/e3sconf/201913901075
14. 10.1088/1757-899X/883/1/012102

16. A.Taslimov, M.Melikuziev, O.Matchonov, M.Ruzinazarov and M.Nasirov. Development of standard cable cross-sections of rural electrical networks // Conmechydro – 2020. IOP Conf. Series: Materials Science and Engineering. 883 (2020) 012105. doi:10.1088/1757-899X/883/1/012105.

17. Tuychiev F.N. Three-phase pole-changing winding for asynchronous machines on the fan drive // AIP Conference Proceedings 2552. 2023. PP, 030013, 1-8. https://doi.org/10.1063/5.0112349.

18. Hoshimov F.A., Rakhmonov I.U., Niyozov N.N., Omonov F.B. Main regularities of change in energy indicators of energy consuming facilities of industrial enterprises // AIP Conference Proceedings 2552. 2023. PP, 030025, 1-7. https://doi.org/10.1063/5.0112388.

19. Bobojanov M.K., Mahmutkhonov S.K., Aytbaev S. Investigation of the Problems Non-Sinusoidal of the Voltage Form // AIP Conference Proceedings 2552. 2023. PP, 05011, 1-6. https://doi.org/10.1063/5.0113890.

20. Rismukhamedov D.A., Tuychiev F.N., Shamsutdinov Kh. Development of pole-changing windings for two-speed motors of hoistingtransport mechanisms // AIP Conference Proceedings 2552. 2023. PP, 050013, 1-7. https://doi.org/10.1063/5.0114061.

21. Rasulov A.N. Method for calculating electro-ferro-magnetic circuits // AIP Conference Proceedings 2552. 2023. PP, 05017, 1-6. https://doi.org/10.1063/5.0111530.

22. Rasulov A.N., Usmonov E.G., Melikuziev M.V. Transformer-resonant circuit for battery charging // AIP Conference Proceedings 2552. 2023. PP, 050018, 1-6. https://doi.org/10.1063/5.0111530.

23. Usmonov E.G. Amplitude-phase Characteristics of the Resonant Circuit when Using it in the Control Circuit of Thyristors Voltage Stabilizers // AIP Conference Proceedings 2552. 2023. PP, 050020, 1-7. https://doi.org/10.1063/5.0111537.

24. Melikuziev M.V. Prospects for the introduction of intellectual systems in city electric networks // AIP Conference Proceedings 2552. 2023. PP, 050021, 1-6. https://doi.org/10.1063/5.0112395.

25. Rakhmonov I.U., Najimova A.M., and Reymov K.M. AIP Conference Proceedings **2647.** 030010. (2022). https://doi.org/10.1063/5.0104788

26. Taslimov A.D. Analysis of the stability of the total costs to the deviation of the network parameters from their optimal values // AIP Conference Proceedings 2552. 2023. PP, 050023, 1-6. https://doi.org/10.1063/5.0112398.

27. Saidkhodjaev A.G. Modern automatic technology of registrations and calculating electric loading // AIP Conference Proceedings 2552. 2023. PP, 050026, 1-5. https://doi.org/10.1063/5.0111540.

28. Bobojanov M.K. Development and Research of Two Speed Motor with Pole-Changing Winding // AIP Conference Proceedings 2552. 2023. PP, 050034, 1-9. https://doi.org/10.1063/5.0114077.

29. Saidkhodjaev A.G., Nuriddinova Kh.R. Reforming of Energy Sector of Uzbekistan: Transition to Market and Increasing Generation // AIP Conference Proceedings 2552. 2023. PP, 050037, 1-5. https://doi.org/10.1063/5.0133486.

30. Rakhmonov I.U., Najimova A.M. AIP Conference Proceedings **2647.** 030011. (2022). https://doi.org/10.1063/5.0104791

31. Rakhmonov I.U., Najimova, A.M., Esemuratova Sh.M., Koptileuov T.T. AIP Conference Proceedings **2647.** 070024. (2022). https://doi.org/10.1063/5.0104793

32. Hoshimov F.A., Rakhmonov I.U., Niyozov N.N., Omonov F.B. AIP Conference Proceedings **2647.** 030025. (2023). https://doi.org/10.1063/5.0112388

33. Rakhmonov I.U., Hoshimov F.A., Kurbonov N.N., Jalilova D.A. AIP Conference Proceedings **2647.** 050022. (2023). https://doi.org/10.1063/5.0112391

34. Rakhmonov I.U., Ushakov V.Ya., Niyozov N.N., Kurbonov N.N., Mamutov M. E3S Web of Conferences **289.** 07014. (2021). https://doi.org/10.1051/e3sconf/202128907014

35. Rakhmonov I.U., Ushakov V.Ya., Najimova A.M., Jalilova D.A., Omonov F.B. E3S Web of Conferences **289.** 07013. (2021). https://doi.org/10.1051/e3sconf/202128907013

Design of a pole changing winding for asynchronous machines drived on conveyors using the ANSYS Maxwell

Nuralibek Rashidov[1,*], *Khamza* Rozmetov[2], *Sabit* Rismukhamedov[1], *Moldagali* Peysenov[1]

[1]Tashkent State Technical University named by Islam Karimov, Tashkent, 100095, Uzbekistan

[2]Tashkent State Technical University Almalyk Branch, 45 Mirzo Ulugbek str., Almalyk, Uzbekistan

Abstract. The article considers the issues of increasing energy efficiency by using multi-speed motors, and also considers the problems of creating pole-changing windings. The procedure for constructing a new two-layer three-phase pole-changing winding with a pole ratio of 4/6, made in 36 stator slots on the basis of the basic switching scheme YYY/YYY, is presented. Based on the results of the analysis of the electromagnetic properties, the optimal winding pitch was selected. An experimental motor with a new pole-changing winding was created on the basis of a magnetic circuit of a serial asynchronous machine of the 4A132M4 type. The analysis of the properties of the electric motor was carried out using the ANSYS/Maxwell program. The results of experimental tests of a new two-speed motor in static and dynamic modes are also presented.

1 Introduction

Today, in the global mining industry, it is important to increase the energy efficiency of conveyor belts and use electricity economically. In this regard, much attention is paid, including optimal control of asynchronous motors of belt conveyor transport, rationing of electricity consumption, the search for combined methods of transporting minerals from deep horizons of open deposits and improving the reliability of conveyor equipment [1].

In the global mining industry, asynchronous motors with short-circuited rotors are usually used for electric drives of belt conveyors up to 3000 meters long and with a power of up to 100 kW. Simple design and low cost the advantage of such electric drives is the high starting torque and the associated appearance of large traction forces, as well as increased belt slippage, their disadvantage. There is a need to regulate the speed of the conveyor when the load current excavated in mines and mine workings increases or decreases [2].

In world practice, there are several methods of regulating the rotational speed of asynchronous motors, which include an asynchronous ventilated cascade, a thyristor frequency controller, and two-speed asynchronous motors are widely used.

The number of poles in the stator does not correspond to the additional change in energy provided in multi-speed electric drives based on asynchronous motors, in which there is a variable block, and is the only adjustment method in which additional waste is not spent on sliding, as a result of which their useful operating coefficient can be high

ANSOFT Maxwell is one of the leading electromagnetic field modeling software and is used to create and research projects of two-and three-dimensional models of electric motors, transformers, as well as various electrical and electromechanical devices [2]. The Maxwell program is based on the method of finite elements (Finite Element Method - FEM) and calculates static, electromagnetic and electric field harmonics, as well as, moreover, transient processes with very high accuracy [3]. With this in mind, instead of real experiments, it is possible to use computer models that are cost-effective.

Rotational Machine Expert (RMxprt) - it is a program that speeds up the processes of design and optimization of electrical machines. The calculation of the working characteristics of RMxprt electric machines uses classical analytical theories and equivalent methods of magnetic circuits. The following starting data of the electric motor are entered into the program: the type of windings and the connection scheme, the characteristics of the stator and rotor materials, geometric parameters, power supply, load, etc.k. Thus, this software tool allows you to significantly speed up the process of manufacturing an electric machine with a standard configuration [3]. As an object of study, let's consider an asynchronous electric motor with a short-circuit rotor of type 4A132M4. Figure 1 shows the basic dimensions of the asynchronous electric motor 4A132M4 using the Ansys Maxwell RMxprt software module.

* Corresponding author: nuralibekrashidov@gmail.com

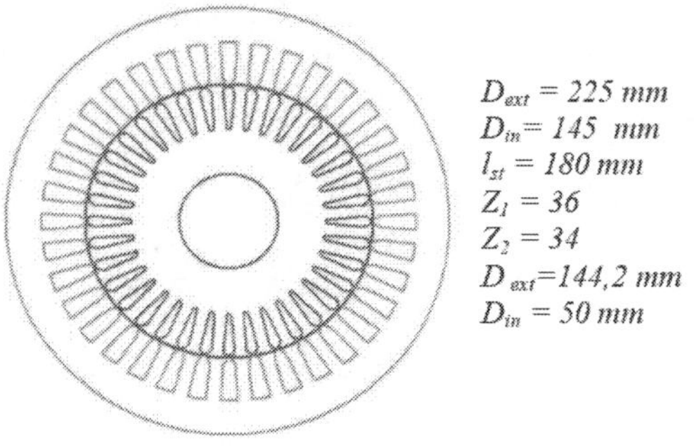

Fig.1. The main dimensions of the electric motor 4A132M4

2 Calculation of the winding factor for a variable number of poles

The number of Poles is considered to be variable winding special chulgam, and the winding step will be uniform for both polarization sides. The only step taken from a technological point of view may be reduced for one pole and extended for the other. The listed factors significantly affect the value of the winding factor.

The value of the winding factor depends on the specifics of the winding scheme. The value of the first harmonic coefficient (n=1) is slightly less than unity, which leads to a slight decrease in the first harmonic of EDF (MDF). Higher harmonics (n = 5, 7, 11, 13, etc.) have a winding factor value much less than unity and its value is close to zero, as a result of which high EDF harmonics (MDF) are significantly reduced [4-15, 16].

The winding factor can be calculated separately for each harmonic using the generalized method [17-20] and the matrix method.

The use of the structural matrix of three–phase winding makes it possible to study winding EDF using the method of matrix algebra and determine the winding factor for any ν-harmonics. For the study of alternating current emission EDF by The Matrix method, the multiphase winding can be written in the following form by the α-ordered harmonic k digital phase EDF vector [4]:

$$\dot{E}_{\phi k\nu} = E_{\phi k} \cdot e^{j\varphi_{k\nu}} \tag{1}$$

here, the $E_{\phi k}$ -EDF vector module; $\varphi_{k\nu}$ – $\dot{E}_{\phi k\nu}$ is the angle of rotation of the vector relative to the positive horizontal axis.

For ν -order harmonics, if the unit vectors EDF are brought to the matrix $[\dot{e}_z]_\nu$, they are written in the following form

$$[\dot{e}_z]_\nu = \begin{matrix} 1 \\ 2 \\ 3 \\ .. \\ i \\ .. \\ z \end{matrix} \begin{bmatrix} \cos\nu 0 & \sin\nu 0 \\ \cos\nu\gamma & \sin\nu\gamma \\ \cos\nu 2\gamma & \sin\nu 2\gamma \\ \dots\dots & \dots\dots \\ \cos\nu(i-1)\gamma & \sin\nu(i-1)\gamma \\ \dots\dots & \dots\dots \\ \cos\nu(z-1)\gamma & \sin\nu(z-1)\gamma \end{bmatrix} \cdot \begin{bmatrix} 1 \\ j \end{bmatrix} = [\dot{e}'_z]_\nu + j[\dot{e}''_z]_\nu \tag{2}$$

here, $\dot{e}'_{z\nu}=cos\ \nu(i–1)\gamma$ and $\dot{e}'_{z\nu}=sin\ \nu(i–1)\gamma$ – $[\dot{e}'_z]_\nu$ and $[\dot{e}''_z]_\nu$ matrix *i*–the number of elements, the number of slot matrix the number to the left ($i = 1...z$) recorded.

the ν-ordered harmonic can be written in the form of the following matrix expression when determining the *k*-digit $E_{\phi\kappa\nu}$ phase EDF and its winding factor

$$\left.\begin{aligned} \dot{E}_{\phi k\nu} &= [C_\phi]^{\tau}_k \cdot [\dot{e}_z]_\nu \\ E_{\phi k\nu} &= \sqrt{([C_\phi]^{\tau}_k \cdot [\dot{e}'_z]_\nu)^2 + ([C_\phi]^{\tau}_k \cdot [\dot{e}''_z]_\nu)^2} \\ tg\varphi &= \frac{[C_\phi]^{\tau}_k \cdot [\dot{e}''_z]_\nu}{[C_\phi]^{\tau}_k \cdot [\dot{e}'_z]_\nu} \\ k_{win.\nu} &= \frac{E_{\phi k\nu}}{N} \end{aligned}\right\} \tag{3}$$

where, $[C_\phi]^{\tau}_k$ – k is the transponized matrix of the digital phase, N – k is the number of digital phase conductors, equal to the modulus sum of the elements of this $[C_\phi]_k$ phase Matrix.

The number of poles based on the basic "YYY/YYY" circuit with 36 equal stator slots and a pole ratio of 2:3 in the circuit is variable, with a pitch of y=6 for the pole side 2p=4, in the MDF form, except for the first harmonic, 2nd, 5th, 7th, 8th, 10th, 11th, 13th, 14th, 16th, 17th harmonics are present and their amplitude is 15.1%, 0.4 %, 0.6%, 0. 3%, 0.2%, 0.4%, 0.2%, 0.2%, 1.9%, 4.5%.

In addition to the first harmonic, there are 2nd, 7th, 11th, 14th, 16th, and 17th harmonics in the form of MDF when the winding step is y=7, and their amplitude is 7.9% as a percentage of the total harmonics, taking into account the winding factor, respectively. 0.4%, 0.3%, 0.3%, 1%, 5%.

In addition to the first harmonic, there are 2nd, 5th, 7th, 8th, 10th, 11th, 13th, 14th, 16th, 17th harmonics in the form of MDF with a winding step u=8, and their amplitude is taken into account. by winding factor 2.2%, 0.2%, 0.1%, 0.4%, 0.3%, 0.1%, 0.1%, 0.3%, 1.9 as a percentage of the amount harmonics respectively % is 5.4%.

Except for the first harmonica in the form of MDF when 2p=6 pole side of this winding is the winding step y=6 5-, 7-, 8-, 11-, 13-, 17-there are Harmonics, the amplitude of which, when taking into account the winding factor, is in the percentage score relative to the general harmonic, respectively 1,2%, 0,9%, 7,6%, 6,4%, 0,4% makes up the.

With the exception of the first harmonica in the form of a MDF, when the winding step is y=8 2-, 5-, 7-, 8-, 10-, 11-, 13-, 14-, 16-, 17- there are harmonics whose amplitude, when taking into account the winding factor, is in percentage the score relative to the common harmonic, respectively 9,6%, 1%, 0,7%, 0,8%, 1,9%, 6,5%, 5,5%, 1,4%, 0,4%, 0,3% makes up.

The winding MDF curve forms a stepped curve around the circumference of the stator, the magnitude of each step in the wedge is equal to the full current of the wedge, and its direction changes in the direction of the current.

By allowing you to determine the MDF for any point of the circle through the myuk integral curve and find its maximum value, you can distribute them along a harmonic line using well-known graphical methods and determine the amplitudes of the MDF harmonics [5].

It is very convenient to construct the Georges diagram for three-phase *m*- and *2m*-zone circuits using the triangular mesh tool, because the currents for the *2m*-zone consist of six-beam symmetrical stars with a turning angle between them of 60°. Their geometric summation can be performed on a triangular grid, the sides of which describe the spatial zone current vectors, and the value is conditionally assumed to be equal to one [21-27, 3, 2].

The Georges diagram allows to study the circuit composition of multi-phase electric machines. If the Georges diagram is symmetric about the axis of the center of gravity, then the negative half-wave of the curve of the MDF should repeat the positive half-wave, then the MDF will consist of only odd harmonic constituents, when the Georges diagram is symmetric, even harmonic constituents will also be present.

The Georges diagram allows us to quantify the level of composition of high harmonics on the MDF curve.

Differential Scattering Factor — is an electromagnetic parameter that characterizes the windings of AC electrical machines. The composition of various harmonics on the MDF curve can be estimated using the differential scattering coefficient.

Figure 2 shows a diagram of Gyorges when the number of stator poses is 36 and the stride step is y=7. The differential scattering factor for this winding step is $2p_1=4$ for the side $\sigma 0=4.6\%$, $2p_2=6$ for the side - $\sigma_0=5.4\%$.

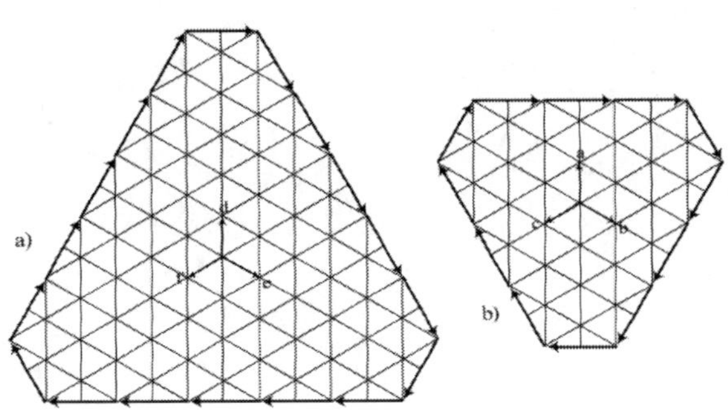

Fig.2. Georges diagram according to the YYY/YYY scheme with Z = 36 and y = 7: a) for $2p_1=4$ sides
b) for $2p_2=6$ sides

The calculation of the winding factor for each harmonic was performed in three steps of the winding. The values of the coefficient of variation for each harmonic are presented in Table 1

Table.1. Winding factor values

Z		*36*					
BC		*YYY/YYY*					
p		*p=for 2 sides*			*p=for 3 sides*		
y		*6*	*7*	*8*	*6*	*7*	*8*
Number of harmonics	1	0.801	0.87	0.911	0.837	0.808	0.724
	2	0.623	0.46	0.246	0.000	0.217	0.375
	3	0.000	0.21	0.375	0.000	0.000	0.000
	4	0.115	0.131	0.086	0.000	0.217	0.217
	5	0.097	0.019	0.072	0.224	0.058	0.194
	6	0.000	0.217	0.217	0.000	0.000	0.000
	7	0.227	0.201	0.09	0.224	0.058	0.194
	8	0.141	0.056	0.161	0.000	0.217	0.217
	9	0.000	0.000	0.000	0.000	0.000	0.000
	10	0.141	0.056	0.161	0.000	0.217	0.375
	11	0.227	0.201	0.09	0.837	0.808	0.724
	12	0.000	0.217	0.217	0.000	0.000	0.000
	13	0.097	0.019	0.072	0.837	0.808	0.724
	14	0.115	0.131	0.086	0.000	0.217	0.375
	15	0.000	0.217	0.375	0.000	0.000	0.000
	16	0.623	0.463	0.246	0.000	0.217	0.217
	17	0.801	0.87	0.911	0.224	0.058	0.194

3 Experimental studies of the working and mechanical properties of a two-speed electric motor

As a mechanical (working) characteristic of an asynchronous motor, in a narrow sense it is understood that the speed of rotation or slip depends on the given power, and in a broad sense a number of other values important for operation, that is, such parameters as input power, current, useful working coefficient and power coefficient, are understood at the invariant value of the given voltage and [5, 6].

Experimental studies of the new 4A132M4/6 engine show that $p_1=2$ on the side $P_2=9$ kW, $\eta=87\%$, $\cos\varphi=0{,}87$, $I_1=18$ A, M=58 nM, $p_2=3$ on the side $P_2=6$ kW, $\eta=85\%$, $\cos\varphi=0.51$, $I_1=20$ A, M=58 nM.

4 Conclusion

Based on the analysis of the results of the experimental studies presented above, it can be concluded that the number of Poles of a new type allows the introduction of a variable-flow-based two-speed motor instead of the single-speed motors available in the transport mechanisms of cargo transportation.

A study of experimental studies of working characteristics of new motors with variable two-speed poles showed that the energy performance of these motors fully corresponds to the required values.

References

1. M.Bobojanov. AIP Conference Proceedings 2552. 050034. (2023). https://doi.org/10.1063/5.0114077
2. D.Rismukhamedov, M.Bobojanov, F.Tuychiev, K.Shamsutdinov. E3S Web of Conferences 264. 03057. (2021). https://doi.org/10.1051/e3sconf/202126403057
3. M.Bobojanov, D.Rismuxamedov, F.Tuychiev, K.Shamsutdinov, K.Magdiev. E3S Web of Conferences 216. 01164. (2020). https://doi.org/10.1051/e3sconf/202021601164
4. M.K.Bobojanov, O.E.Ziyodulla, M.T.U.Ismoilov, E.I.U.Arziev, G.Z.Togaeva. E3S Web of Conferences 177. 03023. (2020). https://doi.org/10.1051/e3sconf/202017703023
5. 4E Electric Motor Systems Annex (EMSA). Policy Guidelines for Motor Driven Units – Part 2, 2018.
6. E Abduraimov, M. Peysenov. E3S Web of Conferences 289. 07026. (2021). https://doi.org/10.1051/e3sconf/202128907026
7. Rakhmonov I.U., Najimova A.M., and Reymov K.M. AIP Conference Proceedings **2647.** 030010. (2022). https://doi.org/10.1063/5.0104788
8. Rakhmonov I.U., Najimova A.M. AIP Conference Proceedings **2647.** 030011. (2022). https://doi.org/10.1063/5.0104791
9. Rakhmonov I.U., Najimova, A.M., Esemuratova Sh.M., Koptileuov T.T. AIP Conference Proceedings **2647.** 070024. (2022). https://doi.org/10.1063/5.0104793
10. Hoshimov F.A., Rakhmonov I.U., Niyozov N.N., Omonov F.B. AIP Conference Proceedings **2647.** 030025. (2023). https://doi.org/10.1063/5.0112388
11. Rakhmonov I.U., Hoshimov F.A., Kurbonov N.N., Jalilova D.A. AIP Conference Proceedings **2647.** 050022. (2023). https://doi.org/10.1063/5.0112391
12. Rakhmonov I.U., Ushakov V.Ya., Niyozov N.N., Kurbonov N.N., Mamutov M. E3S Web of Conferences **289.** 07014. (2021). https://doi.org/10.1051/e3sconf/202128907014
13. Rakhmonov I.U., Ushakov V.Ya., Najimova A.M., Jalilova D.A., Omonov F.B. E3S Web of Conferences **289.** 07013. (2021). https://doi.org/10.1051/e3sconf/202128907013
14. Rakhmonov, I., Berdishev, A., Niyozov, N., Muratov, A., Khaliknazarov, U. IOP Conference Series: Materials Science and Engineering, 2020. 883(1). 012103. DOI: 10.1088/1757-899X/883/1/012103
15. Rakhmonov, I.U., Niyozov, N.N. E3S Web of Conferences, 2019. **139.** 01077. DOI: 10.1051/e3sconf/201913901077
16. Rakhmonov, I.U., Reymov, K.M. Journal of Physics: Conference Series, 2019. 1399(5). 055038. DOI: 10.1088/1742-6596/1399/5/055038
17. R.Karimov. AIP Conference Proceedings 2552. **030014**. (2022). https://doi.org/10.1063/5.0111533
18. R.Karimov. AIP Conference Proceedings 2552. **050012**. (2022). https://doi.org/10.1063/5.0111524
19. S.Dzhuraev, R.Karimov, and ofters. ElConRus, pp. 1166-1169. (2022). doi: 10.1109/ElConRus54750.2022.9755782
20. R.Karimov, A. Kuchkarov, and ofters. Journal of Physics: Conference Series 2094. **052050**. (2021). doi:10.1088/1742-6596/2094/5/052050
21. R.Karimov, N.Kurbanova, and ofters. Journal of Physics: Conference Series. 2094(5). **052042**. (2021). doi:10.1088/1742-6596/2094/5/052042
22. Usmanov E.G., Khusanov B.M. RSES 2020. E3S Web of Conferences **216.** 01161. (2020). https://doi.org/10.1051/e3sconf/202021601161
23. Erejepov, M., Novikov, A.N., Khusanov, B.M., Seytmuratov, B., Sayimbetov, Z. E3S Web of Conferences. **289.** (2021). 07018. https://doi.org/10.1051/e3sconf/202128907018
24. Khakimov, H.T., Shayumova, Z.M., Kurbanbaeva, Z.K., Khusanov, B.M. E3S Web of Conferences **139.** 01076. (2019). https://doi.org/10.1051/e3sconf/201913901076
25. Melikuziev M.V., Nematov L.A., Novikov A.N., Baymuratov K.K. E3S Web of Conferences **289.** 07016 (2021). https://doi.org/10.1051/e3sconf/202128907016
26. A.D.Taslimov, A.S.Berdishev, F.M.Rakhimov, Melikuziev M.V. E3S Web Conf. **Vol. 139**. 2019. https://doi.org/10.1051/e3sconf/201913901081.
27. A.D.Taslimov, A.S.Berdishev, F.M.Rakhimov, Melikuziev M.V. E3S Web Conf. **Vol.139**. 2019. https://doi.org/10.1051/e3sconf/201913901082

E3S Web of Conferences **384**, 01044 (2023) https://doi.org/10.1051/e3sconf/202338401044
RSES 2022

Development of two-speed asynchronous electric motors for the undercarriage of mine self-propelled cars

Furkat Tuychiev[1*], *Alim* Haqberdiev[1]

[1]Tashkent State Technical University named by Islam Karimov, Tashkent, 100095, Uzbekistan

Abstract. The paper presents a new double layer three phase 3/6 pole-changing winding, with 72 slots and 6 terminals with "Y/YY with additional branches" и "Y-Δ/YY" switching circuit. Using the discretely specified spatial function method, a two-speed asynchronous electric motor with a pole-switched winding was designed for self-propelled mine wagons, which were tested in static and dynamic modes of operation. The application areas of two-speed asynchronous motors with pole-switched windings are given. Currently, dynamic modes have not been investigated in the used two-speed engines. As the object of the study, a two-speed asynchronous motor of the AVTM - 6/12 brand of the self-propelled mine wagons drive was adopted in operating modes with no load on the shaft and when starting with a load of 500 Nm and switching to a second speed.

1 Introduction

At the mines of the Tyubegatan potash Deposit (Uzbekistan), wheeled self-propelled mine wagons (SPMW) are used to deliver the rock mass beaten off by a tunneling combine to ore outlets or directly to conveyor bunkers at distances of up to 200 meters [1]. In order to increase the efficiency of self-propelled mine wagons, it is necessary to regulate the speed of rotation of the engines, therefore, three-speed motors of the AVT15-4/6/12 type are used to drive them [2, 3].

The problem of creating pole-changing windings (PCW) for multi-speed electric motors was dealt with by many scientists from around the world [4, 5]. Series two-speed motors with a pole pair ratio of 1/2 have a pole-changing Dahlander winding in the stator [6, 7]. Pole changing method is one of the main methods of the speed control of an induction motor. The number of stator poles can be changed by pole amplitude modulation (PAM) method [8, 9] The method is used both for close ratio as well as for wide ratio [10]. Two-speed motors with pole ratio 2/3 and 3/4 are made with pole-changing windings obtained by the phase modulation method [11, 12].

To simplify the process of constructing a winding circuit, a new way of presenting the current distribution in the form of a discretely specified spatial function, in short, DSSF, was introduced, from which the method was called the "DSSF method" [13, 14].

The purpose of this article is to develop schemes of pole-changing windings and analyze a multi-speed electric motor on the drive of self-propelled mine wagons.

2 Pole changing winding method

The development of schemes for pole-changing windings with a close structure to normal windings, i.e. with improved electromagnetic properties is possible, using the modernized method discretely specified spatial function. The development of pole-changing windings in pole ratios $p_1/p_2=3/6$ based on basic circuits (BS) is peculiar, the use of asynchronous motors with a basic circuit of the type "Y/YY with additional branches" and "Y-Δ/YY" is appropriate. Including, on the drives of the undercarriage of mine self-propelled cars, when obtaining PCW based on schemes like "Y/YY with additional branches" and "Y-Δ/YY", one has to synthesize a suitable one for one of the pluses of the DSSF [15].

As an example, consider the process of obtaining a PCW circuit for a pole ratio of $p_1/p_2=3/6$ with the number of slots $Z_1=72$. A normal m-zone winding with the number of poles $p_2=6$ was taken as the "initial" winding, and a normal 2m-zone winding with $p_1=3$ was taken as a "typical" one (Table 1).

Table 1. Synthesis of windings with pole $p_1=3$

1	2	3	4	5	6	7	8	9	10	11	12	13	Slots
a	a	a	a	$\underline{c}$	$\underline{c}$	$\underline{c}$	$\underline{c}$	b	b	b	b	$\underline{a}$	$p_1=3$ typical
a	a	a	a	c	c	c	c	b	b	b	b	a	$p_2=6$ initial
a	a	a	a	$\underline{c}$	$\underline{c}$	$\underline{c}$	$\underline{c}$	b	b	b	b	$\underline{a}$	$p_1=3$ synt.
14	15	16	17	18	19	20	21	22	23	24		72	Slots
$\underline{a}$	$\underline{a}$	$\underline{a}$	c	c	c	c	$\underline{b}$	$\underline{b}$	$\underline{b}$	$\underline{b}$	…	$\underline{b}$	$p_1=3$ typical
a	a	a	c	c	c	c	b	b	b	b	…	b	$p_2=6$ initial
$\underline{a}$	$\underline{a}$	$\underline{a}$	c	c	c	c	$\underline{b}$	$\underline{b}$	$\underline{b}$	$\underline{b}$		$\underline{b}$	$p_1=3$ synt.

* Corresponding author: tuychievfn@gmail.com

For example, if the position of the DSSF conductor in the original winding is equal to «b», and for a typical DSSF in the same phase as «a», which is known from the location of the three-phase system, currents that are close to «a» «-b», not «b» (direction of instantaneous currents corresponds), therefore, the synthesis winding is written «-b» for DSSF; if the DSSF of the original winding is «c», the DSSF of the typical winding is «-c», then «-c» is written in the DSSF of the synthetic winding [16].

The resulting windings on both sides of the poles are absolutely symmetrical with respect to the power source, the EMF vectors of the same branches of each phase are symmetrical with respect to each other, that is, they are equal in amplitude and rotated in phase by an angle of 2π/3 el.rad. For the basic circuit "Y-Δ/YY" with a winding step y=6 on the side of the poles $2p_1$ and $2p_2$, the winding factors are equal to k_{w1}=0.691 and k_{w2}=0.717, respectively, and for the basic circuit "Y/YY with additional branches" it is equal to k_{w1}= 0.757 and k_{w2}=0.88.

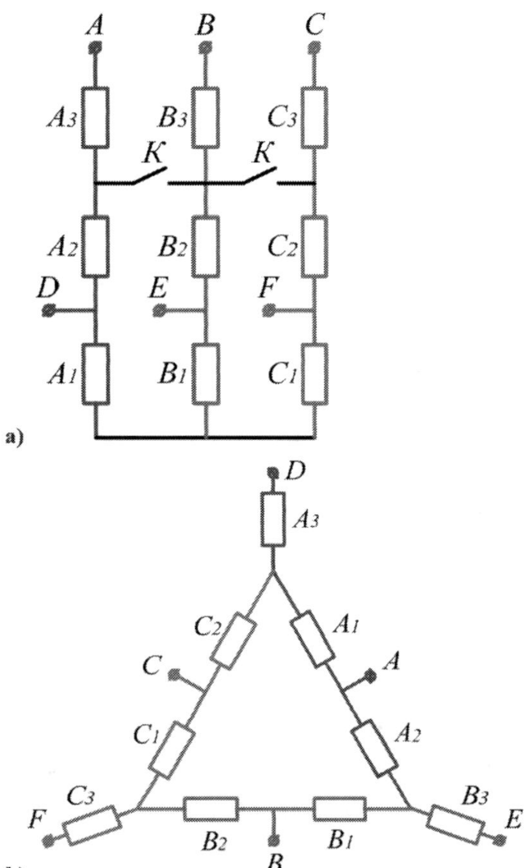

Fig.1. Scheme of switching for PCW:
a) "Y/YY with additional branches"; b) "Y-Δ/YY"

3 Results and discussion

For experimental confirmation of theoretical studies, prototypes of pole-changing two-speed electric motors of the AVTM15-6/12 type were manufactured, which were tested in static and dynamic modes of operation as part of the electric drive of the corresponding mechanisms. The results of experimental studies have shown that for the pole p_1=3, the useful power P_2=55 kW, efficiency η=87.5%, power factor cosφ=0.83, with stator current I_1=65 A and rotor speed n=973 rpm, for the pole p_2=6, the useful power P_2=26 kW, efficiency η=80%, power factor cosφ=0.59, with stator current I_1=48.8 A and rotor speed n=474 rpm were obtained (Fig.2).

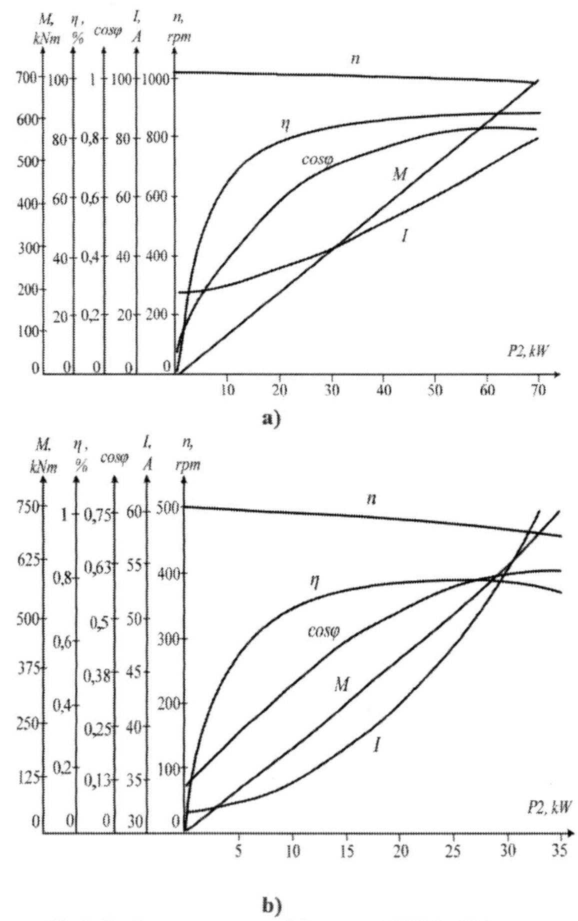

Fig.2. Performance curves of the motor AVTM -6/12:
a) for p_1=3; b) for p_2=6

A common disadvantage of a two-speed asynchronous motor with a squirrel-cage rotor is an additional shock torque, which creates large mechanical stresses and shocks in gears during start-up and switching from one speed to another.

Experimental tests were carried out to study the operation of a newly developed electric motor for self-propelled mine wagons in a pole-changing mode and a curve was obtained for changing the stator current, speed, torque and voltage of a two-speed motor of self-propelled mine wagons when a load of 500 Nm was applied to the shaft.

As an object of research, the operating modes of operation during start-up and when switching to the second speed of a two-speed asynchronous motor AVTM-6/12 in electric drives of the undercarriage of self-propelled mine wagons are studied.

Initially, transient processes were studied and curves were obtained as a function of time, changes in stator current and torque during 6- and 12-pole starts. Figures 1 and 2 show the change curves of the above parameters (I=f(t), M=f(t)).

From these curves it can be seen that the steady state of the engine is very fast, i.e. after 320 ms for $p_1=3$, after 420 ms for $p_2=6$.

As can be seen from the curve in Figure 1, the amplitude value of the starting current of the electric motor is 650 A, the output to the steady state operation of the motor occurs after 0.5 s, and the amplitude value of the steady current is 85 A. The rated motor speed is 974 rpm, the rated motor torque is 450 Nm, and the maximum torque is 2 kNm.

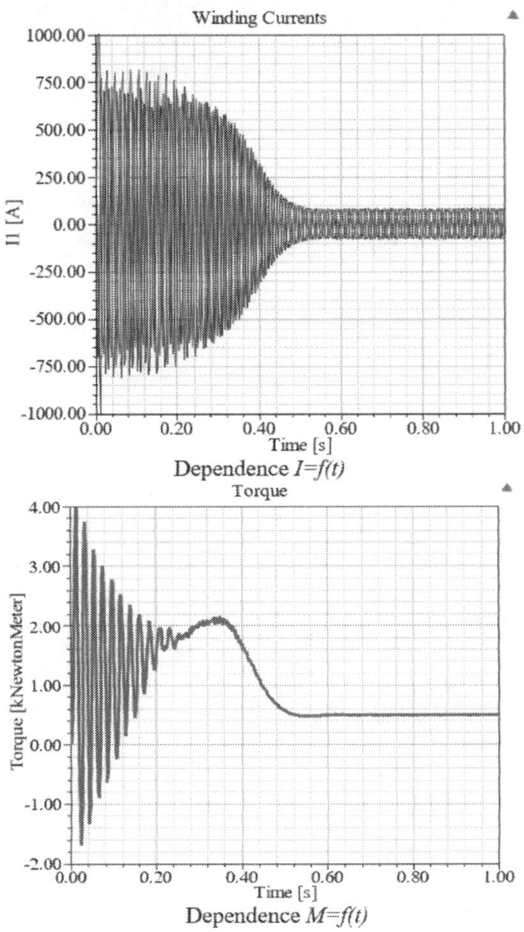

Fig.3. AVTM–6/12 electric motor start from $p_1=3$

As can be seen from the curve in Fig. 4, the amplitude value of the starting current of the electric motor is 210 A, the output to the steady state operation of the engine occurs after 0.75 s, and the amplitude value of the steady current is 60 A. The rated motor speed is 480 rpm, the rated motor torque is 500 Nm, the maximum torque is 1 kNm.

In the case of changing the speed of two-speed variable-pole induction motors from high to low, that is, the self-propelled mine wagons moves along the main underground freight road at a speed of 1000 rpm, when loading or unloading the transition to a speed of 500 rpm, a curve of change in time, motor current (I=f(t)), torque (M=f(t)).

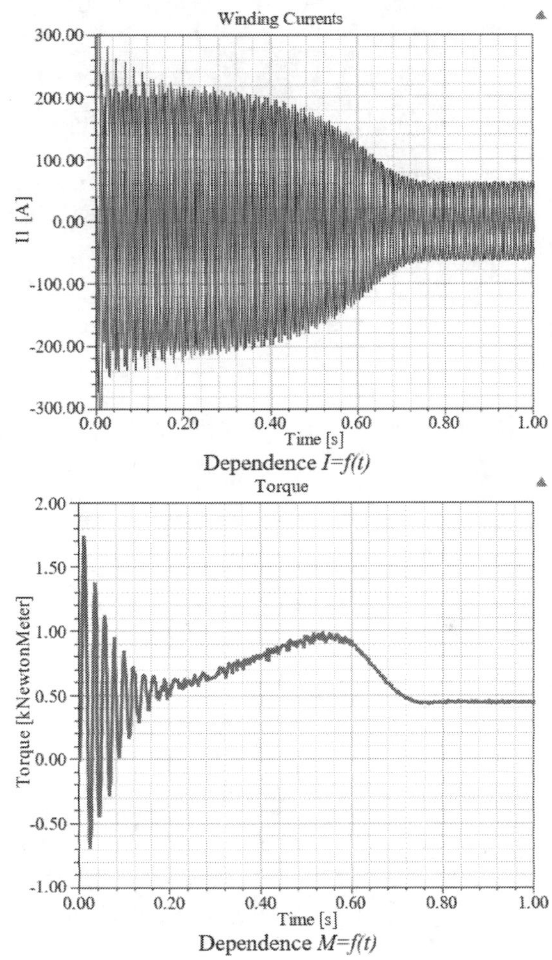

Fig.4. AVTM–6/12 electric motor start from $p_2=6$.

As can be seen from the curve in Fig. 5, the amplitude value of the starting current of the electric motor is 230 A, the output to the steady state operation of the engine occurs after 0.3 s, and the amplitude value of the direct current is 62 A. The rated engine speed is reduced from 970 rpm to 480 rpm. The motor is in braking mode, the maximum braking torque is 1.15 kN, then the maximum torque is 0.66 kNm, the rated torque is 450 Nm.

In the case of a change in the speed of two-speed asynchronous motors with a variable number of poles from high to low, that is, a mine car moved during loading or unloading at a speed of 500 rpm, along the main underground freight road of the transition to a speed of 1000 rpm, a curve of change in time, motor current (I=f(t)), torque (M=f(t)).

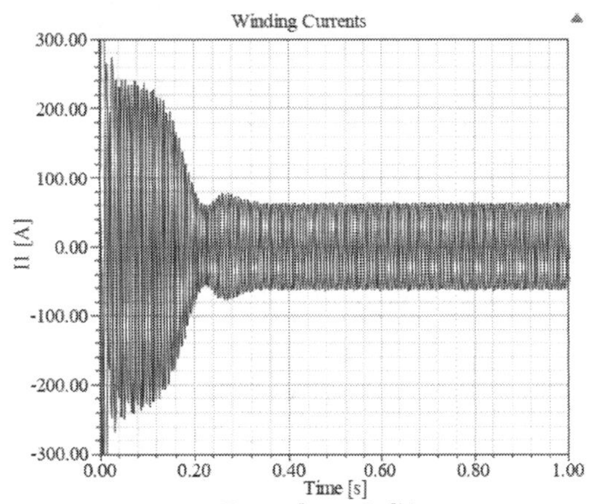

Dependence $I=f(t)$

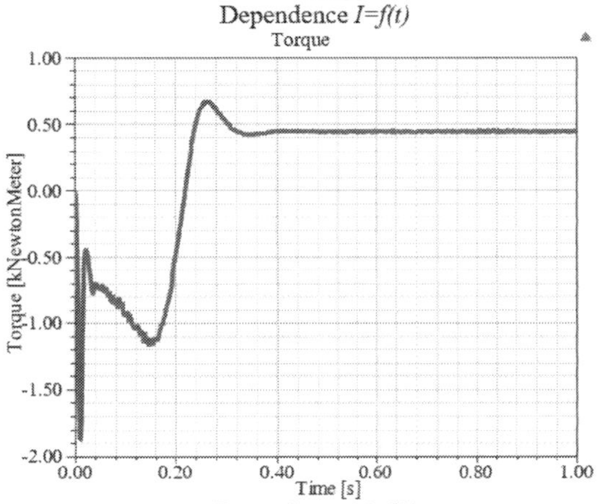

Dependence $M=f(t)$

Fig.5. Switching motor poles ABTM -6/12 from p_1=3 to p_2=6 at a load of 500 Nm

As can be seen from the curve in Fig. 6, the amplitude value of the starting current of the motor is 500 A, the output to the steady state operation of the motor occurs after 0.3 s, the amplitude value of the steady current is 80 A. The rated engine speed has increased from 480 rpm to 970 rpm. The rated torque of the motor is 500 Nm and also, the maximum torque is 2 kNm.

From all the results of studies of the dynamic mode of operation of an electric drive with an AVTM15-6/12 motor, it can be concluded that the transition time of the start-up process, the transition from one speed to another, the level of noise and vibration in the engine were in all cases within the allowable range.

This means that electric drives based on new variable pole asynchronous motors can indeed be created and applied to self-propelled mine wagons operated in underground mines, and as a result, the rational use of electricity and natural resources is ensured.

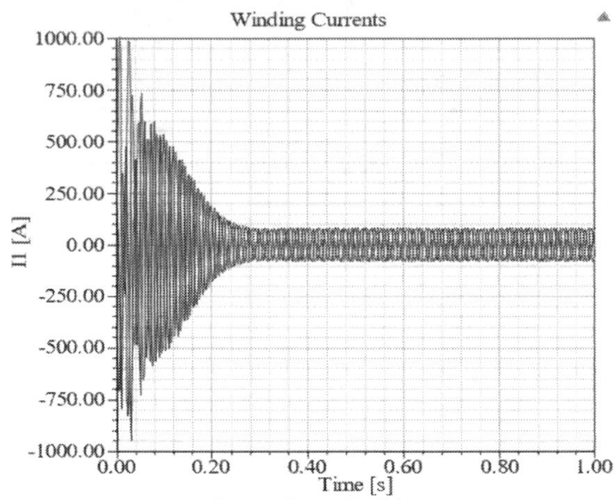

Dependence $I=f(t)$

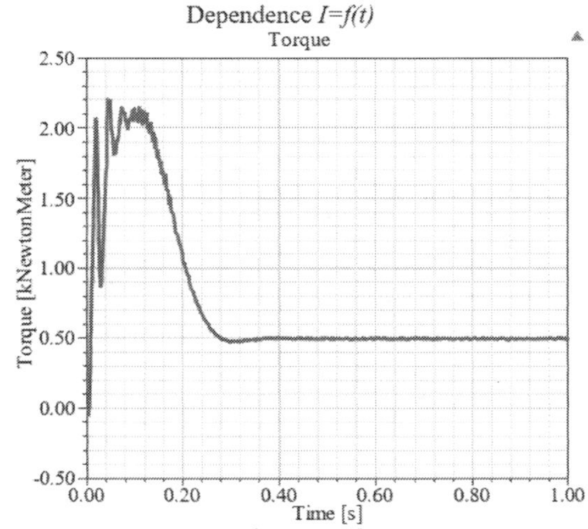

Dependence $M=f(t)$

Fig.6. Switching motor poles ABTM -6/12 from p_2=6 to p_1=3 at a load of 500 Nm

Thus, the experiment carried out on self-propelled mine wagons operating in the harsh conditions of potash mines showed that the electric drives of the undercarriage of a mine self-propelled car based on a two-speed asynchronous motor with a variable number of poles satisfy all basic requirements.

4 Conclusions

As the results of the analysis showed, the new pole switching winding scheme has electromagnetic properties that are as close as possible to traditional two-layer windings.

Using the DSSF method, a new circuit of a pole-switched winding for a pole ratio of 3/6 with a switching circuit "Y/YY with additional branches" and "Y-Δ/YY" with a number of stator slots equal to 72 has been developed, for which a patent for the invention has been received.

The new pole switching winding was tested on an asynchronous motor built on the basis of the AVTM-6/12 motor, developing a rated power of 55 kW/26 kW for 3/6 poles, respectively.

References

1. Toshov Dj.B., Rismukhamedov D.A., Tuychiev F.N., Khakberdiev A.L. Experimental study of dynamic modes of operation of electric motors of running gear of mine self-propelled wagons. Universum: technical sciences: electron. scientific magazine 2022. 3(96). URL:https:// 7universum.com/ru/tech/archive/item/13289.
2. Khakberdiev A.L. Causes of failure of underground self-propelled cars and measures to eliminate them // Mining Bulletin of Uzbekistan - 2015. No. 2. - S. 77-79.
3. Haqberdiev A.L, Toshov J.B. Analysis of the control system of electric motors of the running gear of self-propelled mine cars used in complex mining and technological conditions. E3S Web of Conferences 216, 01135 (2020) RSES 2020. https://doi.org/10.1051/ e3sconf/ 202021601135.
4. Auinger, H. (1978), "Polumschaltbare Dreiphasenwicklung mit 6 Klemmen Ubersicht zum Stand der Technik". Bulletin des Schweizerischen Elektrotechnischen Vereins, Vol. 69, No. 17, pp.926-932.
5. Bogatyrev, N.I., Kreymer, A.S., Vanurin, V.N. and Dganibekov K.A. (2016), "Modulated stator windings of the motor for fan drives", Polythematic Online Scientific Journal of Kuban State Agrarian University, Vol. 118 No.4, pp.1441-1458.
6. Kovacs, C. (2018), "Pole-Changing Windings for Close Ratio and 1:N Ratio Using the 3//Y/3//Y method" Advances in Science, Technology and Engineering Systems Journal, Vol. 3 Iss. 4, pp.241-253.
7. van der Giet, M., & Hameyer, K. (2007). Induction motor with pole-changing winding for variable supply frequency. 2007 IEEE International Electric Machines & Drives Conference, 2, pp.1484-1489.
8. Melcescu, L.M., Cistelecan, M.V., Craiu O. and Cosan B. (2010), "A new 4/6 pole-changing double layer winding for three phase electrical machines", The XIX International Conference on Electrical Machines (ICEM), Rome, Italy, pp.1-6.
9. Broadway, A.R.W. and Ismail, K.S. (1986), "Phase modulated 3-phase pole changing windings", IEE Proceedings B (Electric Power Applications), vol. 133, Iss. 2, pp.61-70.
10. Popov, D.A. and Popov, S.D (1994), "Three-phase pole-changing winding with a pole pair ratio p_1:p_2=1:2". RF patent №2012981, 15 May.
11. Alwash, J.H., Ismail, K.S. and Eastham J.F. (2000), "A novel 16/6 phase modulated winding", IEEE Transactions on Energy Conversion, Vol. 15 Iss. 2, pp.188–190.
12. Liu, H., Wang, J. and Zhang Z. (2016), "Performance analysis of variable speed multiphase induction motor with pole phase modulation", Archives of electrical engineering, Vol. 65 No. 3, pp.425-436.
13. Karimov, Kh.G. and Tupoguz, A. (1987) "Method for Designing Electrically Aligned Windings in AC Machines" Electricity, Vol 9, pp.29-38.
14. Rismukhamedov, D., Bobojanov, M., Tuychiev, F., Shamsutdinov, Kh. (2021). "Development and research of pole-changing winding for a close pole ratio" E3S Web of Conferences 264, (CONMECHYDRO–2021), Tashkent, pp.1-7
15. Tuychiev F. AIP Conference Proceedings **2552.** 030013. (2022). doi.org/10.1063/5.0112349
16. Bobojanov, M., Rismuxamedov, D., Tuychiev, F., Shamsutdinov, Kh. and Magdiev, Kh. (2020), "Pole-changing motor for lift installation", E3S Web of Conferences 216 (RSES 2020), Kazan, Russia, pp.1-5.

Saving electrical energy by using induction motors with pole changing windings in the water supply system

Makhsud Bobojanov[1], *Sardor* Torayev[2,*]

[1]Tashkent State Technical University named by Islam Karimov, Tashkent, 100095, Uzbekistan

[2]"TIIAME" NRU Bukhara Institute of Natural Resources Management, Bukhara, 105009, Uzbekistan

Abstract. This article provides information on the level of provision of clean drinking water to the population of Republic of Uzbekistan and the work carried out in the field, as well as the deficiencies that arise in the water supply enterprise and measures to eliminate them. The operation modes of the enterprise's pump units, the analysis of the main parameters and the possibility of saving electricity and water consumption due to the application of two-speed asynchronous motors with pole change winding for such equipment are presented.

1 Introduction

The world's demand for electricity is increasing at a high rate, the main reasons for which are the growth of the world's population, the development of industry and agriculture.

According to the International Energy Agency, electric motors consume 46% of the electricity produced in the world. This indicator makes up about 70% of the electricity consumed by the industry. As a result, the demand for energy-efficient and high-efficiency electric motors has increased [1].

In order to further accelerate the ongoing work on providing the residents of our republic with clean and high-quality drinking water and improving wastewater services, to create decent living conditions and maintain their health by increasing the coverage of drinking water supply and wastewater services.

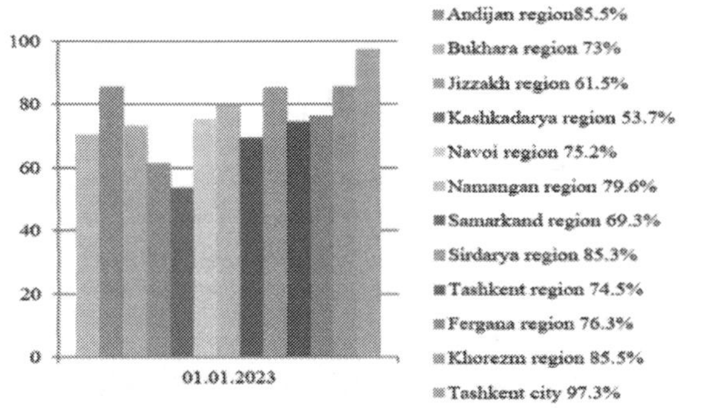

Fig.1. The level of provision of clean drinking water to the population of Uzbekistan

The development strategy of the President of New Uzbekistan for the period of 2022-2026 basically sets the tasks of bringing the level of drinking water supply to the population of the republic to 87%, updating wastewater systems in 32 large cities and 155 district centers.

If we pay attention to the pictures given above, it can be seen that the level of provision of clean drinking water in some regions and cities of our Republic is low. Conducting research in the field and creating modern energy-saving technologies is an urgent issue of today.

2 The current state of the problem

The provision of centralized clean drinking water to the population of Bukhara region is 69% as of January 1, 2022, and 73% as of January 1, 2023.

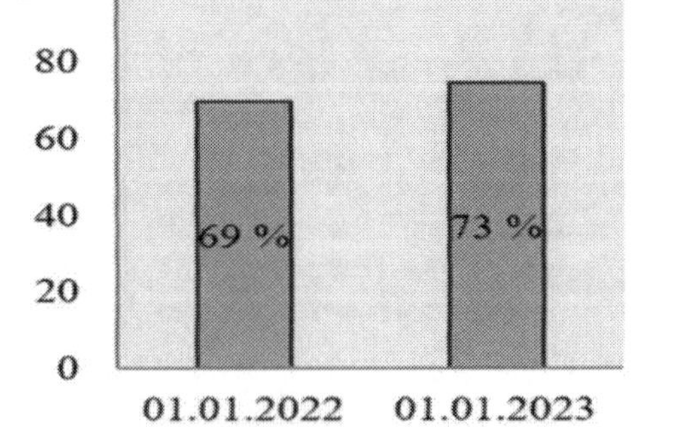

Fig.2. Clean drinking water consumption in the period of 2022-2023 of Bukhara region

* Corresponding author: torayevs43@gmail.com

It can be seen that providing the population with drinking water is an urgent issue today. Here it is worth noting that repairs of pump aggregates of currently operating water supply enterprises, extending their service life, saving electricity, reducing water wastage, protecting the environment and preventing emergency situations, it is important to provide timely and quality service, use of automatic tools.

In the course of conducting research, one-speed electric motors, malfunctions in water pipes due to pressure changes, several times connecting and disconnecting motors to the network due to changes in water consumption, and a number of other problems may be encountered in the water supply company. Such problems lead to a decrease in the efficiency of the water supply company.

The power consumption graph of the electric motors of the city water supply company is variable, due to the increase and decrease of the demand for clean drinking water during the day, the motors work with full load and without load. When the motors work without load, the waste of electric energy increases and efficiency decreases. If we take the Bukhara water supply company as a research object. There are 2 asynchronous motors for 800 kW and 400 kW with a total power of 1200 kW are operating at the pump station. Water consumption mainly occurs in June, July and August. At these times, an additional 400 kW electric motor is activated, and the total power consumption reaches 1600 kW.

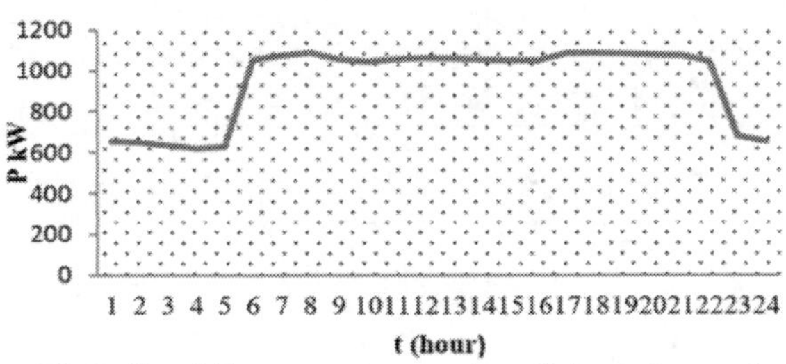

Fig.3. The daily power consumption of the water supply company

It can be seen from the above figure that the water consumption during a day is variable, mainly between 24:00 in the evening and 5:00 in the morning, as the pressure in the water pipes increases and the consumption decreases. As a result, electric motors run without load and waste power.

With aim to find the power consumed by the water supply company, may be determine the average power consumption and determine the electricity consumed during the year [1-5].

We determine the average power consumption of pump units using the following formula.

$$P_{avg} = \frac{\sum_{1}^{i=24} P}{i}; \quad (1)$$

Average power consumption based on the above values

$$P_{avg} = 950\ kW$$; is equal to

Can be determine the electricity consumption of asynchronous motors without changing the rotation speed is equal to

$$W = P_{o'r} \cdot T = 950 \times 8760 = 8322000\ kW \cdot \text{h} \quad (2)$$

According to the pictures obtained from the water supply company in Bukhara, the electric energy consumed by electric motors in one year is 8 322 000 $kW \cdot$ h.

3 New proposed idea, obtained results

Speed control of single-speed asynchronous motors, which are used in practice at pumping stations, due to changes in water consumption, starting and determining their main parameters in these processes is the main factor of economy.

In world practice, there are several ways to control the speed of asynchronous motors, including asynchronous valve cascade, thyristor frequency converter, as well as multi-speed asynchronous motors with changing poles. The method of speed control with multi-speed pole-changing asynchronous motors is the simplest and relatively inexpensive method.

In addition, the voltage can drop by more than 30% as a result of direct starting of motors that receive power from weak mains.

In this case, first of all, the torque developed by the electric circuit decreases, and secondly, the electronic and microprocessor devices that record the voltage drop disconnect the electric circuit from the network.

The use of pump units powered by a two-speed asynchronous motor facilitates the step-by-step start-up process of high-power motors [3-11, 12].

There are several advantages to applying adjustable electrical drives to water supply pumps.

In this case, opportunities are created to optimize technological processes, reduce the number of pumping units, save water consumption, adjust the pressure in water pipes, facilitate service, and save electricity. Also, during the start-up of high-power single-speed electric motors in the water supply company, malfunctions occur in the pipes as a result of the increase in pressure in the water pipes. Based on the above information, there is a need to apply adjustable electrical circuits to the pumps in the water supply company. Instead of single-speed asynchronous motors, the power consumption of two-speed motors with a ratio of 4/6 poles is shown in the graph below [13-20].

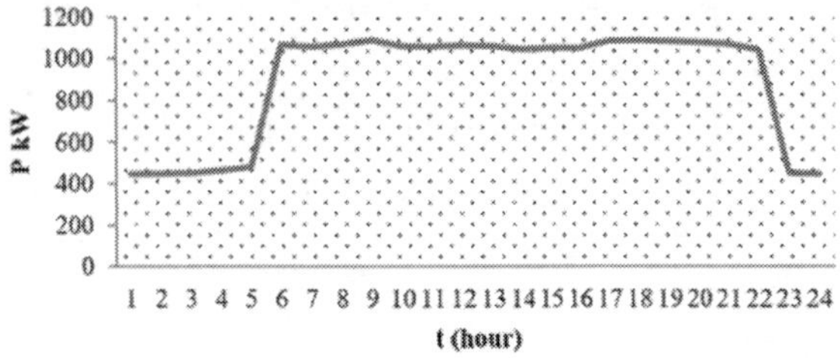

Fig.4. The graph of power consumption in the case of applying two-speed asynchronous motors with changing poles to the pumps of the water supply company.

It can be concluded from the graph that in times of low water consumption, the speed of the electric motor is reduced, i.e. it is changed to 1000 rev/min. As a result, power consumption of motors is reduced, and electricity saving is achieved.

Electric energy consumed by the use of two-speed asynchronous motors with alternating poles for water supply pumps [4 - 8].

The rotation frequency of an asynchronous electric motor is determined by the following formula.

$$\omega = \frac{2\pi f(1-s)}{p}; \quad (3)$$

In here; f - industrial frequency; s- slip; p- number of pairs of poles

We take a slip to s=0,03 in the nominal operating mode of the electric motor, the result of this, angular velocity of the electric motor $\omega_{p=2} = 152,3 \, rad/s$; is equal to;

When the speed of the pump units is changed, the water consumption is determined as follows.

p_1=2 namely n=1500 rpm;

$$\frac{Q_1}{Q_2} = \frac{\omega_1}{\omega_2}; \quad (4)$$

$$\frac{1500}{Q_2} = \frac{157}{152,3};$$

$$Q_{p=2} = 1455 \frac{m^3}{h} = 0,0404 \frac{m^3}{s}; \text{ is equal to.}$$

$$p = 3; \; \omega_{p=3} = 101,5; \text{ is equal to.}$$

$$\frac{Q_1}{Q_3} = \frac{\omega_1}{\omega_3}; \quad (5)$$

$$Q_{p=3} = 1000 \frac{m^3}{h} = 0,28 \frac{m^3}{s};$$

The method of finding the discharge height of the pump unit is as follows.

$$\frac{H_1}{H_2} = (\frac{\omega_1}{\omega_2})^2; \; H_2 = 61m; \quad (5)$$

$$\frac{H_1}{H_3} = (\frac{\omega_1}{\omega_3})^2; \; H_{p=3} = 28m \quad (6)$$

The power consumed by the pump is determined by the following formula.

$$N_{p=2} = \frac{\rho \cdot Q \cdot H}{102 \cdot \eta} = \frac{1000 \cdot 0,404 \cdot 61}{102 \cdot 0,74} = 330 kW; \quad (7)$$

$$P_{p=2} = \frac{N}{\eta} = \frac{330}{0,88} = 375 kW; \quad (8)$$

$$N_{p=2} = \frac{\rho \cdot Q \cdot H}{102 \cdot \eta} = \frac{1000 \cdot 0,28 \cdot 28}{102 \cdot 0,72} = 103 kW; \quad (9)$$

$$P_{p=2} = \frac{N}{\eta} = \frac{103}{0,72} = 143 kW; \quad (10)$$

Above are the results obtained when the rotation speed of the 400 kW asynchronous motor is changed from 1500 rpm to 1000 rpm, and the above formulas were used to change the parameters of the 800 kW motor as a result of the speed change [21-25]. The results of recalculation of two-speed asynchronous motors are presented in the table below.

Table 1. The results of changing the parameters of the pump unit due to the adjustment of the rotation speed of the asynchronous motor.

№	n rpm	P kW	Q m³/sec	H m	N kW
1	1500	P_1=375	0,42	65	330
	1000	P_2=146	0,28	28	103
2	1500	$P_1' = 772$	0,89	65	687
	1000	$P_2' = 300$	0,58	28	220

$$W_2 = [(P_1 + P_1') \cdot t_1 + (P_2 + P_2') \cdot t_2] \cdot 365 \quad (11)$$

Here P_1 and P_1' are the power consumption of the first P_2 and P_2' second electric motors at 1500 and 1000 rpm

t_1 is the time of increased water demand

t_2 is the time of reduced water consumption

$W_2 = [(375 + 772) \cdot 8 + (146 + 300) \cdot 16] \cdot 365 = 5953880 \; kW \cdot h$

$$\Delta W = W_1 - W_2 = 8322000 - 5953880 = 2368120 \; kW \cdot h \quad (12)$$

2368120 $kW \cdot h$ electric energy was saved in one year due to the use of economical two-speed motors with alternating poles instead of asynchronous motors of the water supply company's pumping station [1-11].

4 Conclusion

Summarizing the above research work on rational use of electrical energy and water resource through using of motor with pole-changing windings in the power supply system, can conclude the following:

- the use of two-speed asynchronous motors of the water supply enterprise allows to obtain economic benefits of 710 436 000 sums (around 6,3 million dollar) in one year, as well as to use drinking water efficiently.
- the use of high-power two-speed electric motors with alternating poles facilitates the start-up process and reduces power consumption.
- that allows to adjust the pressure in the water pipes and prevent malfunctions.
- it give the opportunities for rational use of water due to changes in water consumption. It is advisable to use such two-, three- and four-speed asynchronous motors with alternating poles for water supply pump electric drives.

References

1. Vishnu K., Arun Sh., Subramaniam U., Shanmugam P., Norbert H., A comprehensive review on energy efficiency enhancement initiatives in centrifugal pumping system, Applied Energy, **Volume 181**, 2016, Pages 495-513, ISSN 0306-2619,
2. Decree of the President of the Republic of Uzbekistan dated January 28, 2022 PF-60 "On the Development Strategy of New Uzbekistan for 2022-2026".
3. M.Bobojanov. AIP Conference Proceedings **2552**. 050034. (2023). https://doi.org/10.1063/5.0114077
4. Bobojanov M.K., Karimov R.Ch., Qosimov T.H. E3S Web of Conferences **289**. 07012. (2021). https://doi.org/10.1051/e3sconf/202128907012
5. Rismukhamedov D., Bobojanov M., Tuychiev F., Shamsutdinov K. E3S Web of Conferences **264**. 03057. (2021). https://doi.org/10.1051/e3sconf/202126403057
6. M.Bobojanov, D.Rismuxamedov, F.Tuychiev, K.Shamsutdinov, K.Magdiev. E3S Web of Conferences, **216**. 01164. (2020).
7. https://doi.org/10.1051/e3sconf/202021601164
8. R.Karimov, M.Bobojanov, E3S Web of Conferences **216**. 01162 (2020).
9. https://doi.org/10.1051/e3sconf/202021601162
10. Bobojanov M.K., Ziyodulla O.E.Ismoilov M.T.U,.Arziev E.I.U, Togaeva G.Z. E3S Web of Conferences **177**. 03023. (2020). https://doi.org/10.1051/e3sconf/202017703023
11. Usmanov E.G., Rasulov A.N, Bobojanov M.K., Karimov R.C. E3S Web of Conferences **139.** 01079. (2019). https://doi.org/10.1051/e3sconf/201913901079
12. Rakhmonov I.U., Najimova A.M., and Reymov K.M. AIP Conference Proceedings **2647**. 030010. (2022). https://doi.org/10.1063/5.0104788
13. Rakhmonov I.U., Najimova A.M. AIP Conference Proceedings **2647.** 030011. (2022). https://doi.org/10.1063/5.0104791
14. Rakhmonov I.U., Najimova, A.M., Esemuratova Sh.M., Koptileuov T.T. AIP Conference Proceedings **2647.** 070024. (2022). https://doi.org/10.1063/5.0104793
15. Hoshimov F.A., Rakhmonov I.U., Niyozov N.N., Omonov F.B. AIP Conference Proceedings **2647.** 030025. (2023). https://doi.org/10.1063/5.0112388
16. Rakhmonov I.U., Hoshimov F.A., Kurbonov N.N., Jalilova D.A. AIP Conference Proceedings **2647.** 050022. (2023). https://doi.org/10.1063/5.0112391
17. Rakhmonov I.U., Ushakov V.Ya., Niyozov N.N., Kurbonov N.N., Mamutov M. E3S Web of Conferences **289.** 07014. (2021). https://doi.org/10.1051/e3sconf/202128907014
18. Rakhmonov I.U., Ushakov V.Ya., Najimova A.M., Jalilova D.A., Omonov F.B. E3S Web of Conferences **289.** 07013. (2021). https://doi.org/10.1051/e3sconf/202128907013
19. Usmanov E.G., Khusanov B.M. RSES 2020. E3S Web of Conferences **216.** 01161. (2020). https://doi.org/10.1051/e3sconf/202021601161
20. Erejepov, M., Novikov, A.N., Khusanov, B.M., Seytmuratov, B., Sayimbetov, Z. E3S Web of Conferences. **289.** (2021). 07018. https://doi.org/10.1051/e3sconf/202128907018
21. Khakimov, H.T., Shayumova, Z.M., Kurbanbaeva, Z.K., Khusanov, B.M. E3S Web of Conferences **139.** 01076. (2019). https://doi.org/10.1051/e3sconf/201913901076
22. Taslimov A.D. AIP Conference Proceedings **2552.** (2023). 050023. https://doi.org/10.1063/5.0112398
23. Melikuziev M.V., Nematov L.A., Novikov A.N., Baymuratov K.K. E3S Web of Conferences **289.** 07016 (2021). https://doi.org/10.1051/e3sconf/202128907016
24. A.D.Taslimov, A.S.Berdishev, F.M.Rakhimov, Melikuziev M.V. E3S Web Conf. **Vol. 139**. 2019. https://doi.org/10.1051/e3sconf/201913901081.
25. A.D.Taslimov, A.S.Berdishev, F.M.Rakhimov, Melikuziev M.V. E3S Web Conf. **Vol.139**. 2019. https://doi.org/10.1051/e3sconf/201913901082

Methods for determining static characteristics in industrial electrical networks

Khilola Nazirova*

Tashkent State Technical University named by Islam Karimov, Tashkent, 100095, Uzbekistan

Abstract. The article considers the influence of voltage quality on the efficiency of electrical networks and electrical receivers. The economic characteristics of consumers and load nodes and methods of their construction are given. The construction of static characteristics of load nodes is considered by analyzing the mathematical model of the node and using the results of a passive experiment and the possibility of estimating the stability reserve, total power and energy losses and constructing economic characteristics by voltage from them.

1 Introduction

The static characteristics of the load nodes *(UN)* are the dependences of their input active and reactive capacities on voltage. These characteristics are determined in the range of permissible voltage changes at the terminals of U_N $U = (0.9\text{-}1.1)\ U_n$. The form of these characteristics depends on the configuration of the power supply network, the number of energy transformations and the composition of consumers, lighting, etc. Static characteristics *P(U), Q(U)* can be set graphically and analytically, and often have a V-shaped, nonlinear shape, contain valuable information about the *UN*. They are necessary for determining the active and reactive capacities in the node and their losses, for constructing economic characteristics, calculating the static stability of the *UN*, etc [1-6, 8]. Static characteristics of *UN* can be determined by the following methods: analytical; method of active experiment, method of passive experiment. In a particular case, the *UN* can be bipolar, i.e. it has a one-way power supply. The analytical method is based on the known composition of the node load-the scheme of the power supply system and the number of energy consumers. At the same time, a replacement scheme of the *UN* is drawn up, for which the dependences of active and reactive power on voltage are determined, according to which it is possible to build the most graphs [7-15, 18]. The analytical method also makes it possible to study the parameters of the design scheme, as well as the quantitative composition of energy consumers for static characteristics *P(U), Q(U)*. This method is applicable at the operational level, if it is possible to determine the composition of the load with acceptable accuracy.

2 The current state of the investigated problem

To construct the static characteristics of the *UN* by the method of active experiment, it is necessary to be able to regulate an independent variable, the voltage at the input terminals of the load node in the permissible range (0.9-1.1) U_n, and for each value of U to record the steady-state values of *P* and *Q*. When conducting experiments, it is necessary to observe the conditions so that all other factors affecting *P, Q* are absent or minimal compared to the influence of U. To weaken the influence of side factors, experiments are carried out during the period of practical stationary operation of the *UN*. The period of stationary capacity *P, Q* can be determined by a static survey of load schedules *P(t), Q(t)* for typical days (working and non-working days). Thus, the influence of all other factors on *P, Q*, except voltage, is excluded, and 8-10 values of *P, Q* are recorded when adjusting U in the interval (0.9-1.1) U_n. These points have a random spread around the desired actual dependence *P(U), Q(U)*, and therefore finding a static characteristic for these experimental points is performed by methods of mathematical statistics, for example, the least squares method. At the same time, "smoothing" is performed at the beginning and then an analytical expression of the desired curve is given, the coefficients of which are determined by the least squares method or by methods of minimizing a nonlinear function with an estimate of the approximation error [16-22, 9, 10]. The construction of

* Corresponding author: ruzinzarov88@bk.ru

statistical characteristics of the *UN* by the active method is complicated by the regulation of voltage at the input terminals in the interval (0,9-1,1) U_{nom} and difficulties in constructing characteristics for all necessary modes and structure of the load node. When constructing the static characteristics of the *UN* by the method of passive experiment, there is no need to adjust the voltage at the input terminals of the UN. At the same time, changes in *U, P, Q* are isolated and recorded in the normal stationary mode of operation of the unit. Studies show that stresses with a period of several minutes are created by random phenomena in the system that are external to the *UN*, so they can be considered as disturbances at the terminals of the *UN, P, Q*-as a reaction of the *UN* to these disturbances. Thus, by measuring averaged over a time interval of values of ΔU, ΔP, ΔQ, we find the regulatory effects $k\_p=\Delta P/\Delta U$ and $k\_Q=\Delta Q/\Delta U$. Quantitatively, ΔU, ΔP, ΔQ is about one percent of the basic values of these quantities, and their slow changes practically do not cause dynamic processes in the *UN*. According to the known values of the powers P, Q and the regulating effects *k_p, k_Q*, it is possible to determine the static characteristics of the UN by determining the parameters of the substitution scheme of the UN by synthesis methods or using the known regression equations [23-25, 5] k_p, Q_p с k'_p, k'_Q.

where $k'_p = \frac{d^2p}{dU^2}$; $k'_Q = \frac{d^2Q}{dU^2}$

As noted, the shape of the experimental curves of static characteristics *P(U)*, *Q(U)* is similar to V-shaped lines. Analytically, these curves in the range of permissible voltage variation $(0,9\div1,1)\cdot U_n$ can be approximated by: second-order polynomials; dependencies corresponding to the selected structure of the substitution scheme of the UN. Second - order polynomials:

$$P = P_0(a_0 + a_1k_u + a_2k_u^2) \quad (1)$$

$$Q = Q_0(b_0 + b_1k_u + b_2k_u^2) \quad (2)$$

where $k_u=U/U_0$; U_0, P_0 〖Q〗 *0* are the values of voltage, active and reactive power at the input terminals of the *UN* in the initial normal mode; P, Q are the current power values corresponding to the current voltage values.

For a particular *UN*, the coefficients of the polynomials are determined based on the experimentally obtained static characteristics by the least squares method. To satisfy the conditions for passing polynomials through the points $P(U_0)=1$ *u* $Q(U_0)=1$, it is necessary to observe the conditions of restriction $a_0+a_1+a_2= 1$, $b_0+b_1+b_2=1$. Substituting these conditions in (1), (2), we get

$$\frac{P}{P_0} = 1 - a_1(1-k_u) - a_2(1-k_u^2);$$

$$\frac{Q}{Q_0} = 1 - b_1(1-k_u) - b_2(1-k_u^2).$$

Using the experimental points of static characteristics, we find the coefficients a_1 a_2, b_1 b_2 of these expressions, and then a_0 and b_0. The standard error of the approximating curves (1), (2) is determined from the expressions.

$$\sigma_p = \sqrt{\frac{\sum_1^n(P_e-P_p)^2}{n}}; \qquad \sigma_Q = \sqrt{\frac{\sum_1^n(Q_e-Q_p)^2}{n}},$$

where *P_e,Q_e,P_p,Q_p* are experimental and calculated power values; n is the number of experimental points.

Based on the values *σ_p* and *σ_Q*, a conclusion is made about the admissibility of using approximating expressions (1), (2). Below are some recommended values of the coefficients a_i 〖,b〗 _i [3] Approximate values:

$$P = P_0(0{,}83 - 0{,}3k_u + 0{,}47k_u^2); \quad (3)$$

$$Q_{NOkV} = Q_0(3{,}7 - 7k_u + 4{,}3k_u^2); \quad (4)$$

$$Q_{6kV} = Q_0(4{,}9 + 10{,}1k_u + 6{,}2k_u^2). \quad (5)$$

$$Q = Q_0(5{,}6 - 12{,}3k_u + 7{,}7k_u^2););\quad (6)$$

$$Q = Q_0(6{,}7 + 15{,}3k_u + 9{,}55k_u^2) \quad (7)$$

when $\cos\varphi_0=0{,}85$. The polynomial approximation of static characteristics is the discrepancy of such an expression to the physical content of processes in the *UN*. Extrapolation of these characteristics by other voltage values, e.g. lower and higher for tuning purposes anti-accident automation, as well as for the decomposition of capacities into the main components for economic calculations by polynomials, is not possible. Analytical expressions of static characteristics corresponding to the physical meaning of the processes are more suitable in *UN*.

3 Definition of critical elements in energy systems

In most cases, the complex load of the *UN* consists of a combination of asynchronous, synchronous motors and passive load (furnaces, lighting, welding, etc.). These three equivalent elements can be taken as a replacement scheme (Fig. 1).

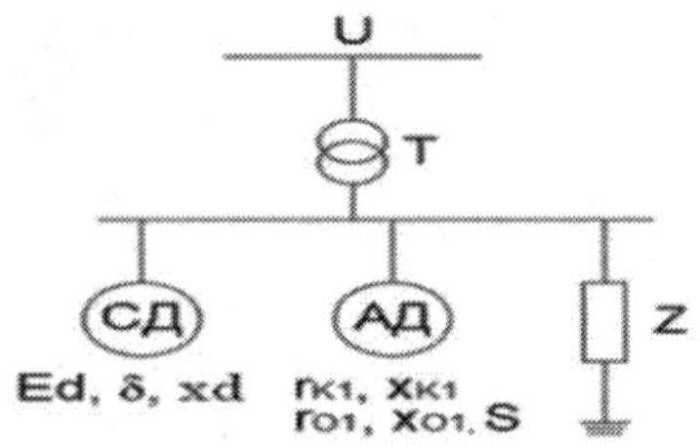

Fig.1. Scheme of asynchronous and synchronous motors and passive load

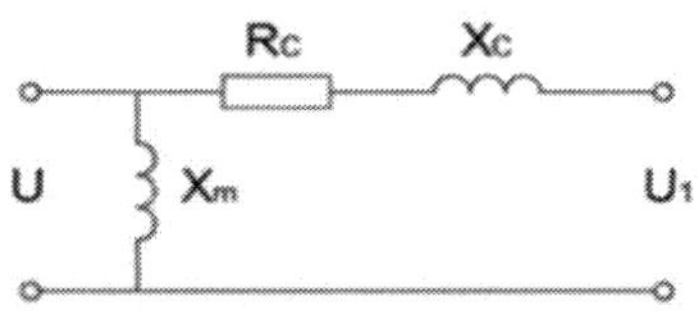

Fig.2. Scheme branch magnetization network transformers

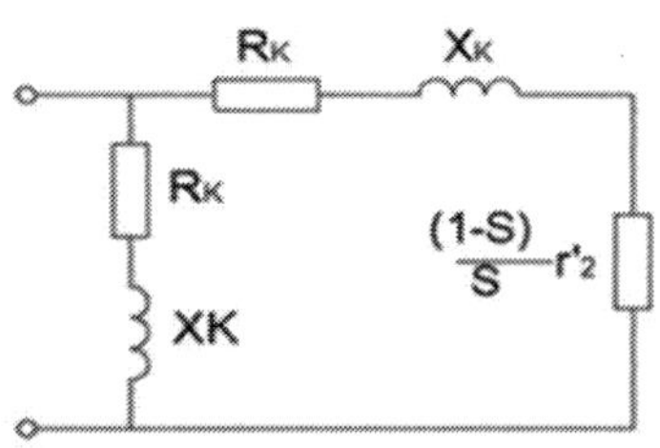

Fig.3. Scheme of asynchronous load with " G" figurative equivalent circuit

In general, the supply network has a complex configuration with many stages of energy transformation and can be represented by an L-shaped circuit with a longitudinal equivalent resistance $Z_c=R_c+jX_c$ and a transverse reactance X_m, which is a branch of magnetization of network transformers (Fig.2).

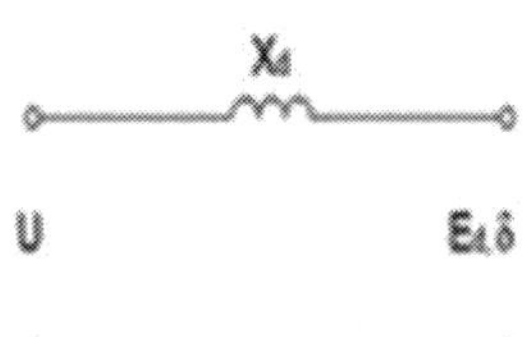

Fig.4. Scheme with synchronous resistance X_d and EDF E_d

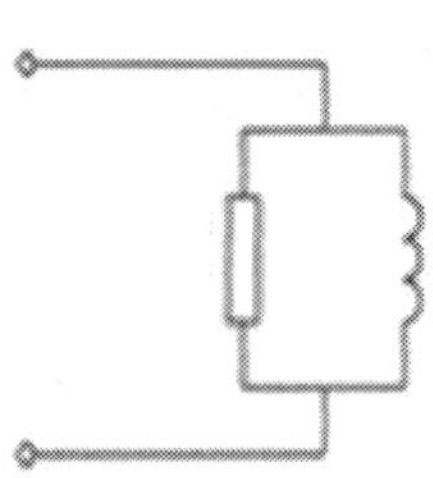

Fig.5. Scheme Circuit passive resistance load Z_n

In turn, the asynchronous load is represented by an L-shaped replacement circuit (Fig.3), the synchronous load is represented by resistances x_d and EMF E_d (Fig.4), and the passive load is represented by resistance Z_n, consisting of parallel connected $R_{(n)}$ and X_p (Fig.5). Then the equivalent substitution scheme (Fig.1) for the UN will correspond to the following analytical expressions of the powers P,Q at the output terminals of the UN.

$$P = P_{\text{пом}}\left(\frac{1-\frac{S_{\text{пом}}}{k_u^{\alpha}}}{1-S_{\text{пом}}}\right)^{\beta_{\text{н}}} + P_{\text{сн}} + P_n k_u^{\alpha} + \frac{\Delta P_{In}}{k_u^2} + \Delta P_{On} k_u^{\alpha_1}; \quad (9)$$

$$Q = Q_{On} k_u^4 - Q_{1C}\left(\frac{k_u^4-b^2}{1-b^2}\right)^{0,5} + Q_{2C} k_u^4 + Q_n k_u^2 + \frac{\Delta Q_{In}}{k_u^{\alpha}} + +\Delta Q_{On} k_u^{\alpha_1}. \quad (10)$$

Where *P_IH, P*_sn are the useful power on the shaft of equivalent asynchronous and synchronous motors at rated voltage; *P_n, Q_n* are the active and reactive power consumed by an equivalent passive element at rated voltage; *P_OH,Q_OH* - active and reactive magnetization power of an equivalent asynchronous motor at rated voltage; Δ*P_OH,*Δ*Q_OH* - active and reactive load losses caused by the magnetization current of an equivalent asynchronous motor at rated voltage; Δ*P_IH,*Δ*Q_IH*- active and reactive load losses caused by the load currents of equivalent asynchronous and synchronous motors; *Q_1C,Q_2C*- reactive power of an equivalent synchronous motor generated and consumed at rated voltage; α,α1- coefficients; *α=2:4,α_1=2:6; β_n* is the coefficient characterizing the moment of resistance of the driven mechanism; b is the coefficient taking into account the influence of the degree of loading and the excitation current of the synchronous load, is the sliding of an equivalent asynchronous motor at rated voltage. All powers are expressed relative to their values at U_n [4].

4 Conclusion

The advantage of the load node models in the form of (9) and (10) is the possibility of determining the component capacities depending on the voltage, the possibility of synthesizing the parameters of equivalent elements replacement circuits for calculating the dynamic characteristics of the load node. All this is very important when solving problems in the field of optimizing the quality of the voltage of the network (node) and finding critical voltages for analyzing the stability of the *UN* operation. It is obvious that for a given static characteristic of *UN*, if necessary, it is possible to switch from the analytical form of polynomials to dependencies (9) and (10).

References

1. A.Rasulov. AIP Conference Proceedings 2552, **050017**, (2023), https://doi.org/10.1063/5.0111530
2. A.N.Rasulov, M.R.Ruzinazarov, and ofters. E3S Web of Conferences 289, **07007**, (2021), https://doi.org/10.1051/e3sconf/202128907007
3. A.N.Rasulov, G.R.Rafikova, and ofters. E3S Web of Conferences 289, **07006**, (2021), https://doi.org/10.1051/e3sconf/202128907006
4. Zykin F.A., Kakhanovich V.S. Measurements and accounting of electric energy. Moscow: Energoatomizdat, 1982.
5. Pospelov G.E. Electrical systems and circuits. G.E. Pospelov, V.T. Fedik, P.V. Pychev. Minsk: Technoprint, 2004. 720s.
6. Fursanov M.I. Determination and analysis of electricity losses in electrical networks of energy systems / M.I. Fursanov. - Minsk.: Belenergosberezhenie 2005. - 207p.
7. A.Rasulov, E.Usmonov, and M.Melikuziev. AIP Conference Proceedings 2552, **050018**, (2023), https://doi.org/10.1063/5.0111531
8. Rakhmonov I.U., Najimova A.M., and Reymov K.M. AIP Conference Proceedings **2647**. 030010. (2022). https://doi.org/10.1063/5.0104788
9. Rakhmonov I.U., Najimova A.M. AIP Conference Proceedings **2647.** 030011. (2022). https://doi.org/10.1063/5.0104791
10. Rakhmonov I.U., Najimova, A.M., Esemuratova Sh.M., Koptileuov T.T. AIP Conference Proceedings **2647.** 070024. (2022). https://doi.org/10.1063/5.0104793
11. Hoshimov F.A., Rakhmonov I.U., Niyozov N.N., Omonov F.B. AIP Conference Proceedings **2647.** 030025. (2023). https://doi.org/10.1063/5.0112388
12. Rakhmonov I.U., Hoshimov F.A., Kurbonov N.N., Jalilova D.A. AIP Conference Proceedings **2647.** 050022. (2023). https://doi.org/10.1063/5.0112391
13. Melikuziev M.V. AIP Conference Proceedings **2552.** (2023). 050021. https://doi.org/10.1063/5.0112395
14. Rasulov A.N., Usmonov E.G., Melikuziev M.V. AIP Conference Proceedings **2552.** (2023). 050018. https://doi.org/10.1063/5.0111530
15. Taslimov A.D. AIP Conference Proceedings **2552.** (2023). 050023. https://doi.org/10.1063/5.0112398
16. Melikuziev M.V., Nematov L.A., Novikov A.N., Baymuratov K.K. E3S Web of Conferences **289.** 07016 (2021). https://doi.org/10.1051/e3sconf/202128907016
17. A.D.Taslimov, A.S.Berdishev, F.M.Rakhimov, Melikuziev M.V. E3S Web Conf. **Vol. 139**. 2019. https://doi.org/10.1051/e3sconf/201913901081.
18. A.D.Taslimov, A.S.Berdishev, F.M.Rakhimov, Melikuziev M.V. E3S Web Conf. **Vol.139**. 2019. https://doi.org/10.1051/e3sconf/201913901082
19. Taslimov A.D., Melikuziev M.V., Najimova A.M., Alimov A.A. E3S Web of Conferences **216**. 01159 (2020). https://doi.org/10.1051/e3sconf/202021601159
20. Saidkhodjaev A.G., Nuriddinova Kh.R. AIP Conference Proceedings **2552.** 050037. (2023). https://doi.org/10.1063/5.0133486
21. R.Karimov. AIP Conference Proceedings 2552. **030014**. (2022). https://doi.org/10.1063/5.0111533
22. R.Karimov. AIP Conference Proceedings 2552. **050012**. (2022). https://doi.org/10.1063/5.0111524
23. S.Dzhuraev, R.Karimov, and ofters. ElConRus, pp. 1166-1169. (2022). doi: 10.1109/ElConRus54750.2022.9755782
24. Rakhmonov I.U., Ushakov V.Ya., Niyozov N.N., Kurbonov N.N., Mamutov M. E3S Web of Conferences **289.** 07014. (2021). https://doi.org/10.1051/e3sconf/202128907014
25. Rakhmonov I.U., Ushakov V.Ya., Najimova A.M., Jalilova D.A., Omonov F.B. E3S Web of Conferences **289.** 07013. (2021). https://doi.org/10.1051/e3sconf/202128907013

A study of a hybrid type stand-alone 3 kW photovoltaic system of Karakalpak state university

Reymov *K.M.* [1*], Esemuratova *Sh.M.*[1], *Khusanov B.M.*[2], Mytnikov *A.V.*[3]

[1]Karakalpak State University, Nukus, 230101, Uzbekistan

[2]Tashkent State Technical University named by Islam Karimov, Tashkent, 100095, Uzbekistan

[3]Tomsk Polytechnic University, Tomsk

Abstract. This article examines the analysis of a 3 kW solar photovoltaic plant installed at the Karakalpak State University. It studied the location of the station and the analysis of sunny days on it. The efficiency of the generating capacity of this photovoltaic plant was also determined during the day, month and all year in comparison with its location.

1 Introduction

It is well known that Uzbekistan is geographically located in favorable climatic conditions for the development of solar energy. The summer climate is cloudless with a sufficiently large number of sunny days, such as more than 300 days, which leads to the use of all types of solar installations [1-6, 11]. On the other hand, despite abundant natural energy sources such as gas, oil, and coal, electrical energy is insufficient in Uzbekistan. It is also known to cause massive power loss due to the aging of power plants and transmission and distribution power facilities. For these reasons, the government has recently decided to increase the renewable energy ratio to 25% by 2030 and build three 100 MW solar power plants by 2021 [7-10, 12].

Currently, photovoltaic power plants' cost is drastically reduced due to technological developments such as modules and inverters. However, to maintain performance for many years as 20 years suitably, various technologies such as technical verification of modules and inverters, optimal design of power plants using such as PVsyst, Solarius PV, and EasyPower, and installation technologies are required [13-20, 8, 7].

For this study, the authors reviewed the installed 3 kW photovoltaic system and analyzed the server's data. In addition, using the Solarius - PV software, the authors simulated the 3 kW system and compared recorded generated electric energy yields with simulation results.

2 Review the 3 kW photovoltaic system

The 3 kW hybrid-type photovoltaic system was installed in January 2020 and currently in operation at Karakalpak state University in Nukus city (42° 45' 32" N, 59 ° 62' 68" E) such as photo 1. The module is a ground-mounted type with a 35° tilt but a 20° azimuth south-west for harmony with the university's main building.

Fig.1. 3 kW photovoltaic system

* Corresponding author: kamal_tstu@mail.ru

The system consists of a hybrid type inverter, polysilicon type modules, and batteries, as shown in Table 1, such as the following. The modules consist of three strings with three parallels, a total of nine modules, and four batteries were series-connected [21-26, 9, 14].

Table 1. Technical specifications of the 3 kW system components

Hybrid Inverter	Module	Battery
Manufacturer: WE Energy Model: WI-3000(48VDC) Hybrid Max. PV Array DC Input: 145V	Manufacturer: Trina Solar Co., Ltd Model: TSM-350PE-15H	Manufacturer: Guangzhou Jalon PowerCo., Ltd Model: NP200-12
AC Input/Output: 1 Phase 220V	Rating: 350W	Rating: 12V 200AH
Capacity: 3 kW	Vmp: 37,9V	
	Imp: 9,23A	

Figure 2 is the main screen of the monitoring system, which is operated through the Internet. The data server is located in Korea and periodically stores data related to the inverter, modules, and batteries. Operators can analyze present and past periods using the monitoring program's functions or save data sets in Excel format to use various software to interpret them.

Fig.2. Main monitoring interface screen

Figure 3 shows two representative daily power patterns. Left is a sunny day without clouds, but the right is a clean day with passing clouds fast.

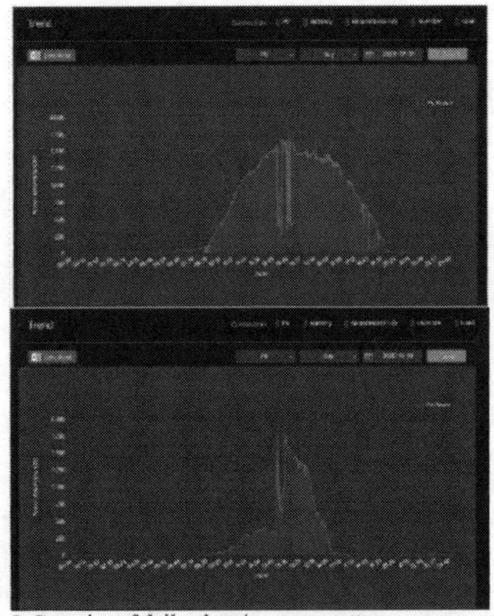

Fig. 3. Samples of daily electric power patterns

The authors use the Solarius - PV simulation software of the ACCA Company to simulate the 3 kW photovoltaic system [4]. Figure 4 shows the database of used inverter and module and modules' tile and azimuth.

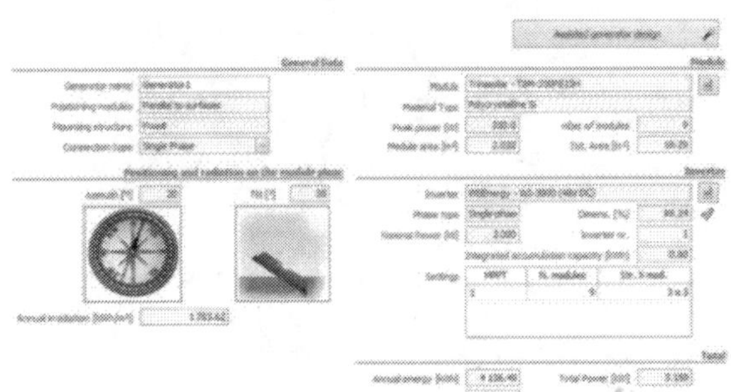

Fig.4. An interface screen of the Solarius - PV design and simulation software Simulation and Analysis

As a result, the software configured three strings and three parallel circuits automatically. Also, it showed some simulation results, such as annual solar radiation 1753.62 kWh/m2 for the modules for given tilt and azimuth, and yearly 4136.49 kWh generated electric energy.

The software shows more detailed monthly energy generation, such as the following graph 1. It indicated that July produces the most 450.12 kWh, and December generated the least 186.93 kWh.

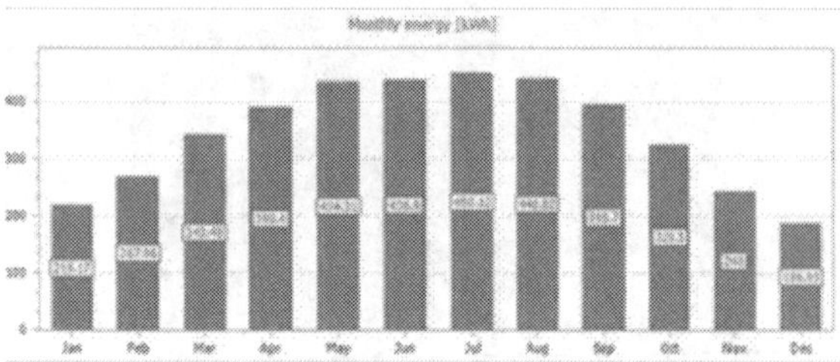

Fig.5. Simulation results for monthly generated electric energy

Figure 6 below compares the Solarius - PV simulation based on real input data and recorded server computer data from February to October 2020. From February to June, simulated and generated energy results are similar, but the difference was slightly larger in other months. According to the recorded data, the generated energy is zero for two days in February and March, thirteen days in August, and four days in September. The authors have a plan to analyze some reasons in detail.

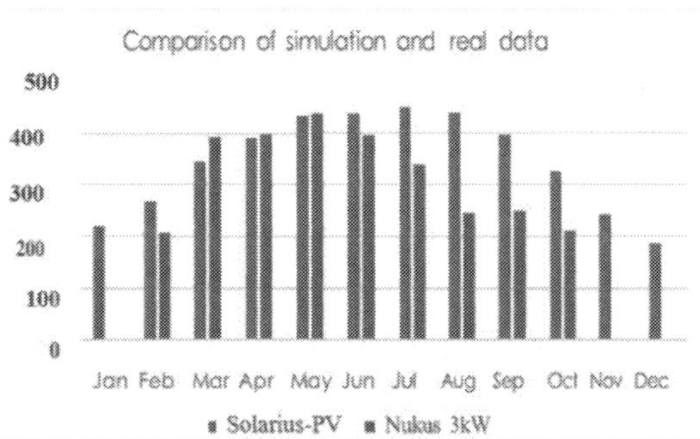

Fig.6. Simulated and real generated electric energy

3 Conclusion

The results of the installation and operation of the 3 kW on-grid hybrid type photovoltaic system (storage system: GEL type 12V 200AH 4ea) installed in Nukus city, Karapalpakstan Republic of Uzbekistan, are as follows.

1. After nine months of operation of the 3kW photovoltaic system, it was confirmed that it

Operates well in the climate and power system environment.

2. Through the real-time remote monitoring system, the authors confirmed that the operating conditions (power generation, the status of each part, environmental temperature, etc.) of the system installed at a distance could be usefully utilized anywhere the Internet is connected.

3. As a result of comparing the power generation record and the simulation result of Solarrius- PV software, four months were similar. But the remaining five months showed a little deviation. The authors will analyze the different reasons in a future thesis.

The authors thank KEA (Korea Energy Association), Hanbek Construction, and WE Energy for support the 3 kW photovoltaic system. Also, thank the ACCA Company for allowing authors to use the Solarius-PV software.

References

1. Abdullaev D.A. Operational processing of measurement information in automated control systems. - Tashkent, Sub, 1985, 128 p.

2. Venikov V.A., Venikov G.V. The theory of similarity and modeling (in relation to the problems of the electric power industry). - M.: Higher school, 1984, 439 pp.

3. Nikiforov G.V., Oleinikov V.K., Zaslavets B.I. Energy saving and energy management in metallurgical production. M: Energoatomizdat, 2003.- 480 p.

4. Novikov SS Methods of operational planning and control of power consumption of large-capacity electric steel-smelting furnaces when operating on the wholesale electricity market: Avtorefer. dis. ... Cand. those. sciences. - M., 2008.- 18 p.

5. Resolution of the Cabinet of Ministers of the Republic of Uzbekistan dated January 12, 2018 N 22 "On additional measures to improve the procedure for the use of electrical energy and natural gas."

6. Decree of the President of the Republic of Uzbekistan no. PP-4477 of April 10, 2019 "On approval of the strategy for the transition of the Republic of Uzbekistan to the green economy for the period 2019—2030". http://www.1ex.uz/docs/4539506.

7. https://www.accasoftware.com/en/solar-design-software/ photovoltaic software (Solarius-PV) site Accessed a Rakhmonov I.U., Najimova A.M., and Reymov K.M. AIP Conference Proceedings **2647.** 030010. (2022). https://doi.org/10.1063/5.0104788

8. Rakhmonov I.U., Najimova A.M. AIP Conference Proceedings **2647.** 030011. (2022). https://doi.org/10.1063/5.0104791

9. Rakhmonov I.U., Najimova, A.M., Esemuratova Sh.M., Koptileuov T.T. AIP Conference Proceedings **2647.** 070024. (2022). https://doi.org/10.1063/5.0104793

10. Hoshimov F.A., Rakhmonov I.U., Niyozov N.N., Omonov F.B. AIP Conference Proceedings **2647.** 030025. (2023). https://doi.org/10.1063/5.0112388

11. Rakhmonov I.U., Hoshimov F.A., Kurbonov N.N., Jalilova D.A. AIP Conference Proceedings **2647.** 050022. (2023). https://doi.org/10.1063/5.0112391

12. Rakhmonov I.U., Ushakov V.Ya., Niyozov N.N., Kurbonov N.N., Mamutov M. E3S Web of Conferences **289.** 07014. (2021). https://doi.org/10.1051/e3sconf/202128907014

13. Rakhmonov I.U., Ushakov V.Ya., Najimova A.M., Jalilova D.A., Omonov F.B. E3S Web of Conferences **289.** 07013. (2021). https://doi.org/10.1051/e3sconf/202128907013s of Oct. 2019

14. Rakhmonov, I.U., Reymov, K.M. E3S Web of Conferences **216.** 01167. (2020). DOI: 10.1051/e3sconf/202021601167

15. Rakhmonov, I.U., Hoshimov, F. E3S Web of Conferences **209**. 07018. (2020). DOI: 10.1051/e3sconf/202020907018

16. Rakhmonov, I., Berdishev, A., Niyozov, N., Muratov, A., Khaliknazarov, U. IOP Conference Series: Materials Science and Engineering, 2020. 883(1). 012103. DOI: 10.1088/1757-899X/883/1/012103

17. Rakhmonov, I.U., Niyozov, N.N. E3S Web of Conferences, 2019. **139.** 01077. DOI: 10.1051/e3sconf/201913901077

18. Rakhmonov, I.U., Reymov, K.M. Journal of Physics: Conference Series, 2019. 1399(5). 055038. DOI: 10.1088/1742-6596/1399/5/055038

19. A.D.Taslimov, A.S.Berdishev, F.M.Rakhimov, Melikuziev M.V. E3S Web Conf. **Vol. 139**. 2019. https://doi.org/10.1051/e3sconf/201913901081.

20. A.D.Taslimov, A.S.Berdishev, F.M.Rakhimov, Melikuziev M.V. E3S Web Conf. **Vol.139**. 2019. https://doi.org/10.1051/e3sconf/201913901082

21. Taslimov A.D., Melikuziev M.V., Najimova A.M., Alimov A.A. E3S Web of Conferences **216**. 01159 (2020). https://doi.org/10.1051/e3sconf/202021601159

22. Saidkhodjaev A.G., Nuriddinova Kh.R. AIP Conference Proceedings **2552.** 050037. (2023). https://doi.org/10.1063/5.0133486

23. Saidkhodjaev A.G. AIP Conference Proceedings **2552**. 050026. (2023). https://doi.org/10.1063/5.0111540

24. Usmonov E.G. AIP Conference Proceedings **2552.** 050020. (2023). https://doi.org/10.1063/5.0111537

25. Rakhmonov,I., Berdishev,A., Khusanov, B.M, Khaliknazarov, U., Utegenov, U. IOP Conference Series: Materials Science and Engineering **883(1)**. 012104. (2020). DOI 10.1088/1757-899X/883/1/012104

26. I.U. Rakhmonov, T.H. Khakimov, I.I. Bakhadirov, B.M. Khusanov. International Journal of Advanced Science and Technology **Vol. 29**. (2020). pp. 1525-1533.

 https://doi.org/10.1051/e3sconf/202338401048

Critical evaluation of energy use in industrial enterprises

Bobojonov *Y.M.*[1,*], Saidkhodjaev *A.G.*[2]

[1]Karakalpak State University, Nukus, 230101, Uzbekistan

[2]Tashkent State Technical University named by Islam Karimov, Tashkent, 100095, Uzbekistan

Abstract. The paper analyzes the use of energy resources in industrial enterprises. Based on the results of the work carried out, the main points of the irrational use of energy resources are given. An analysis of the state of accounting and control over the consumption of energy resources and an overview of the current situation with energy saving at industrial enterprises is given.

1 Introduction

The presented review of literary sources shows that at present the problems of energy saving in industry are solved with varying degrees of completeness, with a predominance of simplified methods. These methods are based mainly on the quantitative indicators of industrial enterprises in the production of final products, without studying the underlying processes of industrial production. As a result, in some cases, only those reserves of energy savings that are on the surface and reserves from the implementation of measures that do not require experimental and methodological guidelines can be identified. Deeper, comprehensive studies of the processes of power consumption are still carried out at a very limited number of enterprises. Naturally, such a situation does not always make it possible to present the true value of energy indicators [1-6, 11, 12].

In addition to the above, it should be noted that the system of analysis and control of energy indicators adopted at the present time is rather primitive. So, for example, in many industries, analysis and control of energy indicators is carried out periodically, by means of special measurements for various periods of time. Naturally, such measurements cannot reflect the entire dynamics of changes in the calculated parameters and do not make it possible to identify the pattern of their changes.

Our studies at a number of industrial enterprises in various industries have shown that with the existing organization of accounting for the consumption of energy carriers and energy resources, it is impossible to assess the real efficiency of their use and reasonably normalize the consumption of energy resources [7-10, 13].

An analysis of the state of accounting and control over the consumption of energy resources at industrial enterprises showed that, with rare exceptions, this issue remains practically insufficiently resolved. So, for example, the unsatisfactory situation that characterizes the majority of enterprises takes place with the instrumentation of the entire energy supply system. As a rule, all enterprises have commercial electricity metering; as for the metering of other types of energy, in many cases it is absent.

Accounting for energy resources in certain industries, production units, energy-intensive units and processes, also does not exist at all enterprises. In addition, as a rule, there is no accounting for individual components of the technological process, such as compressed air, nitrogen, hydrogen, water, etc. This leads to the loss of opportunities for the correct distribution, accounting and control of energy consumption.

A big disadvantage is the weak introduction in industry of automated systems for accounting and control of energy resources, which are installed at single industrial facilities and are used very inefficiently. As a rule, these systems take into account and control only the following parameters: power consumption, energy consumption and indicators of other energy carriers, and thus are used mainly as informational ones.

However, these functions are clearly insufficient for energy saving management, since the obtained information on the consumption of all types of energy resources cannot be used to solve the main energy saving tasks necessary for optimal energy consumption management [15-19, 14].

* Corresponding author: kamal_tstu@mail.ru

2 The current state of the investigated problem

The above review of the current situation with energy saving in industry shows that this problem is in most cases solved without in-depth research and is devoted to solving particular problems of rationalizing energy use processes. In addition, the solutions obtained for individual private tasks cannot be copied for other enterprises, since each of them has its own specifics in terms of the mode of operation (single-shift, two-shift, round-the-clock), products (homogeneous, heterogeneous, multi-assortment, etc.) etc.

At the same time, it should be noted that when developing priority measures, one should not be limited to identifying and eliminating the causes that lie on the surface (overstating the power of equipment, the presence of idle operation of this equipment, an irrational mode of electric lighting, insufficient thermal insulation of equipment, etc.).

The energy intensity of industrial products, as the main indicator of the energy saving process, is a function of many variables associated with electrical, mechanical, thermal, organizational, physical and chemical, as well as other production factors that affect this indicator. Based on this, in order to manage the energy saving process, it is necessary to develop an adaptive control system that is self-adjusting to any changes, both internal and external factors, which allows minimizing the energy component in the cost of products.

The norms existing at the enterprises of the cotton and silk industry, obtained by experimental statistical methods, not only do not stimulate the rationalization of electricity consumption, but do not even stimulate the correct organization of accounting and control over the consumption of electricity. The development of reasonable standards can be carried out only if the entire set of data necessary for this has been identified. In addition to the correct organization of electricity metering, a detailed analysis of electrical loads and electricity consumption by units, workshops and the enterprise as a whole, factors affecting the specific electricity consumption should be identified, energy characteristics should be built, experimental data should be mathematically processed, etc. This work, as V. I. Lapitsky correctly points out, should be carried out on the basis of a systematic approach with the solution of problems of optimizing the energy economy and rational use of electricity.

Naturally, the implementation of such, in some cases, a significant amount of research, requires certain financial outlays on the part of the enterprise to pay for the work of a research or other specialized organization, or even when these works are performed by the energy services of the enterprises themselves. Despite the fact that electricity rationing is a prerequisite for the correct organization of production, that this is a necessary prerequisite for the implementation of all measures to save energy, proper planning and forecasting, such a general assessment of the need for rationing is still insufficient. An operating self-supporting enterprise requires a specific assessment of the economic efficiency of each such work. This assessment should be based on the fact that the process of establishing scientifically based norms is associated, as mentioned above, with the optimization of power consumption modes, i.e. identifying the main, potentially possible measures to reduce losses and unproductive costs of electricity.

3 Definition of critical elements in energy systems

Considering that the studies related to the development of standards require many years for their implementation, all costs associated with the regulation should be brought to the year preceding the beginning of the estimated year of implementation.

The issue of the efficiency of instrument installation, energy metering and the development of standards should be addressed comprehensively, as a single event.

However, speaking about the organization of accounting for all types of energy, it follows, in our opinion, to single out electrical energy from the general problem.

The fact is that the authors of proposals for evaluating the effectiveness of instrument metering are based mainly on examples of installing meters for steam, water, air, gas, etc. These rather expensive devices require significant additional installation and installation costs, especially in those cases where they were not provided for by the project. Therefore, proposals on the economic justification of these works are legitimate.

The situation is different for electricity. Estimated electricity metering is installed at the inputs of all, without exception, industrial enterprises, and thus, we can only talk about justifying the installation of technical (control) metering. This accounting, with rare exceptions, is already available in all shops and energy-intensive units. In addition, at all substations for protection purposes and measurements, there are measuring current and voltage transformers, which means that if there is no technical metering of electricity anywhere, then there are all conditions for installing a meter, and the costs for these purposes are reduced to purchasing a meter [20-24, 8].

The effect from the introduction of measures to save electricity, identified in the process of research and development of standards, was adopted by us as a criterion for the expediency of developing the whole range of issues of regulation, and not just electricity metering.

Very often, energy savings can be obtained as a result of purely technological measures (improvement of technology, installation of more productive machines, etc.). In this case, the electricity component of the total savings should be taken into account.

4 Conclusion

It should also be noted that the proposal to compensate for the costs of rationing by savings from a decrease in specific electricity costs cannot always be used, because. a correctly set rate may be higher than the actual rate due to the introduction of energy-intensive technology. In addition, the specific power consumption may be close in value to the limit of the potential value. Then there will either be no savings, or its magnitude will be very insignificant.

References

1. Han F., Zio E., Kopustinskas V., Praks P. Risk, Reliability and Safety: Innovating Theory and Practice, pp. 2565-2571. (2016). DOI: 10.1201/9781315374987-389.
2. Su H., Zio E., Zhang J. Li X. Reliability Engineering & System Safety, **Volume 175**, Pages 79-91, (July 2018). DOI: 10.1016/j.ress.2018.03.006.
3. Jonsson H., Johansson J., Johansson H. Proceedings of the Institution of Mechanical Engineers, Part O: Journal of Risk and Reliability. **Vol. 222**. No. 2. P. 235-243. (2008). DOI: https://doi.org/10.1243%2F1748006XJRR138.
4. Li T., Eremia M., Shahidehpour M., IEEE Transactions on Power Systems, **vol. 23**, p 1817-1824, (2008). DOI: 10.1109/TPWRS.2008.2004739.
5. Dokic S.B., Rajakovic N.Lj. Energy (2018), DOI: 10.1016/j.energy. 2018.04.165
6. Thompson, J. R., Frezza, D., Necioglu, B., Cohen, M. L., Hoffman, K., Rosfjord, K. International Journal of Critical Infrastructure Protection **Volume 24**, p. 144-165. (2019).
7. Kai L., Ming W., Weihua Z., Jinshan W., Xiaoyong Y. International Journal of Critical Infrastructure Protection **Volume 23**, p. 79-89. (2018).
8. Tichy L. International Journal of Critical Infrastructure Protection, (2019). DOI: 10.1016/j.ijcip.2019.01.003.
9. Tsavdaroglou M., Al-Jibouri S. H.S., Bles T., Halman J. I.M. International Journal of Critical Infrastructure Protection **Volume 21**, p. 57-71. (2018).
10. Praks P., and Kopustinskas V. "CRITIS 2018, LNCS 11260, pp. 3–16. (2019). DOI: 10.1007/978-3-030-05849-4_1
11. Rakhmonov I.U., Najimova A.M., and Reymov K.M. AIP Conference Proceedings **2647.** 030010. (2022). https://doi.org/10.1063/5.0104788
12. Rakhmonov I.U., Najimova A.M. AIP Conference Proceedings **2647.** 030011. (2022). https://doi.org/10.1063/5.0104791
13. Rakhmonov I.U., Najimova, A.M., Esemuratova Sh.M., Koptileuov T.T. AIP Conference Proceedings **2647.** 070024. (2022). https://doi.org/10.1063/5.0104793
14. Hoshimov F.A., Rakhmonov I.U., Niyozov N.N., Omonov F.B. AIP Conference Proceedings **2647.** 030025. (2023). https://doi.org/10.1063/5.0112388
15. Rakhmonov I.U., Hoshimov F.A., Kurbonov N.N., Jalilova D.A. AIP Conference Proceedings **2647.** 050022. (2023). https://doi.org/10.1063/5.0112391
16. Rakhmonov I.U., Ushakov V.Ya., Niyozov N.N., Kurbonov N.N., Mamutov M. E3S Web of Conferences **289.** 07014. (2021). https://doi.org/10.1051/e3sconf/202128907014
17. Rakhmonov I.U., Ushakov V.Ya., Najimova A.M., Jalilova D.A., Omonov F.B. E3S Web of Conferences **289.** 07013. (2021). https://doi.org/10.1051/e3sconf/202128907013
18. R.Karimov. AIP Conference Proceedings 2552. **030014**. (2022). https://doi.org/10.1063/5.0111533
19. R.Karimov. AIP Conference Proceedings 2552. **050012**. (2022). https://doi.org/10.1063/5.0111524
20. S.Dzhuraev, R.Karimov, and ofters. ElConRus, pp. 1166-1169. (2022). doi: 10.1109/ElConRus54750.2022.9755782
21. Melikuziev M.V. AIP Conference Proceedings **2552.** (2023). 050021. https://doi.org/10.1063/5.0112395
22. Rasulov A.N., Usmonov E.G., Melikuziev M.V. AIP Conference Proceedings **2552.** (2023). 050018. https://doi.org/10.1063/5.0111530
23. Taslimov A.D. AIP Conference Proceedings **2552.** (2023). 050023. https://doi.org/10.1063/5.0112398
24. Melikuziev M.V., Nematov L.A., Novikov A.N., Baymuratov K.K. E3S Web of Conferences **289.** 07016 (2021). https://doi.org/10.1051/e3sconf/202128907016

Optimization of electricity generation mode, energy-intensive units, operating and various regimes production

Hoshimov, *F.A.*[1,*], Najimova *A.M.*[2], Rafikova *G.R.*[1], Bakhadirov *I.I.*[1]

[1]Tashkent State Technical University named by Islam Karimov, Tashkent, 100095, Uzbekistan

[2]Karakalpak State University, Nukus, 230101, Uzbekistan

Abstract. The article provides an analysis of the optimization of the modes of electrical loads of energy-intensive units operating in various production modes, in particular, the mode of continuous operation of the unit, the mode of continuous operation, the mode of continuous operation of the unit with a constant hourly output, the mode of continuous operation of the unit with varying productivity, the operation of the unit in variable mode. And also measures are proposed to reduce power consumption, taking into account the technological process.

1 Introduction

An analysis of the work of enterprises in the cotton and silk industries shows that they have significant reserves of energy savings that require the most complete identification and rational use. Their implementation largely depends on the correct organization and feasibility of rationing electricity consumption. The scientifically based rate of specific electricity consumption per unit of output creates the basis for calculating the electricity needs of production sites, workshops and the enterprise as a whole [1-5, 16, 17].

2 The current state of the investigated problem

The work of industrial enterprises is carried out on the basis of progressive rationing, the purpose of which is to determine the consumption of electricity per unit of output for specific production conditions, ensure the rational and economical use of electricity in the production process, and establish the initial value for determining the need for production in electricity.

One of the most important issues to be solved in the development of energy saving technology for power-consuming industrial facilities is to ensure the minimum level of power consumption. At the same time, it should be borne in mind that the nature of the power consumed is strictly subject to the technological regulations of the unit operating in both constant and variable load modes. The complexity of solving the problem is also in the fact that under operating conditions there is a significant deviation of electrical loads due to the influence of various random factors [6-12, 18, 19].

The technical literature discusses methods for constructing "typical graphs" of electrical loads based on identification classification methods that allow the use of mathematical models to solve optimization problems.

However, the use of a set of "typical graphs" of electrical loads and their probabilistic characteristics does not provide the desired accuracy in the calculation of electrical loads and their optimization in the conditions of operation of energy-intensive equipment, since the gap between the calculated and actual loads of industrial enterprises is estimated at 200% [13-15, 20-24, 2].

Studies carried out from the standpoint of a systematic approach make it possible to develop methods for optimizing electrical loads in determining clear dependencies between the operation of an electrical wire and their energy performance.

Below we consider the solution of the problem for the most typical operating conditions of the units.

3 The mode of continuous operation of the unit.

The continuous operation mode of the unit has two characteristic features [25-27, 8, 14]:

a) continuous operation with constant hourly output;

b) continuous operation mode with varying performance.

* Corresponding author: ilider1987@yandex.ru

$$P = P_{id} + P_{aux.mech} + P_u \quad (1)$$

When the units are operating in the and modes, the power consumption can be expressed by the following formula: where is the P_{id} -idle power; $P_{aux.mech}$ - power of auxiliary mechanisms; P_u - useful power of the unit. In those cases when, under the influence of random factors, there is a fluctuation in the value of P, the average value is taken as the calculated load. Power, as a rule, can be taken as a constant value. An exception can be made for units, idling power, which may undergo a certain change during the billing period. An example of this is cement tube ball mills, whose grinding media lose their weight over time, as a result of which the idle power consumption is significantly reduced [28-33, 21, 9].

The mode of continuous operation of the unit with a constant hourly output. In this mode, the power consumption is practically unchanged, provided:

$$P_{id} = const; P_{aux.mech} = const; P_l = const; A = const$$

Under the condition, $P_{id} = \text{var}; P_{aux.mech} = const$ the resulting power for the billing period varies within certain limits.

The mode of continuous operation of the unit with varying performance. Under operating conditions, there may be various disturbances in the rhythm of the supply of processed products, which is associated with a change in productivity and, accordingly, power consumption. In this case, $P = f(A)$ the nature of the change in power consumption can be expressed.

4 Operation of the unit in variable mode.

The operation of a number of units can be characterized by the following modes:

1) The mode of alternation of the maximum load with a complete shutdown of the unit.

2) The mode of alternating the maximum load with the idling of the unit.

3) Mode of alternating maximum and reduced load.

In all these modes, the average power is taken into account, which is determined depending on the performance in accordance with formula (1).

A special place in the calculation of the power consumption in the unit when operating in these modes is occupied by the starting costs of the units, which in some cases have a significant share in the total power consumption. The value of starting costs (start and stop mode) has a significant impact on the energy performance of the unit and for each of them it has its own characteristics. The value of these costs in most cases is determined experimentally and should be taken into account in the process of optimizing power consumption modes.

Measures to reduce power consumption. The choice of the most advantageous mode of operation of the unit, providing a reduction in power consumption, can be ensured by the following measures:

- reduction of the idle speed of the unit due to its correct operation, adjustment of the unit,
- timely and high-quality lubrication, modernization and replacement of individual components;
- reduction of idle time when operating in variable mode;
- reducing the load power of the unit by regulating its speed, monitoring the quality of processed products;
- control of the current and voltage regime during the operation of electrothermal units, as well as optimization of their values, taking into account the specific features of the technological process.

The choice of the most advantageous mode of operation of the unit under conditions a and b can be carried out in accordance with the nomogram constructed according to the expression $P = f(A)$. In particular, $0 \rightarrow P_{max}$ with performance A, the mode will be preferable if, according to the conditions of the technology, the specified mode can be allowed. In this case, the average power $P_{id} \rightarrow P_{max}$ for the regime will be reduced by the value $P_A = P_B$.

5 Conclusion

The energy services of textile and cotton enterprises still allow inaccuracies in the rationing, evaluation and distribution of energy resources, since they do not fully take into account the specifics of the production process, the power consumption modes of technological equipment, the features of its operation, therefore, it is necessary to improve methods for analyzing and predicting the power consumption of the plant (industrial Association), which is a self-supporting unit with an independent balance sheet and reporting.

References

1. Han F., Zio E., Kopustinskas V., Praks P. Risk, Reliability and Safety: Innovating Theory and Practice, pp. 2565-2571. (2016). DOI: 10.1201/9781315374987-389.

2. Su H., Zio E., Zhang J. Li X. Reliability Engineering & System Safety, **Volume 175**, Pages 79-91, (July 2018). DOI: 10.1016/j.ress.2018.03.006.

3. Jonsson H., Johansson J., Johansson H. Proceedings of the Institution of Mechanical Engineers, Part O: Journal of Risk and Reliability. **Vol. 222**. No. 2. P. 235-243. (2008). DOI: https://doi.org/10.1243%2F1748006XJRR138.

4. Li T., Eremia M., Shahidehpour M., IEEE Transactions on Power Systems, **vol. 23**, p 1817-1824, (2008). DOI: 10.1109/TPWRS.2008.2004739.

5. Dokic S.B., Rajakovic N.Lj. Energy (2018), DOI: 10.1016/j.energy. 2018.04.165

6. Thompson, J. R., Frezza, D., Necioglu, B., Cohen, M. L., Hoffman, K., Rosfjord, K. International Journal of Critical Infrastructure Protection **Volume 24**, p. 144-165. (2019).

7. Kai L., Ming W., Weihua Z., Jinshan W., Xiaoyong Y. International Journal of Critical Infrastructure Protection **Volume 23**, p. 79-89. (2018).

8. Tichy L. International Journal of Critical Infrastructure Protection, (2019). DOI: 10.1016/j.ijcip.2019.01.003.

9. Tsavdaroglou M., Al-Jibouri S. H.S., Bles T., Halman J. I.M. International Journal of Critical Infrastructure Protection **Volume 21**, p. 57-71. (2018).

10. Praks P., and Kopustinskas V. "CRITIS 2018, LNCS 11260, pp. 3–16. (2019). DOI: 10.1007/978-3-030-05849-4_1

11. Iakubovskii D., Komendantova N., Rovenskaya E., Krupenev D., Boyarkin D. Geosciences (Switzerland). **Vol.9. №1**. ID: 54. (2019). DOI: 10.3390/geosciences 9010054

12. Rakhmonov I.U., Najimova A.M., and Reymov K.M. AIP Conference Proceedings **2647**. 030010. (2022). https://doi.org/10.1063/5.0104788

13. Rakhmonov I.U., Najimova A.M. AIP Conference Proceedings **2647**. 030011. (2022). https://doi.org/10.1063/5.0104791

14. Rakhmonov I.U., Najimova, A.M., Esemuratova Sh.M., Koptileuov T.T. AIP Conference Proceedings **2647**. 070024. (2022). https://doi.org/10.1063/5.0104793

15. Hoshimov F.A., Rakhmonov I.U., Niyozov N.N., Omonov F.B. AIP Conference Proceedings **2647**. 030025. (2023). https://doi.org/10.1063/5.0112388

16. Rakhmonov I.U., Hoshimov F.A., Kurbonov N.N., Jalilova D.A. AIP Conference Proceedings **2647**. 050022. (2023). https://doi.org/10.1063/5.0112391

17. Rakhmonov I.U., Ushakov V.Ya., Niyozov N.N., Kurbonov N.N., Mamutov M. E3S Web of Conferences **289**. 07014. (2021). https://doi.org/10.1051/e3sconf/202128907014

18. Rakhmonov I.U., Ushakov V.Ya., Najimova A.M., Jalilova D.A., Omonov F.B. E3S Web of Conferences **289**. 07013. (2021). https://doi.org/10.1051/e3sconf/202128907013

19. A.V. Edelev, S.M. Enikeeva, S.M. Senderov,

20. *Computational technologies*. – Vol. 4, **№ 5**. – p. 30 – 35. (1999).

21. S.V. Vorobev, A.V. Edelev, *Software products and systems*. – **№3**. – p. 174 – 177. (2014).

22. L.R. Ford, D.R. Fulkerson. *Flows in networks*. Princeton University Press. 276p. (1966).

23. Vorobev S.V., Edelev A.V. Scientific Bulletin of NSTU **Vol. 62, No. 1**, p. 181-194. (2016).

24. Yu.P. Korotaeva, R.D. Margulova, *Extraction, preparation and transport of natural gas and condensate. Reference manual in 2 volumes. Volume II*. Nedra, 288 p., (1984).

25. *Export of the Russian Federation of the most important goods in 2011 - 2016 (according to the Federal Customs Service of Russia)* http://customs.ru/index.php?option=com_ newsfts&view =category&id=52&Itemid=1978&limitstart=60.

26. Hoshimov, F.A., Rakhmonov, I.U., Niyozov, N.N. E3S Web of Conferences **209**. 07017. (2020). DOI: 10.1051/e3sconf/202020907017

27. Rakhmonov, I.U., Reymov, K.M. E3S Web of Conferences **216**. 01167. (2020). DOI: 10.1051/e3sconf/202021601167

28. Rakhmonov, I.U., Hoshimov, F. E3S Web of Conferences **209**. 07018. (2020). DOI: 10.1051/e3sconf/202020907018

29. Rakhmonov, I., Berdishev, A., Niyozov, N., Muratov, A., Khaliknazarov, U. IOP Conference Series: Materials Science and Engineering, 2020. 883(1). 012103. DOI: 10.1088/1757-899X/883/1/012103

30. Rakhmonov, I.U., Niyozov, N.N. E3S Web of Conferences, 2019. **139**. 01077. DOI: 10.1051/e3sconf/201913901077

31. *Ministry of Energy of the Russian Federation. Statistics*. http://minenergo.gov.ru/activity/statistic.

32. S.M. Senderov, V.I. Rabchuk, S.V. Vorobev, *Proc. of the col. of rep. Methodological issues of reliability research of large energy systems. 90th meeting "Reliability of developing energy systems"*. July 1-7, Irkutsk. (2018).

33. Feoktistov A., Gorsky S., Sidorov I., Kostromin R., Edelev A., Massel L. Communications in Computer and Information Science. **Vol. 965**. P. 289-300. (2019).

Current source converter into stabilized voltage source based on electromagnetic ferromagnetic circuit

Ruzinazarov *M.R.* *

Tashkent State Technical University named by Islam Karimov, Tashkent, 100095, Uzbekistan

Abstract. The article considers a series ferroresonant circuit connected in parallel with a linear capacitance, having an "N" - shaped current-voltage characteristic with a wide zone of the incident section. By compensating this section with a linear inductance characteristic connected in series with the resulting circuit, it is possible to create a device that converts a current source into a stabilized voltage source.

1 Introduction

In a modern automation system, telemechanics and in the power circuit of various devices, electroferromagnetic circuits are widely used as current and voltage stabilizers, phase and frequency converters, they converted a current source into a stabilized voltage source, since devices created on the basis of electroferromagnetic oscillatory circuits have high reliability in operation, resistance to mechanical overload, rather high efficiency, power factor, manufacturing limit from several W to tens of kW, low cost [1-6, 12].

2 The current state of the investigated problem

With the rapid development of some branches of electrical engineering, there is an increasing need to create devices that convert a constant voltage system into a constant current system and vice versa.

The main part of the current source converter into a stabilized voltage source (CSCISVS) is a series ferroresonant circuit, the equivalent circuit of which is shown in Fig. 1. Consider the current-voltage characteristic of this circuit, shown in Fig.2. (curve 3) when powered from a voltage source. The value of the capacitance C1 is chosen such that its characteristic crosses the current-voltage characteristic of the nonlinear inductance in the saturation region. With an increase in the input voltage from U1=0 to the value U1=Umin, the current in the circuit increases in proportion to the voltage. The core of the non-linear coil is not saturated, and its voltage is greater than the capacitor voltage, the current in the circuit lags behind the input voltage by an angle $\pi/2$. When U1= Umin is reached, the current increases sharply, and its phase changes, and its phase changes by an angle π. The "ab" section of the characteristic is unstable. The voltage drop on the inductance leads, and on the capacitance C1 lags in phase from the current by an angle $\pi/2$. Therefore, the "oab" section of the circuit characteristic is inductive in nature, and "bs" is capacitive. A further increase in the input voltage leads to a proportional increase in the current in the circuit. This property of the circuit can be used to generate control signals for power thyristors [7-10, 13, 11, 14].

When a serial ferroresonant circuit is connected to a current source, the current-voltage characteristic of the circuit has the form of an “oabs” curve (Fig. 2). In this case, the section "ab" is stable. The form of this characteristic depends on the ratio of parameters in the circuit. If the falling section is observed in a small range of current changes, then in order to expand it, it is necessary to connect the capacitor C2 in parallel to the ferroresonant circuit with a characteristic passing tangentially to the “oa” curve. (curve 5) of such a circuit (Fig. 3) is built according to the characteristics of the resonant circuit (curve 3) and capacitance C1 (straight line 4) by summing the currents for the same values of the input voltage [15-20].

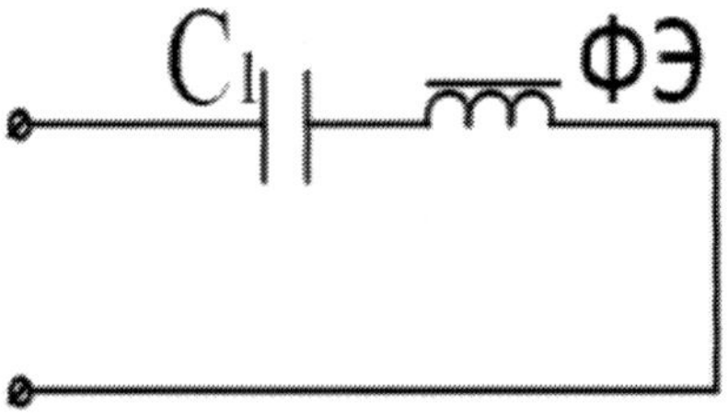

Fig.1. Series ferroresonant circuit

* Corresponding author: ruzinzarov88@bk.ru

 https://doi.org/10.1051/e3sconf/202338401050

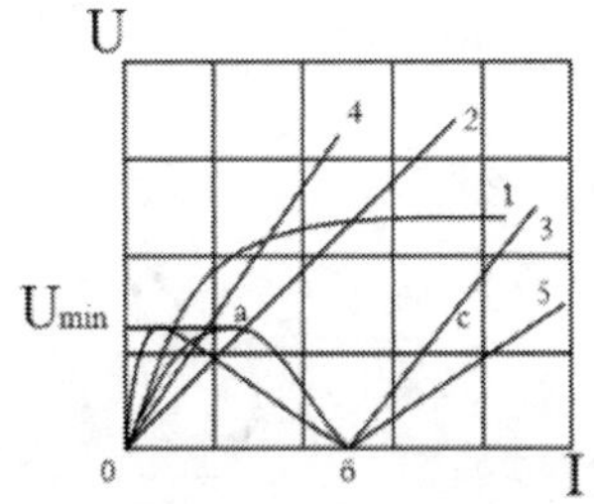

Fig.2. Volt-ampere characteristic of ferroresonant circuit

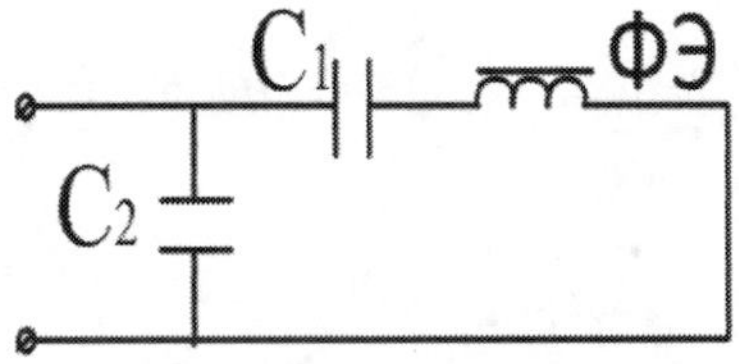

Fig.3. Series ferroresonant circuit with capacitor connected in parallel

This takes into account that in the pre-resonant mode the ferroresonant circuit is inductive, and in the post-resonant mode it is capacitive. As can be seen from Fig.2. with the inclusion of capacitor C1, the zone of the falling section of the characteristic expands in current. The length of the falling section of the characteristic strictly depends on the ratio of the parameters.

In order to compensate for the falling section and obtain the effect of voltage stabilization, we connect in series to the circuit shown in Fig.3., a linear inductance L0. The inductance characteristic must be strictly selected. The current-voltage characteristic of a two-coil ferroresonant circuit (Fig. 4.) is shown in Fig. 5. (curve 3) is obtained by summing curve 1 (current-voltage characteristic of a linear inductance) and curve 2 (volt-ampere characteristic of a ferroresonant circuit shown in Fig. 3). When constructing curve 3, it was taken into account that before resonance, the ferroresonant circuit (Fig. 3) has an inductive character, and after resonance, it is capacitive. As seen from Fig. 5, at certain ratios of the circuit parameters, a pronounced voltage stabilization zone appears. (av). The stabilization effect can also be achieved in the post-resonant mode, if the section "sun" of curve 2 is parallel to the linear inductance characteristic (curve 1). It is obvious that the overall dimensions and energy performance of the device will be better when operating in the pre-resonance mode, than in the post-resonance mode, since the effect of stabilizing the output voltage in the first case is obtained by summing, and in the second by subtracting the characteristics of the circuit elements [21-26, 18].

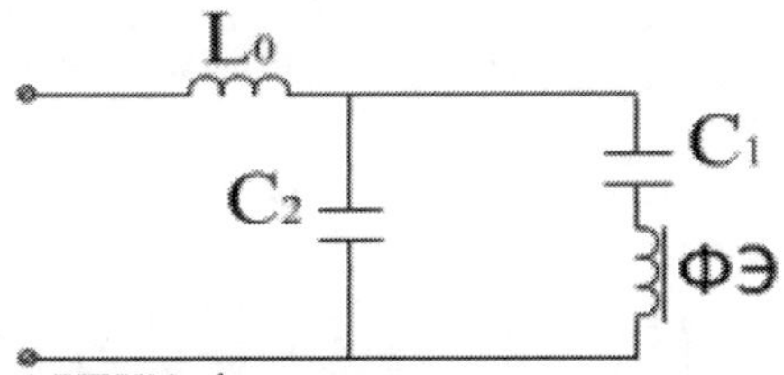

Fig.4. PITSIM scheme

In addition, the shape of the stabilized voltage curve in the pre-resonant mode is close to sinusoidal, and when the device is operated in the post-resonant mode, it is strongly distorted, since the ferromagnetic element is in the deep saturation zone [27-30, 15, 12].

Thus, connecting a series ferroresonant circuit to a current source calls the appearance of a stable falling section on the current-voltage characteristic, by compensating for which it is possible to achieve the transformation of a current source into a voltage source.

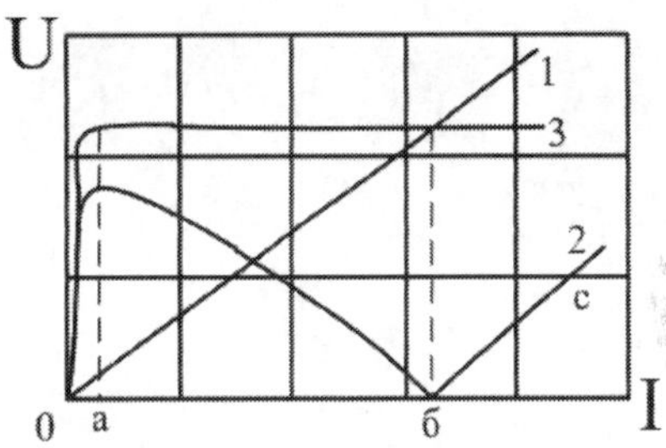

Fig.5. Volt-ampere characteristic of CSCISVS

In the theoretical analysis of ferroresonant circuits, their equivalent circuit plays an important role. Since the presence of a nonlinear ferromagnetic element in the circuit complicates the analytical solution, in order to obtain the basic mathematical expressions, it is necessary to make a number of assumptions that simplify the theoretical analysis. So, to analyze the static modes of the ferroresonant circuit (Rim. 4), we accept the following assumptions [31-33, 15, 1]:

1. The magnetization curve of a non-linear element is approximated by a power function in the form $i=K\psi2$.
2. We neglect the active resistance and leakage inductance of the windings of a ferromagnetic element, since the effect on electromagnetic processes of ferroresonant circuits is not significant.
3. Losses in capacitances and in the core of the linear inductance are not taken into account due to their extreme smallness.
4. A non-linear ferromagnetic element is represented by an equivalent circuit consisting of a non-linear inductance and active resistance connected in series to it.

The PITSIN equivalent circuit is shown in Fig. 3-6. Accepted notation:

U is the circuit voltage;

I is the current flowing through Lo;

I1 is the current flowing through the serial ferroresonant circuit;

I2 is the current flowing through the capacitor C2;

Ψ is the flux linkage of the non-linear element.

For the circuit under consideration, the following relations are valid;

$$i = i_1 + i_2 ; \quad (1)$$

$$U = L_0 \frac{di_1}{dt} + L_0 \frac{dt_2}{dt} + \frac{1}{C_2} \int L_2 dt ; \quad (2)$$

$$O = \frac{d\psi}{dt} + \frac{1}{C_1} \int i_1 dt - \frac{1}{C_2} \int i_2 dt + i_1 r ; \quad (3)$$

There is

$$i_1 = K\psi^2 \quad (4)$$

From equation (3) we determine the current in C2

$$i_2 = C_2 \frac{d^2\psi}{dt^2} + C_2 r \frac{di_1}{dt} + \frac{C_2}{C_1} i_1 ; \quad (5)$$

Then, taking into account (4), (5), we write the expression for the loop current as follows:

$$i = C_2 \frac{d^2\psi}{dt^2} + C_2 rK \frac{d\psi^2}{dt} + K\psi^2 (1 + \frac{C_2}{C_1} ; \quad (6)$$

Introducing the replacement of the displaced

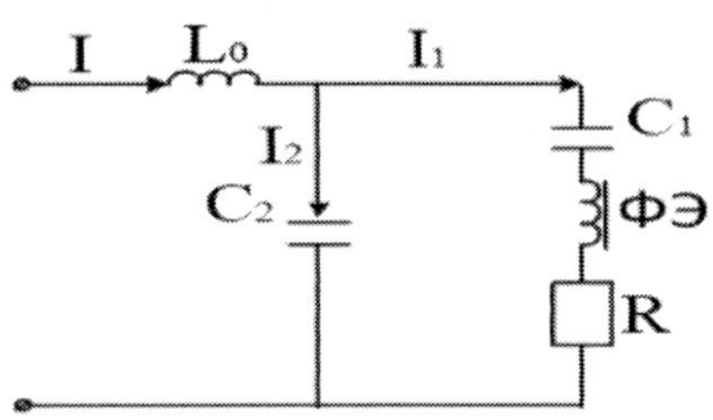

Fig.6. PITSIN equivalent circuit

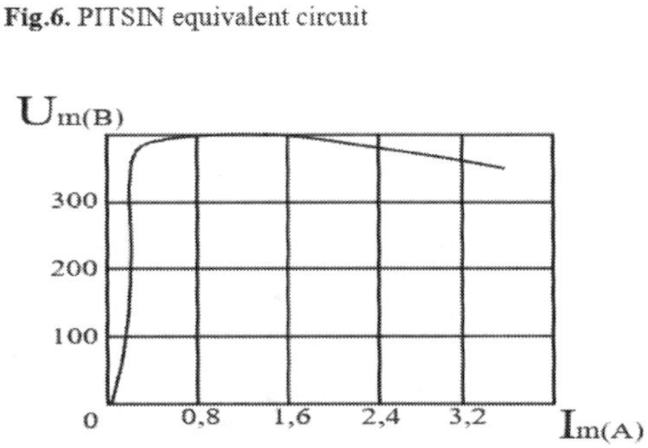

Fig.7. Volt-ampere characteristic of PITSIN

$$Y = \frac{i}{i_0}; \qquad x = \frac{\psi}{\psi_0}; \qquad \tau = \omega t$$

we will rewrite expression (6) in the form:

$$i_0 Y = C_2 \omega^2 \psi_7 \frac{d^0 x}{d\tau^2} + C_2 rK\omega\psi_\delta^7 \frac{dx^7}{d\tau} + K\psi_\delta^7 x^7 (\frac{C_0}{C_1} + 1) \quad (7)$$

We multiply the right and left sides of equation (7) by $\frac{1}{C_2 \omega^2 \psi_0}$; and get:

$$\frac{i_\delta}{C\omega^2 \psi_\delta} Y = \frac{d^2 x}{d\tau^2} + \frac{rK\psi_\delta^6}{\omega} \cdot \frac{dx^7}{d\tau} + \frac{K\psi_\delta^6}{C_2 \omega^2} (\frac{C_2}{C_1} + 1); \quad (8)$$

Let us introduce into equation (8) the notation

$$i_\delta = C_2 \omega^2 \psi_\delta ; \quad d = \frac{64}{35} r\omega C_1 ; \quad \beta = \frac{64(C_1 + C_2)}{35 C_2} ; \quad \psi_\delta^6 = \sqrt{\frac{\omega^2 C_1 64}{35K}} ;$$

Taking into account the accepted notation, we rewrite equation (8) as follows [9]:

$$Y = \frac{d^2 x}{d\tau^2} + d \frac{dx^2}{d\tau} + \beta x^7 \quad (9)$$

Solution (9) will be sought in the form:

$$x = x_m \sin(\tau + 4) \quad (10)$$

$$y = y_m \sin \tau \quad (11)$$

Derivatives of X have the form:

$$\frac{dx}{d\tau} = x_m \cos(\tau + \varphi) \quad (12)$$

$$\frac{d^2 x^2}{d\tau^2} = 2 \frac{dx_m}{d\tau} \cos(\tau + 4) - x_m \sin \tau + 2x_m \frac{d\varphi}{d\tau} \sin(\tau + \varphi) \quad (13)$$

$$x^7 = \frac{35}{64} x_m^7 \sin(\tau + \varphi) \quad (14)$$

$$\frac{dx^7}{d\tau} = \frac{35}{64} x_m^7 \cos(\tau + \varphi) \quad (15)$$

Substituting the values (11) - (15) into equation (9) and grouping the coefficients for the same trigonometric functions, we determine the value of Ym.

$$Y_m = \sqrt{(\beta \frac{35}{64} x_m^7 - x_m)^2 + (d \frac{35}{34} x_m^7)^2} \quad (16)$$

To build the current-voltage characteristic of the circuit, it is now necessary to determine the dependence U=f(ψ). Let us differentiate equation (2).

$$\frac{dU}{dt} = L_0 \frac{di_1}{d\tau^2} + L_0 \frac{di_2}{dt^2} + \frac{i_2}{C_2} \quad (17)$$

Taking into account (4) and (5), we rewrite the expression in the form:

$$\frac{dU}{dt} = L_0 C_2 \frac{d^4\psi}{dt^4} + L_0 K C_2 r \frac{d^3\psi^7}{dt^3} + (\frac{C_2}{C_1}+1)\cdot L_0 K \frac{d^2\psi^7}{dt^2} + \frac{d^2\psi}{dt^2} + \quad (18)$$
$$+ Kr\frac{d\psi^7}{dt} + \frac{K}{C_1}\psi^7;$$

Let's make a change of variables

$$\upsilon = \frac{U}{U_\delta}; \quad x = \frac{\psi}{\psi_\delta}; \quad \tau = \omega t;$$

then (18) will take the following form:

$$\omega U_\delta \frac{d\upsilon}{d\tau} = L_0 C_2 \psi_\delta \omega^4 \frac{d^4 x}{d\tau^4} + L_0 K C_2 r \omega^8 \psi_\delta^7 \frac{d^3 x^7}{d\tau^3} + \omega^2 \psi_\delta \frac{d^2 x}{d\tau^2} + \quad (19)$$
$$+ (\frac{C_2}{C_1}+1)\cdot L_0 K \psi_\delta^7 \omega^2 \frac{d^2 x^7}{d\tau^2} + K\psi_\delta^7 r\omega \frac{dx^7}{d\tau} + \frac{K}{C_1}\psi_\delta^7 x^7;$$

If we multiply the right and left parts of equation (19) by, then we can write $\frac{1}{L_1 C_2 \psi_\delta \omega^4}$

$$\frac{U_\delta}{L_0 C_2 \psi_\delta \omega^3}\cdot\frac{d\upsilon}{d\tau} = \frac{d^4 x}{d\tau^4} + \frac{K\psi_\delta^6 r}{\omega}\cdot\frac{d\,x^7}{d\tau^3} + (\frac{C_2}{C_1}+1)\cdot\frac{K\psi_\delta^6}{C_0\omega^2}\cdot\frac{d^2x^7}{d\tau^2} + \quad (20)$$
$$+ \frac{1}{L_0 C_2 \omega^2}\cdot\frac{d^2 x}{d\tau^2} + \frac{K\psi_\delta^6 r}{L_0 C_2 \omega^2}\cdot\frac{dx^7}{d\tau} + \frac{K\psi_\delta^6}{L_0 C_1 C_2 \omega^2}\cdot K^7$$

We accept the notation $U_\delta = L_0 C_2 \psi_\delta \omega^6, \quad d = \frac{64}{35} C_1 \omega r;$

$$\gamma = \frac{1}{L_0 C_2 \omega^2}, \quad \eta = \frac{64}{35}\cdot\frac{C_1 r}{L_0 C_2 \omega}, \quad \delta = \frac{64}{35 L_0 C_2 \omega^2};$$

Taking into account which (20) takes the form:

$$\frac{d\upsilon}{d\tau} = \frac{d^4 x}{d\tau} + d\frac{d^3x^7}{d\tau^3} + \beta\frac{d^2x^7}{d\tau^2} + \gamma\frac{d^2x}{d\tau^2} + \eta\frac{dx^7}{d\tau} + \delta x; \quad (21)$$

Solution (21) will be sought in the form
Then

$$x = x_m \sin(\tau+\varphi); \quad \upsilon = \upsilon_m \sin(\tau+\theta); \quad (22)$$
$$\frac{d\upsilon}{d\tau} = \upsilon_m \cos(\tau+\theta)$$

Derivatives of x are equal

$$\frac{d^2x}{d\tau^2} = -x_m \sin(\tau+\varphi); \quad (23)$$

$$\frac{d^4x}{d\tau^4} = -x_m \sin(\tau+\varphi); \quad (24)$$

$$\frac{d^2x^7}{d\tau^2} = -x_m \sin(\tau+\varphi)\cdot\frac{64}{35}; \quad (25)$$

$$\frac{d^3x^7}{d\tau^2} = -\frac{35}{64}\cdot x_m^7 \cos(\tau+\varphi); \quad (26)$$

Taking into account (14), (15) and (22) - (26), equation (21) for the steady state will take the form:

$$\upsilon_m \cos(\tau+\theta) = x_m \sin(\tau+\varphi) - d\frac{35}{64}x_m^7\cos(\tau+\varphi) - \beta\frac{35}{34}x_m^7\cdot \quad (27)$$
$$\cdot\sin(\tau+\varphi) - \gamma x_m \sin(\tau+\varphi) + \eta\frac{35}{64}x_m^7\cos(\tau+\varphi) + \gamma x_m^7 \sin(\tau+\varphi);$$

We perform simple trigonometric transformations and group the coefficients for the same trigonometric functions, and after simple transformations we obtain

$$\upsilon_m = \sqrt{[x(1-\gamma 0 + x_m^7(x - \frac{35}{64}\beta)]^2 + [x_m^7\frac{35}{64}\cdot(\eta-\alpha)]^2}; \quad (28)$$

In Fig.7. the current-voltage characteristic of PITSIN is shown, built on the basis of expressions (16) and (28) for the following parameters: L0=1.4 H, C1=5 μF, C2=1 μF, r=100 Om. Here, at certain ratios of parameters, voltage stabilization is observed in a wide range of input current changes (0.2 ÷ 2A)

3 Conclusion

The proposed PITSIN scheme based on an electroferromagnetic circuit, which has a section on the amplitude characteristic, works without a trigger effect at a certain ratio of parameters. To expand the falling section "N" - a shaped current-voltage characteristic of a series ferroresonant circuit, it is proposed to use a capacitor, which must be connected in parallel to it.

References

1. A.N.Rasulov, T.M.Kodirov Electromagnetic circuits in stabilization and regulation modes. Tashkent, ed.ToshGTU, 2014. - P.199.
2. V.V. Gubanov Power semiconductor converters with output stabilizers. –L.: Energy. 1972. -S 133.
3. T.M.Kodirov Electroferromagnetic circuits with a wide supply section on the amplitude and phase characteristics. // News of universities. Electromechanics, No. 7. 1988. -S. 26-30.
4. T.M.Kodirov, A.N.Rasulov Ferroresonant current stabilizer with sinusoidal current curve. //Automation and telemechanics, №11. 1977. -S. 197-200.
5. Fursanov M.I. Determination and analysis of electricity losses in electrical networks of energy

systems / M.I. Fursanov. - Minsk.: Belenergosberezhenie 2005. - 207p.

6. Golovkin P.I. Energy system and consumers of electrical energy. Moscow: Energoatomizdat, 1986.
7. Mikhailov V.V. Tariffs and modes of power consumption. Moscow: Energoatomizdat, 1986.
8. R Karimov. AIP Conference Proceedings **2552**. 050012. (2022). https://doi.org/10.1063/5.0111524
9. S Dzhuraev, R Karimov and ofters, Study and Analysis of Power Quality in the Electrical Networks of the Outdoor Lighting of the Dushanbe City (2022 Conference of Russian Young Researchers in Electrical and Electronic Engineering (ElConRus), 2022), pp. 1166-1169. doi: 10.1109/ElConRus54750.2022.9755782.
10. A.N.Rasulov, M.R.Ruzinazarov, N.Toirova, T.Sh.Alibekova. E3S Web of Conferences **216**. 01171. (2020). https://doi.org/10.1051/e3sconf/202021601171
11. Rasulov, A.N., Ruzinazarov, M.R., Novikov, N.L., Aslanova, G.N., Esemuratova, S. E3S Web of Conferences **289**. 07007. https://doi.org/10.1051/e3sconf/202128907007
12. Rasulov, A.N., Rafikova, G.R., Novikov, N.L., Ruzinazarov, M.R., Esemuratova, S. E3S Web of Conferences 289. 07006. (2021). https://doi.org/10.1051/e3sconf/202128907006
13. Rafikova, G.R., Ruzinazarov, M.R., Makhmutkhonov, S.K. Features stabilization of the consumer current using electro-ferromagnetic circuits. DOI: https://doi.org/10.1051/e3sconf/201913901075
14. Rakhmonov I.U., Najimova A.M., and Reymov K.M. AIP Conference Proceedings **2647**. 030010. (2022). https://doi.org/10.1063/5.0104788
15. Rakhmonov I.U., Najimova A.M. AIP Conference Proceedings **2647**. 030011. (2022). https://doi.org/10.1063/5.0104791
16. Rakhmonov I.U., Najimova, A.M., Esemuratova Sh.M., Koptileuov T.T. AIP Conference Proceedings **2647**. 070024. (2022). https://doi.org/10.1063/5.0104793
17. Hoshimov F.A., Rakhmonov I.U., Niyozov N.N., Omonov F.B. AIP Conference Proceedings **2647**. 030025. (2023). https://doi.org/10.1063/5.0112388
18. Rakhmonov I.U., Hoshimov F.A., Kurbonov N.N., Jalilova D.A. AIP Conference Proceedings **2647**. 050022. (2023). https://doi.org/10.1063/5.0112391
19. Rakhmonov I.U., Ushakov V.Ya., Niyozov N.N., Kurbonov N.N., Mamutov M. E3S Web of Conferences **289**. 07014. (2021). https://doi.org/10.1051/e3sconf/202128907014
20. Rakhmonov I.U., Ushakov V.Ya., Najimova A.M., Jalilova D.A., Omonov F.B. E3S Web of Conferences **289**. 07013. (2021). https://doi.org/10.1051/e3sconf/202128907013
21. A.Taslimov, M.Melikuziev, O.Matchonov, M.Ruzinazarov and M.Nasirov. IOP Conf. Series: Materials Science and Engineering **883**. 012105. (2020). doi:10.1088/1757-899X/883/1/012105.
22. Tuychiev F.N. AIP Conference Proceedings **2552.** 030013. 2023. https://doi.org/10.1063/5.0112349
23. Bobojanov M.K., Mahmutkhonov S.K., Aytbaev S. AIP Conference Proceedings **2552.** 05011. 2023. https://doi.org/10.1063/5.0113890.
24. Bobojanov M.K. AIP Conference Proceedings **2552.** 050034. (2023). https://doi.org/10.1063/5.0114077
25. Saidkhodjaev A.G. AIP Conference Proceedings **2552**. 050026. (2023). https://doi.org/10.1063/5.0111540
26. Rismukhamedov D.A., Tuychiev F.N., Shamsutdinov Kh. AIP Conference Proceedings **2552.** 050013. (2023). https://doi.org/10.1063/5.0114061
27. Rasulov A.N. AIP Conference Proceedings **2552.** 05017. (2023). https://doi.org/10.1063/5.0111530
28. Rasulov A.N., Usmonov E.G., Melikuziev M.V. AIP Conference Proceedings **2552.** 050018. (2023). https://doi.org/10.1063/5.0111530
29. Usmonov E.G. AIP Conference Proceedings **2552.** 050020. (2023). https://doi.org/10.1063/5.0111537
30. Melikuziev M.V. AIP Conference Proceedings 2552. 050021. (2023). https://doi.org/10.1063/5.0112395
31. Rakhmonov I.U., Hoshimov F.A., Kurbonov N.N., Jalilova D.A. AIP Conference Proceedings **2552.** 050022. (2023). https://doi.org/10.1063/5.0112391
32. Taslimov A.D. AIP Conference Proceedings **2552.** 050023. (2023). https://doi.org/10.1063/5.0112398
33. Saidkhodjaev A.G., Nuriddinova Kh.R. AIP Conference Proceedings **2552.** 050037. (2023). https://doi.org/10.1063/5.0133486

Efficient protection and control of electric drives using solid state circuits

Erkin Abduraimov*

Tashkent State Technical University named after Islam Karimov, 100095, Uzbekistan, Tashkent, University St. 2A

Abstract. The paper presents the results of a theoretical study of semiconductor circuits carried out for practical use in the development of a non-contact switching and control apparatus for starting control and under voltage protection of AC motors. According to the results of the analysis of semiconductor circuits, it is recommended to use them when creating relay devices for the automation of electrical equipment. Theoretical investigated semiconductor circuits were used in the creation of a non-contact starter of an asynchronous electric drive with protection against operation at minimal voltage drops in the network.

1 Introduction

The problem of creating reliable and economical start-protective and switching devices and devices as elements of electrical equipment for AC circuits is of great importance for ensuring the uninterrupted operation, clarity and speed of operation of electric drives and other energy devices. They especially have the prospect of using semiconductor contactless switching and control devices in difficult climatic conditions, in various industries, in the field of automatic control of electric drives and in power supply systems.

As it is well known, with a decrease in the supply voltage of asynchronous motors, the level of magnetic flux decreases the torque consequently. This increases the current consumption, leading to a decrease in the voltage level in the power grid, which affects the operation of other devices connected to it. In addition, one should not forget about the starting currents generated when starting the motors. Spontaneous start-up, which occurs when the voltage is restored after its disappearance or when the general switch of the main line machine is turned on, etc., for the engines of most mechanisms of industrial enterprises is unacceptable due to the safety of the operating personnel, due to the danger of the mechanism breaking, due to possible product defects and in a number of ways other reasons. Therefore, with a significant decrease in the voltage in the network or its disappearance, the motors, as a rule, should be automatically switched off by special under voltage protection [1-8].

Currently, under voltage protection in motor control circuits is carried out by line contactors and electromagnetic starters or special under voltage relays. Which carry out the inclusion of the machine at a network voltage of at least 80% of the nominal and automatically turn off the machine when the voltage disappears or drops to 50% of the nominal [9-15].

2 Theoretical researches

To start and protect AC motors from under voltage, we propose a circuit of a non-contact semiconductor device for controlling an asynchronous motor, which have been studied by theoretical and experimental studies. Theoretical studies of the components of the proposed device, the following circuits were considered, shown in Figures 1 and 2. [15-20]

Figure 1.a) shows a diode thyristor circuit, where a relay effect is revealed, with the help of which it is proposed to start and protect against under voltage in the control circuits of asynchronous motors. The explanation of the phenomenon of the relay effect can be carried out using the example of a diode-thyristor circuit, when the thyristor *VT* is connected in series to the network through $R_н$ active resistance, and the control circuit is supplied with current from the network through the active resistance *R* and the diode VD (Fig. 1, a) [1-5].

If you slowly increase the input voltage, then at a certain value of $\boldsymbol{U_{вх}}$, the amplitude value of the control signal will be equal to the thyristor opening current, at this moment the thyristor VT opens abruptly at an angle $\alpha=90^0$ (Fig. 1.b). A further increase in voltage leads to a decrease in the angle $\boldsymbol{\alpha}$ almost to zero, i.e. the thyristor will pass a full positive half-wave of current. The voltage value at which the thyristor opens abruptly depends on the value of the parameters *R* and R_L. Thus, the opening of a thyristor with a jump or the phenomenon of a sharp change in voltage or current on the load in the considered circuit can be called a trigger or otherwise a relay effect [15-20].

* Corresponding author: *abduraimoverkin69@gmail.com*

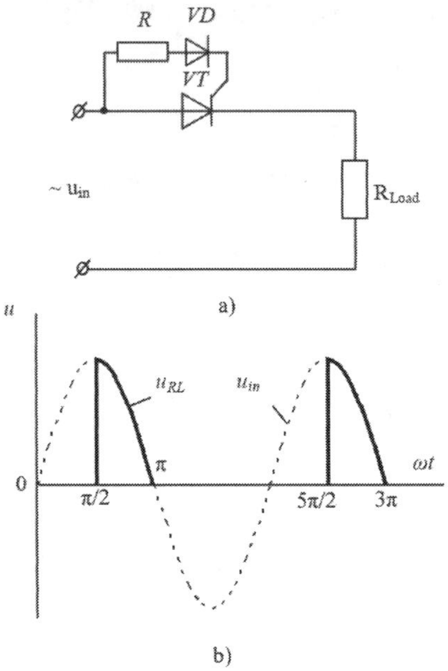

Fig.1. a) Thyristor switching circuit b) Mains voltage and load waveforms

Also, the mode of operation of a circuit consisting of a series-connected thyristor, active resistance and an inductive coil is considered, knowing that the windings of electric motors have an active-inductive nature of the load, and we take into account operation at mains voltages (Fig. 2).

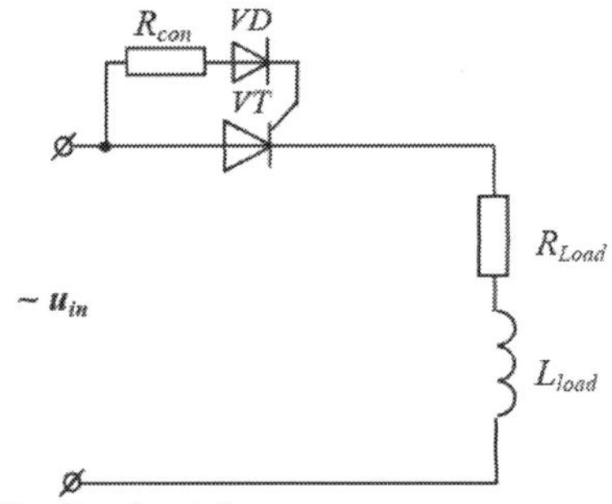

Fig.2. Circuit under study

The equation for this circuit has the following form:

$$u_{in} = u_{thy} + L\frac{di}{dt} + Ri$$

We accept the characteristic of the thyristor as ideal for the open state of the thyristor, while equation will take the form:

$$u_{in} = L\frac{di}{dt} + Ri = U_m \sin(\omega t + \varphi)$$

for

$$\varphi = \frac{\pi}{2}$$

$$L\frac{di}{dt} + Ri = U_m \cos \omega t$$

or

$$\frac{di}{dt} = \frac{U_m \cos \omega t}{L} - \frac{Ri}{L}$$

For different values of t, given the integration step h, we have:

$$i_n = i_{n-1} + \left(\frac{U_m \cos \omega t_{n-1}}{L} - \frac{R}{L} i_{n-1} \right) h$$

Figure 3 shows the voltage and current curves at the terminals of the elements L and R, constructed by solving equation numerical method. As can be seen from this figure, the current gradually increases and the moment of current termination is delayed relative to the moment when the phase voltage passes through zero. It should be noted that the shape of the current curve depends on the ratio of the circuit parameters L and R. The proposed technique allows a qualitative analysis of the steady-state modes and transient processes of the circuit with various variations of the parameters.

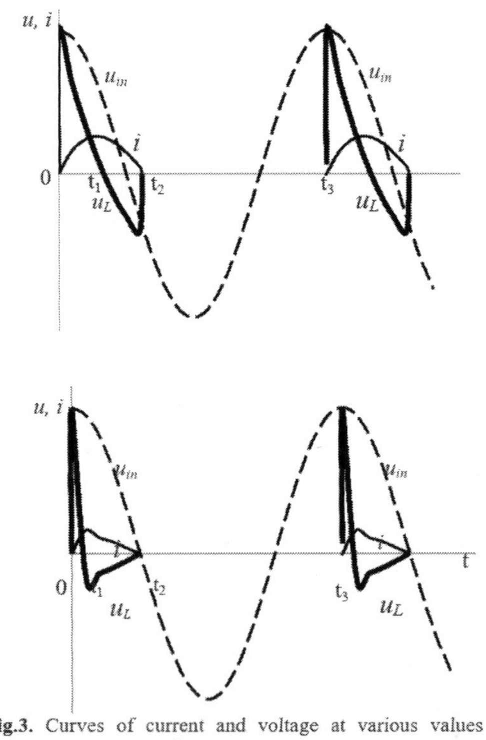

Fig.3. Curves of current and voltage at various values of resistance

Using these investigated circuits, on their basis, a non-contact device was developed to start and protect against the operation of the minimum voltage of AC electric motors, an asynchronous electric drive circuit was proposed (Figure 4). Considering the frequent modes of starting electric motors and operating at low network voltages, a circuit of a non-contact three-phase thyristor starter of an asynchronous electric motor is proposed, consisting of six thyristors, six diodes and resistors, one small-sized intermediate relay and a circuit breaker [21-26].

3 Experimental researches

The engine is started by pressing the SB2 button and energizing the KL relay. At the same time, when the relay KL is activated, it closes its contacts, thereby signals are sent to the control electrodes of the thyristors through the resistor-diode circuit from the anode part of the thyristors themselves. The opening of the thyristors leads to the supply of mains voltage to the motor, and it starts. During operation, if the nominal value of the mains voltage decreases from the allowable value, then the currents in the control circuit of the thyristors, which are supplied from the anode part through the resistor and the diode of the thyristors themselves, also decrease, and this will lead to the closing or disconnection of all thyristors, and, accordingly, to disconnecting the motor from the mains.

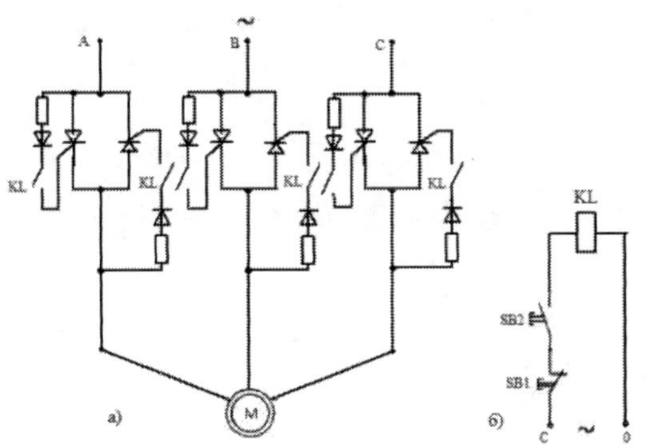

Fig. 4. Contactless thyristor induction motor starter with proposed under voltage protection circuit:
a) - power circuit b) - control circuit

The engine is started by pressing the SB2 button and energizing the KL relay. At the same time, when the relay KL is activated, it closes its contacts, thereby signals are sent to the control electrodes of the thyristors through the resistor-diode circuit from the anode part of the thyristors themselves. The opening of the thyristors leads to the supply of mains voltage to the motor, and it starts. During operation, if the nominal value of the mains voltage decreases from the allowable value, then the currents in the control circuit of the thyristors, which are supplied from the anode part through the resistor and the diode of the thyristors themselves, also decrease, and this will lead to the closing or disconnection of all thyristors, and, accordingly, to disconnecting the motor from the mains.

Experimental studies of the starting mode of the electric motor and the curves of the change in current and voltage of the non-contact relay starter are shown in Fig.5. for one phase. From these graphs it can be seen that the value of the starting current of the rated value is set in two periods and reaches a stable operating mode at permissible values of current and time.

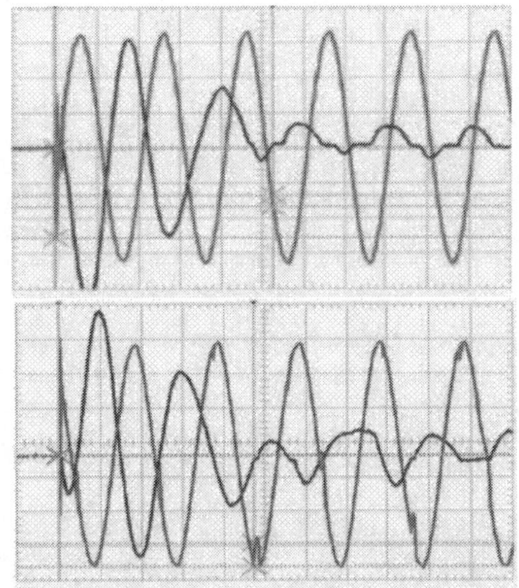

Fig.5. Oscillograms of current and voltage in one phase at the input and output of the starter when starting an asynchronous motor.

4 Conclusions

This Thus, when analyzing the first semiconductor circuit, the opening of a thyristor by a jump or the phenomenon of a sharp change in voltage or current on the load can be called a relay effect and it is used in the creation of the device.

In the following circuit of a connected thyristor, active resistance and inductive coil, the operation of the circuit for an active-inductive load was analyzed, taking into account the nature of the load of the winding of electric motors, a numerical method was used for the analysis, it allows for a qualitative analysis of steady modes and transients.

In the proposed scheme for starting and controlling the protection of an asynchronous electric motor, due to the phenomenon of the relay effect observed on diode-thyristor circuits, it is possible to effectively control the smooth start and at the same time protect the electric motor from operating at low mains voltages, that is, from the minimum voltage.

References

1. E.Abduraimov, M.Peysenov, N.Tairova. AIP Conference Proceedings, 2552, **040012**, (2022), https://doi.org/10.1063/5.0116235
2. E.Kh.Abduraimov, D.Kh.Khalmanov. Journal of Physics: Conference Series, 2094(2), **022072**, (2021), DOI 10.1088/1742-6596/2094/2/022072

3. E.Kh.Abduraimov, D.Kh.Khalmanov, and ofters. E3S Web of Conferences, 289, **07026**, (2021), https://doi.org/10.1051/e3sconf/202128907026
4. E.Kh.Abduraimov, D.Kh.Khalmanov. E3S Web of Conferences, 216, **01106**, (2020), https://doi.org/10.1051/e3sconf/202021601106
5. E.Kh.Abduraimov. E3S Web of Conferences, 216, **01105**, (2020), https://doi.org/10.1051/e3sconf/202021601105
6. E.Kh.Abduraimov, D.Kh.Khalmanov. Journal of Physics: Conference Series, 1515(2), **022055**, (2020), DOI 10.1088/1742-6596/1515/2/022055
7. E.Kh.Abduraimov. Journal of Physics: Conference Series, 072009, **2373**, (2022), DOI 10.1088/1742-6596/2373/7/072009
8. E.Kh.Abduraimov, D.Kh.Khalmanov. Journal of Physics: Conference Series, 072010, **2373**, (2022), DOI 10.1088/1742-6596/2373/7/072010
9. R.Karimov. AIP Conference Proceedings, 2552, **030014**, (2022). https://doi.org/10.1063/5.0111533
10. R.Karimov. AIP Conference Proceedings, 2552, **050012**, (2022). https://doi.org/10.1063/5.0111524
11. S.Dzhuraev, R.Karimov, and ofters. ElConRus, (2022), pp. 1166-1169, doi: 10.1109/ElConRus54750.2022.9755782
12. R.Karimov, A. Kuchkarov, and ofters. Journal of Physics: Conference Series 2094, **052050**, (2021). doi:10.1088/1742-6596/2094/5/052050
13. R.Karimov, N.Kurbanova, and ofters. Journal of Physics: Conference Series 2094(5), **052042**, (2021). doi:10.1088/1742-6596/2094/5/052042
14. K.G.Abidov, O.O.Zaripov, and ofters. AIP Conference Proceedings, 2552, **030023**, (2022), https://doi.org/10.1063/5.0112385
15. K.G.Abidov, O.O.Zaripov, and ofters. AIP Conference Proceedings, 2552, **030022**, (2022), https://doi.org/10.1063/5.0112384
16. K.G.Abidov, O.O.Zaripov, and ofters. E3S Web of Conferences, 289, **07003**, (2021), https://doi.org/10.1051/e3sconf/202128907003
17. K.G.Abidov, and ofters. E3S Web of Conferences, 289, **07004**, (2021), https://doi.org/10.1051/e3sconf/202128907004
18. K.Abidov, N.Khamudkhanova. E3S Web of Conferences, 216, **01111**, (2020), https://doi.org/10.1051/e3sconf/202021601111
19. 19. K.G.Abidov, O.O.Zaripov, and ofters. E3S Web of Conferences, 216, **01110**, (2020), https://doi.org/10.1051/e3sconf/202021601111
20. K.G.Abidov, O.O.Zaripov. E3S Web of Conferences, 139, **01088**, (2019), https://doi.org/10.1051/e3sconf/201913901088
21. A.A.Khashimov, K.G.Abidov. Elektrotekhnika, **7**, pp. 34-38, (2001).
22. A.A.Khashimov, K.G.Abidov. Elektrotekhnika, **10**, pp. 22-26, (2001).
23. 20. K.G.Abidov, A.K.Nuraliev, and ofters. Journal of Advanced Research in Dynamical and Control Systems, **12(7)**, pp. 2167-2171, (2020).
24. A.K.Nuraliev, and ofters. E3S Web of Conferences, 216, **01108**, (2020), https://doi.org/10.1051/e3sconf/202021601108
25. M.Ibadullaev, A.Nuraliev, A.Esenbekov. IOP Conference Series: Materials Science and Engineering, 862(6), **062031**, (2020), DOI 10.1088/1757-899X/862/6/062031
26. S.Begmatov, D.Khalmanov, and ofters. AIP Conference Proceedings, 2552, **040011**, (2022), https://doi.org/10.1063/5.0130666

https://doi.org/10.1051/e3sconf/202338401052

Application of numerical and graphical methods of analysis in nonlinear resistive circuits of electronic devices

Erkin Abduraimov [1*], *Baxtiyor* Nurmatov [1]

[1]Tashkent State Technical University named after Islam Karimov, 100095, Uzbekistan, Tashkent, University St. 2A

Abstract. The article discusses the research issues and methods of analysis of semiconductor circuits used in electronic circuits of contactless switching devices, provides analytical, graphic and numerical methods for the analysis of nonlinear semiconductor resistive circuits. The approximations of the Volt-Ampere characteristics (VACh) of semiconductor elements and the solution of the differential equations of state of electrical circuits by a numerical method using a computer are given. Diode, thyristor semiconductor nonlinear resistive circuits with active and inductive loads are considered. In the analysis of steady-state modes and transient processes, numerical methods for solving differential equations of states by the Euler method were used.

1 Introduction

In connection with the widespread automation of production processes, the introduction of automatic control systems, the requirements for reliability, speed and durability of electrical devices and apparatus have significantly increased. These requirements are largely met by devices based on the use of properties and phenomena inherent in nonlinear resistive circuits. Therefore, the theory of nonlinear resistive circuits and systems is of great importance in the creation of highly efficient electrical devices. Currently, semiconductor circuits are usually used as power switches for switching, regulating and converting devices [1-4]. To improve existing and create new highly efficient and energy-saving technical devices, it is necessary to theoretically and experimentally investigate semiconductor resistive circuits. On the basis of theoretical analyzes and experimental studies of nonlinear resistive circuits, it has been established that in order to ensure high-quality power supply to consumers, it is necessary to use such circuits as power contactless switching devices, current and voltage regulators. Circuits based on nonlinear resistive circuits allow switching power loads at the best dynamic modes, namely, when a sinusoidal current passes through zero, which improves the transient process mode [5-8].

2 Analysis of diode resistive circuits

An uncontrolled semiconductor diode, as an element of an electrical circuit, is a nonlinear asymmetric active resistance, its value depends on the polarity and the magnitude of the potential applied to it. Quite often, when considering the operation of rectifier circuits, in which, as a rule, diodes are used, they use the term "ideal diode", this concept means some kind of asymmetric resistance, the value of which in the positive direction of the current is zero, and in the opposite direction is equal to infinity [6-8]. The VACh of a diode can be obtained experimentally or from reference data for a given semiconductor element. In the analytical study of circuits with valves, an important issue is the choice of the approximating function of the nonlinear element [10]. The VACh of the forward current of a semiconductor diode can be described by a function of the form $i = au_d^2$ (fig.1).

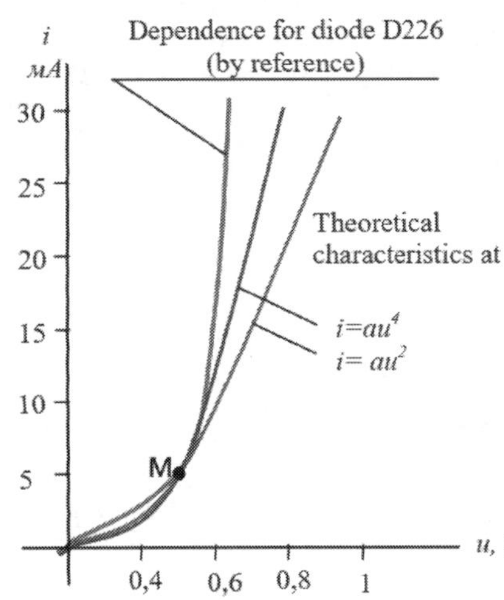

Fig.1. Volt-ampere characteristics

Here, a - coefficient of the approximating function, which is determined by the method of selected points.

* Corresponding author: *abduraimoverkin69@gmail.com*

 https://doi.org/10.1051/e3sconf/202338401052

Taking into account the VAC of the diode, type D226 for the selected point M, we have, a=0,41 [11-14].

Suppose that a diode with a series-connected resistance R (Fig.2) is connected to a network with a voltage $u = U_m Sin\omega t$

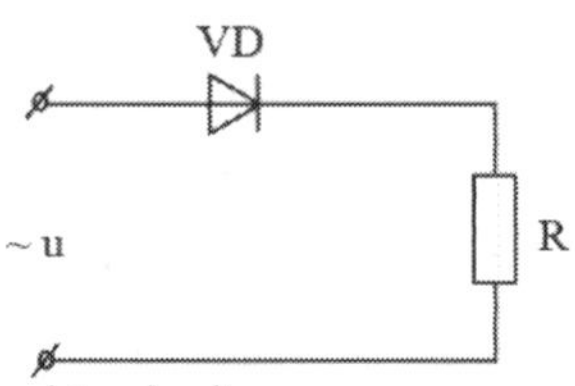

Fig. 2. Diode-resistor circuit

According to the second Kirchhoff's law:

$$u = u_d + Ri \tag{1}$$

in the approximating function, you can write:

$$u = \sqrt{\frac{i}{a}} + Ri \tag{2}$$

After some transformations, we get the following equation:

$$R^2 i^2 - i\left(2Ru + \frac{1}{a}\right) + u^2 = 0$$

where

$$i = \frac{(2Rau + 1) - \sqrt{4Rau + 1}}{2R^2 a} \tag{3}$$

Here the minus sign, taken before the radical, takes into account that under stress u=0 the current will also be zero [15-18]. On the basis of equation (3) with the use of a computer, a graph of the change in current as a function of time is calculated and built. Figure 3. these dependences are presented for various values of the load resistance R for input voltage u=*100Sin wt*.

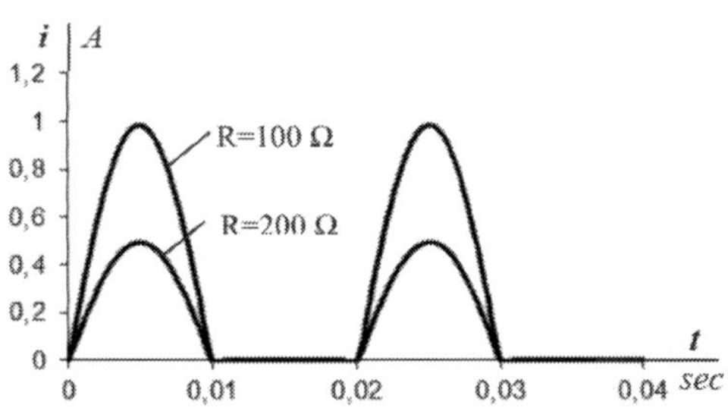

Fig. 3. Current ramps

Let us carry out a theoretical analysis of the circuit shown in Fig. 4, where a diode, inductive and active resistance are connected in series. To analyze this chain, it is proposed to use the numerical solution of the equation of state of the chain by the Euler method [19].

The equations for this chain are as follows:

$$u = u_d + L\frac{di}{dt} + iR \tag{4}$$

Using a more accurate approximating function $i = au_d^4$, where a=0,8 we get [3-4].

$$u = \sqrt[4]{\frac{i}{a}} + L\frac{di}{dt} + iR \tag{5}$$

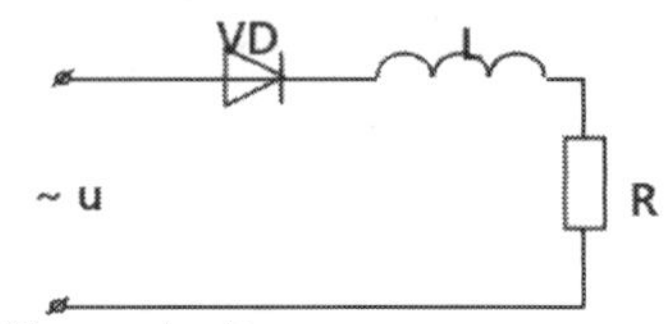

Fig. 4. Half-wave circuit

While $u = U_m Sin\omega t$ we have:

$$\frac{di}{dt} = \frac{1}{L}\left(U_m Sin\omega t - \sqrt[4]{\frac{i}{a}} - iR\right) \tag{6}$$

To solve this equation, according to Euler's method, we take:

$$i_n = i_{n-1} + \frac{1}{L}\left(U_m Sin\omega t_{n-1} - \sqrt[4]{\frac{i}{a}} - Ri_{n-1}\right)h \tag{7}$$

where h-step of integration; $\boldsymbol{i_n}$- instantaneous current value for a step n; i_{n-1} - instantaneous current value for the n-*1* step Figure 5. the current curves are shown, constructed on the basis of solving the equation by a numerical method, using a computer [20-24]. From this dependence it can be seen that there is a relative smoothing of the current curve. The shape of the current curve depends on the ratio of the parameters of the L and R circuit.

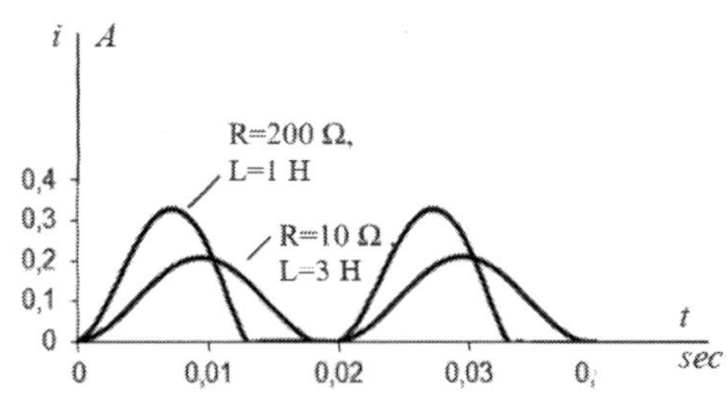

Fig. 5. Current curves

The proposed technique makes it possible to analyze the steady-state modes and transient processes of such circuits for various variations of the parameters L and R.

Let us analyze the circuit shown in Fig. 6, where inductive and active resistances are connected in series to a full-wave diode bridge circuit. To analyze this chain, we propose to use the numerical solution of the equation of state for the chain [25-26].

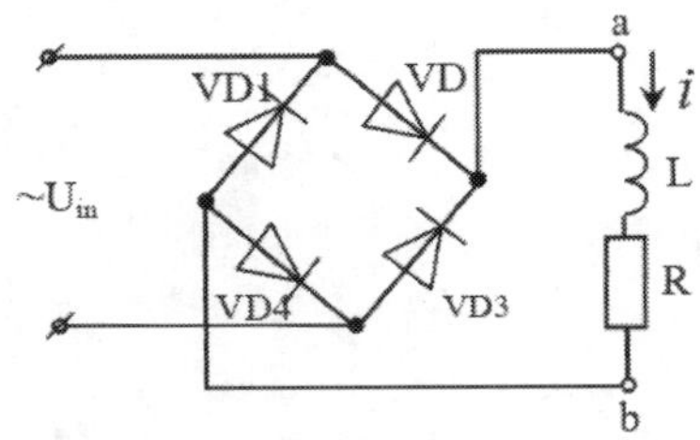

Fig. 6. Bridge rectifier circuit

The equation for this chain is as follows:

$$u_{ab} = L\frac{di}{dt} + Ri \quad (8)$$

Hence

$$\frac{di}{dt} = \frac{1}{L}(u_{ab} - Ri)$$

To solve this equation, we use the numerical Euler method.

$$i_n = i_{n-1} + \frac{1}{L}(U_m Sin\,\omega t_{n-1} - Ri_{n-1})h \quad (9)$$

Based on the results of solving this equation, the current and voltage curves were obtained (Fig. 7).

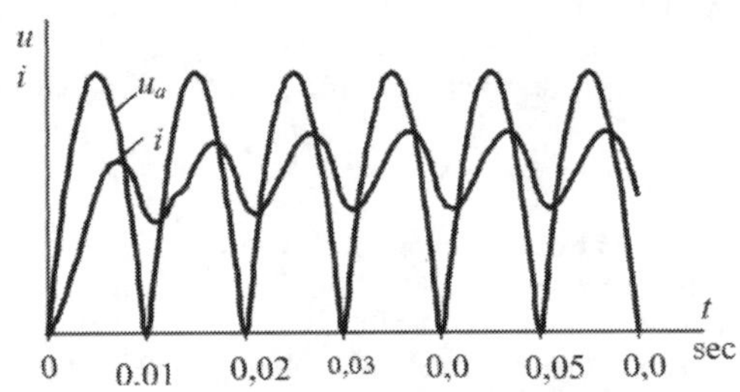

Fig. 7. Curves of current and voltage

When calculating on a computer, the following were taken: u_{ab}=100*Sin wt*; *R*=200 Ohm; *L*=1 H.

3 Analysis of thyristor resistive circuits

Thyristor semiconductor circuits have two stable electrical states (open and closed), have high speed and can switch large load currents. The main property of a thyristor is the ability to delay the moment of its firing in the presence of direct voltage on it. This property of the thyristor makes it possible to create devices with regulation of the value of the output voltage. Using the control current, you can control the moment when the thyristor turns on. The on-state current of the thyristor also flows after the control current is removed. It is possible, among other things, to turn off the anode current and restore the off state of the thyristor by reducing the current in the semiconductor device below the critical value. For thyristors, in the same way as for diodes, there is the concept of "ideal thyristor" [5-7]. Therefore, we can assume that the resistance of an ideal thyristor in the opposite direction, as well as in the forward closed state, is equal to infinity. In the open state, the forward resistance of an ideal thyristor is zero.

The way thyristors are switched by the gate current is of great importance. First, it allows, due to the control signal, to turn on the thyristor at different values of the anode voltage. Secondly, this method makes it possible to switch large currents with a low-power control signal.

We are investigating a circuit consisting of a series-connected thyristor and an active resistance (Fig. 8.) which is connected to an alternating voltage source. To analyze the operation of the considered circuit, you should use the load characteristic and the characteristic of the thyristors [12-14].

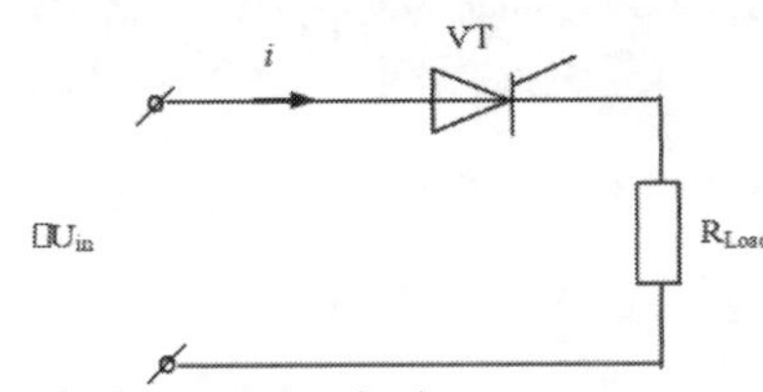

Fig. 8. Thyristor-resistive circuit

An analytical expression of the load characteristic can be obtained based on the second Kirchhoff's law.

$$I = \frac{U_{in}}{R_L} - \frac{U_{thy}}{R_L} \quad (10)$$

This is the equation of a straight line cutting off segments on the coordinate axes $U = U_L$ while $I = 0$ and $I = U_{in}/R_{Load}$ при $U = 0$. (Fig.9.). The points of intersection of the load lines with the thyristor characteristic determine the operating modes of the circuit.

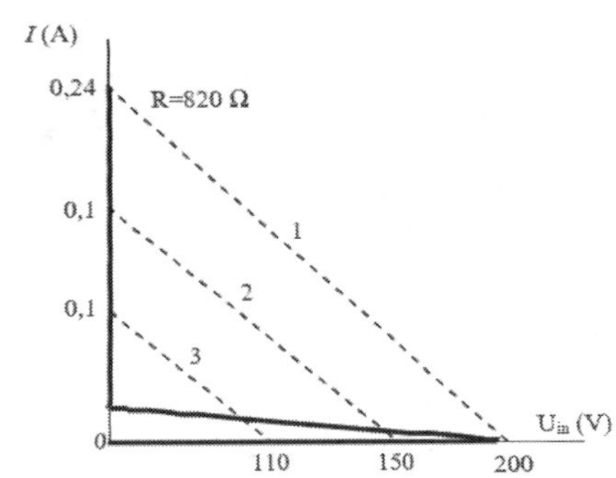

Fig. 9. VACh thyristor and load lines

 https://doi.org/10.1051/e3sconf/202338401052

To display the load line and current-voltage characteristics, consider the operating mode of the KU 202K thyristor with an active load of 820 Ohm. In the absence of a control signal, the thyristor is blocked in both directions and is under the influence of the source voltage. When the voltage of the source reaches a certain value, the thyristor turns on at the moment *1/2*. In this case, the voltage across the resistance will change abruptly to the amplitude value of the alternating voltage [22-23]. The voltage across the thyristor at the moment of switching on changes abruptly to almost zero. The duration of the current through the thyristor and the voltage across the load resistance is a quarter of the period. The values of the currents are determined by the following formulas:

$$i = \frac{U_{m\max}}{R} Sin\omega t = I_m Sin\omega t \qquad (11)$$

for $\alpha < \omega t < \pi$

$$I_{ave} = \frac{1}{2\pi}\int_{\alpha}^{\pi} I_m Sin\omega t\, d\omega t = \frac{1}{2\pi} I_m (1 + Cos\alpha) \qquad (12)$$

in our case for $\alpha = 90^o$ $\quad I_{ave} = \frac{I_m}{2\pi}$

RMS current value:

$$I = I_m \frac{1}{\sqrt{2\pi}} \sqrt{\frac{1}{2}(\pi - \alpha) + \frac{1}{4} Sin2\alpha} \qquad (13)$$

for $\alpha = 90^0$ we get, $\quad I = \frac{I_m}{2\sqrt{2}}$

Thus, the considered circuit can be analyzed using the characteristics of the thyristor and load lines.

Consider the operating mode of a circuit consisting of a series-connected thyristor, an inductive coil and an active resistance (Fig. 10).

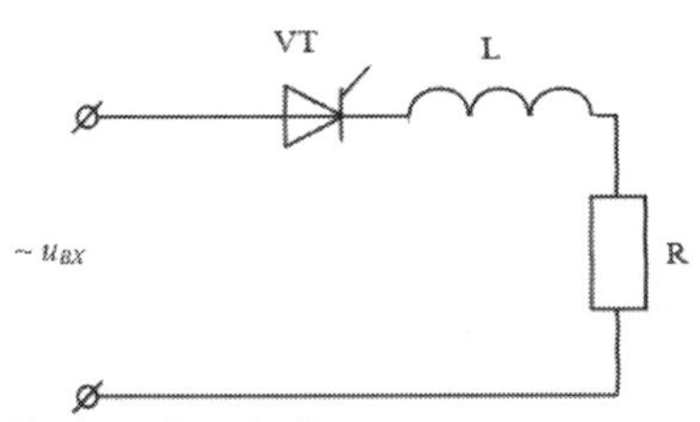

Fig. 10. The investigated scheme

The equation for this chain is as follows:

$$u_{in} = u_{thy} + L\frac{di}{dt} + Ri \qquad (14)$$

We accept the characteristic of the thyristor ideal for the open state of the thyristor, while equation (14) will take the form:

$$u_{in} = L\frac{di}{dt} + Ri = U_m Sin(\omega t + \psi)$$

for $\psi = \pi / 2$

$$L\frac{di}{dt} + Ri = U_m Cos\omega t \qquad (15)$$

Or

$$\frac{di}{dt} = \frac{U_m Cos\omega t}{L} - \frac{R}{L}i \qquad (16)$$

For different values of t, setting the integration step *h*, we have:

$$i_n = i_{n-1} + \left(\frac{U_m Cos\omega t_{n-1}}{L} - \frac{R}{L} i_{n-1}\right)h \qquad (17)$$

The curves of voltage and current at the terminals of the elements L and R and current, constructed by solving equation (17) by a numerical method, are presented.

It can be seen from the figure that the current gradually increases and the moment of termination of the current is delayed relative to the moment of transition of the phase voltage through the zero value. It should be noted that the shape of the current curve depends on the ratio of the circuit parameters *L* and *R* [7-8, 17-18].

4 Conclusions

Thus, a diode-resistor circuit can be investigated analytically, using an even-order power function to approximate the VACh of a diode, using the obtained expression for constructing a change in the load current over time, and a theoretical analysis of a circuit consisting of a series-connected diode, inductance and active resistance is given which is carried out using the fourth-order power function as an approximation of the diode characteristics.

By studying the circuit of a thyristor resistive circuit, it was found that the graphical method of analysis used in calculating transistor circuits can also be recommended for calculating a circuit consisting of a series-connected thyristor and an active resistance. In this case, the VACh of the thyristor is considered to be known for various control currents, and load lines are drawn on it. For the case of active-inductive load, the ideal characteristic of the thyristor is adopted. In this case, the nonlinear differential equations of state of the circuit in all cases were solved by the numerical Euler method using a computer. The proposed methods allow for a qualitative analysis of steady-state modes and transient processes in semiconductor circuits.

References

1. E.Abduraimov, M.Peysenov, N.Tairova. AIP Conference Proceedings, 2552, **040012**, (2022), https://doi.org/10.1063/5.0116235
2. E.Kh.Abduraimov, D.Kh.Khalmanov. Journal of Physics: Conference Series, 2094(2), **022072**, (2021), DOI 10.1088/1742-6596/2094/2/022072
3. E.Kh.Abduraimov, D.Kh.Khalmanov, and ofters. E3S Web of Conferences, 289, **07026**, (2021), https://doi.org/10.1051/e3sconf/202128907026
4. E.Kh.Abduraimov, D.Kh.Khalmanov. E3S Web of Conferences, 216, **01106**, (2020), https://doi.org/10.1051/e3sconf/202021601106
5. E.Kh.Abduraimov. E3S Web of Conferences, 216, **01105**, (2020), https://doi.org/10.1051/e3sconf/202021601105
6. E.Kh.Abduraimov, D.Kh.Khalmanov. Journal of Physics: Conference Series, 1515(2), **022055**, (2020), DOI 10.1088/1742-6596/1515/2/022055
7. E.Kh.Abduraimov. Journal of Physics: Conference Series, 072009, **2373,** (2022), DOI 10.1088/1742-6596/2373/7/072009
8. E.Kh.Abduraimov, D.Kh.Khalmanov. Journal of Physics: Conference Series, 072010, **2373,** (2022), DOI 10.1088/1742-6596/2373/7/072010
9. R.Karimov. AIP Conference Proceedings, 2552, **030014**, (2022). https://doi.org/10.1063/5.0111533
10. R.Karimov. AIP Conference Proceedings, 2552, **050012**, (2022). https://doi.org/10.1063/5.0111524
11. S.Dzhuraev, R.Karimov, and ofters. ElConRus, (2022), pp. 1166-1169, doi: 10.1109/ElConRus54750.2022.9755782
12. R.Karimov, A. Kuchkarov, and ofters. Journal of Physics: Conference Series 2094, **052050**, (2021). doi:10.1088/1742-6596/2094/5/052050
13. R.Karimov, N.Kurbanova, and ofters. Journal of Physics: Conference Series 2094(5), **052042**, (2021). doi:10.1088/1742-6596/2094/5/052042
14. K.G.Abidov, O.O.Zaripov, and ofters. AIP Conference Proceedings, 2552, **030023**, (2022), https://doi.org/10.1063/5.0112385
15. K.G.Abidov, O.O.Zaripov, and ofters. AIP Conference Proceedings, 2552, **030022**, (2022), https://doi.org/10.1063/5.0112384
16. K.G.Abidov, O.O.Zaripov, and ofters. E3S Web of Conferences, 289, **07003**, (2021), https://doi.org/10.1051/e3sconf/202128907003
17. K.G.Abidov, and ofters. E3S Web of Conferences, 289, **07004**, (2021), https://doi.org/10.1051/e3sconf/202128907004
18. K.Abidov, N.Khamudkhanova. E3S Web of Conferences, 216, **01111**, (2020), https://doi.org/10.1051/e3sconf/202021601111
19. 19. K.G.Abidov, O.O.Zaripov, and ofters. E3S Web of Conferences, 216, **01110**, (2020), https://doi.org/10.1051/e3sconf/202021601111
20. K.G.Abidov, O.O.Zaripov. E3S Web of Conferences, 139, **01088**, (2019), https://doi.org/10.1051/e3sconf/201913901088
21. A.A.Khashimov, K.G.Abidov. Elektrotekhnika, **7**, pp. 34-38, (2001).
22. A.A.Khashimov, K.G.Abidov. Elektrotekhnika, **10**, pp. 22-26, (2001).
23. 20. K.G.Abidov, A.K.Nuraliev, and ofters. Journal of Advanced Research in Dynamical and Control Systems, **12(7)**, pp. 2167-2171, (2020).
24. A.K.Nuraliev, and ofters. E3S Web of Conferences, 216, **01108**, (2020), https://doi.org/10.1051/e3sconf/202021601108
25. M.Ibadullaev, A.Nuraliev, A.Esenbekov. IOP Conference Series: Materials Science and Engineering, 862(6), **062031**, (2020), DOI 10.1088/1757-899X/862/6/062031
26. S.Begmatov, D.Khalmanov, and ofters. AIP Conference Proceedings, 2552, **040011**, (2022), https://doi.org/10.1063/5.0130666

Analysis of reactive power consumption "Denav oil extraction plant"

Raxmatillo Karimov[1*], *Dilmurod* Xushvaktov[1]

[1]Tashkent State Technical University named after Islam Karimov, 100095, Uzbekistan, Tashkent, University St. 2A

Abstract. The article deals with the study of non-contact devices, the analysis of reactive power consumption of the "Denav Oil Extraction Plant" and the graphs of reactive power consumption of enterprises. And also, the purpose of the article is from a review of materials on contactless devices for automatic control of the power of capacitor banks and the study of graphs for the consumption of reactive power of enterprises.

1 Introduction

Improving the reliability of power supply systems, developing and implementing energy-saving technologies and ways to reduce power losses is one of the topical issues of today. The reliability of the power supply system is associated with the safe implementation of switching in the system by turning on, off, transferring power to electrical consumers. The efficiency of the power supply system is influenced by many factors, one of the main among them is the optimal start-up and power control of capacitor units [1-5, 15].

Implementation of regulation using non-contact devices improves the reliability of the electrical system and the rational use of electricity [1-5, 15-17].

Ensuring the economical operation of capacitor units is possible by regulating their power when the magnitude and nature of the loads change depending on the voltage at the point of connection of the capacitors [1-5, 16].

Non-contact devices for switching and power control of capacitor units are distinguished by their speed, long service life, quiet operation and other advantages [1-5].

However, existing non-contact devices have a complex system for controlling the operation of thyristors or other semiconductor elements, which creates difficulties in the installation, commissioning and maintenance of such devices and prevents their widespread use [1-5, 17-20].

In addition, devices for power regulation, capacitor banks and switching of electrical equipment are imported from other countries and for the purchase of which foreign currency is required [1-5, 18-19].

Based on the foregoing, it is relevant to develop and implement inexpensive contactless starters with a simplified design for controlling the power of capacitor banks and switching various consumers, including various motors operating in a stressful mode with frequent starts and stops [1-5, 15-20].

2 Analysis of reactive power consumption Denav oil extraction plant

"Denav oil extraction plant" is an oil and fat industry enterprise producing vegetable refined cottonseed and soybean oil, as well as consumer goods, soap, husks, meal, etc. [1, 6-9].

By the nature of consumption of raw materials, the enterprise "Denav Oil Extraction Plant" belongs to the manufacturing industries, that is, it processes raw materials of plant origin [2, 10-12].

According to the purpose of the finished product, the enterprise belongs to a group that sends its products for direct consumption by the population, and sends waste as raw materials to other enterprises [1, 13-15].

According to the duration of work, the enterprise belongs to the group of year-round (permanent) operation. Duration of work during the day - continuity of action [2, 16].

The method of obtaining finished products is the extraction of vegetable oil from oilseeds (cotton, soybeans, sunflower, safflower) [1, 17-19].

According to the level of mechanization and automation of production processes, the enterprise is considered to be complex-automated, that is, all processes of the main and auxiliary production are automated [2, 20-21].

The efficiency of production largely depends on the quality of raw materials supplied for processing (seed oil content), weediness, spoilage, as well as on the degree of

* Corresponding author: raxmatillo82@mail.ru

 https://doi.org/10.1051/e3sconf/202338401053

provision of the enterprise with the necessary raw materials in the right amount [1, 22-25].

The production capacity of the Denav Oil Extraction Plant under the project is 1360 tons per day of processed seeds [2, 26-27].

3 Analysis of the consumption of electric energy

The enterprise is supplied with electricity from the main power grids through the Denov substation with a voltage of 110/10 kV through the feeders GCP-1 (Gas Control Point), GCP-2, GCP-2, Boiler-1, Boiler-2, Water intake-1, Water intake-2, through the Complete Cells Outdoor Switchgear (CCOS) 10 kV [2, 28-31].

The 110/10 kV substation is a deep input substation, i.e. higher voltage (110 kV) is as close as possible to electrical installations of consumers with a minimum number of stages of intermediate transformation and devices. The total installed capacity of 10/0.4 kV transformers is 14300 kVA. The substation is located on the territory of the enterprise, and the connection is made via cable and overhead power transmission lines [32-33].

Commercial metering devices (Automated Electricity Control and Accounting System - AECAS) are installed at the outlets of the Denov substation with a voltage of 110/10 kV along the feeders GCP-1, GCP-2, GCP-3, Boiler-1, Boiler-2, Vodozabor-1 and Vodozabor-2 [34].

To analyze the dynamics of changes in electricity consumption, a summary table was prepared based on the report provided by the department of the chief power engineer [1, 35-38].

Based on the data from the reporting of the enterprise, the total electricity consumption amounted to (see table 1): active energy – 5497990 kWh, reactive energy – 1497220 kVAr·h. The weighted average power factor is cosφ=0.98 [2, 39-41].

For detailing, a separate analysis of power consumption was performed for feeders GCP-3 and Boiler-1. Power consumption data are summarized in Tables 1 and 3. The dynamics of changes in active, reactive energy and power factor are also indicated there. When analyzing the graphs (see figure 2) of the change in power factor, it can be seen that at GCP-3 and Boiler-1 there is a sharp fluctuation of reactive power [42-44].

The main reason for this is the manual control of reactive power compensation. To solve this problem, it is required to install a microprocessor controller for automatic control of reactive power compensation [45].

Table 1. Summary table of active energy consumption

Months	Active Energy, kWh
January	389210
February	1283810
March	391110
April	424110
May	247310
June	251000
July	274810
August	276910
September	232610
October	1727110
Total	5497990

Table 2. Electricity consumption at GCP-3

Months	Active Energy, kWh	Reactive Energy, kVAr·h	Power factor, *cosφ*
January	156000	36000	0,97
February	697200	302400	0,92
March	176400	34800	0,98
April	168000	96000	0,87
May	60000	24000	0,93
June	60000	4800	1,00
July	78000	7200	1,00
August	62400	26400	0,92
September	38400	14400	0,94
October	945600	472800	0,89
Total	2442000	1018800	0.92

Table 3. Electricity consumption at Boiler-1

Months	Active Energy, kWh	Reactive Energy, kVAr·h	Power factor, *cosφ*
January	31400	20000	0,84
February	95800	65400	0,83
March	31000	16600	0,88
April	36000	22000	0,85
May	10400	10200	0,71
June	5500	2440	0,91
July	5800	200	1,00
August	8200	2200	0,97
Total	224100	139040	0,85

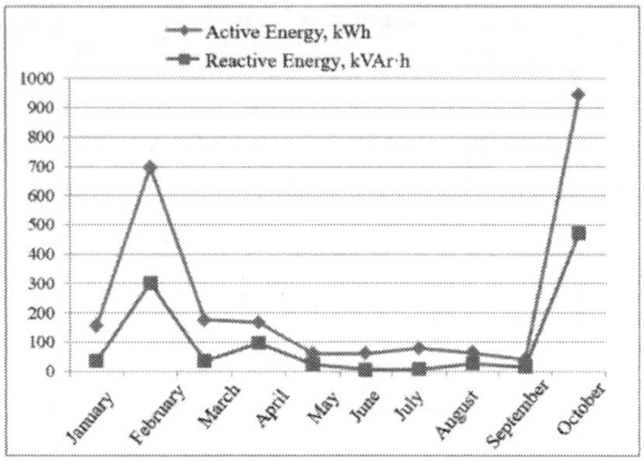

Fig.1. Dynamics of electricity consumption at GCP-3

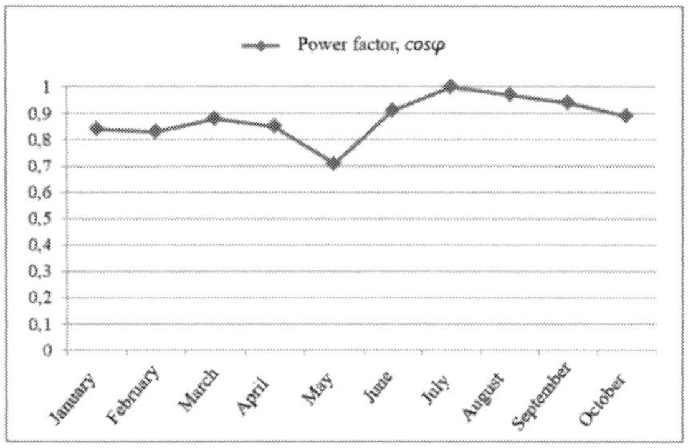

Fig.2. Power factor GCP-3

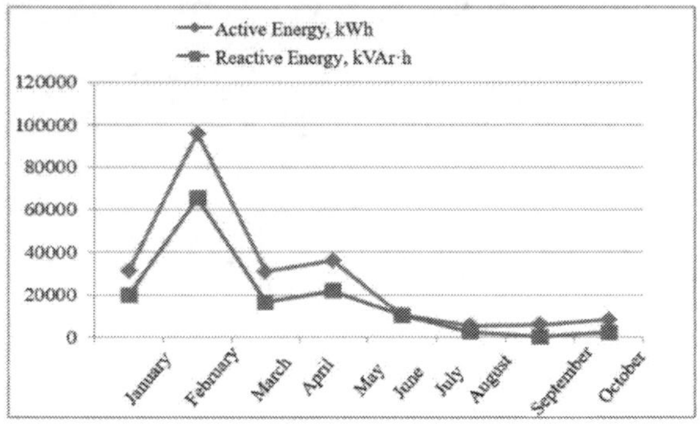

Fig.3. Dynamics of electricity consumption at Boiler-1

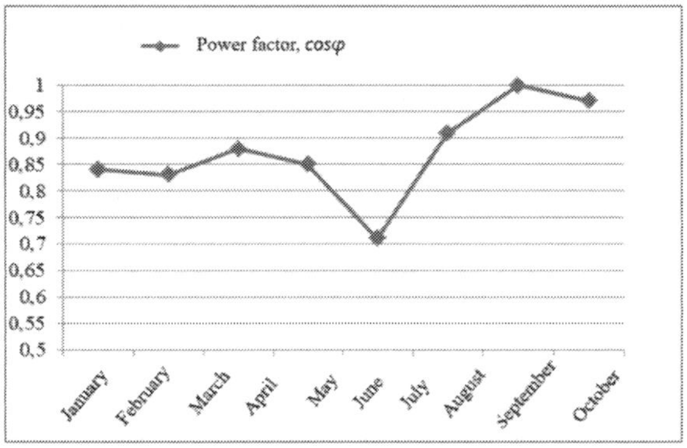

Fig.4. Power factor of Boiler-1

4 Conclusion

As part of the implementation of the article on the topic "Analysis of the consumption of reactive power "Denav Oil Extraction Plant"" for 2021, the following results were obtained:

The study and analysis of graphs of reactive power consumption at the "Denav Oil Extraction Plant" makes it possible to have an idea of the magnitude of the consumed power and understand the importance of constantly maintaining the power factor within the required limits, because undercompensation and overcompensation of reactive power takes place.

In general, for automatic control of the power of capacitor banks without arc formation, it is necessary to develop non-contact devices based on semiconductor elements. Now contactless devices are purchased for currency, i.e. are among the goods imported from abroad, therefore, the solution of the problem of development and implementation of such devices with a simple design and reasonable prices, along with a technical problem, would solve the problem of import substitution, which justifies the relevance of the purpose of this study and the need to continue research in this direction.

References

1. R.Karimov. AIP Conference Proceedings, 2552, **030014**, (2022). https://doi.org/10.1063/5.0111533
2. R.Karimov. AIP Conference Proceedings, 2552, **050012**, (2022). https://doi.org/10.1063/5.0111524
3. S.Dzhuraev, R.Karimov, and ofters. ElConRus, (2022), pp. 1166-1169, doi: 10.1109/ElConRus54750.2022.9755782
4. R.Karimov, A. Kuchkarov, and ofters. Journal of Physics: Conference Series 2094, **052050**, (2021). doi:10.1088/1742-6596/2094/5/052050
5. R.Karimov, N.Kurbanova, and ofters. Journal of Physics: Conference Series 2094(5), **052042**, (2021). doi:10.1088/1742-6596/2094/5/052042
6. K.G.Abidov, O.O.Zaripov, and ofters. AIP Conference Proceedings, 2552, **030023**, (2022), https://doi.org/10.1063/5.0112385
7. K.G.Abidov, O.O.Zaripov, and ofters. AIP Conference Proceedings, 2552, **030022**, (2022), https://doi.org/10.1063/5.0112384
8. K.G.Abidov, O.O.Zaripov, and ofters. E3S Web of Conferences, 289, **07003**, (2021), https://doi.org/10.1051/e3sconf/202128907003
9. K.G.Abidov, and ofters. E3S Web of Conferences, 289, **07004**, (2021), https://doi.org/10.1051/e3sconf/202128907004

10. K.Abidov, N.Khamudkhanova. E3S Web of Conferences, 216, **01111**, (2020), https://doi.org/10.1051/e3sconf/202021601111
11. K.G.Abidov, O.O.Zaripov, and ofters. E3S Web of Conferences, 216, **01110**, (2020), https://doi.org/10.1051/e3sconf/202021601111
12. K.G.Abidov, O.O.Zaripov. E3S Web of Conferences, 139, **01088**, (2019), https://doi.org/10.1051/e3sconf/201913901088
13. E.Kh.Abduraimov, D.Kh.Khalmanov. Journal of Physics: Conference Series, 2373(7), **072010**, (2022), DOI 10.1088/1742-6596/2373/7/072010
14. E.Kh.Abduraimov. Journal of Physics: Conference Series, 072009, 2373, (2022), DOI 10.1088/1742-6596/2373/7/072009
15. E.Abduraimov, M.Peysenov, N.Tairova. AIP Conference Proceedings, 2552, **040012**, (2022), https://doi.org/10.1063/5.0116235
16. E.Kh.Abduraimov, D.Kh.Khalmanov. Journal of Physics: Conference Series, 2094(2), **022072**, (2021), DOI 10.1088/1742-6596/2094/2/022072
17. E.Kh.Abduraimov, D.Kh.Khalmanov, and ofters. E3S Web of Conferences, 289, **07026**, (2021), https://doi.org/10.1051/e3sconf/202128907026
18. E.Kh.Abduraimov, D.Kh.Khalmanov. E3S Web of Conferences, 216, **01106**, (2020), https://doi.org/10.1051/e3sconf/202021601106
19. E.Kh.Abduraimov. E3S Web of Conferences, 216, **01105**, (2020), https://doi.org/10.1051/e3sconf/202021601105
20. E.Kh.Abduraimov, D.Kh.Khalmanov. Journal of Physics: Conference Series, 1515(2), **022055**, (2020), DOI 10.1088/1742-6596/1515/2/022055
21. K.G.Abidov, A.K.Nuraliev, and ofters. Journal of Advanced Research in Dynamical and Control Systems, **12(7)**, pp. 2167-2171, (2020).
22. A.K.Nuraliev, and ofters. E3S Web of Conferences, 216, **01108**, (2020), https://doi.org/10.1051/e3sconf/202021601108
23. M.Ibadullaev, A.Nuraliev, A.Esenbekov. IOP Conference Series: Materials Science and Engineering, 862(6), **062031**, (2020), DOI 10.1088/1757-899X/862/6/062031
24. S.Begmatov, D.Khalmanov, and ofters. AIP Conference Proceedings, 2552, **040011**, (2022), https://doi.org/10.1063/5.0130666
25. Kh.G.Karimov, M.K.Bobozhanov. Elektrichestvo, **1**, pp. 27-32, (1996).
26. M.K.Bobojanov, S.Mahmutkhonov, and S.Aytbaev. AIP Conference Proceedings 2552, **050011**, (2023), https://doi.org/10.1063/5.0113890
27. M.Bobojanov. AIP Conference Proceedings 2552, **050034**, (2023), https://doi.org/10.1063/5.0114077
28. E.Usmanov. AIP Conference Proceedings 2552, **050020**, (2023), https://doi.org/10.1063/5.0111537
29. A.Rasulov. AIP Conference Proceedings 2552, **050017**, (2023), https://doi.org/10.1063/5.0111530
30. A.Rasulov, E.Usmonov, and M.Melikuziev. AIP Conference Proceedings 2552, **050018**, (2023), https://doi.org/10.1063/5.0111531
31. A.N.Rasulov, M.R.Ruzinazarov, and ofters. E3S Web of Conferences 289, **07007**, (2021), https://doi.org/10.1051/e3sconf/202128907007
32. A.N.Rasulov, G.R.Rafikova, and ofters. E3S Web of Conferences 289, **07006**, (2021), https://doi.org/10.1051/e3sconf/202128907006
33. E.G.Usmonov, D.S.Akhmetbaev, M.Mamutov, and ofters. E3S Web of Conferences 289, **07015**, (2021), https://doi.org/10.1051/e3sconf/202128907015
34. M.V.Melikuziev. AIP Conference Proceedings 2552, **050021**, (2023), https://doi.org/10.1063/5.0112395
35. M.V.Melikuziev, and ofters. E3S Web of Conferences 289, **07016**, (2021), https://doi.org/10.1051/e3sconf/202128907016
36. A.D.Taslimov. AIP Conference Proceedings 2552, **050023**, (2023), https://doi.org/10.1063/5.0112398
37. A.D.Taslimov, T.S.Mamarasulova, and ofters. E3S Web of Conferences, 289, **07011**, (2021), https://doi.org/10.1051/e3sconf/202128907011
38. A.D.Taslimov, Kh.Aminov, and ofters. E3S Web of Conferences, 289, **07010**, (2021), https://doi.org/10.1051/e3sconf/202128907010
39. I.U.Rakhmonov, A.M.Najimova, and K.M.Reymov. AIP Conference Proceedings 2647, **030010**, (2022), https://doi.org/10.1063/5.0104788
40. I.U.Rakhmonov, and A.M.Najimova. AIP Conference Proceedings 2647, **030011**, (2022), https://doi.org/10.1063/5.0104791
41. I.U.Rakhmonov, A.M.Najimova, Sh.M.Esemuratova, and T.T.Koptileuov. AIP Conference Proceedings 2647, **070024**, (2022), https://doi.org/10.1063/5.0104793
42. F.A.Hoshimov, I.U.Rakhmonov, N.N.Niyozov, and F.B.Omonov. AIP Conference Proceedings 2552, **030025**, (2023), https://doi.org/10.1063/5.0112388
43. I.U.Rakhmonov, F.A.Hoshimov, N.N.Kurbonov, and D.A.Jalilova. AIP Conference Proceedings 2552, **050022**, (2023), https://doi.org/10.1063/5.0112391
44. K.Reymov, S.K.Makhmuthonov, G.Turmanova, and Q,Uzaqbaev. E3S Web of Conferences 289, **07023**, (2021), https://doi.org/10.1051/e3sconf/202128907023
45. T.Gayibov, K.Reymov, and A.Aytbaev. E3S Web of Conferences 289, **07024**, (2021), https://doi.org/10.1051/e3sconf/202128907024

Prospect for the use of the solar updraft tower project for the generation of electrical energy in the republic of Uzbekistan

Almuhan Nuraliyev [1*], *Yalkin* Adilov [2], *Azamat* Esenbekov [3]

[1] Tashkent State Technical University named after Islam Karimov, 100095, Uzbekistan, Tashkent, University St. 2A
[2] Tashkent University of Irrigation and Agricultural Mechanization Engineers, 39 Kari Niyazi str., Tashkent, Uzbekistan
[3] Karakalpak State University named by Berdakh, 1 Ch.Abdirov str. Nukus, Uzbekistan

Abstract. Fulfilling the decision of the Paris Agreement on Climate Change (COP-21 of the UNFCCC) dated December 12, 2015, the government of the Republic of Uzbekistan has developed and adopted an energy development plan of the Republic for 2020-2030. The plan provides for the construction of 3 GW of wind and 5 GW of solar power plants annually. RES will generate electrical energy in parallel with the generation of energy by thermal stations and transfer it to the general energy system. The main disadvantage of the joint operation of RES and thermal power plants is the need to have standby generating capacity due to the fact that the sources of solar and wind energy are fickle by nature. System costs rise with the increasing share of renewable energy sources. The Solar Updraft Tower project is a renewable energy option that does not have the main disadvantage of other types of renewable energy sources - the variability of electricity production and therefore is more acceptable for operating conditions in a single energy system. The article discusses and evaluates the Solar Updraft Tower (SUT) project for the generation of electrical energy, taking into account the conditions of Uzbekistan.

1 Introduction

Today's generation of electricity from coal, oil and natural gas is environmentally damaging, unsustainable in terms of pricing, and many developing countries cannot afford these energy sources [1-4]. A reasonable technology for the widespread use of renewable energy sources should be:

a) simple and reliable;

b) affordable for technologically less developed countries;

c) should not require additional water resources for cooling;

d) do not produce waste and use environmentally friendly production from renewable or recyclable materials.

The Solar Updraft Tower (SUT) project meets these conditions and makes it possible to take a decisive step towards a global solar energy economy. Economic evaluations based on experience and knowledge accumulated to date [2-6] have shown that large-scale SUTs (≥ 100 mW) are capable of generating energy at prices close to those of conventional power plants. sharply increasing demand, will soon balance the difference in value today, and tomorrow will completely change it. Another argument in favor of the NSC is the global climate change that began in 2018-2020 and requires an immediate solution to the problem of replacing the energy production system from hydrocarbons with energy obtained from alternative sources, regardless of the price of oil. Our opinion is that in the very near future the SUT project will be able to take part in solving one of today's main tasks: the production of sustainable, inexhaustible and affordable energy. According to WEA today: in Australia, under the project of S.N.B, a station for the production of electricity with a capacity of 50 MW was built. Under the plans of the Enviromission Commission, two versions of 200 MW each are currently under construction, using 32 6.25 MW turbines with a collector area of 10 square kilometers under a tower 730 meters high in the Arizona desert. The thermal mass - possibly salt ponds - under the collector zone means power generation will continue into the night. A 50 kW prototype plant of this design operated in Spain in 1982-89. China is building a 27.5 MW solar rising tower in Jinshawan [3-7]. Estimates of the cost of electricity generated range from €0.05 per kilowatt-hour (kWh) to €0.25 (US$0.07 to $0.34 per kWh), depending on land value and financing scenario. By comparison, a conventional gas-fired power plant can produce electricity for as little as 0.05 EUR/kWh. Start projects also enable the development of other applications such as agriculture or horticulture, to extract or distill water, or to improve air pollution in cities.

* Corresponding author: raxmatillo82@mail.ru

 https://doi.org/10.1051/e3sconf/202338401054

2 Research Method

The technological principle of generating electrical energy using SUT is quite simple: under a large glass roof, the sun heats the air [4-8]. Under the influence of the greenhouse effect, the air under the glass roof (collector) heats up more than the air outside and an excess pressure appears, which in turn creates an air flow that is sucked in by a central vertical cylindrical pipe - a chimney. The air flow drives the turbines - generators installed in the chimney, which generate electricity.

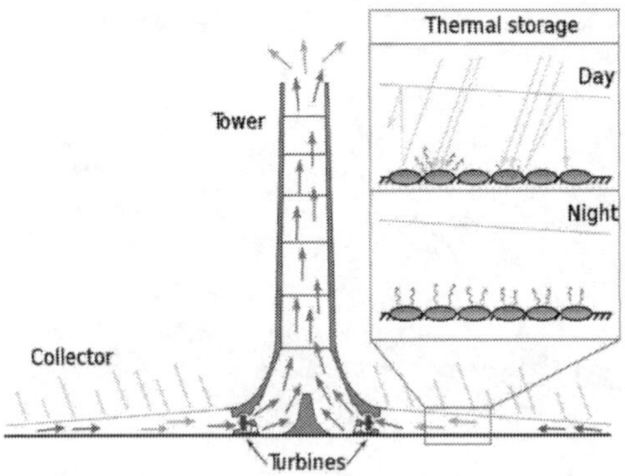

Fig 1. Basic parameters of the SUT

Because the soil below the collector acts as a natural heat storage system, SUTs can run 24 hours on pure solar power with some power reduction at night. Simple water pipes mounted on the ground increase the total heat capacity of the SNB by heating the water in these pipes, during the day and at night this heat warms the air and maintains pressure under the collector.

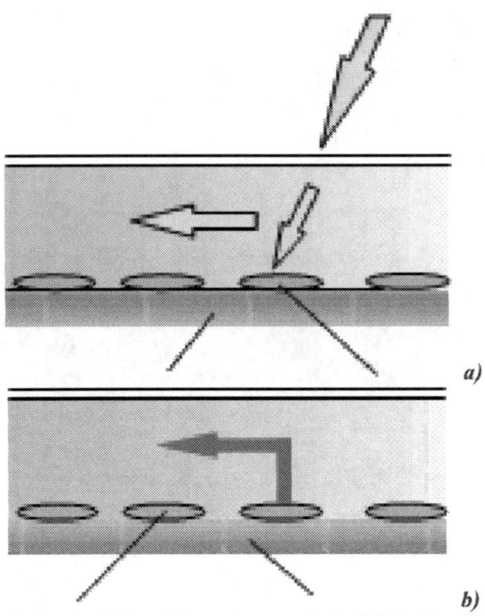

Fig 2. Accumulation of thermal energy in the SUT system

During the day, under the action of solar radiation, heat is accumulated in the water (Fig. 2a), at night, on the contrary, heat is transferred from the water into the air (Fig. 2b). The volume of water in the tubes is chosen to fit a water layer with a depth of 5 to 20 cm depending on the required power output characteristics.

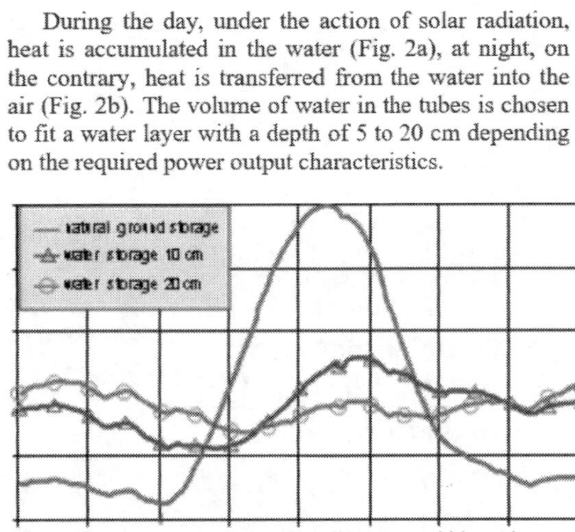

Fig. 3 Change in the power generated by SUT within 24 hours

The electrical output of a solar ascending tower is proportional to the air volume in the tower and the collector area (Figure 1). In other words, one can increase the collector area and decrease the height of the tower, and the same result can be obtained by increasing the height of the tower and decreasing the collector area.

3 Production of electrical energy

The energy contained in the updraft is converted into mechanical energy using stepped turbines which the base of the tower and into electrical energy using conventional wind turbines [5, 9-12].

Using wind turbines, we obtain mechanical power in the form of rotational energy from the air flow in the tower. The turbines in the solar ascending tower do not operate in step speed control, as in a conventional wind power converter, but as a smoothly controlled hydro-generator in which static pressure is converted into rotational energy [6, 13-16].

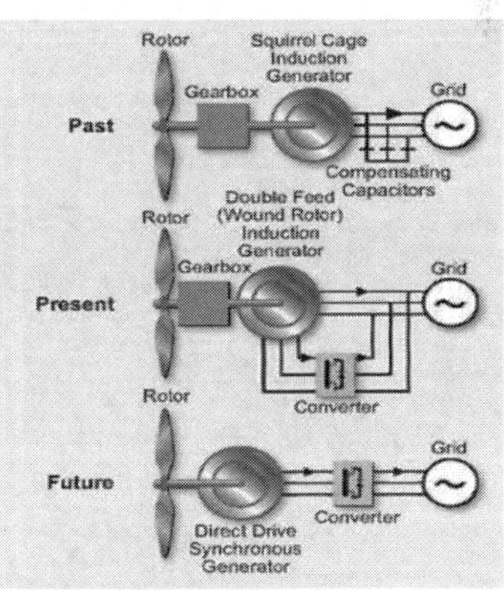

Fig. 4. Types of wind turbines used in SUT

The specific power output (power per area covered by the rotor) in a solar pressure tower is about an order of magnitude higher than that of a conventional Wind Turbine Generator (WTG). The air flow rate before and after the turbine is approximately the same. In order to maximize energy yield, the goal of the turbine control system is to maximize electrical energy yield under all operating conditions [7, 17-19]. To this end, the pitch of the blade is adjusted during operation to adjust the power output according to changing airspeed and airflow.

4 WTG subsystem model

The aerodynamics of the WTG generates the power as a function of wind speed given by:

$$PWTG=12\rho ArCpVw3 \quad (1)$$

where, ρ represents air density having unit kg/m3, generally constant for particular location of installation, Ar and Cp are the swept area (m2) of blades of a particular turbine and the coefficient of power. Vw represents wind speed. The transfer function (i.e. in first order form) may also be given as a lag compensator as:

$$GWTGx(s)=KWTG\backslash(1+sTWTG)=\Delta PWTG\backslash\Delta PW \quad (2)$$

where, KWTG is WTG 's gain constant and TWTG is WTG's time constant.

The angle of rotation of the turbine blades is controlled depending on the operating mode of the frequency inverter (Fig. 4 models "present&future").

5 Results and Discussion

Using SUT in agriculture. An important side effect of placing a large transparent membrane above the ground is to capture the evaporated groundwater and return it back to the topsoil [8, 20-22]. This localized increase in ground moisture can make the soil below the collector suitable for agricultural use by effectively creating a partial greenhouse. In some cases, the land below the collector would not be viable for agriculture without the presence of a membrane. This means that some badlands can be put to productive use, making this energy production strategy more cost effective while building agricultural capital.

The clearance height below the collector can easily accommodate agricultural equipment, and the collector supports can be far enough apart to allow the land to be farmed. The farming system, combined with the updraft tower, can be applied both as a sustainable economic strategy and for the recovery of agriculture in developing countries.

Using SUT with PV panels. Use with other solar technologies, such as plants that use solar radiation to produce EE using photovoltaic (PV). The main disadvantage of PV panels is their strong heating during the generation of EE [9, 23-24]. The use of a solar ascending tower in combination with photovoltaic arrays can increase the efficiency of both systems. The constant wind flow can cool the PV collectors with air while increasing the power output per area of land used, which makes the solar rising tower more efficient.

Combining these systems creates: 1. the loss of some direct solar radiation as a result of its deflection by the membrane. 2. Additional cooling of PV elements due to constant blowing.

Optimization of SUT usage. The quality and quantity of electrical energy received from the solar ascending tower, as well as the economic efficiency of the project, depend on the following factors: 1. the intensity of global solar radiation. 2. SUT parameters (collector area and tower height). 3. Cost of land to be used for the SUT project. 4. The cost of materials and labor that will be used in the implementation of the project [10, 15-17].

a)

b)

Fig. 5. General scheme of the electrical energy and agricultural production based on the SUT.
a) scheme of production of electrical energy;
b) general view of agricultural SUT

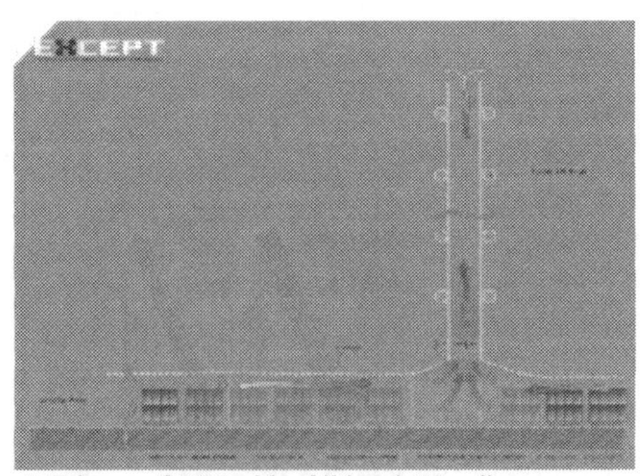

Fig. 6. Scheme of SUT with additional PV collector

Table 1. Typical dimensions of SNB and power generation

SUT power		50 mW	100 mW	200 mW
Height of tower	m	750	1000	1000
Towers diameter	m	90	110	120
Collectors diameter	m	3750	4300	7000
The amount of electricity with solar radiation 2300kWh/m2+	GW in hour	153	320	680
The amount of electricity with solar radiation 1800kWh/m2+	GW in hour	120	250	532
Towers Cost	mln€	72	192	192
Labor cost		18€	18€	5€
Collectors cost	mln€	142	430	390
Turbine + generator cost	mln€	56	146	146
The cost of electricity with solar radiation 2300 (kWh \ m2 year) A	€\κV hour	0.19	0.15	0.12
The cost of electricity with solar radiation 1800 (kWh \ m2 year) A	€\κV hour	0.25	0.2	0.16

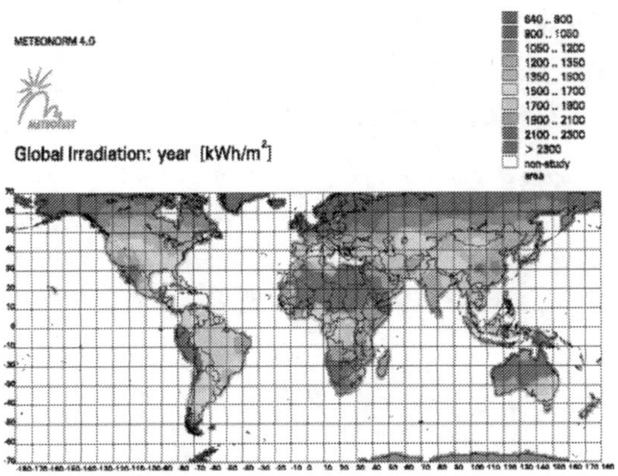

Fig. 7. Map of global solar radiation

Many areas and countries in the Earth's solar belt, that is, a zone around 35° north and south of the equator, are suitable for solar rising towers. As a general rule, annual global horizontal radiation must equal or exceed 1800 kWh/m². There really is no optimal physical size for SUT power plants [11, 25-27]. Optimal dimensions can only be calculated by including the unit costs of the components (collector, column, turbines) for individual sites. Thus, plants with different optimal key dimensions will be built for different sites, but always at the optimal price: if the collector area is cheap and concrete is expensive, then the collector will be large, and the tower is relatively small, and if the land under the collector is expensive, then its area will be smaller but the tower is higher.

For a quick overview, typical dimensions for selected capacities of a solar ascending tower are shown in Table 1. These figures are based on typical material and construction costs [12-16].

Conclusion

The possibility of using the SUT project in the conditions of Uzbekistan. According to the map of solar radiation Fig. 7, Uzbekistan is located in a climatic zone where the level of solar radiation is (1900-2100) kWh/m2 year, and the SUT project is recommended for use. In economic calculations, indicators of the cost of land and labor are of great importance, which will be used for the economic justification of the use of the SUT complex. square meter (1-1.5) $. As for the cost of using labor, in the Republic there is a large surplus of it and, accordingly, prices are much less than 5 € indicated in Table 1. If we add here the relative cheapness of building materials that will be used to build the SUT, then the cost of the project can be reduced by (25- 30)% compared with the data in Table 1. Taking into account the above, we propose to implement the SUT project with a unit capacity of 100 MW in Yangiarik district, Khorezm region, Uzbekistan. 110.5 thousand people permanently live in the region, concentrated in the south of the territory. Electric energy is consumed in the amount of 8 million. kWh per month or 96 million kWh per year. In the district, as a result of the survey, there is an area of 200 hectares unused for agriculture, covered with sand (photo Fig. 8).

Fig. 8. Photo of the site for the placement of the SUT complex in the Yangiaryk district, Khorezm region

Currently, active work is underway on the project, taking into account the standards developed by the Ministry of Innovation of the Republic of Uzbekistan.

References

1. Saad Mekhilef, and ofters. Journal of Thermal Analysis and Calorimetry, **141**, p.649-668, (2020).
2. K.G.Abidov, A.K.Nuraliev, and ofters. Journal of Advanced Research in Dynamical and Control Systems, **12(7)**, pp. 2167-2171, (2020).
3. A.K.Nuraliev, and ofters. E3S Web of Conferences, 216, **01108**, (2020), https://doi.org/10.1051/e3sconf/202021601108
4. M.Ibadullaev, A.Nuraliev, A.Esenbekov. IOP Conference Series: Materials Science and Engineering, 862(6), **062031**, (2020), DOI 10.1088/1757-899X/862/6/062031
5. R.Karimov. AIP Conference Proceedings, 2552, **030014**, (2022). https://doi.org/10.1063/5.0111533
6. R.Karimov. AIP Conference Proceedings, 2552, **050012**, (2022). https://doi.org/10.1063/5.0111524
7. S.Dzhuraev, R.Karimov, and ofters. ElConRus, (2022), pp. 1166-1169, doi: 10.1109/ElConRus54750.2022.9755782
8. R.Karimov, A. Kuchkarov, and ofters. Journal of Physics: Conference Series 2094, **052050**, (2021). doi:10.1088/1742-6596/2094/5/052050
9. R.Karimov, N.Kurbanova, and ofters. Journal of Physics: Conference Series 2094(5), **052042**, (2021). doi:10.1088/1742-6596/2094/5/052042
10. K.G.Abidov, O.O.Zaripov, and ofters. AIP Conference Proceedings, 2552, **030023**, (2022), https://doi.org/10.1063/5.0112385
11. K.G.Abidov, O.O.Zaripov, and ofters. AIP Conference Proceedings, 2552, **030022**, (2022), https://doi.org/10.1063/5.0112384
12. K.G.Abidov, O.O.Zaripov, and ofters. E3S Web of Conferences, 289, **07003**, (2021), https://doi.org/10.1051/e3sconf/202128907003
13. K.G.Abidov, and ofters. E3S Web of Conferences, 289, **07004**, (2021), https://doi.org/10.1051/e3sconf/202128907004
14. K.Abidov, N.Khamudkhanova. E3S Web of Conferences, 216, **01111**, (2020), https://doi.org/10.1051/e3sconf/202021601111
15. K.G.Abidov, O.O.Zaripov, and ofters. E3S Web of Conferences, 216, **01110**, (2020), https://doi.org/10.1051/e3sconf/202021601111
16. K.G.Abidov, O.O.Zaripov. E3S Web of Conferences, 139, **01088**, (2019), https://doi.org/10.1051/e3sconf/201913901088
17. E.Kh.Abduraimov, D.Kh.Khalmanov. Journal of Physics: Conference Series, 2373(7), **072010**, (2022), DOI 10.1088/1742-6596/2373/7/072010
18. E.Kh.Abduraimov. Journal of Physics: Conference Series, 072009, 2373, (2022), DOI 10.1088/1742-6596/2373/7/072009
19. E.Abduraimov, M.Peysenov, N.Tairova. AIP Conference Proceedings, 2552, **040012**, (2022), https://doi.org/10.1063/5.0116235
20. E.Kh.Abduraimov, D.Kh.Khalmanov. Journal of Physics: Conference Series, 2094(2), **022072**, (2021), DOI 10.1088/1742-6596/2094/2/022072
21. E.Kh.Abduraimov, D.Kh.Khalmanov, and ofters. E3S Web of Conferences, 289, **07026**, (2021), https://doi.org/10.1051/e3sconf/202128907026
22. E.Kh.Abduraimov, D.Kh.Khalmanov. E3S Web of Conferences, 216, **01106**, (2020), https://doi.org/10.1051/e3sconf/202021601106
23. E.Kh.Abduraimov. E3S Web of Conferences, 216, **01105**, (2020), https://doi.org/10.1051/e3sconf/202021601105
24. E.Kh.Abduraimov, D.Kh.Khalmanov. Journal of Physics: Conference Series, 1515(2), **022055**, (2020), DOI 10.1088/1742-6596/1515/2/022055
25. S.Begmatov, D.Khalmanov, and ofters. AIP Conference Proceedings, 2552, **040011**, (2022), https://doi.org/10.1063/5.0130666
26. K.Reymov, S.K.Makhmuthonov, G.Turmanova, and Q,Uzaqbaev. E3S Web of Conferences 289, **07023**, (2021), https://doi.org/10.1051/e3sconf/202128907023
27. T.Gayibov, K.Reymov, and A.Aytbaev. E3S Web of Conferences 289, **07024**, (2021), https://doi.org/10.1051/e3sconf/202128907024

E3S Web of Conferences 384, 01055 (2023) https://doi.org/10.1051/e3sconf/202338401055
RSES 2022

Improving the reliability of the operation of reclamation pumping stations by using self-starting electric motors

Kudrat Abidov*

Tashkent State Technical University named after Islam Karimov, 100095, Uzbekistan, Tashkent, University St. 2A

Abstract. The results of full-scale studies of unsteady processes during self-starting of electric motors at pumping stations are presented. The simulation of transient processes in the modes of self-starting of an asynchronous electric drive of a pumping unit is considered in order to identify the nature of the change in flow, pressure and moment of resistance on the pump shaft in the noted modes. The transient processes in the run-down modes when the pump loses the drive, as well as the self-starting of the asynchronous electric drive of the pumping unit, are investigated. In order to improve the technical and operational performance of reclamation pumping stations, the necessity of using the self-starting mode of electric drives of pumping units is substantiated.

1 Introduction

Solving the issue of self-starting of electric motors of pumping units is very important for preventing mass shutdown of consumers and ensuring the uninterrupted operation of modern large pumping stations during short-term power outages. Successful implementation of self-starting of electric motors of critical mechanisms after a short interruption of power supply and a deep voltage drop will minimize damage and ensure reliable operation of the station [1-4]. It is convenient and economical to study the main characteristics of the processes of self-starting of electric motors of pumping units by means of mathematical models on a computer [2-6], taking into account changes in the initial parameters. Field studies of the self-starting process of electric motors are more laborious and expensive, however, they are necessary to assess the reliability of the accepted mathematical model. In this regard, this paper presents the results of field studies of unsteady processes during self-starting of electric motors at pumping stations [7-9].

2 Experimental research

To solve the issue of self-starting, experiments were carried out at typical Amu-Zang pumping stations in the Surkhandarya region of the Republic of Uzbekistan, where asynchronous electric motors of the DAZO-15-59-YUY1 type, U_N=6kV, P_N=630kW, n_N=595 rp/m, I_N=80 A.

At these stations, during short-term power outages, all 16 motors are switched off by relay protection. The valves do not close and the pump units run in coast mode [8-10].

This negatively affects many units of the units (gland packings fail, fastening units relax, etc.), which leads to a large expenditure of time and money to restore the working condition of the units.

The experiments were carried out on one pumping unit No.11 at different values of the time delay, which varied within 1.5-14.5 s. The frequency of rotation on the pump shaft, the stator current, the overrun time, and the self-start time were oscillographed [11-13].

Data processing oscillograms self-start motor DAZO are shown in Table 1.

Table 1.

n=595 rp/m, I_n =80A	t_{off},c	t_{self},c	n_{self}, rp/m	I_{self}, A
1	1,5	0,4	379	573
2	2,2	0,56	333	640
3	7,5	1,04	153	660
4	11,5	1,16	117	680
5	14,5	1,5	54	693

The pumping station under consideration consists of 16 pumps of the 24 NDS type and asynchronous motors of the DAZO type. All engines of the pumping station are powered by one substation. It can be seen from the oscillogram that the multiplicity of the starting current and the duration of self-starting increase with increasing voltage off time. Oscillograms of the self-starting process for a pause time of 1.5 and 2.2 s are shown in Fig.1.

Although non-return valves are installed on these units, increasing the holding time leads to an increase in the starting current and starting duration. It was agreed to re-close the oil circuit breaker at the input of the substation of transformer T_1, which is 1.5 s, taking this

* Corresponding author: raxmatillo82@mail.ru

into account, the delay time of the self-starting device was set to 2 s.

In the course of the experiment, an experiment of self-starting on an open valve was carried out. The self-starting process of pumping unit №6 was carried out with a power interruption t_{off}=2.2s. Self-start time t_{self}=2.4 s. The current surge during self-starting was 6.3 I_{nom}. The rotational speed at the moment of self-starting is n_{self}=380 rp/m. The duration of the transient process from the moment the voltage is off is 4.6 s.

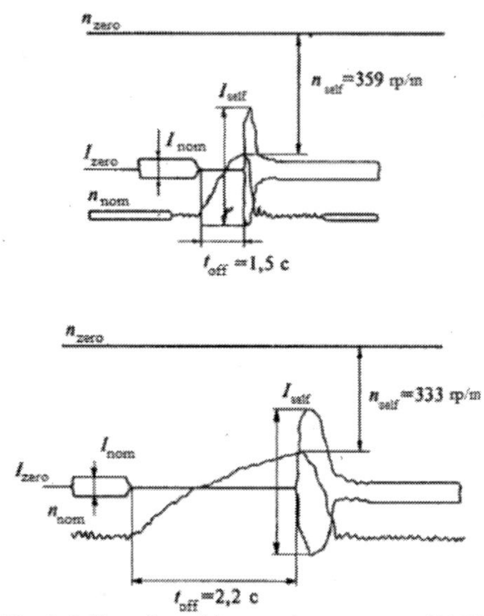

Fig. 1. Self-starting of the asynchronous motor DAZO-15-59-10U1

The maximum pressure value during self-starting on the discharge pipeline is 1.6 H_{nom}. When self-starting, the pressure pulsation on the pressure pipeline is set in 7.3 s. In addition to unit №6, tests were carried out on units № 1,2,4,9, the results of which are given in Table. 2.

Table 2.

Tested pumping unit	n_{self}, rp/m	I_{self}^*	t_{off} c	t_{self} c	H^*
№6	380	6,3	2,2	2,4	1,6
№6	190	6,4	4,5	3,5	1,6
№1	480	5,5	1,2	1,5	
№2	467	5,9	1,3	1,7	
№4	380	6,3	2,2	2,4	
№9	190	6,4	4,8	4,1	
* Multiple of nominal value					

To study the self-starting process of the pumping unit, as well as to clarify the nature of the flow of hydromechanical and electromechanical processes, an experimental stand was created with a 1.5K-6 pump and a KAM-30 electric motor. Oscillograms of self-starting of the electric motor of the pumping unit are shown in fig. 2, and the data corresponding to the experimental study - in table. 3.

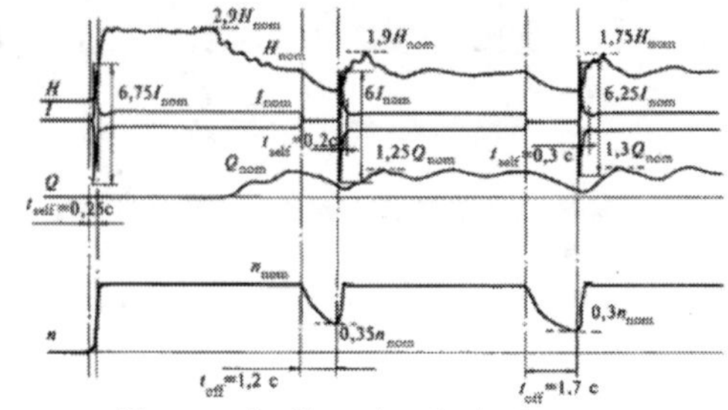

Fig. 2. Oscillogram of self-starting obtained on an experimental stand

Table 3.

Mode	t_{off}, c	t_{off}, c	n_{self} multiple	I_n, multiple	H_n, multiple	Q_n, multiple
Closed valve start	-	0,25	-	6,75	2,9	-
Self start	1,2	0,2	0,35	6	1,9	1,25
Self start	1,7	0,3	0,3	6,25	1,75	1,3
self start	3,3	0,32	0,04	6,33	1,8	1,17
Self start from turbine mode	16,7	0,4	0,3 back hijacking	0,4	1,8	1

The value of the pressure of the hydraulic shock is less than the manometric pressure of the pump developed when it is operated on a closed valve. This is due to the fact that the hydraulic resistance in the pumping unit during self-start does not change instantly. It depends on the starting speed of the machine. Therefore, the hydraulic shock is incomplete. In the experimental setup, the length of the pressure pipeline is relatively short; during self-start, the fluid flow rate changes slowly [13-15].

This paper also considers the simulation of transient processes in the modes of self-starting of an asynchronous electric drive of a pumping unit in order to identify the nature of the change in flow, pressure and moment of resistance on the pump shaft in the noted modes. In these modes, hydraulic, mechanical, electromagnetic transients occur. The results of the study show that the hydromechanical time constant is much greater than the electromagnetic one. Therefore, the electromagnetic time constant of the windings of an asynchronous motor in the self-starting mode is not taken into account.

The hydraulic transient process, taking into account the elasticity of water and the walls of pressure conduits, can be described in the form [14-17], where the relationship between the change in pressure and the velocity of water in the pipeline is determined by the formula:

$$\Delta H = \frac{a}{g}\Delta \vartheta; \tag{1}$$

The hydraulic shock process has a wave character and is described by partial differential equations:

$$\begin{cases} -\dfrac{\partial \vartheta}{\partial t} = g - \dfrac{\partial H}{\partial x} \\ \dfrac{\partial \vartheta}{\partial x} = \dfrac{g}{a^2}\dfrac{\partial H}{\partial t} \end{cases} \tag{2}$$

The general solution of the system has the form:

$$\begin{aligned} H &= H_0 + \varphi - \psi; \\ \vartheta &= \vartheta_0 - \frac{g}{a}(\varphi + \psi); \end{aligned} \tag{3}$$

where H - pressure; ϑ - is the speed of water movement in the pipeline; t - is the time since the occurrence of hydraulic shock; a - is the propagation velocity of shock waves; g- is the acceleration due to gravity; x - is the distance from the origin; φ - equivalent waves of pressure increase; ψ - equivalent pressure reduction waves.

In addition, the hydraulic transition process is described by the equation of unsteady motion of an incompressible fluid:

$$H = H_g + h_M + h_L + h_i \tag{4}$$

where H -is the pump head; H_g - geometric pressure; h_i is the inertial head; $h_M + h_L$ head loss.

The equation of the mechanical transient process is described by the equation of motion of the unit in an unsteady mode:

$$M_{el} - M_g - M_f = \frac{GD^2}{375}\frac{dn}{dt} \tag{5}$$

where M_{el}.-torque of the engine, is determined by the starting characteristic of the engine; M_g - hydraulic moment of the pump; M_f - the moment spent on friction in the seals and bearings of the unit.

The torque of an asynchronous motor is determined by the well-known expression [5, 18-20]:

$$M_{el} = \frac{3(I_2')^2 R_2'}{\omega_0 S}$$

where I_2'-is the reduced rotor current; R_2' - reduced active resistance of the rotor; ω_0 - synchronous angular speed of the motor; S - engine slip.

To determine the hydraulic moment M_g, a four-square characteristic of the pump is used in the form of dependencies between the reduced flow rate Q, rotational speed n, hydraulic moment M:

$$\bar{n} = f(\bar{Q});\quad \bar{M} = f(\bar{Q}). \tag{6}$$

Self-starting will be successful if the motor accelerates to rated speed after voltage recovery and the following conditions are met:

1) normal voltage in the pump impeller;
2) heating of the motor windings at a normal level;
3) the torque of the motor ensures the acceleration of the motor to the nominal speed;
4) allowable value of pressure in pipelines.

3 Research results

Mathematically simulating the process of self-starting and setting the duration of a power interruption at different time intervals, we determine the maximum duration at which self-starting is impossible. In this case, one of the conditions for a successful self-start is no longer fulfilled. Knowing this time, you can correctly set the automatic operation mode when the voltage is restored. To accurately determine the break time limit, it is necessary to simulate a self-start simulation, taking into account all the accepted conditions.

Figure 3 shows the characteristic of the process with a break interval t_{of}=2 s, built according to the calculated data $H=f(t)$, $Q=f(t)$, $n=f(t)$, $M=f(t)$ for an asynchronous electric motor of the type DAZO-15-59-10У1, U_N=6 kV, P_N=630 kW, n=595 rp/m, I_N=80 A, and centrifugal pump type 24НДС [8, 21-22]. At the moment the engine is turned off, M_{el} instantly drops to zero. The hydraulic moment of resistance of the pumping unit, M_g, and the frictional moment, M_g, shown as M in the timing diagram, are saved. This leads to a decrease in the rotational speed n of the pump and the developed pressure H, due to which the flow Q decreases. Although the parameters n, H, Q decrease, the unit mode remains "pumping". In this time interval, when the electric motor is turned on, the second stage begins, self-starting. Under the action of the torque M_{el} of the engine, the pumping unit is accelerated to a steady state.

The results of full-scale research and computer simulation in terms of speed parameters n (Fig. 1 and Fig. 3) coincide.

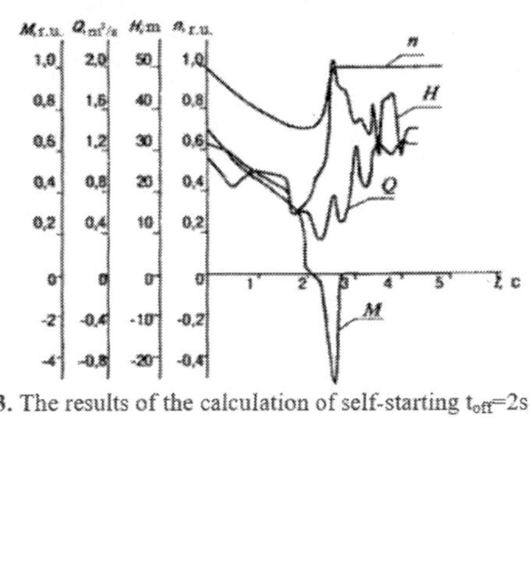

Fig.3. The results of the calculation of self-starting t_{off}=2s

https://doi.org/10.1051/e3sconf/202338401055

Figure 4 shows the characteristics of the transient process of self-starting at a break interval t_{of} =5 s. The pumping unit operating in the steady state is disconnected from the network - this is the first stage of the calculation and the torque of the M_{el} engine instantly drops to zero, while the hydraulic moment of resistance of the M_g pumping unit and the friction moment M_f are preserved.This leads to a decrease in the rotational speed n of the pump and the developed pressure H, due to which the flow Q decreases.At point b (Fig.4), the rotation frequency decreases so much that the pump flow becomes equal to zero, then the water begins to move along the conduits and through the impeller from the upstream to the downstream, i.e. in the turbine direction.But the direction of rotation is kept "pumping". Since in this mode the throughput of the pumping unit drops sharply, the pressure in the spiral chamber, the pressure H, the hydraulic moment M_g increase. At point c, the electric motor is switched on - the second stage begins. Under the action of the engine torque M_{el}, the pumping unit is accelerated to a steady state. From the point with the difference M_{el}-M_g-M_f=M_{din} creates a dynamic accelerating moment, under the influence of which the pump flow slows down and begins to decrease in the "turbine" mode [5-10, 23].

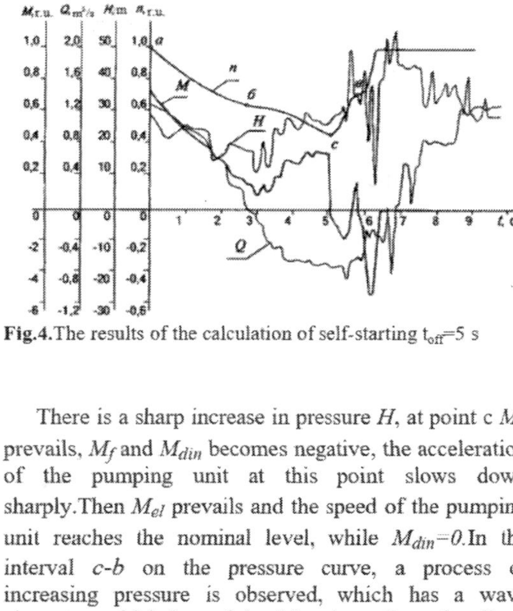
Fig.4.The results of the calculation of self-starting t_{off}=5 s

There is a sharp increase in pressure H, at point c M_g prevails, M_f and M_{din} becomes negative, the acceleration of the pumping unit at this point slows down sharply.Then M_{el} prevails and the speed of the pumping unit reaches the nominal level, while $M_{din}=0$.In the interval c-b on the pressure curve, a process of increasing pressure is observed, which has a wave character, which is explained by the action of a direct wave of hydraulic shock in the pressure pipeline.

4 Conclusion

1. The response time of the automatic reclosure of the electrical network for self-starting is determined from the mode run-out of a specific pumping unit, while it is necessary to control the recovering voltage to a value that ensures successful self-start [11-20].

2. The self-starting system of electric motors must take into account technological limitations, the actions of technological protections and automatic restoration of the operation of auxiliary mechanisms. When designing and determining the conditions for their operation, it is necessary to take into account the influence of various types of transients, especially those caused by load shedding and drive shutdown [21-23].

3. An algorithm for calculating the self-starting mode of a pumping unit with an asynchronous electric drive has been compiled.

References

1. K.G.Abidov, O.O.Zaripov, and ofters. AIP Conference Proceedings, 2552, **030023**, (2022), https://doi.org/10.1063/5.0112385
2. K.G.Abidov, O.O.Zaripov, and ofters. AIP Conference Proceedings, 2552, **030022**, (2022), https://doi.org/10.1063/5.0112384
3. K.G.Abidov, O.O.Zaripov, and ofters. E3S Web of Conferences, 289, **07003**, (2021), https://doi.org/10.1051/e3sconf/202128907003
4. K.G.Abidov, and ofters. E3S Web of Conferences, 289, **07004**, (2021), https://doi.org/10.1051/e3sconf/202128907004
5. K.Abidov, N.Khamudkhanova. E3S Web of Conferences, 216, **01111**, (2020), https://doi.org/10.1051/e3sconf/202021601111
6. K.G.Abidov, O.O.Zaripov, and ofters. E3S Web of Conferences, 216, **01110**, (2020), https://doi.org/10.1051/e3sconf/202021601111
7. K.G.Abidov, O.O.Zaripov. E3S Web of Conferences, 139, **01088**, (2019), https://doi.org/10.1051/e3sconf/201913901088
8. K.G.Abidov, A.K.Nuraliev, and ofters. Journal of Advanced Research in Dynamical and Control Systems, **12(7)**, pp. 2167-2171, (2020).
9. R.Karimov. AIP Conference Proceedings, 2552, **030014**, (2022). https://doi.org/10.1063/5.0111533
10. R.Karimov. AIP Conference Proceedings, 2552, **050012**, (2022). https://doi.org/10.1063/5.0111524
11. S.Dzhuraev, R.Karimov, and ofters. ElConRus, (2022), pp. 1166-1169, doi: 10.1109/ElConRus54750.2022.9755782
12. R.Karimov, A. Kuchkarov, and ofters. Journal of Physics: Conference Series 2094, **052050**, (2021). doi:10.1088/1742-6596/2094/5/052050
13. R.Karimov, N.Kurbanova, and ofters. Journal of Physics: Conference Series 2094(5), **052042**, (2021). doi:10.1088/1742-6596/2094/5/052042
14. E.Kh.Abduraimov, D.Kh.Khalmanov. Journal of Physics: Conference Series, 2373(7), **072010**, (2022), DOI 10.1088/1742-6596/2373/7/072010

15. E.Kh.Abduraimov. Journal of Physics: Conference Series, 072009, 2373, (2022), DOI 10.1088/1742-6596/2373/7/072009
16. E.Abduraimov, M.Peysenov, N.Tairova. AIP Conference Proceedings, 2552, **040012**, (2022), https://doi.org/10.1063/5.0116235
17. E.Kh.Abduraimov, D.Kh.Khalmanov. Journal of Physics: Conference Series, 2094(2), **022072**, (2021), DOI 10.1088/1742-6596/2094/2/022072
18. E.Kh.Abduraimov, D.Kh.Khalmanov, and ofters. E3S Web of Conferences, 289, **07026**, (2021), https://doi.org/10.1051/e3sconf/202128907026
19. E.Kh.Abduraimov, D.Kh.Khalmanov. E3S Web of Conferences, 216, **01106**, (2020), https://doi.org/10.1051/e3sconf/202021601106
20. E.Kh.Abduraimov. E3S Web of Conferences, 216, **01105**, (2020), https://doi.org/10.1051/e3sconf/202021601105
21. E.Kh.Abduraimov, D.Kh.Khalmanov. Journal of Physics: Conference Series, 1515(2), **022055**, (2020), DOI 10.1088/1742-6596/1515/2/022055
22. A.K.Nuraliev, and ofters. E3S Web of Conferences, 216, **01108**, (2020), https://doi.org/10.1051/e3sconf/202021601108
23. M.Ibadullaev, A.Nuraliev, A.Esenbekov. IOP Conference Series: Materials Science and Engineering, 862(6), **062031**, (2020), DOI 10.1088/1757-899X/862/6/062031

 https://doi.org/10.1051/e3sconf/202338401056

Improving the quality of 0.4 kV electricity in household appliances due to voltage regulation

Raxmatillo Karimov*

Tashkent State Technical University named after Islam Karimov, 100095, Uzbekistan, Tashkent, University St. 2A

Abstract. The article discusses a way to improve the quality of electricity up to 0.4 kV household appliances through voltage regulation. As an actuator, a booster transformer with two secondary windings connected in accordance with the primary winding is used. High-speed non-contact voltage relays are used as a device that is sensitive to changes in the input voltage, as well as a device that switches the secondary windings of a booster transformer.

1 Introduction

Electricity, as a special type of product, has certain characteristics that make it possible to judge its suitability for various production processes.

The set of characteristics under which power receivers are able to perform their functions, united by the general concept of power quality, along with reliability, safety, is one of the mandatory requirements for power supply systems. The quality of electricity is characterized by a combination of properties and indicators. Ensuring the required quality of electrical energy is a problem that is solved during its generation, transmission, distribution and consumption. The main parameter and indicator of the quality of electrical energy is considered to be voltage and its quality. The task of managing the quality of electrical energy is to provide technically acceptable values of quality indicators at the terminals of electrical receivers of electrical energy. Deviation of these indicators from acceptable values leads to a disruption in the normal operation of power receivers, a reduction in their service life, the occurrence of product defects, a decrease in productivity in industry, i.e. to various damages [1-5].

Significant results have been obtained in the field of electric energy quality management, corresponding to the achieved level of means for measuring electric energy quality indicators. It was the means of measuring voltage quality indicators that made it possible to take the next step in solving the problem of improving the quality of electrical energy [3-7].

To quantify the quality of electricity, such a system of single or generalized (integral) indicators is needed so that the quality is measurable, comparable and accessible for control and management.

Power plants produce electricity of a sufficiently high quality, and the deterioration of the quality of electricity occurs in the process of its transmission and consumption as a result of the influence of electrical receivers. The characteristic properties of electricity, necessary to determine the requirements for the system of indicators of the quality of electricity, are as follows [8-10]:

- with symmetry and sinusoidality of the three-phase voltage system and voltage and frequency values equal to or close to the nominal values for electrical equipment, the requirements formulated above are almost completely satisfied. The greatest economic efficiency can be achieved with certain deviations from the specified conditions;
- an adverse effect on consumers can manifest itself both constantly, through the accumulation of irreversible changes (marriage or undersupply of products, aging of insulation, etc.), or abruptly (failures or malfunctions in the operation of automatic devices, explosions of capacitor banks, etc.). This circumstance indicates the need to limit the permissible values by lowering the quality of electricity, which determine these negative impacts;
- electromagnetic and other characteristics of electrical systems and power supply systems of consumers change over time, as a rule, according to probabilistic laws, therefore, changes in the quality of electricity are random variables.

One of the reasons for the deterioration in the quality of electricity is the so-called "dips" of voltage, which are observed when switching powerful loads: semiconductor converters, etc. Voltage dips lead not only to a deterioration in the operation of electrical receivers at such enterprises, but also to a complete stop of the entire technological process. For example, up to 0.4 kV of household appliances causes great damage [10-14].

There are a large number of works devoted to voltage regulation using transformers and autotransformers by changing the transformation ratio. However, the switching of transformer winding taps occurs due to electromechanical contacts, which reduces their speed [15-17].

* Corresponding author: raxmatillo82@mail.ru

Devices that could compensate for voltage dips in the supply network must meet the following requirements: speed and sensitivity to changes in the voltage value at the input of this device [18-19].

2 Research contactless device

Such devices include booster transformers with non-contact switching of secondary windings. The executive body that switches the secondary windings of booster transformers, which meets the above requirements, is a non-contact voltage relay developed at the Department of Electrical Engineering of the Tashkent State Technical University, the diagram of which is shown in Fig.1.

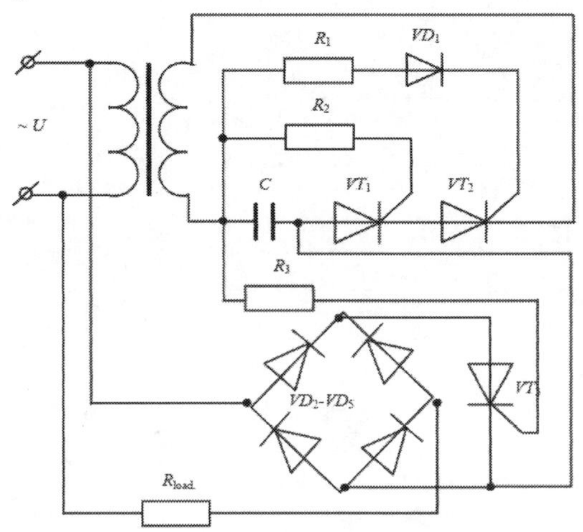

Fig.1. Schematic diagram of a non-contact thyristor voltage relay

It is known that, by controlling the thyristor triggering moment, it is possible to influence the shape of the load current curve. If the phase shift φ between the beginning of the "positive" half-cycle of the anode voltage and the beginning of the forward current flow is zero, the shape of the load current curve will be sinusoidal [20-23].

In the article, the circuit of a thyristor voltage relay with a non-sinusoidal shape of the voltage curve on the load is investigated. For many electrical installations, this is not the nominal mode of operation. To ensure the sinusoidal shape of the current and voltage curves on the load, it is necessary to achieve the opening of the thyristor when the current passes through zero [21-23].

Figure 1 shows a circuit diagram of a non-contact voltage relay with a sinusoidal voltage across the load [24-27].

Non-contact voltage relay works as follows. When a certain value of the input voltage is reached, the unlocking signal on the control electrode of the thyristor VT_2 will be sufficient to open with an angle of 90^0. After the opening of the thyristor VT_2, the thyristor VT_1 opens and the capacitor C is charged to the voltage of the secondary winding. At the same moment, from the plates of the capacitor C, a control signal pulse is applied to the power thyristor VT_3, which has the shape shown in Fig.2. Since a DC signal is applied to the control electrodes of the thyristor VT_2, it remains constantly open, and a sinusoidal current will flow through the load R_4. The moment of operation of the thyristors VT_2 is controlled by selecting the parameter of the resistor R_1 [28-30].

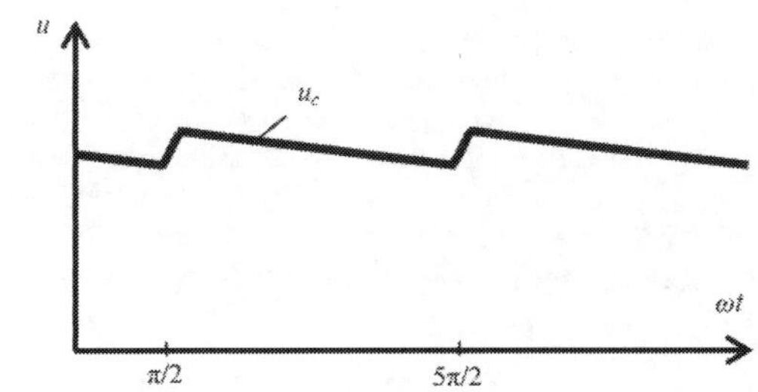

Fig. 2. The shape of the voltage curve on the capacitance

When testing, thyristors of the VT_1 - *KU202I*, VT_2 - *KU202I*, VT_3 - *KU202R* types were used as thyristors, respectively, as a *KTs402B* diode bridge, as active resistances R_1, R_2, R_3, R_4, respectively, R_1=5.6 kOhm, R_2=160 Ohm, R_3=390 Ohm, R_4=2.4 kOhm, as a capacitance C, a capacitor with a capacitance of 30 microfarads, as a transformer, a single-phase transformer with a voltage of 220/20 V was used. Experimental studies have shown that the load R_4=2.4 kOhm was connected to the network at a voltage of 200 V [31-32].

We know the contact circuit for turning on and off the windings of a booster transformer. The disadvantage of such devices is the use of a ferromagnetic element and an electromechanical key in the control unit, an increase in weight and size indicators, and a relatively low reliability of operation [25]. A non-contact device that is built according to a booster circuit has its own advantage - the switching process is carried out without interrupting the current through the load during switching of power switches [33-34].

By using non-contact voltage relays, a control system for a booster transformer was assembled, which provides the rated voltage at the load when the voltage in the supply network decreases (Fig. 3).

In Fig.3. shows a circuit diagram based on a non-contact voltage relay turning on and off the winding of a booster transformer. The device consists of two non-contact relays based on optothyristors. When a certain value of the input voltage is reached, the I-relay is activated and a control signal is given to open the power thyristor VT_4, which turns on the winding of the booster transformer to the network. With a further increase in the input voltage, the II-voltage relay is activated, shunting the diode circuit VU_1 of the optocoupler of the I-relay, with its thyristor circuit of the optocoupler. Thus, terminating the access of the control signal to the power thyristor, i.e. the switching off of the power thyristor is achieved as soon as the load current passes through zero [22, 35-37].

Consider the diagram of a non-contact switching device (Fig. 3) based on a non-contact voltage relay. The winding of the booster transformer is connected to the network by means of a diode bridge VD_3-VD_5, a controlled power thyristor VT_4 is included in the diagonal of the diode bridge. Control signals are fed to the control electrode of the power thyristor through resistor R_4 from the plates of capacitor C_1 (I-relay), which in turn is connected to the secondary winding of a low-power transformer through two low-power controlled thyristors VT_1, VT_2. The control signal for the thyristor VT_1 is supplied from the secondary winding of a low-power transformer through a series-connected R_3. The control signal for the thyristor VT_2 is supplied from the secondary winding of a low-power transformer through the series-connected R_1, R_2 and diode VD_2, as well as through the resistor R_5, since the thyristor VT_3 is closed. Disconnecting the winding of the booster transformer from the network is carried out due to the action of the second voltage relay (II), containing a limiting resistor R_4 through which a signal is supplied to the control electrode of the thyristor VT_3. The opening of the thyristor VT_3 will turn off the control signal of the thyristor VT_2 (I-relay), thereby closing VT_2 and this will turn off the control signal of the power thyristor VT_4. The rest of the relay circuit (II) is carried out in the same way as for relay (I). Variable resistors R_2 in both relays serve to adjust the pickup setting of the relay [2, 37-40].

The developed circuit of a non-contact device for switching on and off a booster transformer has an improvement in weight and size indicators and provides high reliability [31-33].

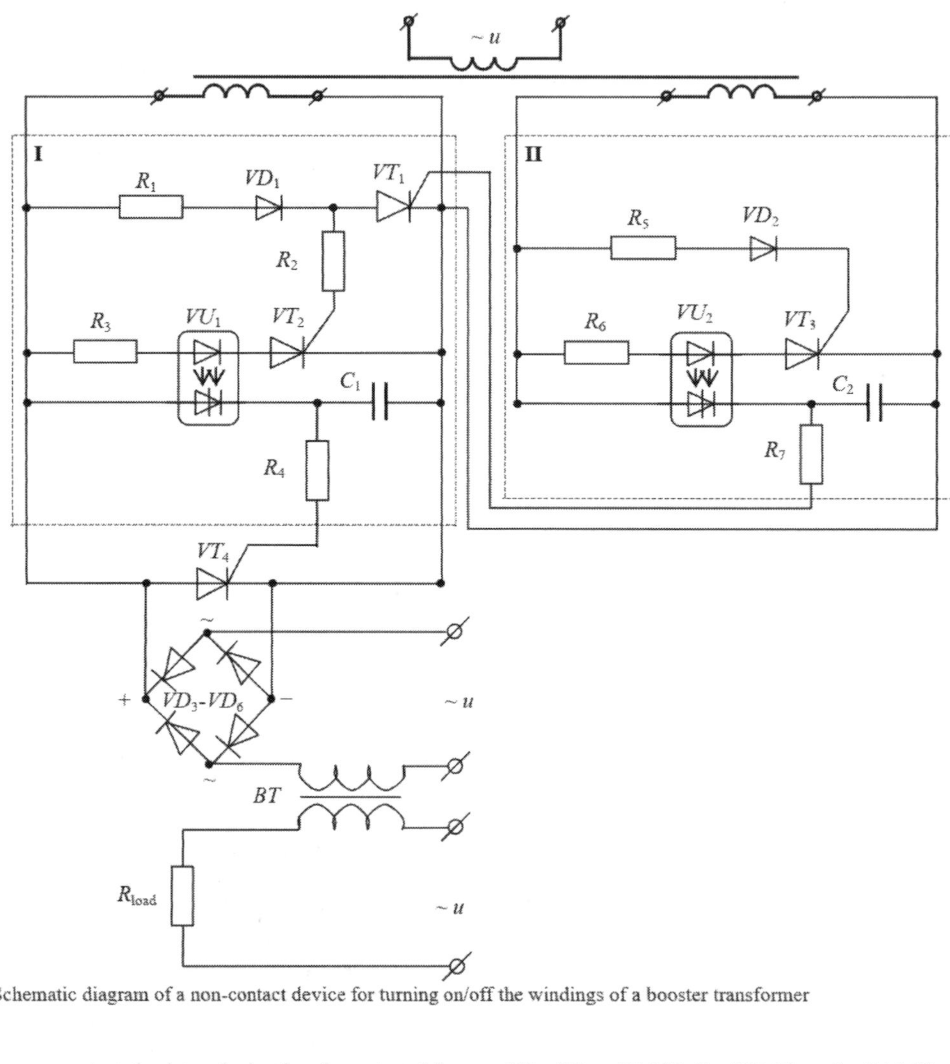

Fig. 3. Schematic diagram of a non-contact device for turning on/off the windings of a booster transformer

This non-contact thyristor device has been tested in the laboratory of the Department of Electrical Engineering. At the same time, thyristors of the following types were used as thyristors VT_1, VT_2, VT_3, VT_4: *KU202Zh*, *KU201K*, *KU202N*, *KU202E*; as diodes VD_1, VD_2 - *D226B*; R_1=820 Ohm, R_2=5,1 kOhm, R_3=1,3 kOhm, R_4=6,8 kOhm, R_5=22 kOhm, R_6=24 kOhm, R_7=6,8 kOhm, R_{load}=20 kOhm, as capacitances C_1 and C_2 - AC capacitors for 500 V. with a capacitance of 2 μF, thyristor optocouplers *MOC* are used as optocouplers

VU_1 and VU_2, diode bridge VD_3-VD_6 - *KTs402V*. Experimental studies have shown that the booster winding of transformer 3 was connected to the network at a voltage of 190 V and turned off at a voltage of 220 V [41-46].

Fig. 4 shows the input-output voltage characteristic of this device.

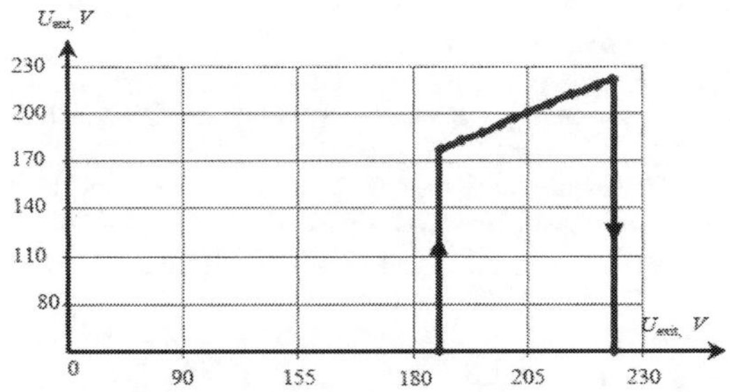

Fig. 4. Characteristic voltage "input-output"

3 Experimental analysis of non-contact voltage relay

In the research laboratory of the Department of "Electrical Engineering" of the Tashkent State Technical University, using an oscilloscope of the LeCroy WaveRunner 64 Xi-A type, experience is given and experimental data for the amplitude value of the voltage are obtained. WaveRunner 64 Xi-A Oscilloscope The 64 MXi-A was developed by LeCroy, the leading US digital oscilloscope manufacturer [47].

Fig. 5 shows the experimentally obtained oscillogram of the voltage change for an active load using a non-contact voltage relay in the control system for switching the windings of a booster transformer.

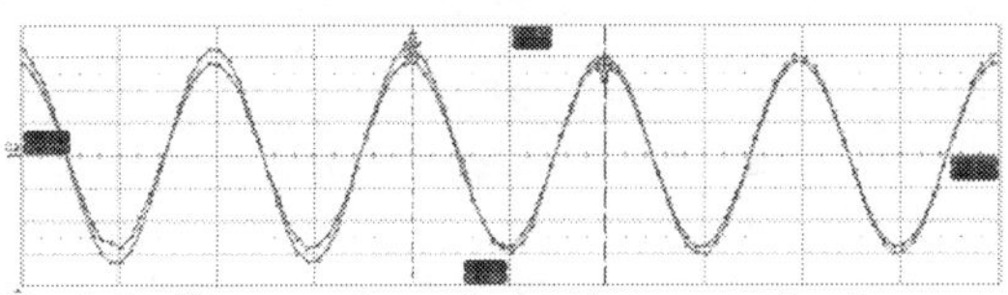

Fig. 5. Oscillogram voltage change "input-output" voltage relay

4 Conclusion

The results of the experiment show that a non-contact voltage relay in the control system will switch the windings of a booster transformer at a control voltage of 18 V in 0.32 seconds.

Thus, the purpose and objectives of the article are the results of the study: *developed non-contact voltage relay; With the use in the control system for switching the windings of a booster transformer, a scheme of a non-contact device is proposed; The application of the device has led to an energy-saving technology; reduction in energy consumption.*

Testing a prototype voltage relay to ensure the voltage deviation is within the allowable range of ±5%, which leads to an improvement in the quality of electricity.

Based on the analysis of the developed voltage relay, a scheme of a device for non-contact switching of the windings of a booster transformer is proposed in the control system.

Obtained from the control of the windings of the booster transformer, contactless switching, the energy characteristics show a high degree of reliability.

Thus, with this device, it is possible to maintain the load voltage stable.

References

1. R.Karimov. AIP Conference Proceedings, 2552, **030014**, (2022). https://doi.org/10.1063/5.0111533
2. R.Karimov. AIP Conference Proceedings, 2552, **050012**, (2022). https://doi.org/10.1063/5.0111524
3. S.Dzhuraev, R.Karimov, and ofters. ElConRus, pp. 1166-1169, (2022), doi: 10.1109/ElConRus54750.2022.9755782
4. R.Karimov, A. Kuchkarov, and ofters. Journal of Physics: Conference Series, 2094, **052050**, (2021). doi:10.1088/1742-6596/2094/5/052050
5. R.Karimov, N.Kurbanova, and ofters. Journal of Physics: Conference Series, 2094(5), **052042**, (2021). doi:10.1088/1742-6596/2094/5/052042
6. K.G.Abidov, O.O.Zaripov, and ofters. AIP Conference Proceedings, 2552, **030023**, (2022), https://doi.org/10.1063/5.0112385
7. K.G.Abidov, O.O.Zaripov, and ofters. AIP Conference Proceedings, 2552, **030022**, (2022), https://doi.org/10.1063/5.0112384
8. K.G.Abidov, O.O.Zaripov, and ofters. E3S Web of Conferences, 289, **07003**, (2021), https://doi.org/10.1051/e3sconf/202128907003
9. K.G.Abidov, and ofters. E3S Web of Conferences, 289, **07004**, (2021), https://doi.org/10.1051/e3sconf/202128907004
10. K.Abidov, N.Khamudkhanova. E3S Web of Conferences, 216, **01111**, (2020), https://doi.org/10.1051/e3sconf/202021601111
11. K.G.Abidov, O.O.Zaripov, and ofters. E3S Web of Conferences, 216, **01110**, (2020), https://doi.org/10.1051/e3sconf/202021601111
12. K.G.Abidov, O.O.Zaripov. E3S Web of Conferences, 139, **01088**, (2019), https://doi.org/10.1051/e3sconf/201913901088
13. E.Kh.Abduraimov, D.Kh.Khalmanov. Journal of Physics: Conference Series, 2373(7), **072010,** (2022), DOI 10.1088/1742-6596/2373/7/072010
14. E.Kh.Abduraimov. Journal of Physics: Conference Series, 072009, 2373, (2022), DOI 10.1088/1742-6596/2373/7/072009
15. E.Abduraimov, M.Peysenov, N.Tairova. AIP Conference Proceedings, 2552, **040012**, (2022), https://doi.org/10.1063/5.0116235
16. E.Kh.Abduraimov, D.Kh.Khalmanov. Journal of Physics: Conference Series, 2094(2), **022072**, (2021), DOI 10.1088/1742-6596/2094/2/022072

17. E.Kh.Abduraimov, D.Kh.Khalmanov, and ofters. E3S Web of Conferences, 289, **07026**, (2021), https://doi.org/10.1051/e3sconf/202128907026
18. E.Kh.Abduraimov, D.Kh.Khalmanov. E3S Web of Conferences, 216, **01106**, (2020), https://doi.org/10.1051/e3sconf/202021601106
19. E.Kh.Abduraimov. E3S Web of Conferences, 216, **01105**, (2020), https://doi.org/10.1051/e3sconf/202021601105
20. E.Kh.Abduraimov, D.Kh.Khalmanov. Journal of Physics: Conference Series, 1515(2), **022055**, (2020), DOI 10.1088/1742-6596/1515/2/022055
21. K.G.Abidov, A.K.Nuraliev, and ofters. Journal of Advanced Research in Dynamical and Control Systems, **12(7)**, pp. 2167-2171, (2020).
22. A.K.Nuraliev, and ofters. E3S Web of Conferences, 216, **01108**, (2020), https://doi.org/10.1051/e3sconf/202021601108
23. M.Ibadullaev, A.Nuraliev, A.Esenbekov. IOP Conference Series: Materials Science and Engineering, 862(6), **062031**, (2020), DOI 10.1088/1757-899X/862/6/062031
24. S.Begmatov, D.Khalmanov, and ofters. AIP Conference Proceedings, 2552, **040011**, (2022), https://doi.org/10.1063/5.0130666
25. Kh.G.Karimov, M.K.Bobozhanov. Elektrichestvo, **1**, pp. 27-32, (1996).
26. M.K.Bobojanov, S.Mahmutkhonov, and S.Aytbaev. AIP Conference Proceedings 2552, **050011**, (2023), https://doi.org/10.1063/5.0113890
27. M.Bobojanov. AIP Conference Proceedings 2552, **050034**, (2023), https://doi.org/10.1063/5.0114077
28. E.Usmanov. AIP Conference Proceedings 2552, **050020**, (2023), https://doi.org/10.1063/5.0111537
29. A.Rasulov. AIP Conference Proceedings 2552, **050017**, (2023), https://doi.org/10.1063/5.0111530
30. A.Rasulov, E.Usmonov, and M.Melikuziev. AIP Conference Proceedings 2552, **050018**, (2023), https://doi.org/10.1063/5.0111531
31. A.N.Rasulov, M.R.Ruzinazarov, and ofters. E3S Web of Conferences 289, **07007**, (2021), https://doi.org/10.1051/e3sconf/202128907007
32. A.N.Rasulov, G.R.Rafikova, and ofters. E3S Web of Conferences 289, **07006**, (2021), https://doi.org/10.1051/e3sconf/202128907006
33. Usmanov E.G., Rakhmonov I.U. Problems of energy and resource saving. **4**, PP. 147-151, (2022).
34. E.G.Usmonov, D.S.Akhmetbaev, M.Mamutov, and ofters. E3S Web of Conferences 289, **07015**, (2021), https://doi.org/10.1051/e3sconf/202128907015
35. M.V.Melikuziev. AIP Conference Proceedings 2552, **050021**, (2023), https://doi.org/10.1063/5.0112395
36. M.V.Melikuziev, and ofters. E3S Web of Conferences 289, **07016**, (2021), https://doi.org/10.1051/e3sconf/202128907016
37. A.D.Taslimov. AIP Conference Proceedings 2552, **050023**, (2023), https://doi.org/10.1063/5.0112398
38. A.D.Taslimov, T.S.Mamarasulova, and ofters. E3S Web of Conferences, 289, **07011**, (2021), https://doi.org/10.1051/e3sconf/202128907011
39. A.D.Taslimov, Kh.Aminov, and ofters. E3S Web of Conferences, 289, **07010**, (2021), https://doi.org/10.1051/e3sconf/202128907010
40. I.U.Rakhmonov, A.M.Najimova, and K.M.Reymov. AIP Conference Proceedings 2647, **030010**, (2022), https://doi.org/10.1063/5.0104788
41. I.U.Rakhmonov, and A.M.Najimova. AIP Conference Proceedings 2647, **030011**, (2022), https://doi.org/10.1063/5.0104791
42. I.U.Rakhmonov, A.M.Najimova, Sh.M.Esemuratova, and T.T.Koptileuov. AIP Conference Proceedings 2647, **070024**, (2022), https://doi.org/10.1063/5.0104793
43. F.A.Hoshimov, I.U.Rakhmonov, N.N.Niyozov, and F.B.Omonov. AIP Conference Proceedings 2552, **030025**, (2023), https://doi.org/10.1063/5.0112388
44. I.U.Rakhmonov, F.A.Hoshimov, N.N.Kurbonov, and D.A.Jalilova. AIP Conference Proceedings 2552, **050022**, (2023), https://doi.org/10.1063/5.0112391
45. K.Reymov, S.K.Makhmuthonov, G.Turmanova, and Q,Uzaqbaev. E3S Web of Conferences 289, **07023**, (2021), https://doi.org/10.1051/e3sconf/202128907023
46. T.Gayibov, K.Reymov, and A.Aytbaev. E3S Web of Conferences 289, **07024**, (2021), https://doi.org/10.1051/e3sconf/202128907024
47. https://www.rlocman.ru/op/tovar.html?di=59773&/WR-64MXi-A

Construction principle of automatic control system adjustable multi-engine drive water lift pump unit

Nargiza Khamudkhanova [1,*], *Mokhira* Idriskhodjaeva [1], *Kholjan* Kholbutayeva [1]

[1]Tashkent State Technical University named after Islam Karimov, 100095, Uzbekistan, Tashkent, University St. 2A

Abstract. The article discusses the issues of the principle of constructing an automatic control system for a multi-motor controlled electric drive with an electrical connection along the rotor chain of asyn-chronous electric motors with phase rotors and a common inverter by a driven network to control the technological process of water supply of pumping units as part of a water-lifting pumping unit. The expediency of constructing a system for automatic control of the frequency of rotation of the drive motors of a pumping unit in the form of a digital electric drive, implemented through a subordinate control system, is disclosed.

1 Introduction

In the practice of operating water-lifting pumping units (PU) equipped with the same type of AC electric drives, the most commonly used options for the structural layout of a group of pumping units (PU) for joint parallel operation in a common pressure pipeline (hydraulic pressure network). At the same time, a reduction in the cost of electrical energy and the rational use of water resources can be achieved with the help of an adjustable multi-engine electric drive, due to a consistent and proportionate change in the speeds (frequencies) of rotation of the pumping station of the water-lifting pump [1-3].

The most promising and fully meeting the requirements for a controlled electric drive system, in relation to water-lifting pumps, can be considered a multi-motor electric drive with electrical coupling (MEDEC) along a rotor circuit with a common grid-driven inverter (CGI), in which the sliding energy of the rotor circuit is transferred (returns) to the network [4].

2 The current state of the investigated problem

In order to provide the PU of the machine water lifting system (MWS) with the required supply, the following options for constructing a closed-loop automatic control system (ACS) for a multi-motor electric drive with electrical communication and a common inverter driven by a network can be applied:

- frequency of rotation of drive motors;
- consumption of each of the functioning pumping units (PU) as a part of pumping station;
- total flow rate of the common pumping system of pumping station;
- the level of the tailwater in the intake structure of the pumping station (PS);
- the options listed above or combinations thereof.

The most simple and convenient in its practical implementation is a closed automatic control system in terms of rotation frequency [5]. However, its application was difficult due to the ambiguity of the relationship between the performance of the NA and its rotational speed, which was expressed in the absence of a specific analytical dependence of the form:

$$Q = f(n, H_{CT}, N, H)$$

The mathematical descriptions of the electromechanical ratios of a multi-motor electric drive with electrical coupling along a rotor circuit proposed in the article in the mode of coordinated rotation of the drive electric motors of the PU make it possible not only to establish the indicated dependence, but also to determine, under the given operating conditions, the optimal value of the speed of rotation of the PS, ensuring that the most energy- and resource-saving modes of operation of PU MWS [6-9].

In our opinion, it is expedient to construct an automatic control system for the frequency of rotation of the drive electric motors of the PU MWS in the form of a digital electric drive (DED) implemented using a slave control system (SCS). The principle of STC is that the external control loop of a certain coordinate (for

* Corresponding author: raxmatillo82@mail.ru

example, speed) is subordinated to the internal control loop (for example, current). Moreover, in one system there can be several subordinate circuits. The advantages of building a system according to the SCS principle in relation to the possibilities of forming the required static and dynamic modes, coordinate restrictions, ease of adjustment are well known, and a number of electric drive systems implemented according to the SCS and their operating modes are widely considered in the literature [10-13].

In our case, it makes sense to use an autonomous digital processor, in which a digital computer machine (DCM) serves only as a source of input information, and digital correction is carried out by a special device that operates autonomously with respect to the digital computer. It is known that in a non-autonomous digital converter, the input receives a control signal, which is a mismatch error and generated by a digital computer, which simultaneously performs the functions of a master device, a comparison element and a digital corrective device [14-17].

However, we chose an autonomous DED as an adjustable electric drive of the PU MWS for a number of reasons. First, the dynamics of the operation of non-autonomous DEDs is directly determined by the frequency of information output from the digital computer, while autonomous DCMs are in a certain sense invariant to this frequency. Secondly, in non-autonomous DEDs, it is more difficult to control the parameters in the process of their control and adjustment, because of which the electric drive system of the PU MWS turns out to be possible to check only in combination with a digital computer. Autonomous DED requires less machine time. In addition, one of the main reasons is the fact that, according to the conditions of the technological process of water supply by the SSW pumping station, the water flow provided by it varies discretely depending on the number of working PUs in the PS, which at its stages, with unit-by-unit control of the supply, can invariably remain quite for extended periods of time, and even throughout the entire work shift. Then it is enough for the duty personnel of the pumping station only, in accordance with the water supply schedule and the energy-saving and resource-saving regime map, of the operation of the PU MWS or, by the product of the current change in the number of functioning pumps, to enter from the keyboard into the computer data on the values of the steady state operation mode of the PU according to the readings of the instruments. After that, the digital computer calculates the value of the optimal rotational speed using the developed software and outputs it in the form of a task signal [18-21].

The proposed ACS consists of digital and analog parts. The functional diagram of the specified ACS, which ensures the stabilization of the speed of rotation of the drive electric motors of the PU in the composition of the PS MWS and the equality of its optimal value, as well as that determined by the task signal, is shown in fig. 1.

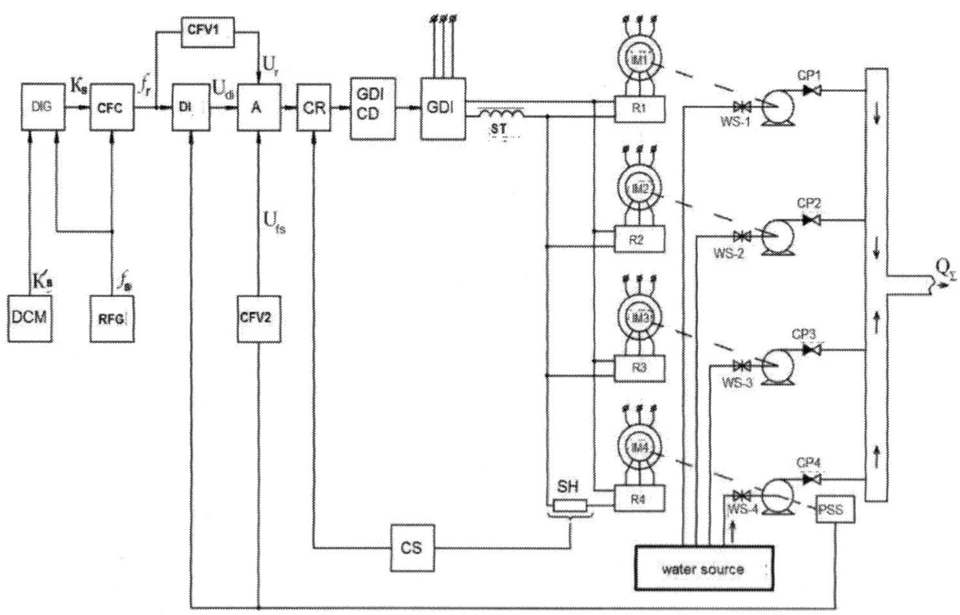

Fig. 1. Functional diagram of the ACS "multi-motor electric drive with electrical connection - pumping unit - pressure network"

Setting the speed (frequency) of rotation of the K'$_s$ in numerical form is fed to the input of the digital intensity generator (DIG), designed to generate signals for the acceleration and deceleration modes of the electric drive. The reference frequency f_r from the reference frequency generator (RFG) is also supplied to the input of the DIG. The DIG has a numerical output short circuit signal, quantized only by level, as a result of which it does not disturb the control system in the steady state. The command of the speed (frequency) of rotation of the K'$_s$ in numerical form is fed to the input of the digital intensity generator (DIG), designed to generate signals for the acceleration and deceleration modes of the electric drive. The reference frequency f_r from the reference frequency generator (RFG) is also supplied to the input of the DIG. The DIG has a numerical output short circuit signal, quantized only by level, as a result of which it does not disturb the control system in the steady state. The principle of operation and the design of the DIG are considered in detail in. The "code-frequency" converter (CFC) converts the number of short circuits into the frequency f_s, which determines the steady rotation speed of the drive electric motors of the PU. A digital integrator (DI) and an analog proportional link combined with an analog adder make up the speed controller (SC). The reference voltage U_r is formed from the frequency f_r by means of a frequency-voltage converter (CFV1). The input of the adder also receives the output voltage from the digital integrator U_r and the feedback signal U_{fs}, which is formed from the frequency f_{fs} of the pulse speed sensor (PSS) using a frequency-to-voltage converter (CFV2). The PSS generates a speed deviation signal at the output in the form of a continuous sequence of pulses, the instantaneous frequency of which is proportional to the instantaneous speed of the sensor. The error level of the IDS is much lower than in tachogenerators, which makes it possible to ensure the most accurate coincidence of the speed of rotation of the drive electric motors of the PU with the given one [22].

The internal loop of the SPU is a current control loop in which the feedback signal is fed to the current regulator (RT) from the shunt (Sh) installed in the circuit of rectified rotor currents of induction motors (AM) by means of an analog current sensor (DT). Since in the considered MEDEC system with a common CGI, the drive electric motors of the PU and the parameters of their rotor circuits are identical, and the rotational speeds of the IM, due to the principle of constructing this system, which ensures their coordinated rotation, are maintained the same, the shunt (SH) can be installed in the circuit rectified rotor current of any of the IMs with an PSS mounted on its shaft [23]. At the same time, this IM will be the drive electric motor of the first-sequence PU, which is constantly in operation as part of the PU MWS. The diagram (Pic. 1) also shows the control unit for the inverter driven by the network (GDICD), the inverter itself driven by the network (GDI), smoothing throttle (ST), centrifugal pumps (R1 - R4) complete with drive electric motors ON (IM1 - IM4) and their uncontrolled bridge rectifiers (CP1 - CP4) [24-28].

3 Conclusion

Thus, the proposed SCS consists of two circuits: the internal one is the current regulation circuit, the external one is the regulation of the speed of rotation of the drive electric motors of the PU. In this case, the speed and current controllers are PI - controllers. The resulting doubly integrating speed control system provides astatic characteristics in terms of control and disturbing influences.

References

1. K.Abidov, N.Khamudkhanova. E3S Web of Conferences, 216, **01111**, (2020), https://doi.org/10.1051/e3sconf/202021601111
2. K.G.Abidov, and ofters. E3S Web of Conferences, 289, **07004**, (2021), https://doi.org/10.1051/e3sconf/202128907004
3. K.G.Abidov, O.O.Zaripov, and ofters. AIP Conference Proceedings, 2552, **030022**, (2022), https://doi.org/10.1063/5.0112384
4. K.G.Abidov, O.O.Zaripov, and ofters. AIP Conference Proceedings, 2552, **030023**, (2022), https://doi.org/10.1063/5.0112385
5. N.Khamudkhanova, and ofters. AIP Conference Proceedings 2402, **060016**, (2021), https://doi.org/10.1063/5.0071557
6. R.Karimov. AIP Conference Proceedings, 2552, **030014**, (2022). https://doi.org/10.1063/5.0111533
7. R.Karimov. AIP Conference Proceedings, 2552, **050012**, (2022). https://doi.org/10.1063/5.0111524
8. S.Dzhuraev, R.Karimov, and ofters. ElConRus, (2022), pp. 1166-1169, doi: 10.1109/ElConRus54750.2022.9755782
9. R.Karimov, A. Kuchkarov, and ofters. Journal of Physics: Conference Series 2094, **052050**, (2021). doi:10.1088/1742-6596/2094/5/052050
10. R.Karimov, N.Kurbanova, and ofters. Journal of Physics: Conference Series 2094(5), **052042**, (2021). doi:10.1088/1742-6596/2094/5/052042
11. K.G.Abidov, O.O.Zaripov, and ofters. E3S Web of Conferences, 289, **07003**, (2021), https://doi.org/10.1051/e3sconf/202128907003

12. K.G.Abidov, O.O.Zaripov, and ofters. E3S Web of Conferences, 216, **01110**, (2020), https://doi.org/10.1051/e3sconf/202021601111
13. K.G.Abidov, O.O.Zaripov. E3S Web of Conferences, 139, **01088**, (2019), https://doi.org/10.1051/e3sconf/201913901088
14. E.Kh.Abduraimov, D.Kh.Khalmanov. Journal of Physics: Conference Series, 2373(7), **072010**, (2022), DOI 10.1088/1742-6596/2373/7/072010
15. E.Kh.Abduraimov. Journal of Physics: Conference Series, 072009, 2373, (2022), DOI 10.1088/1742-6596/2373/7/072009
16. E.Abduraimov, M.Peysenov, N.Tairova. AIP Conference Proceedings, 2552, **040012**, (2022), https://doi.org/10.1063/5.0116235
17. E.Kh.Abduraimov, D.Kh.Khalmanov. Journal of Physics: Conference Series, 2094(2), **022072**, (2021), DOI 10.1088/1742-6596/2094/2/022072
18. E.Kh.Abduraimov, D.Kh.Khalmanov, and ofters. E3S Web of Conferences, 289, **07026**, (2021), https://doi.org/10.1051/e3sconf/202128907026
19. E.Kh.Abduraimov, D.Kh.Khalmanov. E3S Web of Conferences, 216, **01106**, (2020), https://doi.org/10.1051/e3sconf/202021601106
20. E.Kh.Abduraimov. E3S Web of Conferences, 216, **01105**, (2020), https://doi.org/10.1051/e3sconf/202021601105
21. E.Kh.Abduraimov, D.Kh.Khalmanov. Journal of Physics: Conference Series, 1515(2), **022055**, (2020), DOI 10.1088/1742-6596/1515/2/022055
22. K.G.Abidov, A.K.Nuraliev, and ofters. Journal of Advanced Research in Dynamical and Control Systems, **12(7)**, pp. 2167-2171, (2020).
23. A.K.Nuraliev, and ofters. E3S Web of Conferences, 216, **01108**, (2020), https://doi.org/10.1051/e3sconf/202021601108
24. M.Ibadullaev, A.Nuraliev, A.Esenbekov. IOP Conference Series: Materials Science and Engineering, 862(6), **062031**, (2020), DOI 10.1088/1757-899X/862/6/062031
25. S.Begmatov, D.Khalmanov, and ofters. AIP Conference Proceedings, 2552, **040011**, (2022), https://doi.org/10.1063/5.0130666
26. Kh.G.Karimov, M.K.Bobozhanov. Elektrichestvo, **1**, pp. 27-32, (1996).
27. M.K.Bobojanov, S.Mahmutkhonov, and S.Aytbaev. AIP Conference Proceedings 2552, **050011**, (2023), https://doi.org/10.1063/5.0113890
28. M.Bobojanov. AIP Conference Proceedings 2552, **050034**, (2023), https://doi.org/10.1063/5.0114077

E3S Web of Conferences **384**, 01058 (2023)
RSES 2022
https://doi.org/10.1051/e3sconf/202338401058

Transformations of equations with approximation of characteristics of nonlinear elements by power polynomial

Anvar Rakhmatullaev[1*], *Baxtiyar* Abdullayev[1], *Sarvar* Rahmatullayev[1]

[1]Tashkent State Technical University named after Islam Karimov, 100095, Uzbekistan, Tashkent, University St. 2A

Abstract. The article discusses some transformations of the equation associated with approximations in the form of a polynomial, which is quickly transferred to a convenient equation compiled with respect to the main input and output values and gives generalized dependencies for a number of nonlinear systems. As an example, a part of a magnetic circuit with longitudinal and orthogonal magnetization fields is given. Relations are presented that connect the inductions B_1, B_2 as input and output quantities, which in turn are directly related to the corresponding voltages U_1 and U_2. For all three magnetic circuits, dependencies are obtained that convert the input and output currents i_1, i_2 into input and output inductions B_1, B_2 with an approximating function in the form of a polynomial of any degree.

1 Introduction

When studying processes in electrical circuits with non-linear elements, in some cases the problem arises of transforming equations that include approximations of the characteristics of non-linear elements. These transformations are often reduced to obtaining such dependencies (usually connecting the main initial and input values) that are convenient in the study of amplitude, frequency, phase, and other characteristics. However, such transformations often cause great difficulties inherent in the operation of non-linear dependencies [1-4].

The article discusses some transformations of the equation associated with an approximation in the form of a polynomial, which is sufficient to quickly transfer to a convenient equation compiled with respect to the main input and output quantities and give generalized dependencies for a number of nonlinear systems [4-5].

2 The current state of the investigated problem

As an example, three magnetic circuits are given (Fig. 1, 2, 3), which are parts of certain electrical circuits.

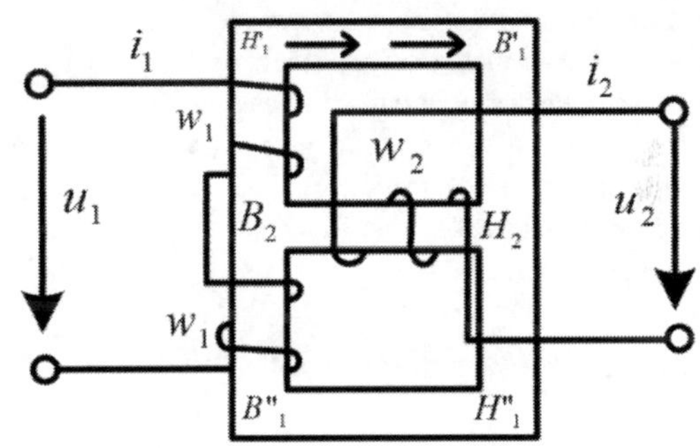

Fig. 1. Magnetic circuit with longitudinal magnetic magnetization circuits

The material of ferromagnetic cores is assumed to be isotropic and the same for all three magnetic circuits, and the dependence of the induction on the magnetic field strength (magnetization curve) is unambiguous and symmetrical relative to the origin of coordinates [6-8]. As is known, such a magnetization curve can be approximated by a polynomial:

$$H = \sum_{23.5} amB^{m} \qquad (1)$$

Where: H – magnetic field strength; m – counted integer; B – magnetic field induction; a_1, a_3, a_5 – approximation coefficients.

* Corresponding author: raxmatillo82@mail.ru

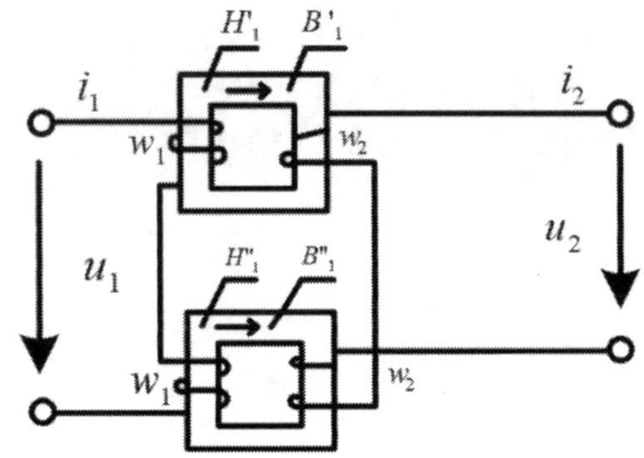

Fig. 2. Double core magnetic circuit

Figures 1, 2 show magnetic circuits with longitudinal magnetic circuits of magnetization. The end rods of a three-rod chain are assumed to be the same. The same condition (identity) is imposed on the cores of a two-core circuit. Figure 3 shows a magnetic circuit with orthogonal magnetic fields of magnetization [9-12].

Equation transformations are made for the general case when any energy sources function in primary and secondary electrical circuits (in one of the electrical circuits, instead of sources, autoparametric excitation of oscillations at any frequency can take place) [13-14].

Let us consider the functional connections of a three-rod magnetic circuit (Fig. 1):

$$i_1 = \frac{l_1}{2W_1}(H_1' + H_1''), \quad (2)$$

$$i_2 = \frac{l_1}{2W_2}(H_1' + H_1'') + \frac{l_2}{W_2}H_2 \quad (3)$$

Where: H_1', H_1'' – strength of the magnetic field of the extreme rods; l_1, l_2 – the lengths of the extreme and rods, respectively.

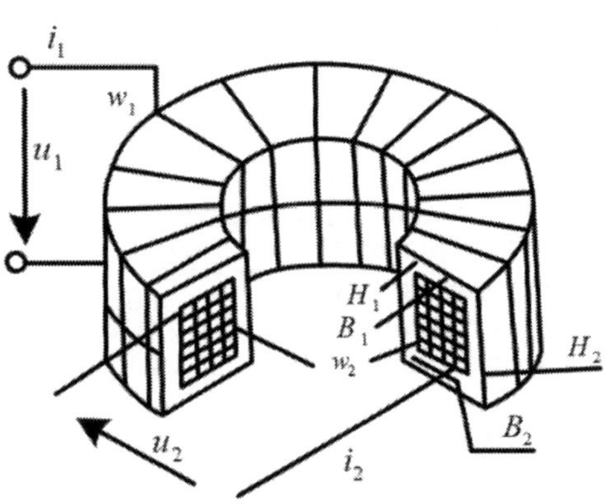

Fig. 3. Magnetic circuit with orthogonal magnetic field of magnetization

Let us introduce approximation (1) into (2) and (3).

$$i_1 = \frac{l_1}{2W_1}\sum_{m=1,3,5} am|(B_1')^m + (B_1'')^m| \quad (4)$$

$$i_2 = \frac{l_1}{2W_2}\sum_{m=1,3,5} am|(B_1')^m + (B_1'')^m| + \frac{l_2}{W_2}\sum am B_2^m \quad (5)$$

The expressions
$(B_1')^m + (B_1'')^m ; (B_1')^m - (B_1')^m$ in (4) and (5) can be converted to, respectively, linking the input and output quantities by induction [15]. For convenience, we accept $B_1' = x, B_1'' = y$.

Let's introduce new variables:

$$a = x + y,\ b = x - y \ \text{ or } \ x = \frac{a+b}{2},\ \ y = \frac{a-b}{2}$$

Then

$$\begin{aligned} x^m \pm y &= \frac{(a-b)^m}{2^m} \pm \frac{(a-b)^m}{2^m} \\ &= \frac{1}{2^m}[(C_m^0 a^m + C_m^1 a^{m-1}b + C_m^2 a^{m\cdot 2}b^2 + \ldots C_m^{m-1}ab^{m-1} \\ &+ C_m^0 a^3 - C_m^1 a^{m-1}b + C_m^2 a^{m-2}b^2 + \ldots + C_m^{m-1}ab^{m-1} \\ &- C_m^m b^m)] = \frac{1}{2^{m-1}}\sum_{k=m-1}^{a} C_m^k (x \pm y)^k (x \pm y)^{m-k} \end{aligned}$$

Where: $K = m-1, m-3, \ldots 2, 0;$ C_m^k – combinations of elements according to *K*.
So,

$$(B_1')^m \pm (B_1'')^m = \frac{1}{2^{m-1}}\sum_{k=m-1}^{0} C_m^k[(B_1') \pm (B_1'')]^k[(B_1') \pm (B_1'']^{m-k} \quad (6)$$

Obviously (Fig. 1)

$$U_1 = W_1 S_1 \frac{dB}{dt} \quad (7)$$

$$B_1 = B_1' + B_1'' \quad (8)$$

$$S_1 B_1' = S_1 B_1'' + S_2 B_2$$

Or

$$B_2 = n(B_1' - B_1'') \quad (9)$$
$$n = S_1/S_2$$

And

$$U_2 = -W_2 S_2 \frac{dB_2}{dt} \quad (10)$$

Taking into account (8) and (9), identity (6) takes the form:

$$(B_1')^m + (B_1'')^m - \frac{1}{2^{m-1}}\sum_{k=m-1}^{0} \frac{C_m^k}{n^k} B_2^k B_1^{m-k}, \quad (11)$$

$$(B_1')^m - \frac{1}{2^{m-1}}\sum_{k=m-1}^{0}\frac{C_m^k}{n^{m-k}}B_1^k B_2^{m-k}. \quad (12)$$

Thus, the expression $(B_1')^m \pm (B_1'')^m$ is converted into relations (11) and (12), which relate the inductions B1, B2 presented here as input and output quantities and, in turn, are directly related to the corresponding voltages U_1 and U_2. [16-18]

Let us introduce identities (11), (12) into (4), (5):

$$i_1 = \sum_{m=1,3,5}\sum_{k=m-1}^{0}\frac{a_m b_1 C_m^k}{2^m n^k W_1}B_2^k B_1^{m-k} \quad (13)$$

$$i_z = \sum_{m=1,3,5...}\sum_{k=m-1}^{0}\frac{a_m l_1 C_m^k}{2^m n^{m-k}\sqrt{2}}B_1^k B_2^{m-k} + \sum_{m=1,3,5...}\frac{a_m l_2}{\sqrt{2}}B_2^m \quad (14)$$

For a two-core magnetic circuit (Fig. 2), functional relationships are valid:

$$i_1 = \frac{C(H_1' + H_1'')}{2W_1} \quad (15)$$

$$i_2 = \frac{C(H_1' - H_1'')}{2W_2} \quad (16)$$

Relations (15) and (16) take the form:

$$i_1 = \sum_{m=1,3,5...}\sum_{k=m-1}\frac{a_m l C_m^k}{2^m W_1}B_2^k B_1^{m-k} \quad (17)$$

$$i_2 = \sum_{m=1,3,5...}\sum_{k=m-1}^{0}\frac{a_m l C_m^k}{2^m W_2}B_1^k B_2^{m-k} \quad (18)$$

For a magnetic circuit with orthogonal magnetic fields (Fig. 3), relations (1, 2) hold:

$$\frac{B_1}{H_1} = \frac{B_2}{H_2} = \frac{B}{H} \quad (19)$$

$$B = \sqrt{B_1^2 + B_2^2}$$

$$H = \sqrt{H_1^2 + H_2^2} \quad (20)$$

Where: B_1, B_2, B - magnetic inductions of the longitudinal, transverse and resulting magnetic fields, respectively; H_1, H_2, H - intensities of the longitudinal, transverse, resulting fields, respectively.

Taking into account (19), (20), (1) and some transformations [19-20], we obtain:

$$H_1 = B_1\frac{H}{B} = \sum_{m=1,3,5...} a_m B_2 B^{m-1} = \sum_{m=1,3,5...}\sum_{k=m-1}^{0} a_m C_{\frac{m-1}{2}}^{1/2} B_2^k B_1^{m-k}$$

$$H_2 = B_2\frac{H}{B} = \sum_{m=1,3,5...} a_m B_2 B^{m-1} = \sum_{m=1,3,5...}\sum_{k=m-1}^{0} a_m C_{\frac{m-1}{2}}^{1/2} B_1^k B_2^{m-k}$$

Where: K=m-1, m-3, … 2,0.

Or

$$i_1 = \frac{l_1}{W_1}H_1 = \sum_{m=1,3,5...}\sum_{k=m-1}^{0}\frac{a_m l_1 C_{\frac{m-1}{2}}^{k/2}}{W_1}B_2^k B_1^{m-k} \quad (21)$$

$$i_2 = \frac{l_2}{W_2}H_2 = \sum_{m=1,3,5...}\sum_{k=m-1}^{0}\frac{a_m l_2 C_{\frac{m-1}{2}}^{k/2}}{W_2}B_1^k B_2^{m-k} \quad (22)$$

If the characteristic of a nonlinear element is not symmetric with respect to the origin, then the approximating function has terms and about even powers [21-25]. And for even degrees, one can obtain the following identities:

$$x^m - y^m = \frac{1}{2^{m-1}}\sum_{k=m-1}^{1} C_m^k (x+y)^k (x-y)^{m-2}$$

Where: K=m-1; m-3; … 5; 3; 1;

$$x^m + y^m = \frac{1}{2^{m-1}}\sum_{k=m}^{0} C_m^k (x+y)^k (x-y)^{m-k}$$

Where: K=m; (m-2); (m-4); … 4;2;0.

3 Conclusion

In conclusion, the following conclusions can be drawn:

1. For all three magnetic circuits, dependences are obtained that convert the input and output currents i_1, i_2 into input and output inductions B_1, B_2 about an approximating function in the form of a polynomial of any degree;

2. Relations (13), (14) of a three-rod magnetic circuit, (17), (18) of a two-core one, (21), (22) - circuits with orthogonal fields differ only in the form of the coefficients (the term $\sum_{m=1,3,5}\frac{a_m l_2}{W_2}B_2^m$ in (14) is combined with the first term at K=0, which makes it possible to obtain generalized dependencies for similar magnetic circuits that are parts of certain electrical circuits

References

1. R.Karimov. AIP Conference Proceedings, 2552, **030014**, (2022). https://doi.org/10.1063/5.0111533
2. R.Karimov. AIP Conference Proceedings, 2552, **050012**, (2022). https://doi.org/10.1063/5.0111524

3. S.Dzhuraev, R.Karimov, and ofters. ElConRus, pp. 1166-1169, (2022), doi: 10.1109/ElConRus54750.2022.9755782
4. R.Karimov, A. Kuchkarov, and ofters. Journal of Physics: Conference Series, 2094, **052050**, (2021). doi:10.1088/1742-6596/2094/5/052050
5. R.Karimov, N.Kurbanova, and ofters. Journal of Physics: Conference Series, 2094(5), **052042**, (2021). doi:10.1088/1742-6596/2094/5/052042
6. K.G.Abidov, O.O.Zaripov, and ofters. AIP Conference Proceedings, 2552, **030023**, (2022), https://doi.org/10.1063/5.0112385
7. K.G.Abidov, O.O.Zaripov, and ofters. AIP Conference Proceedings, 2552, **030022**, (2022), https://doi.org/10.1063/5.0112384
8. K.G.Abidov, O.O.Zaripov, and ofters. E3S Web of Conferences, 289, **07003**, (2021), https://doi.org/10.1051/e3sconf/202128907003
9. K.G.Abidov, and ofters. E3S Web of Conferences, 289, **07004**, (2021), https://doi.org/10.1051/e3sconf/202128907004
10. K.Abidov, N.Khamudkhanova. E3S Web of Conferences, 216, **01111**, (2020), https://doi.org/10.1051/e3sconf/202021601111
11. K.G.Abidov, O.O.Zaripov, and ofters. E3S Web of Conferences, 216, **01110**, (2020), https://doi.org/10.1051/e3sconf/202021601111
12. K.G.Abidov, O.O.Zaripov. E3S Web of Conferences, 139, **01088**, (2019), https://doi.org/10.1051/e3sconf/201913901088
13. E.Kh.Abduraimov, D.Kh.Khalmanov. Journal of Physics: Conference Series, 2373(7), **072010**, (2022), DOI 10.1088/1742-6596/2373/7/072010
14. E.Kh.Abduraimov. Journal of Physics: Conference Series, 072009, 2373, (2022), DOI 10.1088/1742-6596/2373/7/072009
15. E.Abduraimov, M.Peysenov, N.Tairova. AIP Conference Proceedings, 2552, **040012**, (2022), https://doi.org/10.1063/5.0116235
16. E.Kh.Abduraimov, D.Kh.Khalmanov. Journal of Physics: Conference Series, 2094(2), **022072**, (2021), DOI 10.1088/1742-6596/2094/2/022072
17. E.Kh.Abduraimov, D.Kh.Khalmanov, and ofters. E3S Web of Conferences, 289, **07026**, (2021), https://doi.org/10.1051/e3sconf/202128907026
18. E.Kh.Abduraimov, D.Kh.Khalmanov. E3S Web of Conferences, 216, **01106**, (2020), https://doi.org/10.1051/e3sconf/202021601106
19. E.Kh.Abduraimov. E3S Web of Conferences, 216, **01105**, (2020), https://doi.org/10.1051/e3sconf/202021601105
20. E.Kh.Abduraimov, D.Kh.Khalmanov. Journal of Physics: Conference Series, 1515(2), **022055**, (2020), DOI 10.1088/1742-6596/1515/2/022055
21. K.G.Abidov, A.K.Nuraliev, and ofters. Journal of Advanced Research in Dynamical and Control Systems, **12(7)**, pp. 2167-2171, (2020).
22. A.K.Nuraliev, and ofters. E3S Web of Conferences, 216, **01108**, (2020), https://doi.org/10.1051/e3sconf/202021601108
23. M.Ibadullaev, A.Nuraliev, A.Esenbekov. IOP Conference Series: Materials Science and Engineering, 862(6), **062031**, (2020), DOI 10.1088/1757-899X/862/6/062031
24. S.Begmatov, D.Khalmanov, and ofters. AIP Conference Proceedings, 2552, **040011**, (2022), https://doi.org/10.1063/5.0130666
25. M.O.Atajonov, S.J.Nimatov, A.I.Rakhmatullaev, A.B.Sadullaev, AIP Conference Proceedings 2552, **050014,** (2023), https://doi.org/10.1063/5.0112403

https://doi.org/10.1051/e3sconf/202338401059

New pole-changing winding for electric drive of ball mills

Dauletbek Rismukhamedov[1], *Khusniddin* Shamsutdinov[1*], *Sarvar* Ganiev[1], *Khayotullo* Magdiev [1]
Shakhobiddin Khusanov[1]

[1]Tashkent State Technical University named after Islam Karimov, 100095, Uzbekistan, Tashkent, University St. 2A

Abstract. The article substantiates the expediency of using a controlled electric motor in grinding mechanisms in electric drives instead of an unregulated asynchronous motor. The problem of facilitating the start-up of ball mills and ensuring accurate stop of the neck of ball mills using the lowest speed is considered. As the review of the literature showed, speed control in electric drives using two-speed motors is a simple, cheap and reliable option. In many existing two-speed motors, two independent windings are placed in the stator slots; for this reason, the motor magnetic circuit is not effectively used, in turn, the use of one pole-changing winding improves the energy performance of the motor and saves winding copper and insulating materials. Based on the analysis of the operating modes of grinding mechanisms, in terms of facilitating start-up and resource saving (increase the operating time of the engine itself), the expediency of creating a two-speed motor with a pole-changing winding with a pole ratio of 2:5 and improving existing electric drive systems with its use is substantiated.

1 Introduction

In many technological and production processes of mining and processing and ceramic factories, dust preparation systems of thermal power plants, house-building plants, chemical plants, mechanisms for grinding various materials are widely used, which include ball drum mills of periodic action [1].

The presence of large inertial masses and an unbalanced load complicates the acceleration and positioning of the ball mill drum. The maximum value of the load moment is observed when the drum rotates from the vertical axis by 90°, which can lead to its stop if the engine torque is insufficient. In this case, the mill is started by the operator by swinging the drum for 7-10 switching on and off, which leads to rapid wear of the engine and mechanical transmission elements, which determines significant economic losses and requires modernization of the electric drive [2].

2 The current state of the investigated problem

By replacing the single-speed asynchronous motor in this mill with a two-speed pole-changing motor, which has a simple preparation technology mounted on a single stator, it can facilitate the starting process and ensure accurate stopping, as well as drastically reduce the number of mill switching per operation cycle. In this case, the mill is started with a step start, that is, during the start-up process, it is started from the side of low speed and after a full turn of the drum, it is switched to the side of high speed. During mill stop, precise stop is ensured by switching from high speed to low speed.

To build a pole-changing winding at the Tashkent State Technical University, a new method "Discrete-specified spatial functions" was developed. On the basis of this method, many schemes of pole-changing windings were developed for a wide range of pole and phase ratios, close in their properties to serial windings [3].

The basis of the method is a method in which two windings of normal design are taken (one of them is original, the other is typical) with a given number of poles ($2p_1$, $2p_2$) and phases (m_1, m_2) and are simultaneously used in the process of constructing a pole-changing winding, according to the principle approximation of the current distribution and the pattern of the magnetomotive forces of the synthesized winding to the current distribution and the picture of the magnetomotive forces of a typical winding, where each layer of conductors distributed over the grooves with conditional currents is replaced by a discrete element of the winding [4]. A discretely given spatial function of a conventional winding is the state of the conductor and is assigned to the coil side with one conditional conductor, through which a single current of one direction or another flows in a slot belonging to one of the phases and designated identically with this phase [5].

The construction of pole-changing windings is based on special basic circuits that have two groups of terminals and when switching power from one group of terminals to another (Fig. 1), the number of pairs of poles changes (in relation to one group of terminals they are $2p_1$-pole, and in relation to another - $2p_2$-pole) [6].

* Corresponding author: shamsutdinovh83@gmail.com

a)

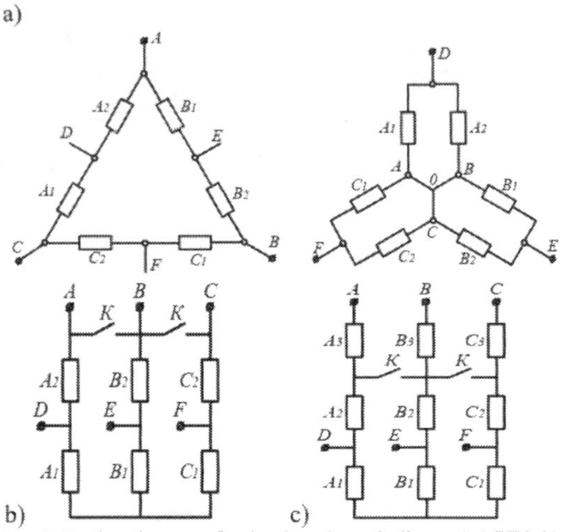

b) c)

Fig 1. Basic schemes of pole-changing windings: a) Δ/YY; b) Y/YY; c) Y/YY with additional branches

The basic scheme may consist of a general and additional parts. In this case, the common part of the basic circuit serves as the basis of the pole-changing winding and participates in the creation of a magnetic field from both poles, and the additional part participates in the creation of a field of only one pole.

Two-speed pole-changing windings are built on the basis of a joint consideration of two multi-phase two-layer loop windings, which are most widely used in the practice of electrical engineering [7, 8].

The development of pole-switchable winding circuits with a fractional pole ratio based on existing basic circuits has its own characteristics, for example, to mitigate the effect of winding fractionality on the pattern of magnetomotive forces of the pole-changing winding according to the basic "Y/YY" or "Δ/YY" circuit, it is advisable to perform windings 2m-2m -zonal, which makes it possible to improve the vibroacoustic characteristics of the engine. In addition, 2m-2m-zone windings have the smallest composition of higher harmonics, which makes it possible to use them in engines with a busy operating mode, in particular, electric drives for grinding mechanisms for step start [9].

When obtaining a pole-changing winding based on the "Y/YY" or "Δ/YY" schemes, it is necessary to synthesize a discretely given spatial function corresponding to one of the poles [10].

As an example, consider the process of obtaining a pole-changing winding circuit for a pole ratio of 2/5 with the number of slots Z=30. In Table 1, a normal 2m-zone winding with the number of poles p_2=5 is taken as the "initial" winding, and a normal 2m-zone winding with p_1=2 is also taken as a "typicalA" winding [11].

Let us write the discretely given spatial function of one layer of these windings under each other:

Table 1. Synthesizing 2p1=4 pole winding

1	2	3	4	5	6	7	8	9	10	11	12	13	14	15	groove
a	$\underline{c}$	b	$\underline{a}$	c	$\underline{b}$	a	$\underline{c}$	b	$\underline{a}$	c	$\underline{b}$	a	$\underline{c}$	b	p_2=5 initial.
a	a	a	$\underline{c}$	$\underline{c}$	b	b	b	$\underline{a}$	$\underline{a}$	c	c	c	$\underline{b}$	$\underline{b}$	p_1=2 typical
a	$\underline{c}$	$\underline{b}$	a	$\underline{c}$	b	$\underline{a}$	$\underline{c}$	b	$\underline{a}$	c	$\underline{b}$	$\underline{a}$	c	$\underline{b}$	p_1=2 synt.

16	17	18	19	20	21	22	23	24	25	26	27	28	29	30	groove
$\underline{a}$	c	$\underline{b}$	a	$\underline{c}$	b	$\underline{a}$	c	$\underline{b}$	a	$\underline{c}$	b	$\underline{a}$	c	$\underline{b}$	p_2=5 initial.
a	a	$\underline{c}$	$\underline{c}$	$\underline{c}$	b	b	$\underline{a}$	$\underline{a}$	$\underline{a}$	c	c	$\underline{b}$	$\underline{b}$	$\underline{b}$	p_1=2 typical
a	$\underline{c}$	b	a	$\underline{c}$	b	$\underline{a}$	c	b	$\underline{a}$	c	$\underline{b}$	a	c	$\underline{b}$	p_1=2 synt.

The discretely given space function of the synthesized winding is obtained by "modulating" the discretely given space function of the original winding with the help of the discretely given space function of the typical winding. This process is based on the principle of approximation of the current distribution to the picture of the magnetomotive force of the pole-changing winding from the pole side 2p1 to the current distribution and picture of the magnetomotive force of a typical winding. The meaning of "approximation" is to determine in each slot the sign of the state of the conductor of a discretely given spatial function of the synthesized winding, depending on the state of the conductor of a discretely given spatial function of a typical winding in this groove and based on the relative position of the phase current vectors in a three-phase system [12].

According to the obtained discretely given spatial function, it is possible to group the coils into the BS branches (Table 2).

Table 2. Grouping coils in the branches of the basic circuit "Δ/YY"

Branches	Coil number
A1	-4, 7, 13, -16, 25
A2	19, -22, -28, 1, -10
B1	-24, 27, 3, -6, 15
B2	9, -12, -18, 21, -30
C1	-14, 17, 23, -26, 5
C2	29, -2, -8, 11, -20

Table 3. Winding data from side p_1=2

Branches of the basic circuit Δ /YY			
Branches	***A-D***	***D-0***	***B-E***
A	6,44	6,44	6,44
k_{win}	0,644	0,644	0,644
φ	42	42	162
Branches	***E-0***	***C-F***	***F-0***
A	6,44	6,44	6,44
k_{win}	0,644	0,644	0,644
φ	162	282	282

Table 4. Winding data from side p_2=5

	Branches of the basic circuit Δ /YY		
	A	*B*	*C*
A	17,32	17,32	17,32
k_{win}	0,866	0,866	0,866
φ	0	120	240

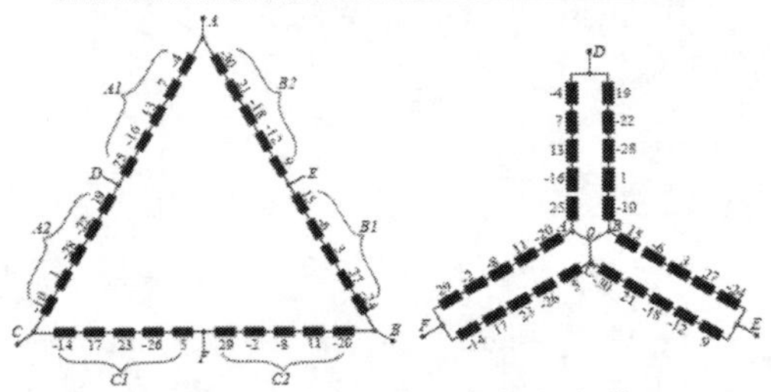

Fig 2. Electric circuit of the pole-switched winding according to the basic circuit "Δ/YY

The resulting winding (Fig. 2) is completely symmetrical with respect to the power source from both poles, the vectors of the electromotive forces of the same branches of each of the phases are symmetrical to each other, i.e. equal in amplitude and shifted in phase by an angle of 2π/3 el. rad., at y=8, the winding coefficients from $2p_1$ and $2p_2$ of the pole side are respectively equal to k_{win1}=0.644 and k_{win2}=0.866 (Table 3 and Table 4) [11].

The resulting pole-changing winding works as follows: When a three-phase power supply is connected to terminals D, E, F (pins A, B, C are combined into a common point), a four-pole rotating magnetic wave occurs in the air gap, all 30 coils are included in the circuit, the electromotive force vectors of each phase are equal in amplitude and shifted along phase by 120^0, that is, the winding on the side of $2p_1$=4 poles is completely symmetrical with respect to the power source.

When a three-phase power supply is connected to terminals A, B, C (terminals D, E, F are free), a ten-pole rotating magnetic wave, all 30 coils are included in the circuit, the electromotive force vectors of each phase are equal in amplitude and shifted in phase by 120^0, that is, the winding on the side of the pole pairs, it is perfectly symmetrical with respect to the power source.

The creation of a two-speed engine, approaching conventional single-speed engines in terms of their energy and weight and size indicators, is one of the urgent problems. The development of such motors makes it possible to improve the existing electric drive with two-speed motors and replace some single-speed motors with two-speed motors in order to save electricity and natural resources in lightly loaded modes, as well as to facilitate the process of starting powerful motors [13].

It should be noted that the task of designing a two-speed motor with one winding is a complex process and can be compared with the simultaneous design of two single-speed motors that have interconnected parameters and cannot be changed unilaterally [14, 18].

3 Results and discussion

Based on the calculated data, «ENERGY MOTORS» LLC manufactured experimental models of a two-speed motor with a developed pole-changing winding (No. IAP 06203, 03/20/2020) for a pole ratio of 4/10 at BS "Δ/YY" based on magnetic cores of serial machines of the АИР112М4 and АИР225L4.

In the research laboratory "Energy-saving technologies in power supply systems" of the Tashkent Technical University, experimental tests were carried out in static modes to determine the losses of idling and short circuit (see tables 5 and 6), as well as the operating and mechanical characteristics of a new two-speed motor with pole-changing winding for the ratio of pairs of poles 2/5 type АИР112М4.

Table 5. Idle experience data

U_1	from p_1=4				
	I_0	P_0	$\cos\varphi_0$	P_{el}	$P_{be}+P_{mag}$
V	A	W		W	W
60	0,9	145	0,62	5,95	139,05
80	1	147	0,43	7,35	139,65
100	1,21	150	0,3	10,76	139,24
140	1,65	160	0,17	20,01	139,99
180	2,2	180	0,11	35,57	144,43
220	3,18	261	0,08	84	177
	from p_2=10				
60	0,95	144	0,4	14,86	126,13
80	1,24	158	0,25	30,44	127,55
100	1,57	176	0,18	48,8	127,19
140	2,3	232	0,12	94,74	137,25
180	3,5	411	0,11	242,5	168,45
220	5,3	780	0,1	556,2	223,8

Table 6. Short circuit experience data

from p_1=4			
U_{1K}	P_K	I_{1K}	$\cos\varphi_K$
V	W	A	
30	160	3,4	0,52
40	280	4,8	0,49
50	440	6,2	0,47
60	650	7,7	0,47
70	940	9,4	0,48
80	1250	10,9	0,48
from p_2=10			
40	240	2,1	0,551
60	420	2,8	0,501
70	525	3,2	0,452
80	650	3,5	0,447
90	790	4	0,423
110	1000	4,6	0,381

The operating characteristics of the electric motor were obtained in the direct load mode (Fig. 3). A DC generator with independent excitation was used as a load machine.

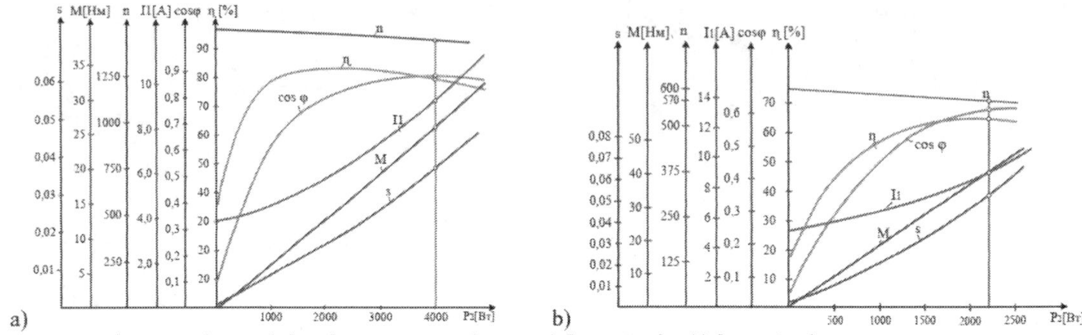

Fig 3. Performance characteristics of a new two-speed motor: a) for p_1=2 poles; b) for p_2=5 poles

The mechanical characteristics of the new two-speed motor are shown in fig. 4

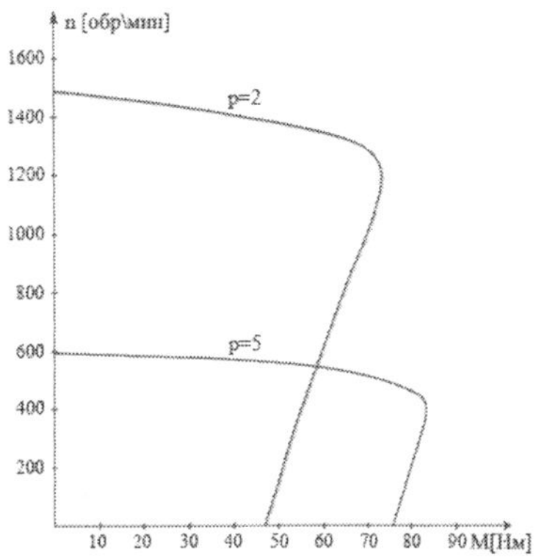

Fig 4. Mechanical characteristics of the new two-speed motor

In connection with the widespread use of motors with a squirrel-cage rotor for ball mills, the behavior of motors during start-up is of great interest.

In the Research and Production Association for the Production of Rare Metals and Hard Alloys of Almalyk MMC JSC, the operation of a ball mill electric drive with a new two-speed motor of the AIR225M10/4 type was studied in the following modes: start-up, operation at a constant speed, switching to the second speed (deceleration) and stop (further all currents are indicated in the amplitude value).

On fig. 5 shows the curve of the stator current as a function of time when starting from standstill on the p_1=2 polarity side. As can be seen from this curve, the onset of the steady state operation of the engine occurs in 800 ms, the starting current is 385 Amperes.

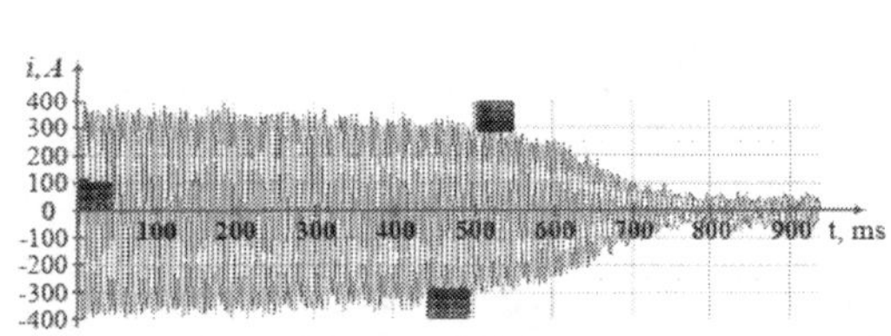

Fig. 5. The curve of the stator current change depending on the time when starting from the state of rest of the electric motor type АИР225M10/4 from the side p_1=2 poles

On fig. 6 shows the curve of the stator current as a function of time when starting from standstill on the p2=5 polarity side. As can be seen from this curve, the onset of the steady state operation of the engine occurs in 400 ms, the starting current is 185 Amperes.

When switching directly from high to low speed, large dynamic torques occur in the drive, which are much higher than the rated torque and can exceed the critical torques in motoring mode. To reduce the dynamic moments in the drive when switching from high to low speed, it is necessary to disconnect the motor from the mains for the duration of braking and, at the moment the synchronous speed corresponding to the low-speed winding is reached, reapply mains voltage to the motor.

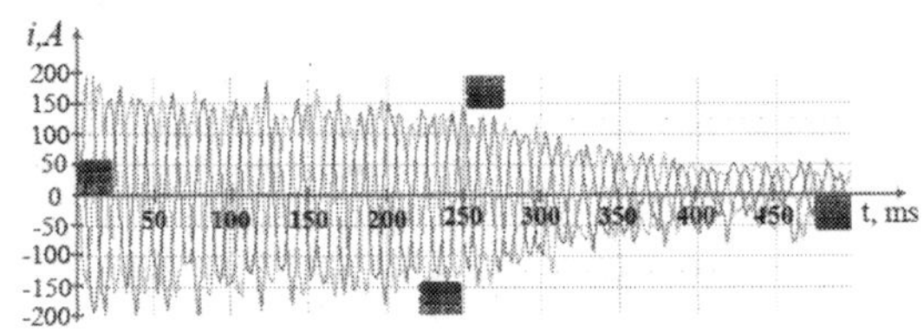

Fig. 6. Curve of stator current change depending on time at starting from a state of rest of an electric motor of the type АИР225M10/4 from the side p_2=5 poles

This method is simpler, but requires a speed or crank angular position sensor with a fairly wide measurement range. During the transition to low speed it is necessary to control the speed of the motor, because due to the variability of the load, it is not possible to pre-calculate the deceleration path, and therefore the free coast of the motor and the switching of the contactors.

On fig. 7 shows the curve of the stator current versus time when starting from standstill on the p_1=2 pole side and switching a two-speed motor to p_2=5 pole. As can be seen from this curve, the onset of the steady state operation of the engine occurs within 800 ms, the starting current is 376 Amperes.

After 2.15 s after the steady state operation on the side p_1=2 poles, the electric motor was switched to p_2=5 poles. As can be seen from this curve, the steady state of the engine operation occurs in 140-150 ms, the starting current is 245 Amperes.

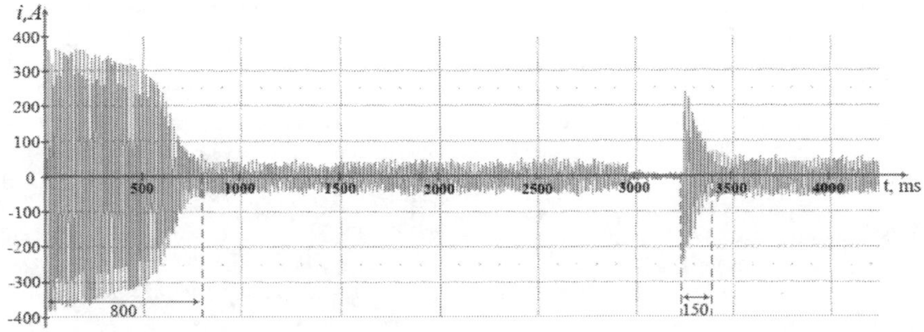

Fig. 7. The curve of the stator current change depending on time when starting from a standstill state of the electric motor of the type АИР225M10/4 from the p_1=2 poles and switching to p_2=5 poles

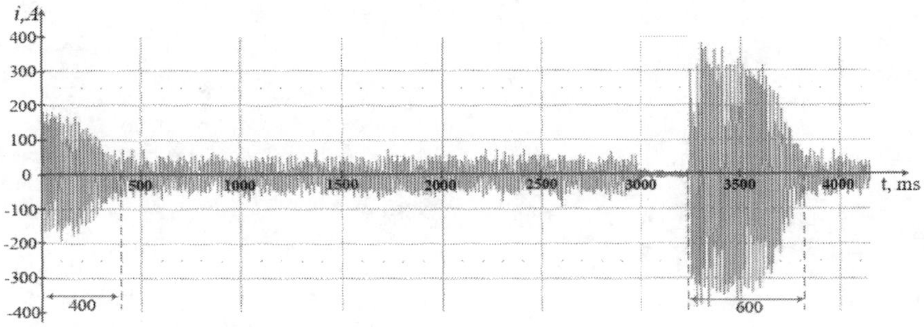

Fig. 8. Curve of stator current change depending on time when starting from a state of rest of an electric motor of the type АИР225M10/4 from p_2=5 poles and switching to p_1=2 poles

On fig. 8 shows the curve of the stator current versus time when starting from standstill on the p_2=5 pole side and switching a two-speed motor to p_1=2 pole

As can be seen from this curve, the onset of the steady state operation of the engine occurs within 400 ms, the starting current is 190 Amperes.

After 2.65 s after the steady state operation on the p2=5 pole side, the electric motor was switched to p1=2 pole. As can be seen from this curve, the steady state of the engine operation occurs within 600 ms, the starting current is 385 Amperes.

Thus, with a direct start, the electric motor reaches a steady state in 800 ms at a starting current of 400 Amperes, and with a stepped start in 600 ms at a starting current of 385 Amperes.

4 Conclusion

As you can see, with a stepped start, the electric motor at a lower starting current reaches the steady state faster, thereby facilitating start-up and resource saving (increasing the operating time of the motor itself and current-carrying parts) due to the lower amount of heat generated by the starting current, and also slows down the process of wear of the mechanical parts of the ball the mill, in addition, it provides an accurate stop of the mechanism without multiple starts, which also affects resource saving in general.

Based on the study of the operating modes of ball mills used in the crushing of various materials in mining, construction, and also in chemical enterprises, it was justified the need to improve the electric drives of these mechanisms by using two-speed electric motors with pole-changing windings.

Using the method of discrete given spatial functions based on the basic circuit "Δ/YY", a new circuit of a pole-changing winding with a pole ratio of 2:5 with high electromagnetic properties was developed and the invention was protected in the form of a patent by the Intellectual Property Agency under the Ministry of Justice of the Republic of Uzbekistan.

References

1. V.A.Perov, E.E.Andreev, L.F.Bilenko. Crushing, grinding and screening of minerals // 4th edition, revised and supplemented. M.: Nedra, p.158, (1990).
2. L.F.Bilenko. Patterns of grinding in drum mills. M.: Nedra, p. 200, (1984).
3. Kh.G.Karimov. Uzbek journal "Problems of Informatics and Energy", - 1992, No. 3/4. - pp. 41-47.
4. Bobojanov M.K., Rismukhamedov D.A., Tuychiev F.N., Shamsutdinov H.F. and Magdiev H.G. Construction and analysis of the pole-changing windings for the pole pairs ratio 5/6 by method discretely specified spatial function// International Journal of Advanced Science and Technology, Vol. 29, No. 11s (2020), pp. 1410-1415.
5. M.K. Bobojanov, D.Rismukhamedov, F.Tuychiev, Kh.F.Shamsutdinov, Kh.Magdiev. Pole-changing motor for lift installation // E3S Web of Conferences, 14 December 2020, Volume 216, 01164 (2020), https://doi.org/10.1051/e3sconf/202021601164.

6. D.Rismukhamedov, Bobojanov M.K., F.Tuychiev, Kh.F.Shamsutdinov. Development and research of pole-changing winding for a close pole ratio // E3S Web of Conferences, 264, 03057 (2021), https://doi.org/10.1051/e3sconf/202126403057.
7. V.I. Popov, New circuits of three-phase windings of electrical machines with improved electromagnetic properties: Monograph. N. Novogorod: VGIPI, 1998. – 116 p.: ISBN 5-88820-044-3.
8. V.N. Vanurin, Stator windings of asynchronous electrical machines: Textbook. - 2nd ed., Rev. and add.- SPb.: «Lan», pp. 2016.-224.
9. D.Rismuhamedov, F.Tuychiev, S.Rismuxamedov, Pole-changing windings for turbomechanism engines. IOP Conf. Series: Materials Science and Engineering, 2020, 883(1), 012140. doi:10.1088/1757-899X/883/1/012140.
10. Dauletbek Rismukhamedov, Furkat Tuychiev and Khusniddin Shamsutdinov. Development of pole-changing windings for two-speed motors of hoisting-transport mechanisms // AIP Conference Proceedings 2552, 050013 (2023); https://doi.org/10.1063/5.0114061.
11. D.Rismukhamedov, F.Tuychiev, Kh.F. Shamsutdinov. Three-phase pole-switching winding with a ratio of pole pairs 2/5 with 30 stator slots // Intellectual Property Agency RepUz. Patent for invention IAP 06203 dated 03/20/2020.
12. D.A. Rismukhamedov, H.G. Karimov, Zh.M. Mavlonov, F.N. Tuychiev, Pole-switchable winding for two-speed electric machine. Patent No. IAP 05385, 23.03.2017.
13. A.O.Di.Tommaso, F. Genduso and R. Miceli, "A new software toolfor design, optimization, and complete analysis of rotating electricalmachines windings". IEEE Transactions on Magnetics, vol. 51, pp. 1–10, April 2015. DOI:10.1109/TMAG.2014.2369860.
14. J. Pyrhönen, T. Jokinen and V. Hrabovcová, Design of Rotating Electrical Machines, John Wiley & Sons, Inc., 2008, p 538. ISBN: 978-0-470-69516-6.
15. M. Caruso, A. Tommaso, F. Marignetti, R. Miceli, G. Ricco Galluzzo. «A general procedure for the construction of Gorges polygons for multi-phase windings of electrical machines», in 2018 Thirteenth International Conference on Ecological Vehicles and Renewable Energies EVER, pp. 1-7, April 2018. https://www.researchgate.net/publication/325350240.
16. 4E Electric Motor Systems Annex (EMSA). Policy Guidelines for Motor Driven Units – Part 2, 2018.
17. Leonard Melcescu, M.V. Cistelecan, O. Craiu, H.B. Cosan. A new 4/6 pole-changing double layer winding for three phase electrical machines. The XIX International Conference on Electrical Machines - ICEM 2010, Italy, 6-8 Sept. 2010. DOI: 10.1109/ICELMACH.2010.5608041.
18. M K Bobojanov, D A Rismukhamedov, F N Tuychiev, and Kh F Shamsutdinov. Development of new pole-changing winding for lifting and transport mechanisms // E3S Web of Conferences 365, 04024 (2023) https://doi.org/10.1051/e3sconf/202336504024
19. M.V. Cistelecan, L.M. Melcescu, H.B. Cosan, M. Popescu. Induction motors with changeable pole windings in the ratio 1:4. International Aegean Conference on Electrical Machines and Power Electronics and Electromotion, Joint Conference. Istanbul, Turkey, 04 April 2013. DOI: 10.1109/ACEMP.2011.6490700
20. M.K.Bobojanov, S.Mahmutkhonov, and S.Aytbaev. Investigation of the Problems Non-Sinusoidal of the Voltage Form. AIP Conference Proceedings 2552, 050011, (2023), https://doi.org/10.1063/5.0113890
21. M.Bobojanov. Development and Research of Two Speed Motor with Pole-Changing Winding. AIP Conference Proceedings 2552, 050034, (2023), https://doi.org/10.1063/5.0114077
22. M.K.Bobojanov, R.Ch.Karimov, T.H.Qosimov, S.D.Zh.Dzhuraev. Development and experimental study of circuits of contactless device for automation of compensation of reactive power of capacitor batteries. E3S Web of Conferences, 289, 07012, (2021), https://doi.org/10.1051/e3sconf/202128907012
23. R.C.Karimov. Non-contact voltage relay for switching windings of a boost transformer. E3S Web of Conferences, 139, 01079, (2019), https://doi.org/10.1051/e3sconf/201913901079
24. Kh.G.Karimov, M.K.Bobozhanov. New pole-changing windings of asynchronous motors. Elektrichestvo, 1, pp. 27-32, (1996).

 https://doi.org/10.1051/e3sconf/202338401060

Ways to use intellectual platforms in the Republic of Uzbekistan

Sarvinoz Salomova*

Tashkent State Technical University named after Islam Karimov, 100095, Uzbekistan, Tashkent, University St. 2A

Abstract. The article is scientifically based on the development of the activities of JSC "Ohangarontsement", a cement producer in Uzbekistan in the digital economy, based on the use of intelligent platforms. Ways to reduce the cost and increase the efficiency of products based on digital information and communication technology are presented.

1 Introduction

Today, the experience of the developed countries of the world shows that the source of economic growth of countries is the development and effective use of intellectual technologies, innovative products and services. Reforms in this area were launched in India 40-50 years ago, and in 2019 the revenue from this sector reached $159 billion [1]. According to the experience of countries such as South Korea, Japan, China, India, Finland, two times more investments were invested than in other countries, they are mainly focused on the production of industrially necessary devices, software, information and communication technology products and intellectual services and their effective application in industries. According to a World Bank study, in 2020, most economically developed countries achieved an increase of 3-3.9% of the country's GDP, 7.2% per capita income through the use of information and intelligent platform technologies in production and services, which led to an increase in the number of Internet users by 15% (compared to 2017) [2].

In world practice, various aspects of the development of industrial enterprises based on the effective use of intelligent information and communication technologies have been carefully studied by foreign economists, such as: huwal M. Davis; Ali Hoshim, Bill Alan, Eduardo Talero, Desai P. He was Jem Dener, Ebenezer Laser, Reimo Suomi, Emilia Gombaz [3-9].

The issues of effective use, management and digitalization of intelligent platforms and information and communication technologies in the industrial sectors and services of the national economy in our country have been studied in detail by such scientists as: A.N. Aripov, B.A. Begalov, A.Sh. Bekmuradov, T. K.Iminov. N.A.Iminova, A.T.Kenzhabaev, B.Sh.Makhkamov, N.M.Makhmudov, M.A.Makhkamova, Sh.I.Otajonov, T.Z.Teshabaev, G.B.Shonazarova, G. A. Khamdamova, S. S. Salomova O. I. Shibarshova [10-25] and a number of others.

2 Methods

In the context of the digitalization of the economy, the organization of production based on an intellectual platform is not only the main source of growth that ensures the development of an individual economic entity, but of the entire system. Today, the diversification of enterprises based on innovative technologies operating in the sectors of the economy of Uzbekistan will strengthen their sources of income and profitability, but also provide an important basis for reaching new heights in the world market of our country. In particular, computerization, robotization and, ultimately, the digitalization of the production processes of industrial enterprises is one of the sources of income growth. In the context of the digitalization of the economy, the organization of new types of production through the use of new information and communication technologies in industries and in all sectors of industry and the application of information and communication technologies to existing production processes will lead to economic and social benefits.

The research used comparative and logical analysis methods.

Based on the study, the author substantiates that the emergence of several new types of services for organizing a digital production process through the use of ICT in the production process of industrial enterprises:

• a qualitatively new logically organized type of product (service);

• a type of intelligent service that is significantly improved in the enterprise's production process and compared to services available in the market;

• new service with some logical and digital improvements in product development;

• Services offered to market segments in manufacturing processes and presented as a broad system of manufacturing facilities enabling digital manufacturing.

* Corresponding author: salomovass87@gmail.com

One of the types of services that form the production process of an enterprise on a new basis is the services of an intelligent platform. The intelligent platform is a network software hardware "superstructure" that allows the use of intelligent technologies to diversify, adapt and organize new production processes through automation, robotization and, finally, digitalization of the production process [26].

In our opinion, the evolutionary transition of production to intellectual services in the formation of an informed society towards a digital society is conditionally carried out in three stages. At the first stage, industrial enterprises felt the need for ICT services8. During this period, the main service was the development of automated data transmission, as well as automated production processes (Fig. 1).

With the development of society, as a result of the increase in information and communication technologies in production processes, the demand for industrial automation began to grow, and there was a need for modern technologies in production. As a result, the process of automation of ICT systems was implemented. Thanks to the computerization and robotization of systems and networks, there has been a transition to a broadband environment, which means that now society has realized the need to implement not only services, but also a programmed production process.

At the second stage, software logical intelligent platforms were used to program the production process. Intellectualization has become dependent on factors such as the spread of ICT types, quality improvement, digitization level, platform level, networking level. The development of smart platform technologies in industrial sectors of the economy and, eventually, the addition of robotic manufacturing and software services has led to the convergence of smart services.

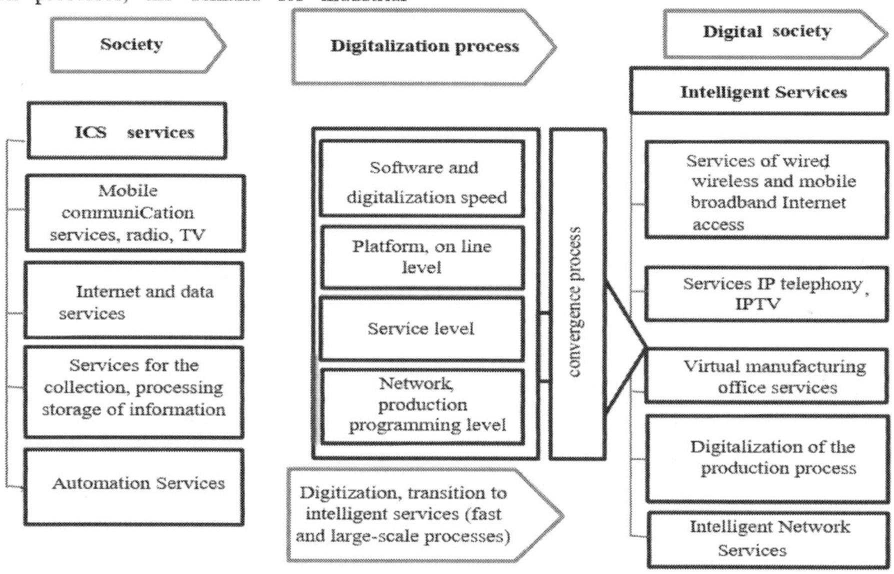

Fig.1. Convergence of the production process and services of the intelligent platform[24]

At the final stage, as a result of globalization, one can observe the transition of the information society to the digital economy. In it, society provides the production process with broadband Internet services, software, intelligent platform technologies and many additional types of digital services, as well as the use of intelligent network services in the production process. In such an environment, the production process in industry is carried out on the basis of intelligent platform technologies, the development of which leads to the formation and development of a separate market for intellectual services in production. In his research work, the author systematized the factors influencing this process in the activities of industrial enterprises (Fig. 2).

There are three main factors influencing the development of the intelligent platform technology market in the industry, which, in turn, can be divided into several parts. The first factor influencing the development of these technologies is the funding factor. The financing factor consists of indicators such as rapid reimbursement of costs for organizing services, cost reduction due to an increase in the size of the business.

The second factor influencing the development of intelligent platform technologies in manufacturing is the increase in the size of the service market. The desire of society for a digital economy will automatically lead to the expansion of the market for intellectual services. This, in turn, makes the services of this platform an integral part of other types of production processes and some services.

The third factor influencing the development of intelligent platform technologies in the real sectors of the economy is technological development.

3 Results and Discussion

There are three main factors influencing the development of the intelligent platform technology market in the industry, which, in turn, can be divided into several parts. The first factor influencing the development of these technologies is the funding factor. The financing factor consists of indicators such as rapid reimbursement of costs for organizing services, cost reduction due to an increase in the size of the business.

The second factor influencing the development of intelligent platform technologies in manufacturing is the increase in the size of the service market. Society's desire for a digital economy will automatically lead to the expansion of the market for intellectual services. This, in turn, makes the services of this platform an integral part of other types of production processes and some services.

The third factor influencing the development of intelligent platform technologies in real sectors of the economy is technological development. The process of technological development consists of such parts as the expansion of the scope of intellectual services (based on digitalization), the introduction of new modern technologies that allow intelligent management of production processes as a result of the globalization of information.

JSC "Akhangarancement" is one of the leading enterprises in the cement industry of the Republic of Uzbekistan. The design capacity of the enterprise for the production of cement-2 is 180,000 tons per year. It also includes the Combine of Building Materials and Structures, which produces a wide range of reinforced concrete structures and concrete mixtures. The main activities of Akhangarancement are the production and sale of cement, the production and sale of concrete and reinforced concrete products, the production of refractory bricks, the exploration and development of minerals, the extraction and processing of mining waste. Processing, transport services, loading and unloading operations in road and rail transport, wholesale trade and foreign economic activity, etc. (Table 1). Since 2004, the State Committee for Architecture and Construction of the Republic of Uzbekistan has been producing high-quality cement based on GOST 10178-85, as well as GOST 31108-2003.

In order to organize the production of high-quality cement and other building materials that meet modern and international standards, Akhangarancement JSC adopted an investment program for 2017-2025. The purpose of this modernization program is to ensure the transition of the enterprise to a new "dry" method of cement production, which involves the use of ICT and software technologies at all stages of production by creating an intelligent platform.

Table 1. Economic indicators of Okhangarocement JSC

	2012	**2013**	**2014**	**2015**	**2016**	**2017**	**2018**	**2019**
Volume of cement production	1533700	1620131	1700501	1771000	1863000,0	1875000,0	1623970,0	1885669,0
Wage Fund (thousand soums)	11611960,5	14514950,7	19366534,4	20740904	22522117,9	26362788,3	30621488,5	47014466,5
Electricity consumption (thousand m^2.)	202672	204829	207663	213200,0	211519,0	203851,0	173098,0	194252,0
Consumed natural gas (thousand.m^2)	244797	263864	265906	266543,0	275910,0	269922,0	239031,0	265934,0
Number of key employees	1201,0	1217,0	1211,0	1223,0	1193,0	1194,0	1124,0	1184,0
The number of auxiliary and additional thread workers	59,0	63,0	61,0	55,0	58,0	56,0	52,0	54,0
Level of technological renewal (thousand soums)	**1090000,0**	**1370000,0**	**1650000,0**	-	**4419210,0**	**2090000,0**	**10000000,0**	**11000000,0**

Today, the state of the production processes of society, equipped with ICT and digital technologies, does not meet the demand (diagram 1).

We know that the basis of any production process is 3 factors: the object of labor, labor tools (ie buildings and structures, equipment and technology) and labor power. Table 2 shows the analysis of the level of renewal of fixed assets (total equipment) of JSC "Ohangarontsement" for 2015-2019, one of these three factors.

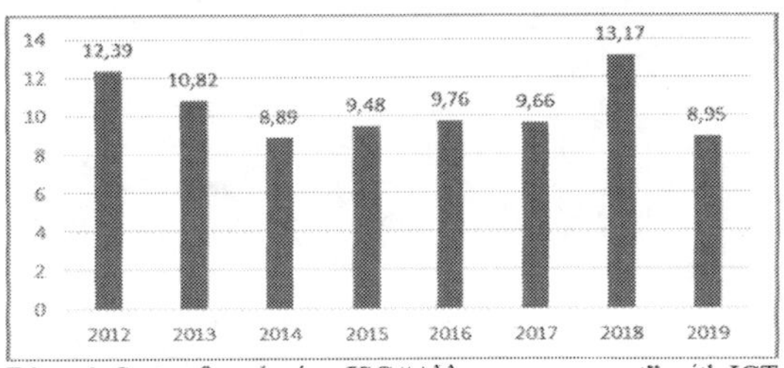

Diag. 1. State of equipping JSC "Akhangarancement" with ICT and digital technologies (%) [24].

Table 2. Analysis of the level of renewal of fixed assets (total technical technologies) of JSC " Akhangarancement " (thousand rubles) [24]

	Years				
	2015	2016	2017	2018	2019
Initial cost of fixed assets	196688284	219862872	254631527	403376478	420881915
Depreciation of fixed assets	132111971	153739198	169226466	267440335	297080236
Residual value of fixed assets	64576313	66123674	85405061	135936143,00	123801679
Fitness of fixed assets,%	32,83	30,07	33,54	33,7	29,41
Cement production volume	1771000	1863000	1875000	1623970	1885669
New fixed tools added	4232210	4419210	2090000	10000000	13000000
Renewal rate of fixed assets%	6,55	6,68	2,45	7,36	10,5

The data show that in 2017, the level of technical and technological renewal in the joint-stock company fell by 2.45%, and in the remaining years, it can be seen as an increase compared to previous years.

The rapid penetration of the digital economy and its capabilities, Internet technologies, computer networks, various digital platforms, mobile communication equipment and electronic services into our lives in recent decades has led to the acceleration and facilitation of "digital production", has established a reorganization of informatization and digitalization capabilities by economic sectors and countries of the world.

In the digital economy, the main factor is the management of production using information and communication technologies and tools, the use of digital data, the processing of large data sets, their quick and easy analysis and use, electronic payments, electronic services, electronic payments, e-commerce, which require reorganization. processing of production processes.

The conceptual model for creating an intelligent platform in industrial enterprises provides a holistic view of the relationship between production processes and the control system, the integration of production and its digital control provides preliminary design and step-by-step implementation of these processes.

The enterprise makes full use of the achievements of modern digital technologies in improving the production process, manage production through programming and digitization, and on this basis ensure integration between departments and workshops. To solve these problems, the author has developed a system of links between problems and solutions for the use of intelligent platform technologies in improving the enterprise management system (Fig. 2).

The analyzes show that, despite the scale of work on the introduction of digital technologies and innovative developments in the country's cement materials industry, the existing financial, scientific and technical potential is not fully used.

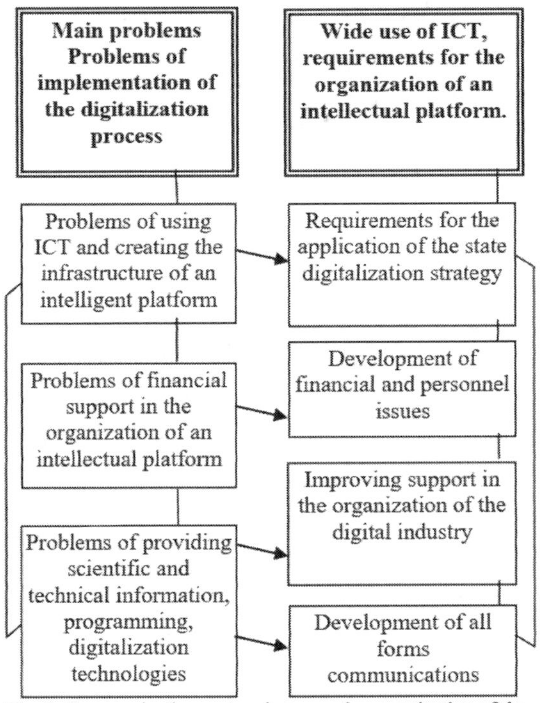

Fig. 2. Communication system between the organization of the intelligent platform and the development of ICT [24]

This, in turn, includes the introduction of digital technologies at cement production enterprises and the management of an intelligent platform, the digitalization of the statistical base and reporting activities, the digitalization of production activities, the digitalization of processes at the subject and system levels, the organization of the integration of micro- and macro-level systems, the implementation of the effective operation of other processes on digitalization.

In this work, a model was developed for the effective organization of the use of digital technologies based on the infrastructure of an intelligent platform in Akhangarancement JSC (Fig. 3.

The production process infrastructure based on the company's intelligent platform includes: digital organization of production processes, process security from raw materials to finished cement, electronic digital signature and identity management system, fact-finding system using blockchain technology, Internet process systems, packaging and transportation of products. In a digital economy, the production process at an enterprise must be carried out with reliable information security. As part of this information integration, information about the production process must be calculated electronically in real time on a daily basis. This will need to be done through a mutually integrated data exchange interface.

The production process infrastructure based on the company's intelligent platform includes: digital organization of production processes, process security from raw materials to finished cement, electronic digital signature and identity management system, fact-finding system using blockchain technology, Internet process systems, packaging and transportation of products. In a digital economy, the production process at an enterprise must be carried out with reliable information security.

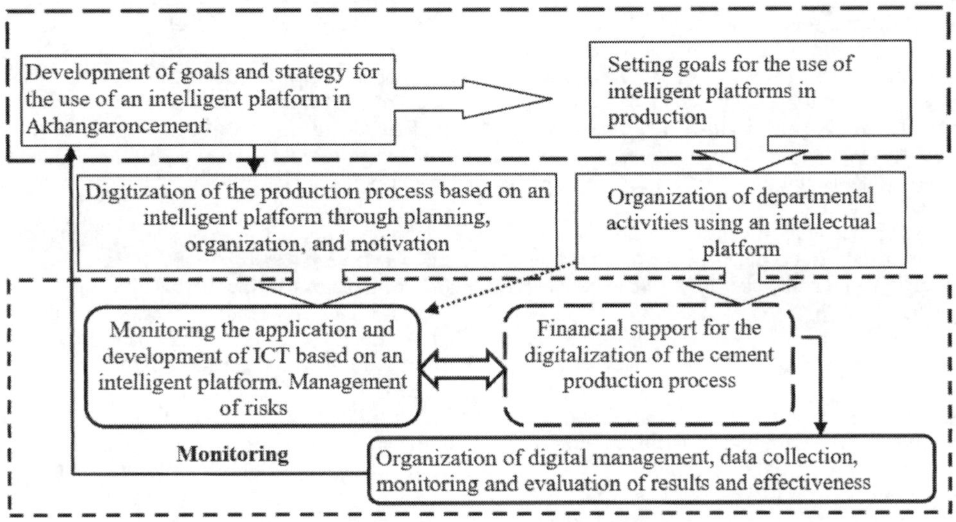

Fig .3. A model for the effective organization of the use of ICT based on the infrastructure of an intelligent platform in Akhangarancement JSC [24]

As part of this information integration, information about the production process must be calculated electronically in real time on a daily basis. This will need to be done through a mutually integrated data exchange interface.

4 Conclusion

Studies on the effective development of industrial enterprises based on an intellectual platform made it possible to draw the following conclusions:

During the COVID-19 pandemic, which has become a global problem in the world, the digitalization of production processes and the organization of new production processes based on the Industry 4.0 intellectual platform have highlighted the relevance and necessity of the digital economy.

The main goal of applying and using technologies of digital and software products through the creation of an intelligent platform in industrial enterprises is, on the one hand, to achieve high profits, on the other hand, to increase the efficiency of users.

The processes of demand and supply formation of digital technologies and intellectual services in industrial enterprises play an important role in the modern market economy. It covers topics such as a comprehensive consumer survey, identifying the various factors that influence demand patterns and supply processes, modeling them, developing application capabilities and content levels, and how the state can influence and encourage digitalization processes.

References

1. http://www.nasscom.in/indian-itbpo-industry - Website of the ICT Agency of India./
2. Minges M. Exploring the relationship between broadband and economic growth.WDR. 2015.
3. Hywel M. Davies; Ali Hashim, Eduardo Talero Information systems strategies for Public Financial Management», World Bank, 1993,
4. Ali Hashim, Bill Allan Treasury reference model», World Bank & IMF, March 2001;
5. Desai P. Introduction to the TSA in Transition Economies: IMF, 1996;
6. Financial Management Information Sidtems and Open Budjet Data: Do Government Report on Where the Money Goes? Washington DC: World Bank, Ebenezer G.Laizer, Reima Suomi. Evaluation of Integrated Financial Management Information Sistem (IFMIS) in Malavi.2017.

7. Proceedings of the 11th European Conference on Information Sistem Management.ISBN:978-1-911218-52-4,
8. Combaz Emilie, Implementing Integrated Financial Management Information Sidtems (GSDRC Helpdesk Research Report, 1229). Birmingham,UK:GSDRC,University of Birmingham.
9. Aripov A.N. Problems of improving the management system in the field of information and communication. Abstract of the dissertation for the degree of Candidate of Economic Sciences. O'zRPHDJQA, 2004 .;
10. Begalov B.A. Econometric modeling of the formation and development trends of the information and communication market. Abstract of the dissertation for the degree of Doctor of Economics. - T .: 2001 .;
11. Bekmurodov A.Sh., Shamshieva N.N. Marketing strategies of telecommunication companies (on the example of mobile companies in Uzbekistan). - Tashkent .: Economy, 2008. - 49 p .;
12. Iminov T.K., Aripov A.N. Issues of management in the field of information and communication technologies in Uzbekistan. Monograph. - T .: Science and technology, 2005. - 300 p .;
13. N.A.Iminova. Risk management in the field of communication. Monograph, T .: Communicator. 2012. - 178b .:
14. Kenjabaev A.T. Problems of formation of the national system of informatization in business activity. ict.fan.dok. Dis. T .: TSTU, 2005. - 321 p .;
15. B.Sh.Maxkamov. Effective development of mobile services in the context of modernization of the economy. I.f.d. (DSc) dissertation abstract. SamISI, 2016 .;
16. Maxmudov N.M. The role of the processing industry in accelerating economic growth: an empirical analysis // Science and Innovative Development. T .: 2018. -№3.-30-36-b.;
17. M.A.Maxkamova. Formation of the organizational and economic mechanism of effective management of innovative activity in the enterprises of the machine-building complex of the Republic of Uzbekistan. I.f.d.diss.-T .: 2004.
18. Otajonov Sh.I. Improving the effectiveness of organizational and economic mechanisms for managing the infrastructure of innovation activities., Diss. i.f.d., T .: 2018y .;
19. Teshabaev T.Z. Improving the management of innovative activities in the higher education system on the basis of information technology. I.f.d.disser.avtorefrati.T .: 2019.
20. Shonazarova G.B. Improving the mechanisms of innovation capacity management in industrial enterprises., Ph.D. dissertation, T .: 2018;
21. Khamdamova G.A., Shibarshova O.I. Development of marketing activities of information and communication technologies. Uch.posobie, T .: Alokachi., 2007.
22. Salomova S.S "Effective development of industrial enterprises on the basis of intellectual platform." Abstract of the dissertation for the degree of Doctor of Philosophy in Economics. T.: 2021
23. Amirov D.M. and others. Explanatory dictionary of information and communication technologies, Tashkent, 2010.575p.
24. Collection of normative legal documents on information and informatization. T .: Ministry of Justice of the Republic of Uzbekistan, 2008, 464p.

AUTHOR INDEX

EDP Sciences – Web of Conferences
17, Avenue du Hoggar
Parc d'Activité de Courtabouf
BP 112
F-91944 Les Ulis Cedex A
France

ISBN 978-1-7138-7247-4

9 781713 872474

CURRAN ASSOCIATES INC.
proceedings.com